ALGEBRAIC CALCULUS

A Radical New Approach to Higher Mathematics for Students of Electronics and Computer Graphics

Roderick Lumsden

M.A.Cantab., Ph.D.

Published by Eusephany Publishing 2016

Published by Eusephany Publishing 2016

ISBN 978-0-9935483-0-7

The cover illustrations, as with the first edition, both show 3D computer graphics images of models developed completely by programming. They were both created to demonstrate that fully programmatic design is genuinely feasible. Conventional architectural software was used to create the final rendered images. The front cover image shows a classic Eighteenth Century form known as an Imperial Staircase, while the back cover takes its inpiration from Frank Lloyd Wright's house built for John Storer in California in 1923.

Preface to the First Edition

This book advocates a radically new approach to the introduction of Higher Mathematics at Freshman level. I adopt a slightly polemical tone because I'm aiming to stimulate debate. The methods, and some of the terminology, that I propose may appear unconventional, but they have sound roots in mathematical history and translate exceptionally well into digital practice, so I'll start by reviewing this background.

The mathematical methods introduced by Élie Cartan the better part of a hundred years ago are now widespread in research-level work. But what is not fully acknowledged is that they can revolutionize the *teaching* of the subject too. All that is needed is a readable, informal account of them. Bringing in these methods, suitably simplified, right at the start, in a simple, engaging style, transforms the clarity and comprehensibility of the subject. The true meaning of so many aspects of intermediate mathematics falls naturally into place.

So I'm doing two things:

- I'm showing that the idea of differential forms, which crystallised around a hundred years ago, allied to the concept of simplexes, suffices as a foundation to develop the entire body of the calculus easily and quickly, and gives a much more coherent line of development.
- I'm putting it in a way that is clear, readable and, hopefully, entertaining. So I have preferred English readability to mathematical formality wherever reasonably possible.

Along the way, I cover in some depth various supporting fields such as vector algebra, with an introduction to the up and coming area of geometric algebra, and I also give a good, but more critical, introduction to the subject of generalised functions, which were more the fashion in Europe in the fifties. And to enrich the readability of the text, there are digressions into fields that are not obviously mathematical, especially if they relate to computer graphics or are particularly relevant to digital practice. I would hope the book's groundbreaking approach will be especially interesting to teachers working in digital applications at this level.

So for those teaching the subject, I'll first give a brief summary of what I see as the salient original features of the book.

1) I introduce differentiation using the exterior derivative on a scalar function to generate a 1-form, so making it multivariate from the start.
2) I *define* integration as a product between a differential form and a simplex.
3) I use the axioms of a group to show that the addition of angles in the circle leads naturally to the idea of complex numbers.
4) The book incorporates geometric algebra into the presentation of vector algebra and analysis from an early stage.
5) Generalised Functions are introduced fully based on differential forms, and this treatment prepares the way for an advanced coverage of Fourier and Laplace transforms.

A word might be in order about the genesis of the work.

Off and on over quite a few years, I've jotted down ideas about mathematics, some of which I explored at the time, and others as possible lines of future exploration. It seemed time to bring all these jottings and ideas together, and the present work is the result. I started thinking of keeping them in note form, but decided it might be better to take the opportunity to put them into a continuous unfolding stream of ideas, and gradually the form of a textbook started to emerge.

There is a consistent theme to the development. This theme is an attempt to see how much of basic calculus could be developed in a purely algebraic way, without recourse to all the conceptual difficulties of infinities, infinitesimals and the formal theory of limits.

The answer seems to be that virtually all of it can be, and that these concepts, surprisingly, do not have any special significance for the calculus, having merely been instrumental in its development by something of a historical accident. I'm anticipating some hostile feedback to this assertion, but nevertheless I believe it to be sound.[1]

This is an interpretation that is highly apposite in the modern world, dominated as it is by digital technology. In computer applications, it is methods of calculation that will work and that will give results that are more important than the philosophical aspects of the subject. As a student, I assumed that all mathematicians were geniuses (I still am in awe of them), and that their teachings were beyond question. The first suspicions that mathematical philosophy might be more flexible, more subtle, more fluid and less rigid, less dogmatic and less sacrosanct were stirred in my mind when I became aware of the wide variety of supposedly "fundamental" models of mathematical logic that appeared in the early days of computing, none of which actually anticipated the multiple-register machine that has come to dominate the world today.

Again, in introducing the basic calculus, one cannot fail to be aware of the deep rift between the thinking of Newton and that of Leibniz, and of the attacks of Bishop Berkeley. The entire discipline has been subject to changes in interpretation all through its history, and there have been several earlier attempts to put it on a more purely algebraic basis, from Euler,[2] and especially from Möbius and Kronecker, and most recently from a team put together by Jerry Marsden at Caltech. My effort combines the concept of simplexes with that of a graded algebra in a way that may not be entirely new, but with which I have perhaps been a little more disrespectful of tradition!

Taking all these strands of thought together, I have decided to be unabashed about presenting this as a personal interpretation or viewpoint. This is how *I myself* see the calculus, and that approach is true to the origins of this work, as a compendium of ways of using mathematics that have worked for me. It's perhaps no bad thing for students to be introduced early to the idea that opinion does have a part in mathematics: the subject is not a perfect sacrosanct whole, but is an evolving discipline subject to the vagaries of fashion of the time that each contributor has made his or her mark. It is perhaps a sad truth that in no other subject does authority and orthodoxy go less challenged.

---

[1] Just as the Fundamental Theorem of Algebra strictly *does* require limiting concepts to define real numbers, so the basics of the calculus can, surprisingly, be set up without them. I keep limits to the margins of the subject right up to Chapter 7.

[2] Euler too believed differentials to be fundamental, so I'm not in bad company. My dissatisfaction with the idea of sets was shared by Kronecker.

So one thing I am trying to get across in this text is that there *are* other ways of looking at most mathematical topics, and the established orthodoxy may be quite the wrong approach for some students. Different people have different ways of thinking, and the methods or the development that may suit some may be needlessly alienating for other perfectly capable people. Over and above that, there may be areas where the orthodoxy is plain wrong, and methods developed to try to incorporate a new concept at the time it was discovered might be much improved upon. This is particularly true of generalized functions, where there is still an undercurrent of dissatisfaction with the way the subject was left by Schwartz and Lighthill some sixty years ago.

Clearly mathematics is not so subjective as a subject like history, or even economics, because integral to it is the fact that it is self-checking: mathematical methods can be downright wrong, and if they are, the answers they give will simply be wrong when put back into the original problem. So I have been at pains to show that the answers my methods give do pass this test, albeit within the limits of my ability. But granted that undeniable constraint, there is still a lot of latitude in the ways of working. Would Newton have recognized his theory of "fluxions" in Cartan's differential geometry?

My belief, perhaps surprising in someone who has never really liked programming, is that the best test for the acceptability of a mathematical concept in the modern world is how well it translates into digital practice. Computers have to work, but philosophy doesn't.

Therefore, central to this presentation has been a strong playing down of all the philosophical garbage that mathematics has accumulated over the centuries. In doing so, I have sometimes adopted a somewhat polemical tone, but I am hoping to entertain as much as to exacerbate when I do so. If mathematics could do with a whiff of opinion, it might well do with a bit of humour too. I have always found the best texts to be the most human.

So the conventional definition→theorem→proof approach is one I avoid. What I try to do is to get across the salient ideas by simple examples, so that readers can move forward into the subject fast. As a student, I made a big mistake in trying to acquire a completely rigorous understanding from the word go. I now believe that seeing the lie of the land from a selection of examples enables one to navigate easily afterwards: trying to map the whole country at 1:10000 from the start gets one hopelessly lost.

I have also brought in topics such as colour vision and music, that are not normally seen as mathematical, but which have interesting properties when viewed in that way.

The text suggests quite a few changes in notation and terminology, most of which will probably not catch on, but a few of which might. I would expect it to be read alongside more conventional presentations: much of its purpose, after all, is simply to put across the idea that there *are* other interpretations of the theory. I would hope that working the exercises in the normal textbooks using the approaches suggested here will be highly fruitful.

Chapter Summary.

1) The first chapter reviews a wide range of topics in basic pre-calculus maths that will be needed for the subsequent work. There are excursions into colour vision and musical scales, and the chapter ends with the elements of the theory of simplexes and a fairly rigorous use of simplex theory to develop simple electrical circuit theory. Colour vision provides an excellent anticipation of the graded algebra of 3-space, with the complementarity between primary and secondary colours, and between black (scalar) and white (pseudoscalar) anticipating Hodge duality. The chapter ends with a look at Euler characteristics.
2) The second chapter sets the tone of the rest of the book, developing basic calculus starting with Cartan's exterior derivative (although I prefer not to call it that at this stage) acting just upon simple scalar functions, giving partial derivatives and ordinary derivatives side-by-side from the start, and suggesting some new terminology. I have never forgotten my early student miseries over partial differentiation, miseries I hope the present methods completely circumvent. Although chapter 2 isn't a long chapter, it introduces many of the key principles of the calculus. The chapter ends with a brief discussion of how this theory will extend to higher grades and touches on the alternating product, but the full development of this is deferred to chapter 6.
3) Chapter 3 uses the theory developed so far to introduce logarithms and exponentials, where again I suggest some new and more compact notation, and ends with the application of these functions to linear differential equations in electrical circuits, building on the section in chapter 1.
4) Chapter 4 develops basic trigonometry with a strong emphasis on the intimate relationship between the circular functions and complex numbers, and stresses that it is much easier to work in complex algebra when dealing with angles. These ideas reappear in chapter 5. I should stress that in digital work this is a much easier way of handling angles, as I know from extensive experience in computer graphics.
5) In the fifth chapter, I introduce vectors and geometric algebra, putting much of the material from chapter 4 into this broader context, touching on quaternions as historically relevant to vector algebra, leading on to the idea of graded algebras and giving a detailed handling of rotations in the context of geometric algebra, with a final section on eigenvalues, eigenvectors and eigenfunctions.
6) Chapter 6 is the crux of the whole book. It develops integration theory using a simple algebraic definition of an integral as a product between a differential form and a simplex, shows how the simplex factor can be extended to manifolds, and how generalised surface integrals fit into this framework, with especial attention to how this all fits into a graded algebra. I have decided to relegate to an appendix a discussion of just why the alternating product is the relevant one for higher differential forms, simply giving a slightly modified account of the conventional method that shows that only the alternating property of the Jacobian determinant will correctly give volumes.
7) Chapter 7 covers Laurent Schwartz's theory of generalised functions, introducing this as an example of using functions, rather than geometric objects, as vectors. Along the way, I develop elementary complex analysis up to the Cauchy integral formula, and cover some of the anomalies of delta functions.
8) The next chapter attempts to give a thorough grounding in Fourier-Laplace transforms and convolution algebra, following directly from chapter 7. I have to confess that I am far from happy with either my own proof, which I therefore relegate to an appendix, or any of the conventional proofs, of the Fourier Inversion Theorem. I have a deep-rooted suspicion that all

these "proofs" subtly contain a definitive element because inverting the order of integration gives an "integral" that can only be construed as a delta function. As with anything based on limiting values, I have reservations about improper integrals, their meaning being strongly *subject to definition*. So I tentatively propose that the Inversion Theorem is best regarded as being as much a definition as a true theorem. Most proofs of it depend on some prior defined property of generalised functions and it seems reasonable to place this definition into the theorem itself.

9) The final chapter moves into quite different territory, giving a basic account of projective geometry, a subject highly relevant to computer graphics. The treatment here leans strongly on the historical origins of the subject and its relation to architectural projections.

Conventions

Throughout the book I use **bold** type to indicate terms used in a formal sense, although I have not been completely rigorous in this in the interests of readability. Too much bold can be disconcerting. *Italics* are used simply for emphasis, and I have avoided putting text into both bold and italic, again in the interests of clarity.

I use appendices to cover topics that are either incidental to the main text or, more particularly, topics where I am less confident of my argument. In chapters 6, 7 and 8 there are quite a lot of these "grey areas" and I would be grateful to anyone who can suggest more rigorous arguments here, as long as they fit into the general line of development of the book.

There are no exercises. Since the main publishing companies have been unwilling to take on the book, I have decided simply to get it out into the world as it is, as much as a discussion platform as the undergraduate textbook that I had at one time hoped it would become. As suggested above, I think students would find it highly instructive to apply the methods presented here to some of the exercises in the standard texts.

I also avoid numbering equations, which may seem a heinous crime, but I have always argued that if an equation is sufficiently important to be referred to again and again, it should be *named*, not numbered, and I have found this rule sufficient in all but a very few cases where I have put an "X", say, against an equation to which I will refer back.

The text has been written entirely in Microsoft Word, and some of the notational changes I propose have been introduced to facilitate working in a conventional word processor. I've never really liked LaTeX, and I think that such changes as replacing the archaic old English "∫" for integration are long overdue, as is the placing of the integral symbol *between* the differential form and the domain of integration.

The tone of the work throughout is informal and exploratory, and yes, occasionally slightly polemical and contentious. I hope it will be seen as bringing a breath of fresh air into a teaching area that has become a little too conventional, and that readers may think some of my suggestions exciting and entertaining.

I would like to thank Professor Stephen Gull at the Cavendish Laboratory in Cambridge for his advice and encouragement in seeing an earlier draft of the book. I know Professor Gull would have preferred a book more oriented towards geometric algebra, but this was a subject completely unknown to me when I assembled these ideas. I suspect he would describe my approach as interesting, if misguided! I might also thank the reviewer with one publisher who lambasted the work mercilessly, apparently interpreting my definition of the differential operator as indicating that I thought the Leibnitz rule defined linearity.

Errors? Well, there are bound to be myriads of errors! I've checked and rechecked as best I can, but it's only too easy to see what you expect to see when going over mathematics, so I am putting a full error-reporting facility into the website version of the book, and I'll attempt to incorporate any corrections subsequently as long as I can understand them!

Roderick Lumsden
January 2012

Preface to the Second Edition

The most significant change in the new edition is that I've finally had the courage to break completely with tradition in the introduction of the Fourier transform and develop it in a way as radical as my approach to basic calculus. Since I feel confident that this approach will be far more acceptable to students, I would hope that the entire tone of this Chapter will be more upbeat and readable.

The ε notation introduced in the treatment of Fourier analysis in the previous edition is now used through much of the text, simplifying the look of a large part of the mathematics so that the ideas involved stand out more clearly from the algebra. I have also added another digression in the introduction of this notation (in Chapter 4) on the concepts of summer and winter time, of right ascension and the use of hours as angular measure, as this is a good example of the sort of mathematically trivial concept that can nevertheless be very confusing. I've also added a brief treatment of hyperbolic functions along similar lines to the main development of angles.

Chapter 5 is also expanded, with a brief treatment of how computer graphics systems distinguish points from vectors and why, and an introduction to the ordinary Gibbs cross product, a much clarified discussion of the relationship between spaces and their duals and the transforms between them, introducing the term "dual transform" to give a formal name to a distinction which occurs repeatedly in advanced work. I wanted to simplify the Section on rotations but ended up making it significantly larger! I still feel it's better this way: because of its deceptive familiarity the concept of rotation has many hidden pitfalls, and students should begin to be aware of them from the start.

Building on the "dual transform" concept, Chapter 6 has a much fuller and more consistent development of the switch from the elementary idea of differentials as numbers to the standard modern interpretation of them as the duals to tangent vectors, and this concept I now develop fully by example. Chapter 6 also has a better treatment of areas and volumes which were so integral to the early development of the calculus, and there's been a slight reduction in the Appendices here. The Appendices to Chapter 8 have now gone completely. Readers of the first Edition have probably gathered that I use the Appendices as a repository of essentially experimental ideas, although they are not exclusively used that way.

I've altered quite a few details of the presentation of electrical theory to make it more consistent, and hopefully more comprehensible. In Chapter 9, I've also developed the Geometric Algebra proof of the Desargues's Theorem fully and shown how the duality in Geometric Algebra translates directly into that of Projective Geometry using David Hestenes's own papers on this.

I've added a fairly large Appendix to the whole book on Methods of Integration, as this was a subject largely ignored in the first, experimental, edition, but is a skill students need to acquire.

The book will hopefully look more professional now too, having a more conventional format and many much better illustrations, and I've also added Exercises to the end of each Chapter. Most of these are conventional problems, and aim to give the student confidence that the methods they're learning in this text are of general applicability. I've also made extensive use of MS Equation rather than trying to lay out equations as tables, but I've retained my own symbolism nonetheless; key equations *are* now numbered as well.

A key hallmark of the text is that I try to tackle head-on the sort of problems that baffle many a student and I try to work out exactly *why* they cause the problems they do. More conventional texts simply don't attempt to do this. An example is the way we're all taught at school that the squares of all numbers are positive, but then we're suddenly introduced to the idea of a square root of a *negative* number even though it's been proved to us that this is impossible. So how? Why? It can't be! To people like myself who are more philosophically inclined than computational, these conflicts are liable to undermine our entire confidence in the subject. It's for such people that I'm writing.

As this edition is in monochrome, I will make full colour versions of selected images available on the accompanying website www.algebraiccalculus.com, where they may be freely downloaded. This new version of the website should be available by the end of April 2016.

Roderick Lumsden
March 2016

Table of Contents

Chapter 1

PRELIMINARIES

1.1 Functional Notation

In mathematics, a **function**[3] is a rule for calculating one value from one or more other, given (or **data**) values. **Functions** can be very simple, like $x + 3$, which for the **datum** $x = 7$ would give the resulting value 10, and for $x = -122.72$ would give the result -119.72. The **data**, written in their generic form as algebraical variables, are called the **arguments**, or sometimes the **parameters**, of the **function**. Thus the **function** $x + y \times z$, which I may also write as $x + y.z$ or $x + yz$,[4] has three **arguments**, *x, y* and *z*. For $x = 4$, $y = 3$ and $z = 10$, this simple **function** returns the value 34. But the general expression $x + y.z$ describes all the vast infinity of possible results associated with all the infinite possible values of *x, y* and *z*.

Necessarily, a **function** gives a unique result for any particular values of the **arguments**. **Functions** cannot be ambiguous.

Very often in mathematical discussions, all the infinity of a *single* **function** is still not enough, and we want to discuss *any* **function** possibly meeting some constraining definition. This is when we use **functional notation**. So we might write:

$$f(x, y, z)$$

to indicate *any* **function** with three numeric **arguments** *x, y* and *z*, whatever the specific form of the **function**. So this notation would include $x + y.z$, but also $ax^2 + bxy + c.z.\mathbf{log}(x + yz^4)$ and an infinite

---

[3] I use **bold** for words that have a special meaning in this text, a brief definition of which usually appears in the Glossary, and *italics* simply for emphasis, although italics may be used occasionally for words with a technical meaning not special to this text. These conventions, however, I use rather loosely as too much bold or italic can be difficult to read.

[4] Here I use the stop "." to indicate multiplication rather than the multiplication sign "×", as this is the normal practice in higher mathematics. Frequently, even the stop is omitted, and simply *juxtaposing* the variables, as in *xy*, defines multiplication. The stop between two *actual* numbers, as in 22.743, is simply the decimal point.

multitude of other more complicated **functions**.[5] A specific value of a **function** can be written replacing the **arguments** with the chosen values. So if $f(x, y) = x + 2y$, $f(3, 4) = 3 + 2 \times 4 = 11$. I may occasionally have cause to write this as $f(x, y)(3, 4) = 11$ to clarify which values are being used for which **arguments**.

If one or more of the arguments must be **whole numbers** or **integers** – something like 1, 22 or 469, or even −18 or 0, but *not* 3.1415926 or 99.999 or anything with a decimal point – it is customary to use the letters of the alphabet from *i* onwards – *i, j, k, l, m, n* – to represent such **arguments** *and to write them as subscripts*. So you may see forms like:

$$g_{i,j}(x, y)$$

which mean that *g* actually has *four* **arguments** *i, j, x, y* but the first two must be **integers**. It is customary in these circumstances to think of the various $g_{i,j}$ forms – like $g_{1,1}$, $g_{1,2}$, $g_{3,1}$, $g_{4,3}$ – as *actually separate* **functions**. This interpretation is most natural when, as is often the case, such **integer arguments** are restricted to a limited range like 1 to 3. In this example, if *i* and *j* were indeed limited to being in the range 1 to 3, we would have all told *nine "g"* functions, which we might write without the commas in the subscripts as:

$$g_{11}(x, y),\ g_{12}(x, y),\ g_{13}(x, y),\ g_{21}(x, y),\ g_{22}(x, y),\ g_{23}(x, y),\ g_{31}(x, y),\ g_{32}(x, y),\ g_{33}(x, y)$$

but obviously we can only omit the commas if the **integers** do not go beyond 9!

This subscripting of **integer parameters** is much used for functions that have no other arguments, such as a 3×3 array of values:

$$\begin{matrix} a_{11} & a_{12} & a_{13} \\ a_{21} & a_{22} & a_{23} \\ a_{31} & a_{32} & a_{33} \end{matrix}$$

which is called a **matrix**, a general term for such two-dimensional arrays defined by two **integer indices**, as the **integer parameters** now tend to be called, and in a case like this it's hard to think of *a* as a **function** at all. It is indeed normal to think of such a 3×3 **matrix** simply as a set of nine separate numbers.

Sometimes one **function** may be associated with, or **derived** from, another, and we would like the **functional notation** to keep track of the new **derived function**'s origin. A common example is the standard **ordinary derivative** of the calculus, where we can write the **derivative** of $f(x)$ as $\partial f(x)$, so the "*f*" is still in there to show the **original function**. More commonly, we need to work with *several* such **derivative functions**, one for each of *several* **arguments** of the original **function**. So, for example, the **function** $g(x, y, z)$ has *three* **partial derivative functions**:

$$\partial_x g(x, y, z) \qquad \partial_y g(x, y, z) \qquad \partial_z g(x, y, z)$$

In this context, you may quite often see forms like $\partial_x g(x)(3)$ to indicate $\partial_x g(x)$ evaluated at the specific value $x = 3$.

---

[5] Again here I've used *both* the stop *and* juxtaposition to indicate multiplication.

Quite often the **arguments** are themselves dependent on **integer** values, and we might have:

$$g(x_1, x_2, x_3, y_1, y_2, y_3)$$

in which case our two subscript notations come into conflict, as most word processors, including this one, only allow one level of subscripting (and sensibly in view of readability). In such cases I will put the **integer arguments** back into normal bracket notation and write:

$$\partial_{x(2)}g(x_1, x_2, x_3, y_1, y_2, y_3) \text{ or } \qquad \partial_{x(i)}g(x_i, y_i)$$

where in the second example $g(x_i, y_i)$ shows a common way of writing forms like $g(x_1, x_2, x_3, y_1, y_2, y_3)$ which might be better written in some form of **list notation** of which a common variant is $g(\{x_i\}, \{y_i\})$. But hardly anyone ever does!

In mathematical texts, the term **mapping** is often used to indicate a **function** *in aggregate*, as for example, when we might speak of a **function** like:

$$x = a\xi + b\psi \qquad y = c\xi + d\psi$$

as **mapping** a rectangular area in ξ and ψ defined by $\xi_1 < \xi < \xi_2$ and $\psi_1 < \psi < \psi_2$ into a *parallelogram* in the plane of x and y. Seen as a **function**, each pair of values (ξ, ψ) determines a unique *point* (x, y) in the xy plane, but we may speak of all the points in the rectangle, *in aggregate*, as being **mapped** into the parallelogram. The term comes from maps in geography, which are just such a representation of a part of the Earth's surface seen in aggregate. Strictly speaking, the terms **function** and **map** or **mapping** are regarded as interchangeable, but in practice they are generally used with this bias, that **maps** and **mappings** refer to aggregates. In these circumstances too, it is often useful to distinguish between the "source" object – here a rectangle – and the "destination" or "result" object. For this purpose, I will use the well-established terms **domain** and **range** of a **function**. The **domain** is actually the set of *all possible* **argument** values for the **function** and the **range** the set of *all possible* result values. In the aggregate context, I can use these terms to talk of the **domain object**, **domain area** or **domain zone** to refer to the **domain** rectangle of the present example, and the **range object**, **range area** or **range zone** to refer to the **range** parallelogram, but in any given example I will use the more specific "**domain** rectangle" and "**range** parallelogram", or even "result(ing) parallelogram".

1.2 Graphs and Chirality

Figure 1.2.1 shows the standard Cartesian configuration of x and y **axes** (the plural of the singular **axis**) that provide the usual way to label any point in a plane. It is named after René Descartes, who devised this standard in the early seventeenth century. The point where the two **axes** meet is called the **origin**, and any point in the plane can be labelled by assigning a number to the x value or x **coordinate** which specifies how far to the *right* of the **origin** the point in question lies, and assigning a number to the y value or y **coordinate** which specifies how far *above* the **origin** the point in question lies. If the particular point is to the *left* of the **origin**, it takes a *negative* x value, if it lies *below* the **origin**, it is assigned a *negative* y value. The labelling is always written with the x value first and the y value second, generally in brackets, so for example, a point P labelled (4, 7) would have $x = 4$ and $y = 7$ and so lie 4 units of distance to the *right* of the **origin** and 7 units *above*, a point Q labelled (−12, 0) would have $x = -12$ and $y = 0$, and so lie 12 units of distance *left* of the **origin** and actually *on* the *x* **axis**, as $y = 0$ means it is neither above nor below the **origin**. The *Figure* shows a few examples.

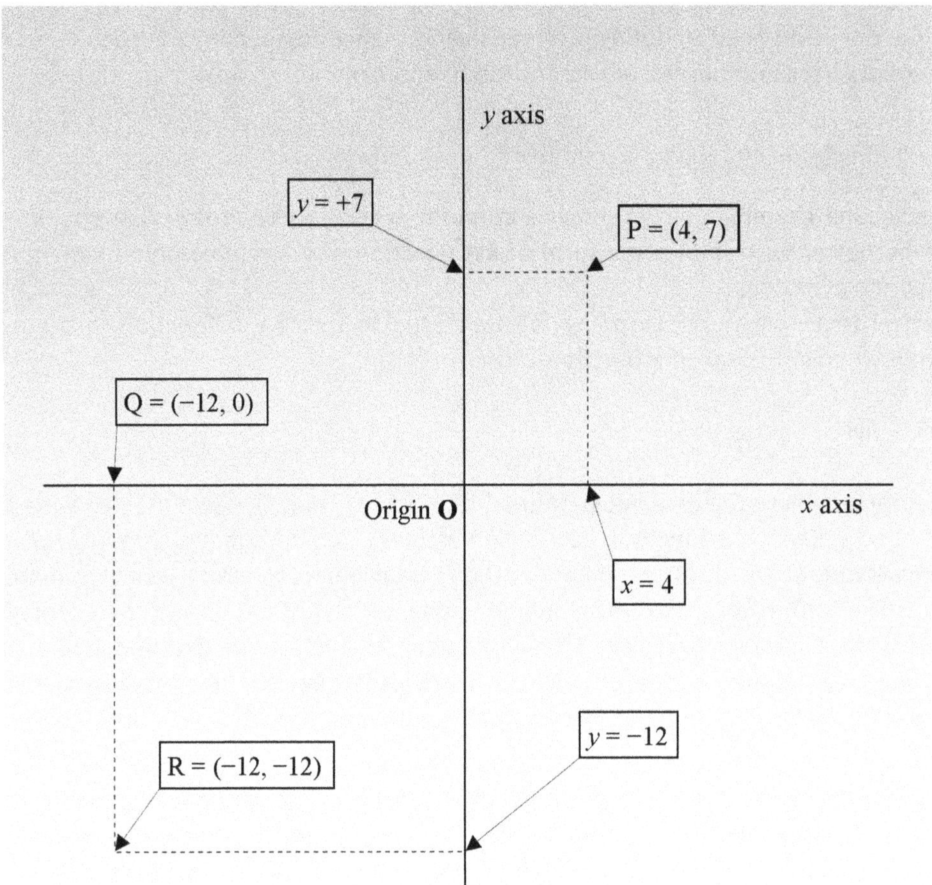

Figure 1.2.1

This is one of the few conventions in mathematics that is *universal*. This layout of the **axes**, the *x* and *y* names for them, and the *x*-first order of labelling are used always, without exception.

It need not have been so. It would have been possible to have the *x* and *y* axes named the other way round, with the *x* vertical, or to have had *negative y* values used for distances *above* the **origin**, or *negative x* values for points to its *right*. All three of these would have involved a **reflection** of the established **axes** in various ways (about a diagonal, about the *x* **axis**, or about the *y*) and would have given *mirror images* of the established system, which could *not* have been rotated *in the plane* into the standard position. Such **axes** would therefore be called **left-handed**, whilst the conventional ones are **right-handed**.

In the plane, **left-handed axes** are never used. But when we come to add a *third* axis, to represent objects in three-dimensional space, *both* **right-handed** and **left-handed** systems *are* used. By convention, the third **axis** is labelled the *z* **axis**, not surprisingly, and if, with the standard **right-handed** layout of *x* and *y*, the *positive z* **axis** comes *towards* you *out of* the plane of the *x* and *y*, the set of all three is called **right-handed**, but if it runs *away from* you, the three are called **left-handed**. Looking at it another way, if you think of the standard *xy* layout as lying on the *ground*, if *z* values *above this plane are positive*, the three **axes** form a **right-handed** system, if *z* values *below this plane are positive*, the three **axes** form a **left-handed** system. These variants are shown in *Figure 1.2.2.*

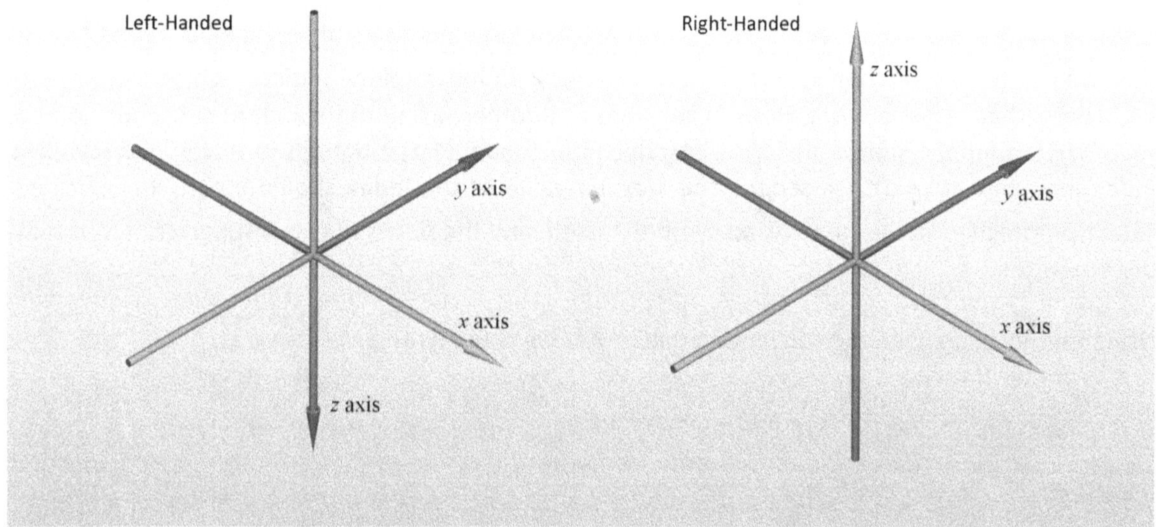

Figure 1.2.2

Again, a **reflection** of a **right-handed** system, this time in any *plane* rather than in a line, changes it into a **left-handed** system and vice versa.

This phenomenon of *handedness* applies not just to **coordinate** systems, but to anything that is not symmetrical in any of the **axes** of the space in question. The general scientific name for such *handedness* is **chirality**, which is just the English word in Greek (I will also use the adjective **chiral**). Mathematicians normally call it **orientation**, and it is often described in ordinary language as the **sense** of something. I will only use the word **sense** in this context in bold, because the word has so many other meanings in English.

Chirality is a very, very strange phenomenon. It is a property of the number of dimensions involved. So the letter "L" is **chiral** in the *plane*, but not in three-dimensional space. Our *xy* **coordinate** system is again **chiral** in the plane, but not in three dimensions, but now the addition of a *z* **coordinate** makes the set of *three* **axes chiral**. Particularly strange is that there is no means of describing unambiguously in *language* which **sense** is which: so there is no way we could tell an alien in the Andromeda Nebula which is our right-hand and which our left. (In the nineteen-sixties it was discovered that one of the four fundamental forces of nature, the "weak" force involved in radioactive decay, appears to be **chiral**, leading to the suggestion that we could refer to this force in talking to our alien, but will this force have the same **sense** in Andromeda?) And why the strange division into just *two* kinds, left and right?

Permutations – re-orderings of a sequence of objects – also show a **chiral** division into two kinds, and so can be used to label the two different senses of objects in space, an approach which I shall discuss briefly in Section 1.13. The ability of **permutations** to describe the **orientation** of **coordinates** also appears in the theory of **determinants**, a field in which perhaps the most famous theoretician was the Revd. Charles Dodgson, or Lewis Carroll. I will briefly discuss **determinants** in Section 1.7, and more fully in Section 1.8.

1.3 Summation

Often in mathematics, we have cause to add together large numbers of very similar **terms**.[6] To denote this, we use the **summation symbol** Σ. It is used with an **index** (plural **indices**) which labels the bits of each **term** that vary from one **term** in the sum to another, using the standard subscript notation for such whole number **indices** that was introduced in Section 1.1, although if *powers* are involved, the **index** may appear as a *superscript*.[7] The *starting* value of the **index** should strictly be specified as a subscript attached to the **summation symbol** Σ itself, and the *final* value as a superscript, but these are often omitted.

Using this notation, with the convention that $x^0 = 1$, we could write $ax^2 + bx + c$ as:

$$\Sigma^2{}_{i=0}\, a_i x^i \equiv a_0x^0 + a_1x^1 + a_2x^2 = c + bx + ax^2$$

if we define:[8] $a_0 = c,\ a_1 = b,\ a_2 = a$

Strictly, the superscript (^{2}) and the subscript ($_{i\,=\,0}$) on the Σ should be vertically aligned, but this is not possible on this word processor. Very commonly, the *starting* and *final* value will conform to a standard range and can be left implied, giving the forms:

$$\Sigma a_i x^i \quad \text{and} \quad \Sigma_i a_i x^i$$

In the second of these, the **index** of the **summation** appears as a subscript on the Σ on its own, without the range values, a usage that is very common as it helps highlight the relevant **index** clearly. All this may seem to be an unnecessary complication with $ax^2 + bx + c$, but it becomes useful if we're writing $a_{10}x^{10} + a_9x^9 + a_8x^8 + a_7x^7 + a_6x^6 + a_5x^5 + a_4x^4 + a_3x^3 + a_2x^2 + a_1x + a_0$, which can again be written just as $\Sigma a_i x^i$ if we know, or don't care about, the number of **terms**. The "don't care about" gives us another *generic* way of writing expressions of the same basic kind, much as we did with **functional notation**, and I will use this in the discussion of **polynomials** in Section 1.9 below.

The **summation notation** enables us neatly to define a *multiplication* for the **matrices** or two-dimensional number **arrays** introduced in Section 1.1. Since we know the range of the summation is always 1 to 3 for three-by-three **matrices**, we can use the short form without the *starting* and *final* values and define the (i, j)'th element of the **product matrix** c_{ij} calculated from *multiplying* $\{a_{ik}\}$ and $\{b_{kj}\}$ as:

---

[6] **Terms** are entities we add together. **Factors** are ones we multiply together.
[7] Another use of *superscript* **indices** will be mentioned in Chapter 5.
[8] I also introduce the symbol "≡" here to indicate "is exactly the same as", or **identity**, a concept defined at the start of Section 1.6. This symbol is also used for **Boolean equivalence**, defined in Section 1.4, which is a special case of **identity**.

$$c_{ij} = \Sigma_k\, a_{ik}.b_{kj} = \Sigma_k\, a_{ik}\times b_{kj}$$

so for example $c_{21} = a_{21}.b_{11} + a_{22}.b_{21} + a_{23}.b_{31}$.

This illustrates another aspect of the **summation notation**, that the actual **summation index** *disappears* in the final result that is on the left. Inside the sum expression, the **indices** *i, j* and *k* all appear, but *k* – the **summation index** itself – has disappeared in the result c_{ij}. *k* has just been used as an auxiliary variable in defining the sum, and doesn't contribute to the result. It is like a working variable in a computer subroutine. Such variables are called **dummy variables** or **bound variables**, the "binding" being to the Σ operation.

1.4 Predicates and Sets

Predicates are **functions** that give only one of two resulting values: 0 or 1, true or false, yes or no. So we might have a **predicate** *Round*(*x*) with *Round*(*Earth*) = 1, *Round*(*Brick*) = 0, and the connection with *language*, with *adjectives* and with logical assertions should be clear.

Because of their "true/false" nature, **predicates** figure very large in theories of logic, and algebras of **predicates** can be formed using the **Boolean operators** "∧" for "*and*" and "∨" for "*or*", which can be defined simply numerically as **functions** of their 0/1 **arguments**, because there are only four possible values for the two **arguments**, (0,0), (0,1), (1,0) and (1,1). So by giving the resulting values for all four combinations, we can completely define the **functions**:

∧:	$1 \wedge 1 = 1$	$1 \wedge 0 = 0$	$0 \wedge 1 = 0$	$0 \wedge 0 = 0$
∨:	$1 \vee 1 = 1$	$1 \vee 0 = 1$	$0 \vee 1 = 1$	$0 \vee 0 = 0$

These **Boolean operators** enable us to form complex expressions like:

$$(P(x) \wedge Q(y)) \vee (R(x) \wedge (x + 3y^2 > 19)) \vee (z = 2)$$

where we mix general **predicates** P(), Q() and R() with algebraic **equations** and **inequalities**, as these too must themselves be **predicates**, because they are either true (≡ 1) or false (≡ 0). The numbers in brackets here use the common convention that 1 corresponds to "true" and 0 to "false". Substituting different values for the variables *x*, *y* and *z* will give different results for this complete expression, but always either 0 or 1. P(*x*) might correspond to "*x* is an **integer** (a whole number)", Q(*x*) might mean "*x* is negative", when we could express it as: $Q(x) \equiv x < 0$.

As it happens, the symbol "≡" which I commonly use for "is identical with" or "is equivalent to" or "is always the same as" is itself often defined as a **Boolean operator**:

≡ :	$(1 \equiv 1) = 1$	$(1 \equiv 0) = 0$	$(0 \equiv 1) = 0$	$(0 \equiv 0) = 1$

where the sense of "sameness" is clear: if the two **arguments** are the same, the value is 1, if different 0. The other important **Boolean operator** is the *unary* operator of **negation**, which acts on a single variable (hence "*unary*"), switching its value:

$$!(0) = 1 \qquad !(1) = 0$$

You should be aware of **de Morgan's Laws**, which state that a **negation** of a "$\wedge$" expression goes down onto its constituents, but switches the "$\wedge$" to an "$\vee$" and *vice versa*:

$$!(P(x) \wedge Q(y)) = (!P(x) \vee !Q(y))$$

$$!(P(x) \vee Q(y)) = (!P(x) \wedge !Q(y)).$$

The **inverse** of a **function** – not to be confused with the **inverse element** of an element of a **group**, as described in the next Section – is a procedure that *reverses* the **function**, working on the *result* of the **function** and returning the **argument** or **arguments** that gave that result. So if $y = f(x) = 4x - 9$, we can define the **inverse** of f as $f^{-1}(y) = (y + 9)/4$ which gives us back x. **Functions** and their **inverses** always have this property, that applied successively, they cancel out:

$$f^{-1}(f(x)) = f^{-1}(y) = (y + 9)/4 = (f(x) + 9)/4 = (4x - 9 + 9)/4 = 4x/4 = x$$
$$f(f^{-1}(y)) = f((y + 9)/4) = 4(y + 9)/4 - 9 = (y + 9) - 9 = y.$$

Very few **functions** have **inverses** – are "**invertible**". **Predicates** in particular are, to all intents and purposes, *never* invertible. It's easy to establish whether something is round, at least approximately, but unthinkable to find a way of discovering all round objects. Nevertheless, mathematics has been profoundly affected throughout the twentieth century by the artificial notion that actually, in some perfect intellectual paradise, *all* **predicates** are **invertible**. The **inverses** of **predicates** are called **sets**, and postulating their existence leads to innumerable logical conflicts. But a huge effort was made, especially in the first half of the twentieth century, to reconstruct *all* mathematics in terms of **set theory**. This is a persistent trend in intellectual thinking: in much the same way, whenever economists refer to something as "real" like "real GDP" or "real interest rates", they actually mean something that exists only in the imagination of economists, and is at best defined through something as arbitrary and unreal as a statistical price index.

Having deep reservations about the validity of **set theory**, I will use the term **set** only in a general and intuitive sense which may suggest a **predicate inverse** as in "the set of everything that's round" or may just reflect ordinary English usage as in "this set of equations".

1.5 A First Look at Axiomatic Systems: Groups

There are two basic ways of introducing mathematical concepts. One is to work out a mechanism or **algorithm** such as the procedure for the multiplication of numbers, and then gradually to work out its properties, such as that the order of the factors in a multiplication doesn't matter, that multiplying by 1 leaves the other factor unchanged, and so on. The other method, which with the notable exception of Euclidean geometry, has only come to be emphasised since the nineteenth century, is to *start* by postulating the properties, and then develop mechanisms that will satisfy them. This is the **axiomatic method**. The **axiomatic method** has proved so fruitful that it has in the twentieth century become the normal method of introducing mathematical concepts, and I will use it extensively. I said *mechanisms*, in the plural, because very often an extraordinary variety of mechanisms or computational devices will satisfy the same **axioms**. One of the most general **axiomatic systems**, and one that has had an enormous influence on mathematical thought since its discovery by E.Galois (1811-32), whose fame was unfortunately guaranteed by his death at the age of twenty in a duel, is the **Group**. A **group** obeys just four basic **axioms** describing an otherwise undefined operation $*$ on the *elements* of the **group**:

Axiom I (Closure): For any two elements $\boldsymbol{a}$ and $\boldsymbol{b}$, $\boldsymbol{a} * \boldsymbol{b}$ is an element of the **group**.

Axiom II (Associativity): For elements $\boldsymbol{a}, \boldsymbol{b}, \boldsymbol{c}$: $(\boldsymbol{a} * \boldsymbol{b}) * \boldsymbol{c} = \boldsymbol{a} * (\boldsymbol{b} * \boldsymbol{c})$

Axiom III (Identity Element): There is a unique **unit** or **zero** element **e** such that for any element $\boldsymbol{a}$:

$$\boldsymbol{a} * \mathbf{e} = \mathbf{e} * \boldsymbol{a} = \boldsymbol{a}$$

Axiom IV (Inverse): For any $\boldsymbol{a}$, there is a unique **inverse element** $\boldsymbol{a}^{-1}$ such that:

$$\boldsymbol{a} * \boldsymbol{a}^{-1} = \boldsymbol{a}^{-1} * \boldsymbol{a} = \mathbf{e}$$

There may also be an **axiom**:

Axiom V (Commutativity): For any two elements $\boldsymbol{a}, \boldsymbol{b}$: $\boldsymbol{a} * \boldsymbol{b} = \boldsymbol{b} * \boldsymbol{a}$

but this may or may not be included. When it is included, the **group** is called a **commutative group**, where, incidentally, the stress on the word *commutative* is on the second syllable: commUtative. Observe too how the **unit** or **zero** element is more generally called the **Identity Element**.

All manner of things fit these **axioms**. Ordinary numbers form a **commutative group** for the operation of *addition* + where $\mathbf{e} \equiv 0$, and the **inverse** of a is $-a$. They also form a **commutative group** with respect to the operation of *multiplication*, and for this $\mathbf{e} \equiv 1$ and the **inverse** of a is $1/a$. (Strictly 0 must be excluded from this second case because 0/0 is undefined).

A more exotic, yet actually simpler, **group** is the set of rotations of the equilateral triangle shown in *Figure 1.5.1*. Here we define the elements of the **group** to be the set of *rotations*:

$e \equiv$ identity, *no* rotation
$p \equiv$ *anticlockwise* rotation about O through 120°
$q \equiv$ *anticlockwise* rotation about O through 240°
$r \equiv$ rotation about a through 180°
$s \equiv$ rotation about b through 180°
$t \equiv$ rotation about c through 180°

Two rotations are equal if they transform the triangle into the same position, and $*$ is "followed by" so that $p * q$ is "rotation p followed by rotation q".[9] So, for example, $r * t = q$.

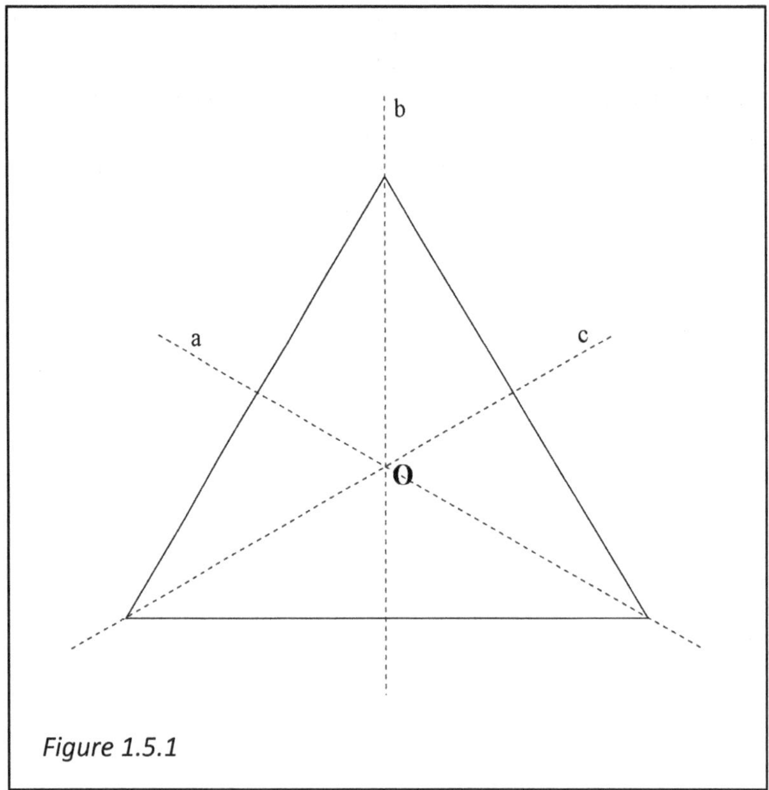

Figure 1.5.1

A more specific set of **axioms** which I will refer to from time to time is the set defining the ordinary **real numbers** that we use all the time. I give the complete listing, in a somewhat condensed form, below, in which you will see the **group axioms** re-appearing for both + and ×.[10] These are called the **axioms** of an **ordered field**. The set of elements of the **field** we refer to as *F*.

Axiom I:	$a + b$ is an element of F	(Closure of +)
Axiom II:	$(a + b) + c = a + (b + c)$	(Associativity of +)
Axiom III:	$a + 0 = 0 + a = a$	(Existence of *unique* **zero**)
Axiom IV:	$a + {}^{-}a = {}^{-}a + a = 0$	(Additive Inverse)
Axiom V:	$a + b = b + a$	(Commutativity of Addition)
Axiom VI:	$a \times b$ is an element of F	(Closure of ×)
Axiom VII:	$(a \times b) \times c = a \times (b \times c)$	(Associativity of ×)
Axiom VIII:	$a \times 1 = 1 \times a = a$	(Existence of *unique* **unit**)
Axiom IX:	$a \times 1/a = 1/a \times a = 1$	(Multiplicative Inverse)

[9] This example comes from C.B.Allendoerfer and C.O.Oakley *Principles of Mathematics*, McGraw-Hill 1963, p.115.
[10] C.B.Allendoerfer and C.O.Oakley *op.cit.*, pp. 57 and 75

Axiom X:	$a \times b = b \times a$	(Commutativity of Multiplication)
Axiom XI:	$a \times (b + c) = (a \times b) + (a \times c)$	(Distribution Law)
Axiom XII:	For each *a*, *b*, one and only of the following is true: $a < b$ $a = b$ $a > b$	(Trichotomy)
Axiom XIII:	If $a > b$ and $b > c$ then $a > c$	(Transitivity of >)
Axiom XIV:	If $a > b$ then $a + c > b + c$	
Axiom XV:	If $a > b$ and $c > 0$ then $a \times c > b \times c$	

The very last **axiom** will be shown in Chapter 4 to lead to a property about which you should be warned. Normally, *applying the same change to both sides of an existing equation (such as $a = b$) or inequality (such as $a < b$) means the same relationship holds, but beware the* **inequality trap**:

If $a > b$ and $c < 0$ then $a \times c < b \times c$

i.e. multiplying both sides of an **inequality *by a negative number*** *changes the sense* of the **inequality**.

Note how the "Identity Element" of a **group** appears as the **zero** of *addition*, but as the **unit** of *multiplication*. In Chapter 4 we will see how *the same* operation has an Identity Element that can be seen as a **zero** in one interpretation, and as a **unit** in another.

1.6 Equations and Inequalities

Statements that are true for *every* value of the variables appearing in them are **identities**. **Axioms** are **identities**: by definition they hold whatever the values of their constituent variables. The term **equation** is generally reserved for **equalities** – statements involving an "=" – that hold *only for certain specific values* of the variables in the expression. This is simply customary usage: the word **equation** originally meant the act of making things equal, and could apply to **identities** like $(a + b) + c = a + (b + c)$ as well. To **solve** an **equation** means to find the specific value or values for which it is true, and these are usually called the **solution(s)** to the **equation**, although again the word **solution** can also mean the act of solving the **equation** and has in recent years acquired a vile salesmen's sense of referring to any software product, and by extension to virtually any product at all.

Solving an **equation** very often entails finding the **inverse** of a **function**, a concept that was discussed in Section 1.4. As an example, we could describe $4x + 7$ as a **function** $f(x)$ and write $y = f(x)$. Now to **solve** the equation:

$$4x + 7 = 0$$

we could formulate the **inverse** $f^{-1}()$ and set it as a **function** of *y*:

$$y = f(x) = 4x + 7 \quad \rightarrow \quad 4x = y - 7 \quad \rightarrow \quad x = (y - 7)/4$$

so that $f^{-1}(y) = (y - 7)/4$, where above I have used the logical **implication** symbol "$\rightarrow$", which is sometimes useful in derivations. So our original **equation** $4x + 7 = 0$ is now **solved** by $f^{-1}(0)$, which equals $-7/4$. To check,

$$4(-7/4) + 7 = -7 + 7 = 0.$$

These two ways of viewing the same idea should be kept in mind.

The solution – refraining from bold now as this is a very general word – to simple **linear equations** like this one (**linear equations** are **equations** containing no higher powers of any of the variables, such as x^2, nor any products of two or more variables, such as xy) become more interesting when several variables appear, and these will be discussed in Section 1.7.

The next step up from **linear equations** like that above is the general **quadratic equation** in one variable, which is an **equation** of the form:

$$ax^2 + bx + c = 0.$$

It should be clear that the solution to *any* form like $ax^2 + bx + c = d$ can be included in this form, simply by replacing c by $c - d$. The general **quadratic** is solved by a trick called "completing the square" which was known in Babylon over three thousand years ago. The idea is to make the left-hand side into a simple squared term of the form $(p + q)^2$, which is called in the same tradition of terminology a "perfect square". We will not need these terms after this section, so I will not put them bold. Expanding $(p + q)^2$ gives:

$$(p + q)\,(p + q) = p^2 + 2pq + q^2$$

and if we rewrite $ax^2 + bx + c = 0$ as $x^2 + (b/a)x + c/a = 0$ we can identify p with x, and so see that the first two terms in the expression match the first two in the $(p + q)^2$ form if we identify q with $b/2a$:

$$p^2 + 2pq \equiv x^2 + 2xq = x^2 + 2x(b/2a) = x^2 + (b/a)x$$

so that $x^2 + (b/a)x + (b/2a)^2$ *would* be a "perfect square":

$$(x + b/2a)^2 = x^2 + (b/a)x + (b/2a)^2$$

Since we can add the same thing to both sides of any **equation** and the result will still be true, we can add $(b/2a)^2 - c/a$ to both sides of $x^2 + (b/a)x + c/a = 0$ to obtain:

$$x^2 + (b/a)x + c/a + (b/2a)^2 - c/a = (b/2a)^2 - c/a$$

so $\quad x^2 + (b/a)x + (b/2a)^2 = (b/2a)^2 - c/a$

so $\quad (x + b/2a)^2 = (b/2a)^2 - c/a$

and this gives the answer:

$$x + b/2a = \sqrt{[(b/2a)^2 - c/a]} \qquad \text{or} \qquad x = -\, b/2a \pm \sqrt{[(b/2a)^2 - c/a]}$$

keeping in mind that the **square root** can have either sign, which we indicate by the symbol "±". This solution is normally expressed as:

$$x = -\, b/2a \pm \sqrt{[(b/2a)^2 - 4ac/(2a)^2]} = -\, b/2a \pm \sqrt{(b^2 - 4ac)}/\sqrt{(2a)^2}$$

or $$x = [-\, b \pm \sqrt{(b^2 - 4ac)}]/(2a)$$

and you should note that the "±" means there are *two* solutions. At this stage, we can only allow for solutions when the **discriminant** $b^2 - 4ac$ is *positive*. In Chapter 4, I will show how the solution takes the form of a **complex number** if $b^2 - 4ac$ is *negative*.

An **inequality** is a statement involving one of the **inequality operators** "<", ">", "≤" or "≥" where the last two mean "is less than *or* equal to" and "is greater than *or* equal to". These **operators** make statements about the **order** of the two numbers or expressions they connect, a statement that has the precise mathematical sense defined by the **axioms of order**, which were **Axioms** XII to XV in Section 1.5. Once again I would like to draw attention to the point that, whereas normally, *applying the same change to both sides of an existing equation (such as $a = b$) or inequality (such as $a < b$) means the same relationship holds, this fails in the one case of the* **inequality trap**:

If $a > b$ and $c < 0$ then $a \times c < b \times c$

i.e. multiplying both sides of an **inequality** *by a negative number* *changes the sense* of the **inequality**.

1.7 Linear Equations

I defined **linear equations** above as **equations** which contain no higher powers of any of the variables, such as x^2, nor any products of two or more variables, such as xy. But **constant** terms are allowed. If such terms are present, it is customary to write them as a single term standing alone on the right-hand side. This leads to the standard formulation of **linear equations**, in forms like:

$$ax + by + cz + dw = k \qquad \text{or} \qquad ax + by = c$$

where all **constant** terms are gathered together on the right, and any minus signs included into the **coefficients,**[11] so that $x - 3y$ can be put in the standard form $ax + by$ with $a = 1$, $b = -3$. Sometimes subscripted variables are used, giving forms like:

[11] The constant multipliers of the variables in a **linear equation** are often referred to as **coefficients**, but this term is used rather loosely.

$$a_{11}x_1 + a_{12}x_2 + a_{13}x_3 = b_1$$
$$a_{21}x_1 + a_{22}x_2 + a_{23}x_3 = b_2$$
$$a_{31}x_1 + a_{32}x_2 + a_{33}x_3 = b_3$$

where in the a_{ij} **coefficients**, the first **index** labels the **equation**, the second the **variable**. This is indeed the standard form of a system of three **linear equations** in three unknowns. If all the **constant** terms b_i are zero, the system of **equations** is called **homogeneous**, if not it is called, perhaps not surprisingly, **non-homogeneous**. Such systems of n **linear equations** in n unknowns – note the "square" pattern – have the curious property that the criterion that must hold for the **homogeneous equations** to have a (non-zero) solution is precisely the condition that must ***not*** hold for the **non-homogeneous** system to have a solution. This criterion involves a single number called the **determinant** of the $\{a_{ij}\}$ **matrix.** If the **determinant** is zero, the **homogeneous equations** will have a solution, but the **non-homogeneous** will not, but if the **determinant** is *non-zero*, the **non-homogeneous** will have a solution, but now the **homogeneous** will not![12]

This is most easily seen in the 2 × 2 case, which we will write as:

$$ax + by = e$$
$$cx + dy = f$$

Suppose $e \neq 0$ and $f \neq 0$.[13] Multiply the first equation through by d and the second by b:

$$adx + bdy = de$$
$$bcx + bdy = bf$$

and take the difference:

$$adx + bdy - bcx - bdy = adx - bcx = de - bf$$

so $\quad x = (de - bf)/(ad - bc)$.

Having solved for x, we can get y by either equation, say: $y = (e - ax)/b$. The point is that we could not have obtained a valid solution for x in the first place if $ad - bc = 0$, because *division by zero is undefined*. $ad - bc$ is the **determinant** of the 2 × 2 system of **linear equations**.

If $e = f = 0$, then the first equation gives:

$$y = -ax/b$$

but if we put this into the *second* equation, we get:

---

[12]As mentioned earlier, **determinants** were the subject of the Revd. Charles Dodgson's research, and there is a charming anecdote that when Queen Victoria read *Alice's Adventures in Wonderland*, she asked the author to send her a copy of his next book, which he did: *The Algebraical Theory of Determinants*.

[13]Here "≠" means "is not equal to". I will later use "≈" to mean "is approximately equal to".

$$cx - dax/b = 0 \quad \text{or} \quad (cb - da)x = 0$$

which entails that either $x = 0$ or $cb - da = 0$ which is the same as $ad - bc = 0$. Now $x = 0$ is a valid solution, when we will also have $y = -ax/b = 0$. But to have a solution with $x \neq 0$, we *must* have the **determinant** $ad - bc = 0$. Note too that when this *is* the case, we can choose *any* value for x, and set $y = -ax/b$ to get y. There is now an *infinity* of solutions!

If there are *more* **equations** *than* **variables**, the system will generally have *no* solution.

If there are *more* **variables** *than* **equations**, we can *choose* the values of the "surplus" **variables** *to be anything we want*, as we could for x in the **homogeneous** system, and these chosen values will effectively create new **constant** terms, so that we reduce the system to a **non-homogeneous** system in the same number of **equations** as variables.

For higher dimensional systems – 3×3, 4×4 and so on – the theory of **determinants** gets rather complicated, involving confusing rules on **permutations**, and I always work just by eliminating one variable using one equation at a time. So above I would use the first **equation** to get, in the **non-homogeneous** case:

$$y = (e - ax)/b$$

and then put this into the second to get:

$$cx + d(e - ax)/b = f$$

or $\quad bcx + d(e - ax) = bf$

so $\quad (bc - da)x = bf - de$

$$x = (bf - de)/(bc - da) = (de - bf)/(ad - bc)$$

as before. This method is known as **Gaussian elimination** and it avoids a lot of hassle.

Linear equations become very important in using the **calculus**, as we will see in Chapter 2, because the calculus is based on a special **operator**, the **differential operator**, which creates from any arbitrary algebraic **equation** a **derived equation** that is **linear** in a new set of variables called **differentials**, and where the original variables now contribute to defining new **coefficients** for these **linear equations**. The effect is that for every set of values satisfying the original equation, we get a new **linear equation** or system of **linear equations**.

This is as good a time as any to mention a notion that will be used repeatedly, that of **linear independence**. It is generally defined for **vectors**, but since I haven't introduced these as yet, I will define it for **functions**. A set of **functions** $\{f_i(x)\}$[14] is **linearly independent**

---

[14] Curly brackets {} are often used to denote *lists* of things, as in the next section.

if $\Sigma a_i f_i(x) = a_1 f_1(x) + a_2 f_2(x) + a_3 f_3(x) + a_4 f_4(x) + \ldots + a_n f_n(x) = 0$ for all x

implies that all the $a_i = 0$.

So if the $\{f_i(x)\}$ are **linearly independent**, $\Sigma a_i f_i(x) \equiv 0$ *only if all the* $a_i = 0$, or, putting it the other way, if *any* $a_i \neq 0$, then $\Sigma a_i f_i(x) \neq 0$. As an example, the powers of x (i.e. x, x^2, x^3, x^4, etc.) are **linearly independent** and I will use this in equating coefficients of like powers of x in Section 1.9.

It is important to realize that this only applies when $\Sigma a_i f_i(x) = 0$ *for all values of* x, so that this is known to be an **identity**. If the equation only holds for a *specific* value or *specific* values of x, then we simply have an **equation** for x. So

$$ax^2 + bx + c = 0$$

is an **equation** for two precise values of x, *unless it is asserted that it holds for all values of* x, when **linear independence** of the powers entails that:

$$a = 0, \quad b = 0, \text{ and } c = 0.$$

The meaning of the concept, which at first seems rather elusive, is that if **linear independence** is *not* so, then one of the $f_i(x)$ **functions** could be expressed in terms of the others, like:

$$f_j(x) = \Sigma_{i \neq j}(a_i/a_j) f_i(x).$$

We *cannot* express the **function** x^2 as $-(b/a)x - c/a$ *precisely because* $ax^2 + bx + c = 0$ is only true for all values of x when $a = b = c = 0$.

1.8 Matrices and Determinants

A **standard vector** of **dimension** m is a **function** of a single **integer** over a finite range of values from 1 to m. It may be written in **list notation** as $\{x_i\}$, as $[x_1, x_2, x_3, \ldots x_m]$ or simply by the name in bold, as, say, $\mathbf{x}$. The **argument** "i" is called the **index**. A **matrix** of **dimension** $m\times n$ (pronounced "m by n") is a **function** of *two* **integers** over finite ranges of values 1 to m and 1 to n. If $m = n$ the **matrix** is called a **square matrix**. **Square matrices** are much the most common kind. **Matrices** are commonly written as a two-dimensional array, with the *first* **argument** or first **index** giving the *row* number, and so often being called the **row index**, the *second* **argument** or second **index** giving the *column* number, and so often being called the **column index**.

So a 3×3 **matrix** may be written:

$$\begin{pmatrix} a_{11} & a_{12} & a_{13} \\ a_{21} & a_{22} & a_{23} \\ a_{31} & a_{32} & a_{33} \end{pmatrix}$$

Like **vectors, matrices** may also be written in **list notation** $\{a_{ij}\}$ or simply as bold capital letters like **A**. The row-column layout is merely a convention in writing: there is no inherent geometrical meaning in it.

Because the ranges of the **indices** are finite, the entire **function** may be specified by explicitly giving its values. So a **vector** of **dimension** 3 – also called a **3-vector** – may be given as [22.7, 13.9, −7.2], and a 3×3 **matrix** may be given as:

$$\begin{pmatrix} 12.9 & -4.23 & 17 \\ 11.41 & 6.22 & 9.04 \\ 7 & 37 & 1 \end{pmatrix}$$

Matrices can be multiplied. The definition of **matrix multiplication** is that if **C** = **A** × **B**, then:

$$c_{ik} = \Sigma_j a_{ij} b_{jk}$$

which is only defined if the **column dimension** of **A** equals the **row dimension** of **B**. So the **product** of two 3×3 **matrices A** and **B** would be:

$$\begin{pmatrix} a_{11}b_{11} + a_{12}b_{21} + a_{13}b_{31} & a_{11}b_{12} + a_{12}b_{22} + a_{13}b_{32} & a_{11}b_{13} + a_{12}b_{23} + a_{13}b_{33} \\ a_{21}b_{11} + a_{22}b_{21} + a_{23}b_{31} & a_{21}b_{12} + a_{22}b_{22} + a_{23}b_{32} & a_{21}b_{13} + a_{22}b_{23} + a_{23}b_{33} \\ a_{31}b_{11} + a_{32}b_{21} + a_{33}b_{31} & a_{31}b_{12} + a_{32}b_{22} + a_{33}b_{32} & a_{31}b_{13} + a_{32}b_{23} + a_{33}b_{33} \end{pmatrix}$$

where the *outer* **indices** in each sum always reflect the position of the element in the resulting **product matrix** and do not change through the sum, so for example, $a_{31}b_{11}+ a_{32}b_{21}+ a_{33}b_{31}$ is the c_{31} element of the **product** and its *outer* **indices** are always 3 for the **row index** of the a_{ij} factor and 1 for the **column index** of the b_{ij} factor.

It looks nasty, but one soon acquires the habit of mentally visualising the appropriate *column* of the *second* **matrix** placed over the appropriate *row* of the first. So as an example:

$$\begin{pmatrix} 3 & 5 \\ 4 & 1 \end{pmatrix} \times \begin{pmatrix} 7 & 9 \\ 2 & 3 \end{pmatrix} = \begin{pmatrix} 3 \times 7 + 5 \times 2 & 3 \times 9 + 5 \times 3 \\ 4 \times 7 + 1 \times 2 & 4 \times 9 + 1 \times 3 \end{pmatrix} = \begin{pmatrix} 31 & 42 \\ 30 & 39 \end{pmatrix}$$

Matrix multiplication is *not* **commutative. B** × **A** ≠ **A** × **B**. For example,

$$\begin{pmatrix} 7 & 9 \\ 2 & 3 \end{pmatrix} \times \begin{pmatrix} 3 & 5 \\ 4 & 1 \end{pmatrix} = \begin{pmatrix} 7 \times 3 + 9 \times 4 & 7 \times 5 + 9 \times 1 \\ 2 \times 3 + 3 \times 4 & 2 \times 5 + 3 \times 1 \end{pmatrix} = \begin{pmatrix} 57 & 44 \\ 18 & 13 \end{pmatrix}$$

which is clearly not the same as the previous result. DON'T FORGET THIS.

Every **square matrix** has a defined **determinant**. For 2×2 **matrices** it is given by the formula:

$$\textbf{Det}\begin{pmatrix} a & b \\ c & d \end{pmatrix} = \begin{vmatrix} a & b \\ c & d \end{vmatrix} = ad - bc$$

as we saw in the previous Section. The vertical bars are a common notation for **determinants**. For 3×3 **matrices**, the formula is:

$$a_{11}(a_{22}a_{33} - a_{23}a_{32}) - a_{12}(a_{21}a_{33} - a_{23}a_{31}) + a_{13}(a_{21}a_{32} - a_{22}a_{31})$$

where you can see by looking at the corresponding **matrix** earlier in this Section, that we take each element of the first row and multiply it by the **determinant** of the 2×2 **matrix** that is obtained if we omit the row and column in which the first row element lies. And the signs swap with alternate terms too! You may just be able to see that this expansion could be written as:

$$\sum_{ijk = \text{Permutation}(123)} \varepsilon_{ijk}\, a_{1i}a_{2j}a_{3k}$$

i.e. every term is of the form $a_{1i}a_{2j}a_{3k}$ for *ijk* running through every **permutation** of 123, with the sign ε_{ijk} being *positive* if the **permutation** is **cyclic** – as 123, 231, 312 – and *negative* if the **permutation** is **anti-cyclic** – as 321, 213, 132. If you can see this, you must be a born mathematician. When I first encountered this permutational concept of **determinants**, I was hopelessly lost. But this model always works: the 2×2 expansion comes out as $\sum_{ij = \text{Permutation}(12)} \varepsilon_{ij}a_{1i}a_{2j} = a_{11}a_{22} - a_{12}a_{21}$. Here there are only two **permutations**, 12 (**cyclic** by default) and 21 (**anti-cyclic**).

Determinants also offer a different way of solving **non-homogeneous linear equations** from the **Gaussian elimination** given above. This is most easily seen for the 2×2 case:

$$a_{11}x_1 + a_{12}x_2 = k_1$$
$$a_{21}x_1 + a_{22}x_2 = k_2$$

where as before, the **row index** on $\{a_{ij}\}$ or **A** identifies the **equation**, the **column index** identifies the unknown variable. The **determinantal** solution, called **Cramer's rule**, is that each unknown is found as a **quotient** of two **determinants**:

$$x_1 = \begin{vmatrix} k_1 & a_{21} \\ k_2 & a_{22} \end{vmatrix} \div \begin{vmatrix} a_{11} & a_{12} \\ a_{21} & a_{22} \end{vmatrix} = (k_1a_{22} - a_{21}k_2)/(a_{11}a_{22} - a_{12}a_{21})$$

$$x_2 = \begin{vmatrix} a_{11} & k_1 \\ a_{21} & k_2 \end{vmatrix} \div \begin{vmatrix} a_{11} & a_{12} \\ a_{21} & a_{22} \end{vmatrix} = (a_{11}k_2 - k_1a_{21})/(a_{11}a_{22} - a_{12}a_{21})$$

In these formulae I use the traditional division symbol "÷" which stands out more clearly.

The **matrix** whose **determinant** gives the **denominator** is just the **A matrix** itself. The **matrix** whose **determinant** gives the **numerator** is formed by replacing the *column* corresponding to the variable being solved by the **vector** of the right-hand side (RHS) constant values. There are no complications with signs.

The 3×3 version of **Cramer's rule** for the set of **linear equations**:

$$a_{11}x_1 + a_{12}x_2 + a_{13}x_3 = k_1$$
$$a_{21}x_1 + a_{22}x_2 + a_{23}x_3 = k_2$$
$$a_{31}x_1 + a_{32}x_2 + a_{33}x_3 = k_3$$

is:

$$x_1 = \begin{vmatrix} k_1 & a_{12} & a_{13} \\ k_2 & a_{22} & a_{23} \\ k_3 & a_{32} & a_{33} \end{vmatrix} \div \begin{vmatrix} a_{11} & a_{12} & a_{13} \\ a_{21} & a_{22} & a_{23} \\ a_{31} & a_{32} & a_{33} \end{vmatrix}$$

$$x_2 = \begin{vmatrix} a_{11} & k_1 & a_{13} \\ a_{21} & k_2 & a_{23} \\ a_{31} & k_3 & a_{33} \end{vmatrix} \div \begin{vmatrix} a_{11} & a_{12} & a_{13} \\ a_{21} & a_{22} & a_{23} \\ a_{31} & a_{32} & a_{33} \end{vmatrix}$$

$$x_3 = \begin{vmatrix} a_{11} & a_{12} & k_1 \\ a_{21} & a_{22} & k_2 \\ a_{31} & a_{32} & k_3 \end{vmatrix} \div \begin{vmatrix} a_{11} & a_{12} & a_{13} \\ a_{21} & a_{22} & a_{23} \\ a_{31} & a_{32} & a_{33} \end{vmatrix}$$

Throughout these equations, I've used the vertical bar notation to indicate the **determinant**.

Although **matrix multiplication** is **non-commutative**, there *is* a **unit matrix** and **matrices** with a non-zero **determinant** do have an **inverse**. The **unit matrix** is that with all the diagonal (top left to bottom right or all the a_{jj}) elements equal to 1, so for 3×3 **matrices** it is:

$$\begin{pmatrix} 1 & 0 & 0 \\ 0 & 1 & 0 \\ 0 & 0 & 1 \end{pmatrix}$$

and it is commonly written **I**. Algebraically, it can also be represented by the **Kronecker delta** symbol δ_{ij}, which has the value 1 if $i = j$, and 0 if $i \neq j$. So the **inverse** of a **matrix A** is the **matrix** $\mathbf{A}^{-1}$ which obeys:

$$\mathbf{A} \times \mathbf{A}^{-1} = \mathbf{A}^{-1} \times \mathbf{A} = \mathbf{I},$$

and that *is* correctly written – **matrices** *do* **commute** with their **inverses**. In terms of the individual number elements, we must have:

$$\Sigma_k\, a_{ik}.(a^{-1})_{kj} = \Sigma_k\, (a^{-1})_{ik}.a_{kj} = \delta_{ij}.$$

This is quite a demanding condition, because we have a complete set of $n \times n$ equations for each *column* of **I**, and so evaluating **inverses** is hard work. I'll illustrate the principle by the 2×2 case. To make things a little clearer, I'll write $(a^{-1})_{ik}$ as x_{ik}, and then we see:

First **I** column ($j = 1$): $\Sigma_k\, a_{ik}.(a^{-1})_{k1} = \Sigma_k\, a_{ik}.x_{k1} = \delta_{i1}$

($i = 1$) $a_{11}x_{11} + a_{12}x_{21} = \delta_{11} = 1$

($i = 2$) $a_{21}x_{11} + a_{22}x_{21} = \delta_{21} = 0$

which gives:

$$x_{11} = \begin{vmatrix} 1 & a_{12} \\ 0 & a_{22} \end{vmatrix} \div \begin{vmatrix} a_{11} & a_{12} \\ a_{21} & a_{22} \end{vmatrix} \qquad x_{21} = \begin{vmatrix} a_{11} & 1 \\ a_{21} & 0 \end{vmatrix} \div \begin{vmatrix} a_{11} & a_{12} \\ a_{21} & a_{22} \end{vmatrix}$$

Second **I** column ($j = 2$): $\Sigma_k\, a_{ik}.(a^{-1})_{k2} = \Sigma_k\, a_{ik}.x_{k2} = \delta_{i2}$

$$(i = 1) \qquad a_{11}x_{12} + a_{12}x_{22} = \delta_{12} = 0$$
$$(i = 2) \qquad a_{21}x_{12} + a_{22}x_{22} = \delta_{22} = 1$$

which gives:

$$x_{12} = \begin{vmatrix} 0 & a_{12} \\ 1 & a_{22} \end{vmatrix} \div \begin{vmatrix} a_{11} & a_{12} \\ a_{21} & a_{22} \end{vmatrix} \qquad x_{22} = \begin{vmatrix} a_{11} & 0 \\ a_{21} & 1 \end{vmatrix} \div \begin{vmatrix} a_{11} & a_{12} \\ a_{21} & a_{22} \end{vmatrix}$$

Evaluating the **determinants** gives:

$$x_{11} = a_{22}/(a_{11}a_{22} - a_{12}a_{21}) \qquad x_{21} = -a_{21}/(a_{11}a_{22} - a_{12}a_{21})$$
$$x_{12} = -a_{12}/(a_{11}a_{22} - a_{12}a_{21}) \qquad x_{22} = a_{11}/(a_{11}a_{22} - a_{12}a_{21})$$

but be careful here: the row and column positions are *switched*. As written, these elements form the **transpose** of the $\mathbf{A}^{-1}$ **matrix**. x_{21} appears top right and x_{12} bottom left. Putting them the right way round gives:

$$\mathbf{A}^{-1} = 1/|\mathbf{A}| \times \begin{pmatrix} a_{22} & -a_{12} \\ -a_{21} & a_{11} \end{pmatrix}$$

which does indeed give:

$$\mathbf{AA}^{-1} = 1/|\mathbf{A}| \times \begin{pmatrix} a_{11} & a_{12} \\ a_{21} & a_{22} \end{pmatrix} \times \begin{pmatrix} a_{22} & -a_{12} \\ -a_{21} & a_{11} \end{pmatrix}$$

$$= 1/|\mathbf{A}| \times \begin{pmatrix} a_{11}a_{22} - a_{12}a_{21} & -a_{11}a_{12} + a_{12}a_{11} \\ a_{21}a_{22} - a_{22}a_{21} & -a_{21}a_{12} + a_{22}a_{11} \end{pmatrix}$$

$$= 1/|\mathbf{A}| \times \begin{pmatrix} |\mathbf{A}| & 0 \\ 0 & |\mathbf{A}| \end{pmatrix} = \begin{pmatrix} 1 & 0 \\ 0 & 1 \end{pmatrix}.$$

and

$$\mathbf{A}^{-1}\mathbf{A} = 1/|\mathbf{A}| \times \begin{pmatrix} a_{22}a_{11} - a_{12}a_{21} & a_{22}a_{12} - a_{12}a_{22} \\ -a_{21}a_{11} + a_{11}a_{21} & -a_{21}a_{12} + a_{11}a_{22} \end{pmatrix}$$

$$= 1/|\mathbf{A}| \times \begin{pmatrix} |\mathbf{A}| & 0 \\ 0 & |\mathbf{A}| \end{pmatrix} = \begin{pmatrix} 1 & 0 \\ 0 & 1 \end{pmatrix}.$$

1.9 Polynomials

A **polynomial** in one variable is an expression – a **function** – of the form $\Sigma a_i x^i$ such as:

$$ax^2 + bx + c \qquad \text{or} \qquad 27x^3 - 3x^2 + 4x + 22$$

There are also **polynomials** in more than one variable, sometimes called **mixed polynomials**, such as:

$$17x^2y - 3xz + 4y \qquad \text{or} \qquad 2.44p^2q^2 - pq$$

but in this section I will be concerned with the single-variable forms. **Polynomials** can be added and subtracted, and multiplied and divided much like ordinary numbers, so for example:

$$(x^2 - 2x + 3) + (x - 4) = x^2 - x - 1$$

terms of like **degree** (the power of x)[15] being added together. Multiplication is a little more subtle:

$$(x^2 - 2x + 3) \times (x - 4) = x^3 - 2x^2 + 3x - 4x^2 + 8x - 12 = x^3 - 6x^2 + 11x - 12$$

but division gets decidedly nastier, so I will deal with that in a minute. The similarity with ordinary numbers is no surprise, as the normal decimal representation of a number is a **polynomial** form. So 349 is simply $3 \times 10^2 + 4 \times 10 + 9$, which is $3x^2 + 4x + 9$ with $x = 10$. But *fractional decimals* cannot be expressed in a strictly **polynomial** form. Still, we can easily extend the **polynomial** concept to cover this case. So 349.627 is:

$$3 \times 10^2 + 4 \times 10 + 9 + 6 \times (1/10) + 2 \times (1/10)^2 + 7 \times (1/10)^3$$

but this involves powers of the *reciprocal* of 10. In Chapter 3, I will show how such powers of reciprocals can be written as *negative* powers, $(1/x)^n = x^{-n}$ and this would enable us to write 349.627 as:

$$3 \times 10^2 + 4 \times 10 + 9 + 6 \times 10^{-1} + 2 \times 10^{-2} + 7 \times 10^{-3}$$

which is much neater, but strays outside the strict definition of a **polynomial** which allows only *positive* powers, together with *constants* which can be regarded as exhibiting a *zero* power, as I will again show in Chapter 3. For now, it will be helpful just to accept as a convention that *anything raised to a power of zero equals one*. So $17^0 = 1$, $10^0 = 1$ and $x^0 = 1$. This finally brings the "9" above into line, giving 349.627 as:

$$3 \times 10^2 + 4 \times 10^1 + 9 \times 10^0 + 6 \times 10^{-1} + 2 \times 10^{-2} + 7 \times 10^{-3}$$

making a smooth series. But keep in mind that strictly a **polynomial** is defined for powers $n \geq 0$ only, and under this definition the stronger analogy is between **polynomials** and **integers** or *whole numbers*.[16]

---

[15] The **degree** of the entire **polynomial** is the highest power of x occurring in it.

So we come to the nasty issue of **polynomial division**, which is generally presented in a way analogous to the notorious long division that caused so may schoolchildren so much misery (myself especially) before calculators were invented. The idea of the method is to use an algorithm – a method of calculating – based on *successive remainders*, where we try a single term like $3x^2$, multiply the **divisor** by this and subtract the result from the **dividend** to get a **remainder** and go round again using the **remainder** as the new **dividend**.

What do all these terms mean? To explain, I'll introduce the specialized form of **functional notation** commonly used for **polynomials**, which is to write them as **functions** with capital letter names. So $P(x)$ and $Q(x)$ might represent arbitrary **polynomials**. Using this convention, I will write a **polynomial division** as:

$$N(x)/D(x)$$

where I choose N and D to represent the "thing being divided" and the "thing doing the dividing" respectively because the standard mathematical terms for these two roles are the **numerator** (the bit on top or on the left) and the **denominator** (the bit underneath or to the right). **Dividend** is an old term for **numerator**, and **divisor** is an old term for **denominator**. The **remainder** needs a little more explanation.

We're trying to find a new **polynomial** $Q(x)$, called the **quotient**, which will satisfy:

$$Q(x) = N(x)/D(x)$$

but this means that $N(x) = Q(x)D(x)$, or expanding $Q(x)$ as $a_nx^n + a_{n-1}x^{n-1} + \ldots + a_1x + a_0$:

$$N(x) = a_nx^n\, D(x) + a_{n-1}x^{n-1}\, D(x) + \ldots + a_1x\, D(x) + a_0\, D(x).$$

It generally turns out that we can't do it precisely, and end up with a bit left over. That bit is the **remainder**. To work out the answer by the **long division algorithm**, we try a form $Q_k(x) = q_kx^k$ first, where q_k is a constant and k is the **degree** of $N(x)$ minus the **degree** of $D(x)$. We first work out $R_k(x) = N(x) - q_kx^k \times D(x)$, which we might call the first **partial remainder**, replace $N(x)$ with $R_k(x)$, and try again with some form $Q_{k-1}(x) = q_{k-1}x^{k-1}$ to get a new **remainder** $R_{k-1}(x) = R_k(x) - q_{k-1}x^{k-1} \times D(x)$ and round again, slowly chipping away at the original **numerator** until we get an $R_p(x)$ whose **degree** (the highest power of x occurring in it) is less than the **degree** of $D(x)$, which is as far as we can go with strict **polynomials** (i.e. without *negative* powers). $R_p(x)$ is the overall **remainder**, that is to say the bit left over.

If it sounds nasty, it is pretty bad, and I was expected to learn this at the age of eight! It gets a bit easier with an example. Let's take $N(x) = 12x^2 - 6x + 3$ and $D(x) = 4x + 2$, so we want to evaluate:

$$(12x^2 - 6x + 3)/(4x + 2)$$

---

[16] As in Solomon Feferman *The Number Systems* Addison-Wesley 1964 Chapter 5.

Now $12x^2/4x = 3x$ so try that as $Q_1(x)$, giving

$$R_1(x) = N(x) - 3x.D(x) = (12x^2 - 6x + 3) - 3x(4x + 2) = 12x^2 - 6x + 3 - 12x^2 - 6x$$

$$= -12x + 3.$$

Now using *this* as the **dividend**, and noting that $(-12x)/(4x) = -3$, try $Q_0(x) = -3$ to get:

$$R_0(x) = R_1(x) - (-3).D(x) = -12x + 3 + 3(4x + 2) = -12x + 3 + 12x + 6 = 9$$

So, including the **remainder** $R_0(x)$, as being "the bit we couldn't get rid of" *divided by* $D(x)$, the answer is:

$$Q(x) = (12x^2 - 6x + 3)/(4x + 2) = Q_1(x) + Q_0(x) + R_0(x)/D(x)$$

$$= 3x - 3 + R_0(x)/D(x) = 3x - 3 + 9/(4x + 2).$$

To check, we must have $Q(x)D(x) = N(x)$, or:

$$(3x - 3 + 9/(4x + 2))\,(4x + 2) = (3x - 3)(4x + 2) + 9 = 12x^2 + 6x - 12x - 6 + 9$$

$$= 12x^2 - 6x + 3$$

which is right!

There's an easier way of doing this. This depends on the fact that different powers of the same variable are **linearly independent** in the sense defined in Section 1.7. Since the **degree** of the **quotient** must be the difference between the **degrees** of **numerator** and **denominator**, and the **remainder** must have a **degree** at most one less than that of the **divisor** $D(x)$, we can simply put undetermined forms for $Q(x)$ and $R(x)$ in the equation:

$$N(x) = Q(x)D(x) + R(x)$$

and solve for the undetermined coefficients by equating coefficients of like powers. Again, an example will clarify things. Using our previous example with **degree**($N(x)$) = 2 and **degree**($D(x)$) = 1, we must have **degree**($Q(x)$) = 1 and **degree**($R(x)$) = 0. So put:

$$(12x^2 - 6x + 3) = (ax + b)(4x + 2) + c = 4ax^2 + 4bx + 2ax + 2b + c$$

so we must have, equating coefficients of x^2: $12 = 4a$ or $a = 3$, equating coefficients of x: $-6 = 4b + 2a$, so $4b = -12$ or $b = -3$, and equating coefficients of x^0: $3 = 2b + c = -6 + c$, so $c = 9$, so again:

$$Q(x) = 3x - 3 + 9/(4x + 2).$$

Could we get rid of the **remainder** if we used *negative* powers as we do for fractions? For example, readers who are familiar with the ordinary numerical form of long division will know that we can get the decimal part of a number like this:

```
        0.05286...
227 )12.0000000
      11.35
       0.65
       0.454
       0.196
       0.1816
       0.0144
       0.01362
       0.00078
```

Surprisingly, we can do this for **polynomials** if we allow negative powers, and using the easier method above. Replacing the **divisor** 227 in the foregoing by $2x^2 + 2x + 7$, and the **dividend** 12 by $1.x + 2$, we can try:

$$(2x^2 + 2x + 7).(ax + b + cx^{-1} + dx^{-2} + ex^{-3} + fx^{-4}) + px^{-3} + qx^{-4} = x + 2$$

which gives:

$$2ax^3 + 2ax^2 + 7ax + 2bx^2 + 2bx + 7b + 2cx + 2c + 7cx^{-1} + 2d + 2dx^{-1} + 7dx^{-2} + 2ex^{-1} + 2ex^{-2} + 7ex^{-3} + 2fx^{-2} + 2fx^{-3} + 7fx^{-4} + px^{-3} + qx^{-4} = x + 2$$

and equating coefficients of like powers again, we get:

x^3:	$2a = 0$	so $a = 0$
x^2:	$2a + 2b = 0$,	so $b = 0$
x:	$7a + 2b + 2c = 1$,	so $c = 0.5$
x^0:	$7b + 2c + 2d = 2$,	so $1 + 2d = 2$ or $d = 0.5$
x^{-1}:	$7c + 2d + 2e = 0$,	so $3.5 + 1 + 2e = 0$, or $e = -2.25$
x^{-2}:	$7d + 2e + 2f = 0$,	so $3.5 - 4.5 + 2f = 0$, or $f = 0.5$
x^{-3}:	$7e + 2f + p = 0$,	so $-15.75 + 1 + p = 0$, or $p = 14.75$
x^{-4}:	$7f + q = 0$,	so $q = -3.5$

so that:

$$Q(x) = 0.5x^{-1} + 0.5x^{-2} + -2.25x^{-3} + 0.5x^{-4} + (14.75x^{-3} + -3.5x^{-4})/(2x^2 + 2x + 7).$$

Now let's put $x = 10$:

$$Q(x) = 0.5/10 + 0.5/100 - 2.25/1000 + 0.5/10000$$

$$+ (14.75/1000 - 3.5/10000)/(227)$$

$= (500 + 50 - 22.5 + 0.5)/10000 + (147.5 - 3.5)/10000 \times 1/227$

$= 528/10000 + (144/227)/10000 = 0.0528634$

agreeing with our previous numerical long division.

Appropriate to this Section is a summary of the idea of **Partial Fractions**, which as so often I'll illustrate with an example. Suppose we have two **polynomials** in a ratio, the **numerator** on top being of lesser **degree** than the **denominator**. If this isn't the case, we need to divide one into the other until we have a **remainder** that *does* fit this requirement. So let's assume that's done and we have, say:

$$\frac{3x-4}{x^2-x-12} = \frac{3x-4}{(x-4)(x+3)}.$$

Partial fractions are the reverse of evaluating a sum of two fractions by putting them over a common **denominator**. We need to find A and B such that:

$$\frac{A}{(x-4)} + \frac{B}{(x+3)} = \frac{3x-4}{(x-4)(x+3)}.$$

Cross-multiplication shows that this will be satisfied if:

$$\frac{A(x+3)+B(x-4)}{(x-4)(x+3)} = \frac{3x-4}{(x-4)(x+3)},$$

or simply

$$A(x+3) + B(x-4) = 3x - 4.$$

Multiply this out to give:

$$(A+B)x + 3A - 4B = 3x - 4.$$

Since x and 1 are **linearly independent**, this will be satisfied for arbitrary x if:

$$A + B = 3 \qquad \text{and} \qquad 3A - 4B = -4.$$

So $B = 3 - A \rightarrow 3A - 12 + 4A = -4$, and so $A = 8/7$, so $B = 13/8$.
To check, put these values back into $A(x + 3) + B(x - 4) = 8/7x + 24/7 + 13/8x - 52/7 = 3x - 28/7 = 3x - 4$.

The method can also work if the **denominator** involves a **quadratic**, when we need to use a **numerator** form like $Ax + B$ rather than plain A or B. For example,

$$\frac{7x-4}{(x^2+1)(3x-2)} = \frac{Ax+B}{(x^2+1)} + \frac{C}{(3x-2)}.$$

Now we need $(Ax + B)(3x - 2) + C(x^2 + 1) = 7x - 4$. So:

$$3Ax^2 + 3Bx - 2Ax - 2B + Cx^2 + C = 7x - 4.$$

Therefore

$$3A + C = 0 \qquad 3B - 2A = 7 \qquad -2B + C = -4,$$

which gives $A = -2/13$, $B = 29/13$, $C = 6/13$. Checking,

$$(-2/13x + 29/13)(3x - 2) + 6/13(x^2 + 1)$$
$$= -6/13x^2 + 4/13x + 87/13x - 58/13 + 6/13x^2 + 6/13$$
$$= 91/13x - 52/13 = 7x - 4.$$

An important word of warning: if there is a repeated factor in the **denominator** like $(x - 1)^2$, we have to use not $A/(x - 1)^2$, but *two* **terms** $A/(x - 1) + B/(x - 1)^2$ just for this factor.

1.10 Infinite Series

It is sometimes useful to postulate the existence of **polynomials**, and indeed of other sums, with an *infinite* number of **terms**. This, like the postulate of **sets**, is something about which I have deep philosophical reservations. But if rather than *infinite*, we try to think of these as *arbitrarily large* sums, then we have a basically workable concept. Such a sum is called a **series**, and its defining concept is that however many **terms** we have, we can always add another.

An outstanding, and a very simple, example, is the **geometric series** Σax^j or:

$$a + ax + ax^2 + ax^3 + ax^4 + ax^5 + ax^6 + ax^7 + ax^8 + ax^9 + \ldots$$

This can be evaluated for any number of terms by a simple dodge. Defining the sum for n terms as:

$$S_n = a + ax + ax^2 + ax^3 + \ldots ax^{n-1}$$

(which does have n terms), we form:

$$xS_n = ax + ax^2 + ax^3 + ax^4 + \ldots ax^n$$

and subtract one from the other to obtain:

$$S_n - xS_n = a - ax^n$$

so

$$S_n = a(1 - x^n)/(1 - x)$$

If $|x| < 1$ (i.e. $x > -1$ and $x < +1$)[17], then as n becomes arbitrarily large, which we write as "$n \longrightarrow \infty$" and say "n tends to infinity", x^n gets smaller and smaller until we can ignore it completely, and say that the **series converges to the limit** $a/(1 - x)$ and so despite its "infinite" origins, the **series** "converges" to a perfectly well-defined **function**.

Note that this "**convergence**" does *not* hold if the **absolute value** of x, $|x|$, is *not* less than 1.

The calculus is normally derived entirely from this difficult notion of a **limit**, a concept which I believe is avoidable. I will begin my attempt to show this in Chapter 2.

One other idea which fits in here is the concept of proof by **mathematical induction**. This is a way of proving an assertion for arbitrary **integer** values n by showing that it holds for $n = 1$ and that if it holds for n, it must also hold for $n + 1$. An example will show what is meant. Say we want to prove that for all **integers** $n \geq 1$, $1 + 2 + 3 + \ldots + n = n(n + 1)/2$.

For $n = 1$, $\quad 1 = 1(1 + 1)/2 = 2/2 = 1$

Now assume that the result is true for n, and evaluate it for $n + 1$. The left-hand side will be:

$$1 + 2 + 3 + \ldots + n + n + 1$$

But, *by hypothesis*, $1 + 2 + 3 + \ldots + n = n(n + 1)/2$, so:

$$1 + 2 + 3 + \ldots + n + n + 1 = n(n + 1)/2 + n + 1 = [n(n + 1) + 2n + 2]/2$$

$$= [n^2 + 3n + 2]/2 = (n + 1)(n + 2)/2$$

which is our original proposition, now expressed, and so proved, for $n + 1$.[18]

Observe that the proposition could be formulated as a **predicate**

$$P(n) = [1 + 2 + 3 + \ldots + n = n(n + 1)/2]$$

so we first show that $P(1)$ = true, and then that $P(n) \longrightarrow P(n + 1)$ where the symbol "$\longrightarrow$" is used to mean **logical implication**: that $P(n)$ **implies** $P(n + 1)$.

[17] $|x|$ is the **absolute value** of a number – its value regardless of its sign. So $|-3.56| = 3.56 = |3.56|$.

[18] This example is from C.B.Allendoerfer and C.O.Oakley, *Principles of Mathematics*, McGraw-Hill 1963, p.91.

1.11 Colour and Pascal's Triangle

Figures 1.11.1− and *1.11.1+* show the familiar idea of the mixing of colours. The *Figures* are logically the same, differing only in the *suggestion* in *Figure 1.11.1+* of moving in from an initial black space, adding more colours until we end up with white, while *Figure 1.11.1−* reflects what happens in printing, where we start with a white page, and add first the rather insipid colours **yellow**, **cyan** and **magenta**, then go onto the stronger **red**, **green** and **blue** and finally end up with black.[19]

The first approach (that in *Figure 1.11.1+*) is called the **additive**, the second the **subtractive**. In the first, we start with black and progressively add more and more light until we attain white, in the second we start with white and progressively *remove* light until we end up with black.

The first thing to understand, something that took centuries to clarify, is that these are *physiological* colours. *They don't exist in the physics, only in our eyes*. We have three receptors in the retina at the back of our eyes, which are sensitive to slightly different parts of the spectrum of light that we see. The "**red**" receptors are most sensitive in the red or long-wavelength part of the spectrum, the "**green**" to slightly shorter wavelengths, the "**blue**" to slightly shorter again, but there's a lot of overlap, and we can sense light to significantly shorter wavelengths than the peak of the sensitivity of the "**blue**".

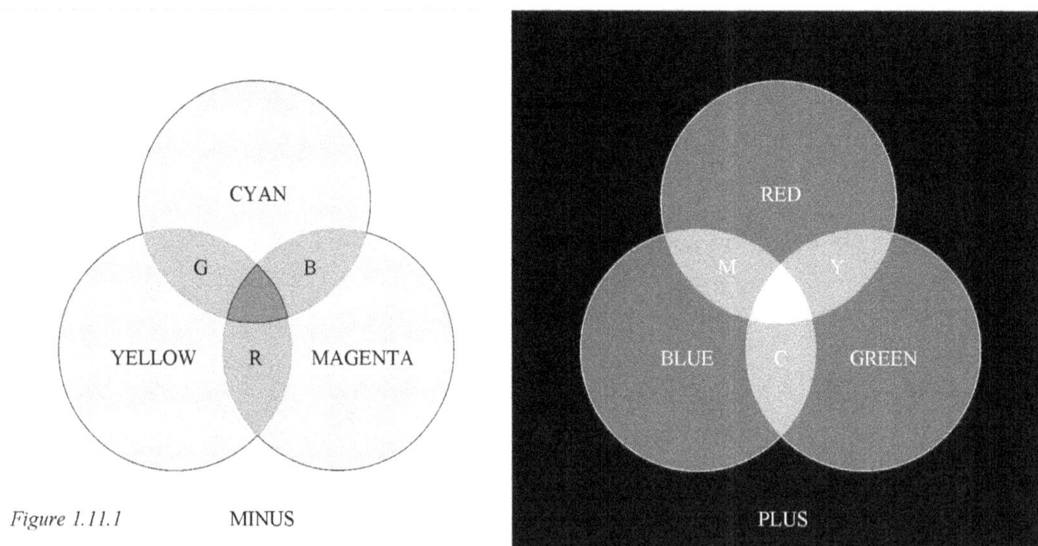

Figure 1.11.1

What we see as a *"colour"* are the three numbers that give the aggregate response of each of the three receptors, so writing the response of the **red** receptor as ρ(R), the **green** as ρ(G) and the **blue** as ρ(B), the three numbers ρ(R), ρ(G), ρ(B) tell us what colour we're seeing. The overlap of the responses means that no actual distribution of light energy can trigger just one receptor, so if we regard ρ() values as running from 0 to 1, there is no *actual* or *physically realizable* light that will give ρ(R) = 1 and ρ(G) = ρ(B) = 0, but the phosphors on the screen of this laptop don't do a bad job, and by showing our eyes a mix of

[19] In the black-and-white edition you'll just have to imagine the colours. Note that the "−" version correctly reflects the fact that CMY printing alone cannot produce a satisfactory black, and so a separate blac**K** is always used, hence the system is called "CMYK". Selected images can be downloaded in full colour from the website www.algebraiccalculus.com.

three light sources that correspond to "**red**", "**green**" and "**blue**", they can create a large part of the possible ρ(R) : ρ(G) : ρ(B) range of ratios that we can see and convince us that we're seeing "full colour". I'll refer to this coding of colours as the **RGB** system, a term that is common in computer work.

Red, **Green** and **Blue** – **R**, **G** and **B** – are called the **primary colours**. When two receptors are equally stimulated but not the third, we see a **secondary colour**. This can only happen in three ways, so there are again three **secondary colours**. They are:

magenta :	$\rho(R) = \rho(B) \approx 1,\ \ \rho(G) \approx 0$
yellow :	$\rho(R) = \rho(G) \approx 1,\ \ \rho(B) \approx 0$
cyan :	$\rho(B) = \rho(G) \approx 1,\ \ \rho(R) \approx 0$

They can also be labelled by the receptor that *isn't* triggered, as respectively, **minus green**, **minus blue**, and **minus red.**[20] The **secondary colours** are also referred to as the **CMY** system and are the colours used in printing inks, because these work *subtractively*. A mathematical model can be built if we put:

$$M = R + B$$
$$Y = R + G$$
$$C = B + G$$

when black becomes **zero** or 0, and white becomes:

$$W = R + G + B$$

And so white is the *only* **tertiary colour**. Now suppose we had only *two* receptors, as do people with **red-green** colour blindness, as these two receptors are mutations of each other and sometimes the gene mutates back. Now we have *two* primaries R and B, and so white now becomes the *only* **secondary**:

$$W = R + B$$

If we had *four* receptors, as I believe many birds do, keeping "R", "G" and "B" and calling the fourth one, say, "Q", we would see:

Four Primaries:	R, G, B, Q
Six Secondaries:	R + G, R + B, R + Q, G + B, G + Q, B + Q
Four Tertiaries:	R + G + B = −Q, R + G + Q = −B, R + B + Q = −G, G + B + Q = −R

and white, R + G + B + Q, would become the only **quaternary colour**. How much more colourful the world would be. No wonder birds sing so much. This pattern of numbers also occurs in the **coefficients** of the products of $(p + q)^n$ as:

$$(p + q)^0 = 1$$
$$(p + q)^1 = p + q$$
$$(p + q)^2 = p^2 + 2pq + q^2$$

[20] "≈" here means "is approximately equal to".

$$(p + q)^3 = p^3 + 3p^2q + 3pq^2 + q^3$$
$$(p + q)^4 = p^4 + 4p^3q + 6p^2q^2 + 4pq^3 + q^4$$
$$(p + q)^5 = p^5 + 5p^4q + 10p^3q^2 + 10p^2q^3 + 5pq^4 + q^5$$

which shows the form:

					1					
				1		1				
			1		2		1			
		1		3		3		1		
	1		4		6		4		1	
1		5		10		10		5		1

and is called **Pascal's Triangle** after Blaise Pascal (1623–1662) although it was known in China at least as far back as the Sung or Song dynasty four hundred years earlier. Its defining characteristic is that each number is the sum of the two immediately above it. Our own pattern of colour vision is given by the fourth line here, which I have highlighted.

1.12 Musical Scales[21]

Figure 1.12.1 shows the standard musical scales as agreed at an international conference in 1939 and in use ever since. This diagram is plotted **logarithmically**, a term which will be more fully explained in Chapter 3. The essence of it is this. Whereas on a normal ruler, equal intervals show *equal differences*, on a **logarithmic** plot, equal intervals show *equal ratios.* So a metre rule will have ten decimetre intervals, each indicating a *difference* of one decimetre in distance, which add up to the metre:

$$1/10 + 1/10 + 1/10 + 1/10 + 1/10 + 1/10 + 1/10 + 1/10 + 1/10 + 1/10 = 1$$

But in *Figure 1.12.1* the twelve equal steps around the circle each indicate a *ratio* of $\sqrt[12]{2}$ each (the twelfth root of 2, or the number that multiplied by itself twelve times equals 2), so altogether they specify a doubling of the sound frequency, which is the definition of an **octave**:

$$\sqrt[12]{2} \times \sqrt[12]{2} \times \sqrt[12]{2} \times \sqrt[12]{2} \times \sqrt[12]{2} \times \sqrt[12]{2} \times \sqrt[12]{2} \times \sqrt[12]{2} \times \sqrt[12]{2} \times \sqrt[12]{2} \times \sqrt[12]{2} \times \sqrt[12]{2} = 2$$

and every time we go round the full circle *clockwise*, the frequency *doubles*, and every time we go fully round *anticlockwise* the frequency *halves*.

So the three o'clock position, marked "C" on the diagram, does not represent 1.25 × the frequency at 12 o'clock (or the point marked "A"), but instead, writing the sound frequency at "C" as f(C):

$$f(\mathrm{C}) = \sqrt[12]{2} \times \sqrt[12]{2} \times \sqrt[12]{2} \times f(\mathrm{A}),$$

[21] References for this section are Sir James Jeans *The Science of Music*, Cambridge University Press,1943 and Harry F. Olson, *Music, Physics and Engineering*, Dover 1967.

likewise writing the frequency at A as *f*(A). The international conference established the reference point of the whole system as *f*(A) = 440 cycles per second (cps), giving the 3 o'clock position the value of 523.251 cps. I have chosen to list the notes on the diagram starting an **octave** lower, at *f*(A) = 220 and *f*(C) = 261.626, so that the **major** scale beginning at C and coming back round to C (3 o'clock) *passes* through A = 440 cps, and the diagram therefore covers two cycles or **octaves**, running from 220 cps through 440 cps to 880 cps. Any **octave** or cycle could have been chosen: three **octaves** up from the standard would have given *f*(A) = 3520 and *f*(C) = 4186.009.

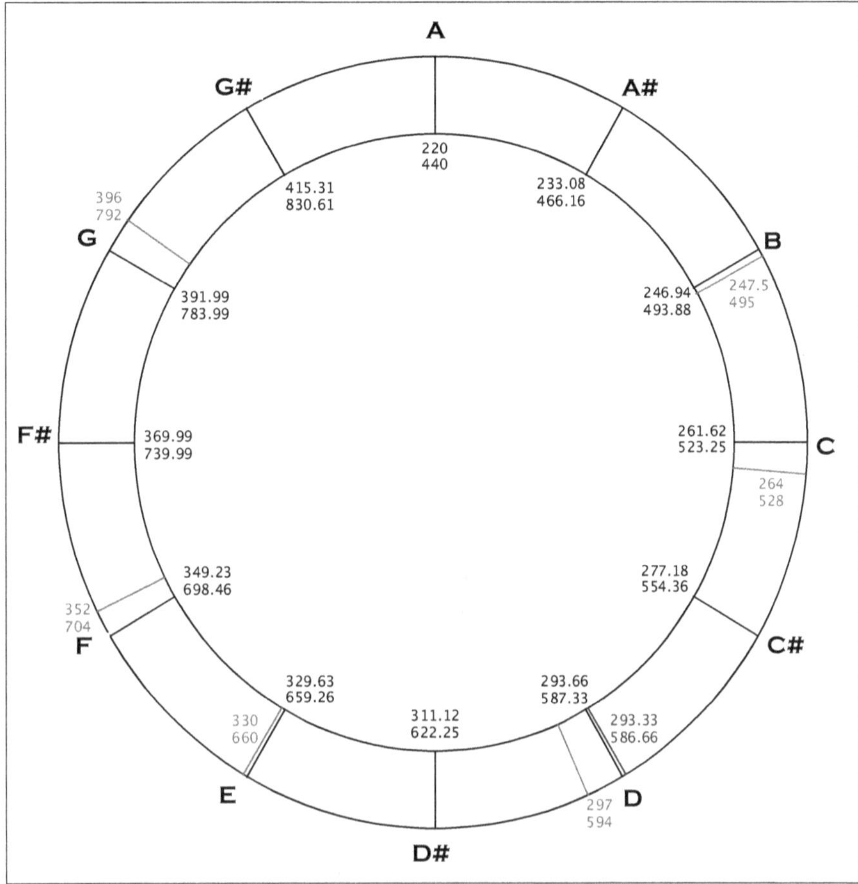

Figure 1.12.1

These twelve equal intervals around the clock face are known as the scale of **equal temperament**, which was a triumph of seventeenth century mathematics and engineering designed to reduce the cost of the new keyboard instruments which use fixed resonators (sound generators whose frequency cannot be changed, at least while playing) and were to evolve into the piano and the church organ. It is a bastard scale, because its intervals are a mathematical compromise and not truly harmonic, giving Western music a somewhat brash flavour, which I am quite at home with, but which is noticeably "off" in the ears of more musically sensitive people. But it simplified the engineering of musical instruments enormously, leading to the huge expansion in music from the early seventeenth century on. The inventor of this scale was no minor player, but a mathematician of the first magnitude – Simon Stevin (1548–1620) – who also extended the decimal notation to fractional numbers, so giving us forms like

0.3267 and 3.1415926, and made major contributions to the theory of **logarithms** and the theory of perspective.

The true musical scales are called **diatonic scales** and are shown in the *Figure* in red for the **major** scale and blue for the **minor**, but in this case only where it departs from the **major**.[22] All three include the A ≡ 440 point. The **diatonic scales** are defined by having the harmonic intervals (the ratios of the frequencies of two notes) defined by simple whole number. These ratios are shown below as multiples of the **key note** – the first note – in the scale:

Major: 1, 9/8, 5/4, 4/3, 3/2, 5/3, 15/8, 2

Minor: 1, 9/8, 6/5, 4/3, 3/2, 8/5, 9/5, 2

although there are variants of the **minor** known as the **harmonic minor** and **ascending melodic minor**. The standard **minor** as shown is also called the **descending melodic minor**. The scales can also be represented in terms of the ratios between *adjacent* notes, so for example the ratio between the fourth and fifth notes in the **major** would be (3/2)/(4/3) = (3/2) × (3/4) = 9/8. It now becomes convenient to put the notes *between* the intervals, so written in this way the two basic scales of **C Major** and **A Minor** would appear as:

C Major: **C** 9/8 **D** 10/9 **E** 16/15 **F** 9/8 **G** 10/9 **A** 9/8 **B** 16/15 **C**

A Minor: **A** 9/8 **B** 16/15 **C** 10/9 **D** 9/8 **E** 16/15 **F** 9/8 **G** 10/9 **A**

although originally **E** and **B** weren't there – a system known as **pentatonic**. The small 16/15 intervals are called **semitones** (written **S**), and all the larger, although they vary a little, are called **tones** (written **T**). The distribution of the two defines the two kinds of scale, **major** having the pattern **TTSTTTS** and the **minor TSTTSTT**. You may spot that these are **cyclic permutations** of each other: if you start at sixth position in the **major**, and go round cyclically, you get the pattern of the **minor**: **TS – TTSTT**.

In the **equal temperament** or "bastard" scale, all the **tones** *are* actually identical ratios, corresponding exactly to *two hours* on our clock face or a ratio of $\sqrt[6]{2} = (\sqrt[12]{2})^2$ = 1.122462, and the **semitones** are exactly half the ratio of the **tones**, corresponding to *one hour* on our clock face or a ratio of $\sqrt[12]{2}$ = 1.059463. Before Stevin's invention, musicians used a variety of scales which were based on choosing a different starting point or **key note** to define a new set of seven notes. So if you had an instrument set to a **major** scale and started at **E** you would get the pattern of ratios between adjacent notes of:

16/15, 9/8, 10/9, 9/8, 16/15, 9/8, 10/9

roughly corresponding to **STTTSTT**. As I understand it, this was the basis of all those Lydian, Phrygian and Dorian scales that are the bane of every music student. Each would have a genuinely different distribution of ratios, and so a quite unique sound. Stevin's system swept all that aside, and because every interval had now a standard basis of one **semitone** equal to a frequency ratio of $\sqrt[12]{2}$, it is easy to

[22] Shown here are, strictly, only the *particular* **minor** scale beginning at A = 440, and the *particular* **major** that begins at C = 264. As noted below, the ratios of **majors** and **minors** are exactly **cyclic permutations** of each other, *except at D* which has an interval of ratio 9/8 *before* it and 10/9 *after* it in the **major**. In the **minor**, these are reversed. So this is the only note that they do not have in common.

reproduce *exactly* the same pattern of ratios starting at any chosen note. So we could start a new **minor** scale at the 1 o'clock position, and taking the notes marked on the circle as: A#, C, C#, D#, F, F#, G#, reproduce the **TSTTSTT** pattern exactly. This is not quite how it's presented in the music texts, and to understand that, we need the concept of **enharmonics** and, getting more mathematical, the **sharp** and **flat operators**. First, the **operators**: the **sharp operator** # moves us one **semitone** *clockwise* around the circle, so for example, on the diagram, #(A) = A#, #(B) = C, #(G#) = A. The **flat operator** ♭ takes us one **semitone** *anticlockwise*, so ♭(A) = G#, ♭(F) = E, ♭(C) = B. The two **operators** are **inverses**, so applied successively, they cancel each other: ♭(#(X)) = #(♭(X)) = X for any note X. This introduces the possibility of writing every note on the **equal temperament** twelve note clock in three ways, as "itself", as the **sharp operator** applied to the note one **semitone** *anticlockwise* round from it, and as the **flat** applied to the note one **semitone** *clockwise* round from it. (In practice, many of these would read as **inverse** pairs like ♭(D#) and such forms are not used.) These alternative codings for the same note are called **enharmonics**. We can use this to write the new scale on A# not as A#, C, C#, D#, F, F#, G#, but as A#, B#, C#, D#, E#, F#, G#, so preserving the A-B-C-D-E-F-G nomenclature.

What the textbooks do is slightly more slick. Taking the **major** sequence **TTSTTTS**, they apply the # **operator** to the note *following* the first **S**, producing the sequence **TTTSTTS**, so if we started with CDEFGAB we get CDE(F#)GAB. Now take the G as the **key note** – the note after the first *new* **S**, and proceed cyclically, so we get again **TTSTTTS**, but this time with GABCDE(F#). Do the same *again* and we get first: GAB(C#)DE(F#) and taking the fifth note – the one after the first *new* **semitone** – as the **key note**, and going round the cycle, we get: DE(F#)GAB(C#). The elegance of this approach – the "circle of fifths" method – is that we add precisely *one* new **sharp** # each time.

We can do a similar dodge with the **flat operator**. This time we apply the **flat** to the note *before* the *second* **semitone** – still working on a **major** scale – to get the pattern: **TTSTTST** and now if we take the *fourth* note as the **key note**, we again recover the **TTSTTTS** series. So starting again with CDEFGAB, we go to CDEFGA(B♭) and choosing F as the **key note**, get FGA(B♭)CDE. Repeating this trick, we go via FGA(B♭)CD(E♭) to (B♭)CD(E♭) FGA, each time now adding just one **flat** and restoring the **TTSTTTS** sequence on the second step where we choose the new **key note**. This is similarly referred to as a "circle of fourths" because it's the *fourth* note at each stage that becomes the new **key note**.

Each new **major** scale so created has an associated **minor** scale obtained by taking the *sixth* note in the **major** as the **key note** of the **minor** and proceeding around the cycle, so **major** CDEFGAB correspond to **minor** ABCDEFG, GABCDE(F#) to E(F#)GABCD and so on. These associated **major** and **minor** scales are called the **relative major** to the **minor**, and the **relative minor** to the **major**.

1.13 Simplexes[23] and Topology

Simplexes are a formalisation of the simplest geometrical objects of any given number of dimensions. A **0-simplex** is simply a point. Two points, or **0-simplexes**, are connected by a line, which is a **1-simplex**. Three lines connecting three points, or three **1-simplexes** connecting three **0-simplexes**, form a triangle in the plane, which is a **2-simplex**. Four points in *space* connected by four lines define four triangles in space, and the roughly tetrahedral volume enclosed by the four triangles is a **3-simplex**. Putting that again: four **0-simplexes** in *space* connected by four **1-simplexes** define four **2-simplexes** in space, and the roughly tetrahedral volume enclosed by the four triangles is a **3-simplex**.

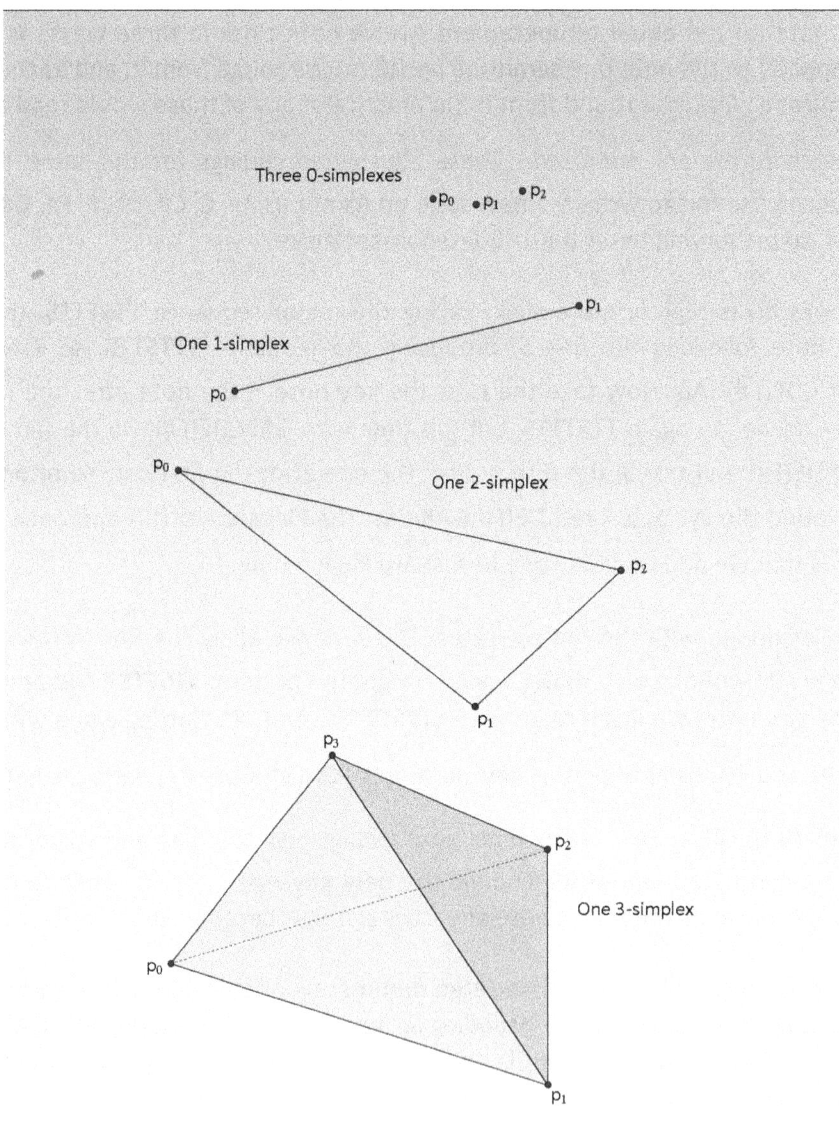

Figure 1.13.1

[23] Those with a classical education will prefer the plural form "simplices", but this is an invented word, so we can choose its plural as we wish, and US usage as here seems to be well established.

Figure 1.13.1 shows the basic idea. The key point is that as we go up through the dimensions, each set of $n + 1$ **($n-1$)-simplexes** forms the **boundary** enclosing the volume of the **n-simplex**. So two *points* form the end-points which are the **boundary** of a *line*, three *lines* form the **boundary** of a *triangle*, and four *triangles* form the **boundary** of **3-simplex** volume.

Note, too, that in a **simplex**, every point is connected to every other point.

The formal definition of an **n-simplex** is: given n+1 points $\mathbf{p}_0, \mathbf{p}_1, \ldots \mathbf{p}_n$, **A** is an **n-simplex** if every point of **A**, **p**, can be expressed uniquely in the form:

$$\mathbf{p} = x_0\mathbf{p}_0 + x_1\mathbf{p}_1 + \ldots + x_n\mathbf{p}_n \qquad \text{where each } x_i \geq 0 \text{ and } x_0 + x_1 + x_2 + \ldots x_n = 1$$

but you will really need to be familiar with **vectors**, the subject of Chapter 5, to understand this. The $\{x_i\}$'s are called the **barycentric coordinates** of the **simplex**.

Simplexes are **orientated**, that is to say they are **chiral** in the general sense introduced in Section 1.2. To put it another way, it turns out to be very useful to give them a **signature**, to make them **signed**. Points can then be marked as +**p** and −**p**, and we can write *lines* or **1-simplexes** as:

$$\mathbf{p}_1\mathbf{p}_2 = -\,\mathbf{p}_2\mathbf{p}_1$$

Simply specifying the *ordered* list of defining points of a **simplex** is the most common way of writing them. So a triangular **2-simplex** is written $\mathbf{p}_1\mathbf{p}_2\mathbf{p}_3$ but here the sign rule operates more subtly. By the previous equation $\mathbf{p}_1\mathbf{p}_2 = -\,\mathbf{p}_2\mathbf{p}_1$, so that every time we swap two points, the whole **simplex** changes sign, so:

$$\mathbf{p}_1\mathbf{p}_2\mathbf{p}_3 = -\,\mathbf{p}_2\mathbf{p}_1\mathbf{p}_3 = \mathbf{p}_2\mathbf{p}_3\mathbf{p}_1 = -\,\mathbf{p}_3\mathbf{p}_2\mathbf{p}_1 = \mathbf{p}_3\mathbf{p}_1\mathbf{p}_2 = -\,\mathbf{p}_1\mathbf{p}_3\mathbf{p}_2$$

these being all the possible **permutations** of the **indices**. [123], [312], [231] are called **cyclic permutations** of each other as we move the elements round in a circular or *cyclic* way to get each from the last, and so are [213], [321] and [132], but we cannot get from one set to the other in this way. So these two sets of orderings do indeed form just *two* distinct classes, and so do represent an **orientation** or a **chiral** division – a "handedness" – of the ordering.

We can use the **signature** of a **simplex** to define a **boundary operator** which generates the **boundary (n−1)-simplex** from any given **n-simplex**. The formal expression is this:

$$\mathbf{b}(\mathbf{p}_0\mathbf{p}_1\mathbf{p}_2\ldots\mathbf{p}_n) = \Sigma_i (-1)^i\, \mathbf{p}_0\mathbf{p}_1\mathbf{p}_2\ldots\{\mathbf{p}_i\}\ldots\mathbf{p}_n$$

where the $\{\mathbf{p}_i\}$ notation indicates that this i'th point is *omitted* from that term (and you have to start the numbering at 0 to get the signs right). It looks obscure but it's really very simple, as examples will show. The **boundary** of a *line* is given by its *signed* end-points:

$$\mathbf{b}(\mathbf{p}_0\mathbf{p}_1) = (-1)^0\mathbf{p}_1 + (-1)^1\mathbf{p}_0 = \mathbf{p}_1 - \mathbf{p}_0 \qquad \text{using } (-1)^0 = 1$$

the **boundary** of a triangle is given by its *signed* sides:

$$\mathbf{b}(\mathbf{p}_0\mathbf{p}_1\mathbf{p}_2) = (-1)^0\mathbf{p}_1\mathbf{p}_2 + (-1)^1\mathbf{p}_0\mathbf{p}_2 + (-1)^2\mathbf{p}_0\mathbf{p}_1 = \mathbf{p}_1\mathbf{p}_2 - \mathbf{p}_0\mathbf{p}_2 + \mathbf{p}_0\mathbf{p}_1$$

and of a **3-simplex** by its *signed* triangular faces:

$$\mathbf{b}(\mathbf{p}_0\mathbf{p}_1\mathbf{p}_2\mathbf{p}_3) = (-1)^0\mathbf{p}_1\mathbf{p}_2\mathbf{p}_3 + (-1)^1\mathbf{p}_0\mathbf{p}_2\mathbf{p}_3 + (-1)^2\mathbf{p}_0\mathbf{p}_1\mathbf{p}_3 + (-1)^3\mathbf{p}_0\mathbf{p}_1\mathbf{p}_2$$

$$= \mathbf{p}_1\mathbf{p}_2\mathbf{p}_3 - \mathbf{p}_0\mathbf{p}_2\mathbf{p}_3 + \mathbf{p}_0\mathbf{p}_1\mathbf{p}_3 - \mathbf{p}_0\mathbf{p}_1\mathbf{p}_2$$

and the elegant significance of the signs is apparent in *Figure 1.13.1*. For example, the **boundary** of the triangle is:

$$\mathbf{p}_1\mathbf{p}_2 - \mathbf{p}_0\mathbf{p}_2 + \mathbf{p}_0\mathbf{p}_1 = \mathbf{p}_1\mathbf{p}_2 + \mathbf{p}_2\mathbf{p}_0 + \mathbf{p}_0\mathbf{p}_1$$

so the points are running in sequence: 1⟶2, 2⟶0, 0⟶1 around the triangle once we set all the signs to + by switching the order of $\mathbf{p}_0\mathbf{p}_2$ to $\mathbf{p}_2\mathbf{p}_0$.
Note too that:

$$\mathbf{b}(\mathbf{p}_1\mathbf{p}_2 + \mathbf{p}_2\mathbf{p}_0 + \mathbf{p}_0\mathbf{p}_1) = \mathbf{p}_2 - \mathbf{p}_1 + \mathbf{p}_0 - \mathbf{p}_2 + \mathbf{p}_1 - \mathbf{p}_0 = 0,$$

an example of the general theorem that: the **boundary** of a **boundary** is zero or **bb** ≡ **0**.

This is a very important result. *The* **boundary operator** *is defined the way it is precisely so that this will be so.*

The subject of **topology** can be reformulated in a way that gives a central role to **boundary operations**. **Topology** is the mathematical discipline which deals with what is *connected* to what in mathematical structures without any specific notion of distance. So London is connected to Manchester by land, but not to Paris, although Paris is connected to Beijing. By *sea*, London would be connected to Sydney and New York, but still not to Paris! Another example of a **topological** concept is that a circle anywhere on or inside a sphere can be progressively shrunk to a point without leaving the sphere, but not a circle looping through the handle of a cup-shaped object. The sphere and the cup have different **Euler characteristics**, a number which can be defined for a connected set of **simplexes** as the number of *faces* (**2-simplexes**) plus the number of *vertices* (**0-simplexes**) minus the number of *edges* (**1-simplexes**). For a closed **polyhedron**, this number is always 2. So a **tetrahedron** has 4 faces, 4 vertices (or *points*), and 6 edges so E = 4 + 4 – 6 = 2. For a **cube**, 6 faces, 8 vertices and 12 edges give 6 + 8 – 12 = 2.

Describing **topology** in terms of **boundary operators** has deep implications, because the **boundary operator** turns out to have profound affinities to the **differential operator** which is the foundation of the **calculus**, which we will begin to look at in Chapter 2.

I will use the common convention of writing **simplexes** of arbitrary **dimension** – whether **0-simplexes** of **dimension** 0, **1-simplexes** of **dimension** 1, **2-simplexes** of **dimension** 2 or whatever – by the Greek letter σ. With this form, an arbitrary sum of **simplexes** such as:

$$a_1\sigma_1 + a_2\sigma_2 + a_3\sigma_3 + a_4\sigma_4 + a_5\sigma_5 + \ldots a_n\sigma_n$$

where the a_i are arbitrary numbers, usually, but not necessarily ±1, and the σ_i are of the same **dimension** *p*, is called a ***p*-chain**.

A ***p*-chain** c_p which satisfies $\mathbf{b}c_p = 0$, i.e. one whose **boundary** is zero, is called a **closed *p*-chain** or alternatively, a ***p*-cycle**. The prefix "*p*-" may be omitted, and we may refer to **chains** and **closed chains** or **cycles**. Thus the **boundary** of any **simplex** is **closed** or is a **cycle**, because $\mathbf{bb}\sigma = 0$.

I will also introduce the term a **strict 1-chain** to indicate a **simple sum**[24] of **1-simplexes** which can be put in an order such that the *end point* of each is the *start point* of the next. I will often omit the "**1-**" and just call this a **strict chain**. Put algebraically, this concept means that there is an ordering of the sum in which, if the *i*'th **simplex** is written $\sigma_i = \mathbf{p}_{i,0}\mathbf{p}_{i,1}$, then for $i = 2$ to n (if there are n terms in the sum):

$$\mathbf{p}_{i,0} = \mathbf{p}_{i-1,1}.$$

So $\mathbf{p}_1\mathbf{p}_2 + \mathbf{p}_2\mathbf{p}_3 + \mathbf{p}_3\mathbf{p}_4 + \mathbf{p}_4\mathbf{p}_5 + \mathbf{p}_5\mathbf{p}_6$ is a **strict chain**. **Strict chains** have the special property that their **boundaries** consist of just the two extreme end points of the **chain**, so:

$$\mathbf{b}(\mathbf{p}_1\mathbf{p}_2 + \mathbf{p}_2\mathbf{p}_3 + \mathbf{p}_3\mathbf{p}_4 + \mathbf{p}_4\mathbf{p}_5 + \mathbf{p}_5\mathbf{p}_6) = \mathbf{p}_2 - \mathbf{p}_1 + \mathbf{p}_3 - \mathbf{p}_2 + \mathbf{p}_4 - \mathbf{p}_3 + \mathbf{p}_5 - \mathbf{p}_4 + \mathbf{p}_6 - \mathbf{p}_5$$

$$= \mathbf{p}_6 - \mathbf{p}_1$$

since all the intermediate points cancel out. If also $\mathbf{p}_{1,0} = \mathbf{p}_{n,1}$ so the extreme end points also match, then the **strict chain** is a **cycle**, and this shows where the term **cycle** comes from: the **1-simplexes** form a *ring*. In the foregoing, this would be so if $\mathbf{p}_6 = \mathbf{p}_1$.

This definition can be generalized to higher **dimensions**. For the **strict chain** defined above has two essential properties:

- Every **0-simplex** in the **boundary** of the **chain** either occurs just once, or occurs twice with opposite signs.
- Every **1-simplex** in the original **chain** has at least one **boundary 0-simplex** that occurs with opposite sign in the **boundary** of another **1-simplex** in the **chain**.

So I will define a **strict *n*-chain** as a **simple sum** of ***n*-simplexes** such that:

- Every **(*n*−1)-simplex** in the **boundary** of the **chain** either occurs just once, or occurs twice with opposite signs.
- Every ***n*-simplex** in the original **chain** has at least one **boundary (*n*−1)-simplex** that occurs with opposite sign in the **boundary** of another **(*n*−1)-simplex** in the **chain**.

[24] I will also informally define a **simple sum** of **simplexes** as one all of whose **coefficients** are +1.

By the foregoing argument, we can define the **strict** or **outer boundary** of a **strict *n*-chain** as the set of **(*n*−1)-simplexes** in the **boundary** of the **chain** that do *not* occur twice with opposite signs. So in the example of a **strict 1-chain** above, $\mathbf{p}_1$ and $\mathbf{p}_6$ constitute the **outer boundary**.

Strict 2-chains are the basic items used to indicate surfaces, and thereby just about anything, in computer graphics, where they are called **meshes**. They are therefore hugely important in the modern world, and I will use them in Chapter 6 to extend the theory of **integration** to the generalized surfaces called **manifolds**. I will also return to **meshes** in Section 1.15.

I will often call a **1-simplex** a **line segment**, as this is a very common concept in mathematics. If a **line segment** lies on one of the **coordinate axes** described in Section 1.2, it is often called an **interval** and is written $[a, b]$ where a and b are the values of the **coordinate** on whose **axis** the **line segment** lies, at $\mathbf{p}_0$ and at $\mathbf{p}_1$ respectively. In other words $\mathbf{p}_0\mathbf{p}_1 = [a, b]$. They are alternative ways of defining the same **interval**.[25]

1.14 Electrical Circuits

Electrical circuits are a standard example for the application of many of the techniques we shall be developing. Their theory is also applicable to many other fields like hydraulics, acoustics and the thermal properties of buildings.

Electrical circuits are closely related to the **simplexes** we have just looked at. They consist of sets of **nodal points** linked by a network of lines. Distributed along the lines are various elements that have specific properties which are well described by **linear equations** between pairs of variables. There may also be elements at the **nodal points**, whose properties are also described by **linear equations**, generally between four variables. These **nodal point** elements are commonly the **active** parts of the circuit – valves, transistors, FET's – while the line elements are commonly **passive** and of three kinds: **resistors**, **capacitors** and **inductors**, so classified according to what kind of linear approximation best suits them.

At any point anywhere in the circuit – that is to say, arbitrary points, not necessarily **nodal points**, a term I'd prefer to reserve for points where three or more lines branch – there will be an **electrical potential** (regard this as an arbitrary term as the physical meaning is not important) defined with respect to an arbitrary "ground" level, just as on Ordnance Survey maps, heights are defined relative to an arbitrary "sea-level" at a place called Newlyn. So between any two points there is a difference in this **potential** called a **voltage**. Along any line in any part of the circuit, there is defined a **current**, the meaning of which was uncertain for much of the early history of electricity, but which turned out to be a *flow*, generally of the negatively charged elementary particles called **electrons**. By this time, however, circuit theory had become established with a choice of sign that would correspond to a flow in the

[25] I avoid here any questions of whether the *actual* end points a and b are themselves included in the **interval**, as such hair-splitting will not be needed for our immediate purposes.

opposite direction, and that's the way it's stayed. But as I will show, this is not as unfortunate as it is traditionally thought to be.

We can formally define the **voltage difference** or **voltage** along a single line element, regarding such as line element as a **1-simplex** $\mathbf{p}_1\mathbf{p}_2$, as:

$$v(\mathbf{p}_1\mathbf{p}_2) = V(\mathbf{b}(\mathbf{p}_1\mathbf{p}_2)) = V(\mathbf{p}_2 - \mathbf{p}_1) = V(\mathbf{p}_2) - V(\mathbf{p}_1)$$

where I use lower-case $v()$ to represent the **voltage difference** and upper-case $V()$ to represent the **potential** at a point, but I also allow the $V()$ **function** to distribute across the points of a **chain**, taking the signs of the points.[26]

Note that under this definition, $v(\mathbf{p}_2\mathbf{p}_1) = -v(\mathbf{p}_1\mathbf{p}_2)$, so the **voltages** are **oriented**: switching the order of the points switches the sign of the **voltage**.

Currents we can take as a **function** of each line element which we may think of as a *flow* from $\mathbf{p}_1$ to $\mathbf{p}_2$, which is therefore *also* **oriented**, so that, writing **currents** with the letter i:

$$i(\mathbf{p}_2\mathbf{p}_1) = -i(\mathbf{p}_1\mathbf{p}_2).$$

The theory of electrical circuits has two fundamental laws called **Kirchhoff's laws**. They are:

Kirchhoff's Current Law (KCL):

> The sum of the **currents** *into* any point in the circuit, or *out of* any point in the circuit, is zero.

Kirchhoff's Voltage Law (KVL):

> The sum of the **voltages**, *consistently oriented*, around any **1-cycle** of lines, however tortuous, is zero.

The difficulty is in getting the *signs* right in the application of these laws, and so I will try to state this very carefully indeed. I speak from bitter experience.

Consider a connected ring of points – **0-simplexes** – connected by lines – **1-simplexes** – such that the end point of each line is the start point of the next, and the ring is **closed**, so the end point of the last line is the start of the first and so the lines form a **1-cycle**.

So we have the points, say five of them, $\mathbf{p}_1$, $\mathbf{p}_2$, $\mathbf{p}_3$, $\mathbf{p}_4$, $\mathbf{p}_5$. By the premises we've stated, the lines must be: $\mathbf{p}_1\mathbf{p}_2$, $\mathbf{p}_2\mathbf{p}_3$, $\mathbf{p}_3\mathbf{p}_4$, $\mathbf{p}_4\mathbf{p}_5$, $\mathbf{p}_5\mathbf{p}_1$. Remember that these are **oriented** so that $\mathbf{p}_1\mathbf{p}_2 = -\mathbf{p}_2\mathbf{p}_1$, and we carry over these **orientations** to the **voltages**. So writing, for example, $v(\mathbf{p}_2\mathbf{p}_3) = V(\mathbf{p}_3) - V(\mathbf{p}_2)$ where as before

[26] In Section 2.6, we will see that this is conceptually an **integral** and in the notation which I will introduce in that Section it would take the form $\mathbf{d}V \mid \mathbf{p}_1\mathbf{p}_2 = V(\mathbf{p}_2) - V(\mathbf{p}_1)$.

$v()$ represents the **voltage** difference *between* points, and $V()$ represents the **electrical potential** *at* a point, and going right round our ring of points, we have:

$$v(\mathbf{p}_1\mathbf{p}_2) + v(\mathbf{p}_2\mathbf{p}_3) + v(\mathbf{p}_3\mathbf{p}_4) + v(\mathbf{p}_4\mathbf{p}_5) + v(\mathbf{p}_5\mathbf{p}_1)$$

$$= V(\mathbf{p}_2) - V(\mathbf{p}_1) + V(\mathbf{p}_3) - V(\mathbf{p}_2) + V(\mathbf{p}_4) - V(\mathbf{p}_3) + V(\mathbf{p}_5) - V(\mathbf{p}_4) + V(\mathbf{p}_1) - V(\mathbf{p}_5)$$

$$= 0$$

because every term appears exactly twice, with opposite signs. Note how the **orientation** is vital: if we'd put $v(\mathbf{p}_4\mathbf{p}_3) = V(\mathbf{p}_3) - V(\mathbf{p}_4)$ rather than $v(\mathbf{p}_3\mathbf{p}_4) = V(\mathbf{p}_4) - V(\mathbf{p}_3)$ we would have:

$$= V(\mathbf{p}_2) - V(\mathbf{p}_1) + V(\mathbf{p}_3) - V(\mathbf{p}_2) + V(\mathbf{p}_3) - V(\mathbf{p}_4) + V(\mathbf{p}_5) - V(\mathbf{p}_4) + V(\mathbf{p}_1) - V(\mathbf{p}_5)$$

$$= 2V(\mathbf{p}_3) - 2V(\mathbf{p}_4) \neq 0.$$

Now look at what is arguably the simplest possible circuit, shown in *Figure 1.14.1*.

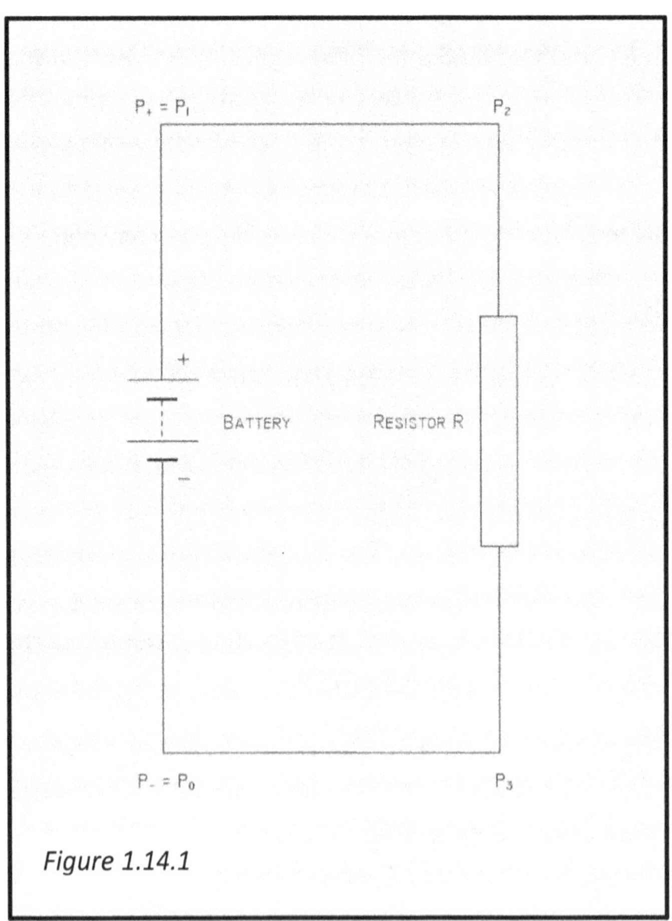

Figure 1.14.1

I give the points $\mathbf{p}_0$ and $\mathbf{p}_1$ the synonyms $\mathbf{p}_-$ and $\mathbf{p}_+$ to emphasise that they correspond to the terminals of the battery. In this circuit, the straight line elements $\mathbf{p}_1\mathbf{p}_2$ and $\mathbf{p}_3\mathbf{p}_0$ by definition represent line elements along which the **voltage difference** is zero: $v(\mathbf{p}_1\mathbf{p}_2) = v(\mathbf{p}_3\mathbf{p}_0) = 0$. Now $v(\mathbf{p}_0\mathbf{p}_1) = V(\mathbf{p}_1) - V(\mathbf{p}_0) = V(\mathbf{p}_+) - V(\mathbf{p}_-)$ represents the *battery* **voltage**, and again by definition, since $V(\mathbf{p}_+) > V(\mathbf{p}_-)$:

$$v(\mathbf{p}_0\mathbf{p}_1) = V(\mathbf{p}_1) - V(\mathbf{p}_0) = V(\mathbf{p}_+) - V(\mathbf{p}_-) > 0.$$

In other words, a battery in the circuit forces the **orientation** of the **voltage** across it.

Now by **KVL**, since the four **line segments** $\mathbf{p}_0\mathbf{p}_1$, $\mathbf{p}_1\mathbf{p}_2$, $\mathbf{p}_2\mathbf{p}_3$, $\mathbf{p}_3\mathbf{p}_0$ form a **1-cycle**, we must have:

$$v(\mathbf{p}_0\mathbf{p}_1) + v(\mathbf{p}_1\mathbf{p}_2) + v(\mathbf{p}_2\mathbf{p}_3) + v(\mathbf{p}_3\mathbf{p}_0) = 0.$$

Therefore, since $v(\mathbf{p}_1\mathbf{p}_2) = v(\mathbf{p}_3\mathbf{p}_0) = 0$, $v(\mathbf{p}_0\mathbf{p}_1) + v(\mathbf{p}_2\mathbf{p}_3) = 0$, but since $v(\mathbf{p}_0\mathbf{p}_1) > 0$, this means that we must have $v(\mathbf{p}_2\mathbf{p}_3) < 0$.

Now let's look at the **currents. KCL** can be expressed algebraically as:

$$\Sigma_k\, i(\mathbf{p}_k\mathbf{p}_0) = \Sigma_k\, i(\mathbf{p}_0\mathbf{p}_k) = 0$$

at any point $\mathbf{p}_0$, where the k=1 to n line elements $\mathbf{p}_k\mathbf{p}_0$ are all the line elements connected to the point $\mathbf{p}_0$. Note how here $\mathbf{p}_0$ is always in the same position through all the **terms** of each sum. So at, say, point $\mathbf{p}_1$ in *Figure 1.14.1*, we have:

$$i(\mathbf{p}_0\mathbf{p}_1) + i(\mathbf{p}_2\mathbf{p}_1) = 0$$

as these are the only two lines connected to $\mathbf{p}_1$, but because $i(\mathbf{p}_2\mathbf{p}_1) = -i(\mathbf{p}_1\mathbf{p}_2)$, this means that:

$$i(\mathbf{p}_0\mathbf{p}_1) = i(\mathbf{p}_1\mathbf{p}_2).$$

Similarly, we can show that $i(\mathbf{p}_0\mathbf{p}_1) = i(\mathbf{p}_1\mathbf{p}_2) = i(\mathbf{p}_2\mathbf{p}_3) = i(\mathbf{p}_3\mathbf{p}_0)$.

So if we choose the sign of $i(\mathbf{p}_0\mathbf{p}_1)$, all the **currents** in the circuit are uniquely determined. We have two possibilities: either we give $i(\mathbf{p}_0\mathbf{p}_1)$ the same sign as $v(\mathbf{p}_0\mathbf{p}_1)$ or the opposite sign. The choice determines the precise form of the law obeyed by the **resistor** element at the right.

A **resistor** is an element defined by a linear relation between **voltage** and **current** called **Ohm's Law**. This is generally expressed as $v = i.R$ where v is the **voltage** between the ends of the **resistor** and i is the **current** through it, and R is the constant of proportionality, which is always taken to be positive: $R > 0$. What is not usually made clear is that we need to use **oriented voltage** and **oriented current**, and here our choice needs to be made. Since $v(\mathbf{p}_2\mathbf{p}_3) < 0$ because $v(\mathbf{p}_0\mathbf{p}_1) > 0$, if we choose $i(\mathbf{p}_0\mathbf{p}_1) > 0$, with the same sign as $v(\mathbf{p}_0\mathbf{p}_1)$, then since $i(\mathbf{p}_2\mathbf{p}_3) = i(\mathbf{p}_0\mathbf{p}_1) > 0$, we obtain:

$$v(\mathbf{p}_2\mathbf{p}_3) = -\, i(\mathbf{p}_2\mathbf{p}_3) \times R.$$

If we choose $i(\mathbf{p}_0\mathbf{p}_1) < 0$, with the *opposite* sign to $v(\mathbf{p}_0\mathbf{p}_1)$, we get the more usual form:

$v(\mathbf{p}_2\mathbf{p}_3) = + i(\mathbf{p}_2\mathbf{p}_3) \times R.$

Both choices are acceptable. It might be thought that if the **current** really consists of electrons flowing round the circuit from the $\mathbf{p}_-$ terminal of the battery to the $\mathbf{p}_+$, and so from $\mathbf{p}_-$ or $\mathbf{p}_0$ to $\mathbf{p}_3$ to $\mathbf{p}_2$ to $\mathbf{p}_1$ or $\mathbf{p}_+$, then we *should* have $i(\mathbf{p}_-\mathbf{p}_+) < 0$, so that the flow $i(\mathbf{p}_+\mathbf{p}_-) > 0$, represents a positive flow back through the battery $\mathbf{p}_+$ to $\mathbf{p}_-$. But the problem there is that *the sign of the electrons, which is negative, is considered to be factored into* $i()$. So $i(\mathbf{p}_-\mathbf{p}_+) < 0$ actually represents a flow of negative electrons from $\mathbf{p}_-$ to $\mathbf{p}_+$!

The historically accepted convention is that:

$$i(\mathbf{p}_-\mathbf{p}_+) = i(\mathbf{p}_0\mathbf{p}_1) > 0$$

consistently with

$$v(\mathbf{p}_-\mathbf{p}_+) = v(\mathbf{p}_0\mathbf{p}_1) > 0$$

i.e. that the **current** and **voltage** through a battery – also referred to as an **emf** or **e.m.f.** or **electromotive force** – are *similarly* **oriented**. In *Figure 1.14.1* this would give a *positive i* going clockwise round the circuit, and *in fact this correctly corresponds to the flow of negative electrons anticlockwise*. This strictly gives the *negative* form of **Ohm's law**:

$$v(\mathbf{p}_2\mathbf{p}_3) = - i(\mathbf{p}_2\mathbf{p}_3) \times R \quad \text{or} \quad v = - i \times R$$

for v and i **oriented** the same way.

The reason we never see it written that way is that if the **current** is defined as going round a circuit in the same sense as the **voltages** – the same way that the **voltages** are **oriented** – then for the simple ring with one battery of E volts and two or three **resistors** R_1, R_2, R_3 as shown at the top in *Figure 1.14.2* we now get from **KVL**:

$$E + v_1 + v_2 + v_3 = E - i.R_1 - i.R_2 - i.R_3 = 0$$

or

$$E = i.R_1 + i.R_2 + i.R_3$$

where *all* the terms – E, i, and all the R's – are *positive*. In other words, putting the "**emf**'s" and the iR's on opposite sides of the equation hides the negative form of the **Ohm's law**. Had we worked with $v = +i.R$, we would still have got an answer, but now the clockwise i would be *negative*. When in Chapter 3 we introduce the other line elements – **capacitors** and **inductors** – we will have to be very careful indeed about these rules.

Often, for a given **strict chain** between **nodal points**, the *opposite choice of* **current orientation** will be made, which here would mean we'd use *anticlockwise* **currents** rather than *clockwise*:

$i(\mathbf{p}_3\mathbf{p}_2) = i(\mathbf{p}_2\mathbf{p}_1) = i(\mathbf{p}_1\mathbf{p}_0) = i(\mathbf{p}_0\mathbf{p}_3) = -i(\mathbf{p}_0\mathbf{p}_1)$, etc.

Now even with the standard convention we would have:

$v(\mathbf{p}_2\mathbf{p}_3) = + i(\mathbf{p}_3\mathbf{p}_2) \times R$, which will appear ambiguously as just $v_R = i.R$.

Because we do not "know" the direction of **current** and can assign it arbitrarily for each **1-chain**, there's a lot of latitude within the constraints that:

$i(\mathbf{p}_m\mathbf{p}_n) = -\, i(\mathbf{p}_n\mathbf{p}_m)$ and that, for any $\mathbf{p}_n$, $\Sigma_k\, i(\mathbf{p}_k\mathbf{p}_n) = 0$.

So beware!

Because all the pitfalls are encapsulated in this most simple of all possible circuits, I'll just recapitulate this argument, using the simplified notation:

$v_{12} := v(\mathbf{p}_1\mathbf{p}_2)$ and $i_{12} := i(\mathbf{p}_1\mathbf{p}_2)$.

Here I use the symbol ":=" to mean that the left-hand side is *defined* as the right. I'll call this **indicial notation** for electric circuits, and although I won't use it in general in this text, it's a great help in getting the signs right. Remember that e.g. $v_{34} = v(\mathbf{p}_3\mathbf{p}_4) = V(\mathbf{p}_4) - V(\mathbf{p}_3)$, the $\mathbf{p}_4$ **potential** minus the $\mathbf{p}_3$. So in our little circuit above we have the battery equation:

$E = v_{01} > 0$ because $v_{01} = V(\mathbf{p}_1) - V(\mathbf{p}_0) = V(\mathbf{p}_+) - V(\mathbf{p}_-) > 0$.

For the conductors we assume:

$v_{12} = v_{30} = 0$.

But by **KVL**, $v_{01} + v_{12} + v_{23} + v_{30} = 0$, because this is the sum of all the $V(\mathbf{p}_i)$ twice, once with a plus sign and once with a minus. Since $v_{12} = v_{30} = 0$, this means:

$v_{01} + v_{23} = 0$,

or

$v_{23} = -v_{01} < 0$.

If we assume the definition $v_{23} = R.i_{23}$ for the same orientation, then since $R > 0$ *by definition*, this means that $i_{23} < 0$ since v_{23} is. Since $i_{01} = i_{12} = i_{23} = i_{30}$ as shown above, this would mean that $i_{01} < 0$ or the current through the battery from the negative *to* the positive pole is negative, which it would be if we defined it as $e \times N$ where N is the number of electrons passing per second, and e is the charge on one electron *which is negative*. That is, if the electrons flowed *clockwise*. But in fact the battery acts as a pump, and though electrons flow freely "downhill" from $V(\mathbf{p}_0)$ to $V(\mathbf{p}_1)$ *outside* the battery, they're being pumped "uphill" from $V(\mathbf{p}_1)$ to $V(\mathbf{p}_0)$ *inside.*

We now also have $E = v_{01} = -v_{23} = -i_{23}.R$.

In fact the standard convention, insofar as there is one at all, is to define:

$$E = i_{01}.R = i_{23}.R \tag{1.14.1}$$

which since $E > 0$ and $R > 0$ by definition, implies $i_{23} > 0$. But because $v_{23} = -v_{01} = -E$, this means that:

$$v_{23} = -i_{23}.R. \tag{1.14.2}$$

Note that this means that equation (1.14.1) is of the form $v = +i.R$, but (1.14.2) is of the form $v = -i.R$. It all depends which *v* you mean!

In general in this text I shall use the definition, as here, that $i_{01} > 0$, or that the current *through the battery* from the – to the + pole is positive. So this means that across an *element* – here a **resistor** – we have $v_{23} = -i_{23}.R$ with a *negative sign*, as $v_{23} < 0$ but $i_{23} > 0$, and $R > 0$ by definition.

This convention means that the current *anticlockwise* in the circuit of *Figure 1.14.1* is actually *negative*, or $i_{32} < 0$. But if we regard this as $i_{32} = e \times N < 0$, or as the number of electrons passing per second *with the negative charge* $e < 0$ *factored in*, this actually *correctly* represents the flow of electrons *anticlockwise* around the circuit from the negative $\mathbf{p}_0$ pole to the positive $\mathbf{p}_1$.

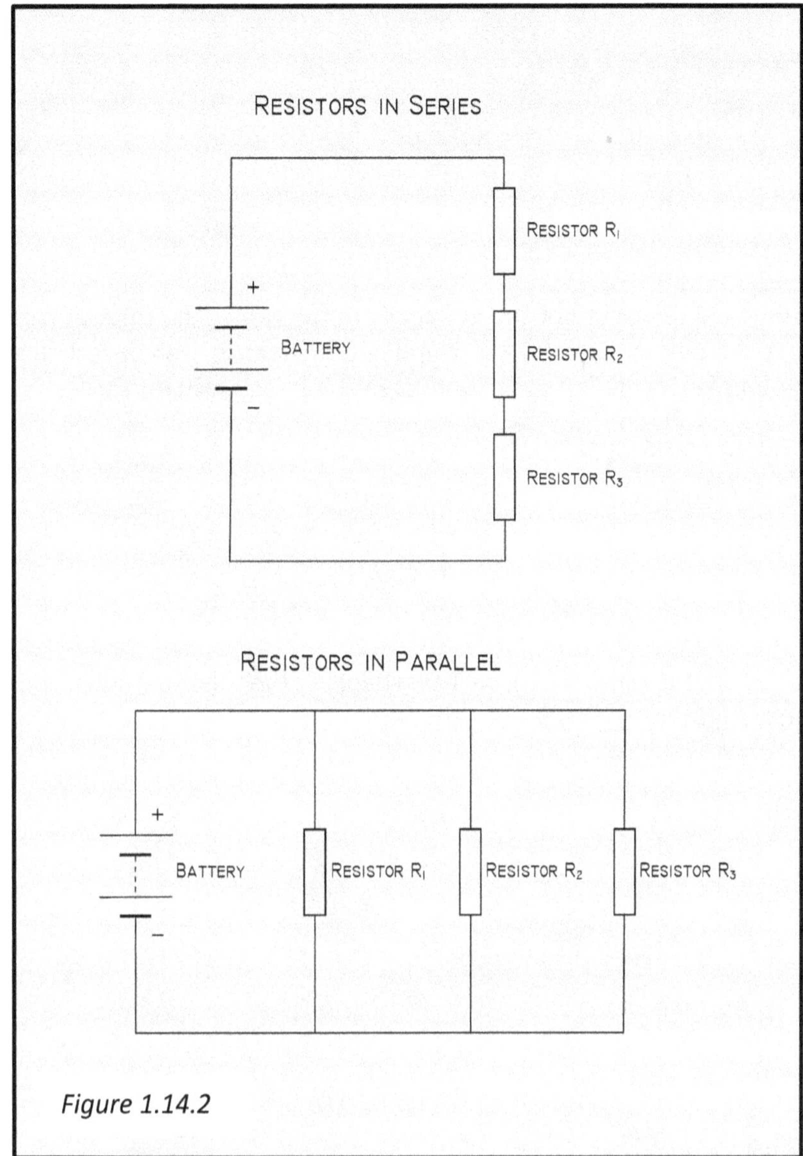

Figure 1.14.2

The theory we've developed so far enables us to determine the rules for **series** and for **parallel resistors**. The **series** case is the one considered above, defined by the requirement that the **current** through all three **resistors** is the same by the simple application of **KCL** to a point **p** lying between any two of them that was given above for $\mathbf{p}_1$.[27] So in this case we can lump the three **resistors** together to give:

$$E = i.R_1 + i.R_2 + i.R_3 = i.(R_1 + R_2 + R_3) = i.R.$$

The complementary case is also shown below in *Figure 1.14.2* and the defining characteristic here is that all three **resistors**, which we may index from $k = 1$ to 3, run between two points $\mathbf{p}_{top(k)}$ and $\mathbf{p}_{bottom(k)}$, where all the $\mathbf{p}_{top}$ **potentials** are the same and so are all the $\mathbf{p}_{bottom}$ ones, and so for each k:

[27] Even this could be waived, because we could choose opposite **orientations** of "*i*" for adjacent segments! It is usual, however, never to do this and to use the same "*i*" for all elements in a **strict chain** between **nodal points** (**branch points**).

$$E + v(\mathbf{p}_{\text{top}(k)}\mathbf{p}_{\text{bottom}(k)}) = 0,$$

so

$$v(\mathbf{p}_{\text{top}(1)}\mathbf{p}_{\text{bottom}(1)}) = v(\mathbf{p}_{\text{top}(2)}\mathbf{p}_{\text{bottom}(2)}) = v(\mathbf{p}_{\text{top}(3)}\mathbf{p}_{\text{bottom}(3)}) = -E.$$

So

$$E = i_1.R_1 = i_2.R_2 = i_3.R_3$$

where now the **currents** are different but the **voltages** are the same. The roles of **current** and **voltage** have reversed. To get the new answer for the combined **resistance** we convert this problem to the same form as the last problem by defining the **conductance** for each **resistor** as:

$$\mu = i \,/\, v = 1 \,/\, R$$

so that:

$$\mu v = i,$$

and since now all the v's are the same (= $-E$), we get, swapping the roles of v and i, and writing i_E for the total current through the battery:

$$i_E = i_1 + i_2 + i_3 = v.(\mu_1 + \mu_2 + \mu_3) = v.\mu$$

or simply:

$$\mu = (\mu_1 + \mu_2 + \mu_3).$$

If we now define an aggregate R as

$$R = 1 \,/\, \mu = 1 \,/\, (\mu_1 + \mu_2 + \mu_3)$$

this can be seen to imply:

$$(1 \,/\, R) = (1 \,/\, R_1) + (1 \,/\, R_2) + (1 \,/\, R_3).$$

This somewhat fundamentalist **simplex**-based development of circuit theory will never really be needed in practice. After a while, one gets the hang of allocating the signs intuitively without difficulty. But this rigorous formulation should give the reader something to fall back on when problems arise.

1.15 Euler Characteristics

In computers, surfaces are stored as data structures called **meshes** which are sets of tiny triangles that approximate the shape of the surface. More formally, if we call **0-simplexes vertices**, **1-simplexes edges** and **2-simplexes** either **faces** or **facets**, we can proceed to define two **faces** as **sharing an edge**, if there is an **edge** that, regardless of its sign, appears in the **boundary** set of both **faces**. Thus in *Figure 1.15.1*(*a*), the two **faces** $\mathbf{p}_0\mathbf{p}_1\mathbf{p}_2$ and $\mathbf{p}_0\mathbf{p}_3\mathbf{p}_1$ **share** the **edge** $\mathbf{p}_0\mathbf{p}_1$, as:

$$\mathbf{b}(\mathbf{p}_0\mathbf{p}_1\mathbf{p}_2) = \mathbf{p}_1\mathbf{p}_2 - \mathbf{p}_0\mathbf{p}_2 + \mathbf{p}_0\mathbf{p}_1 = \mathbf{p}_0\mathbf{p}_1 + \mathbf{p}_1\mathbf{p}_2 + \mathbf{p}_2\mathbf{p}_0$$

and $$\mathbf{b}(\mathbf{p}_0\mathbf{p}_3\mathbf{p}_1) = \mathbf{p}_3\mathbf{p}_1 - \mathbf{p}_0\mathbf{p}_1 + \mathbf{p}_0\mathbf{p}_3 = -\mathbf{p}_0\mathbf{p}_1 + \mathbf{p}_3\mathbf{p}_1 + \mathbf{p}_0\mathbf{p}_3$$

so that these two **faces share** the **edge** $\mathbf{p}_0\mathbf{p}_1$.

Then a (**connected**) **mesh** is a sum of **faces**, each one of which has at least one **edge shared** with another **face** in the sum, but with no **edge shared** by more than two **faces**. A **mesh** is **consistently orientated** if, as above, any **shared edge** has opposite signs in the **boundaries** of the two **faces** sharing it. Under these two criteria, a **mesh** is the same thing as a **strict 2-chain** as defined in Section 1.13.

An actual continuous surface, which is a (two-dimensional) **manifold**, can be construed as a continuum limiting case of the simpler discrete concept of a **mesh** of **simplexes**. This is a somewhat philosophical distinction: traditionally mathematics has regarded **manifolds** as "real", with **meshes** as a coarse "approximation" to the "reality". Since in a computer, only **meshes** are real, with **manifolds** being the artificial intellectual abstraction, it seems to me best to regard the boot as being on the other foot!

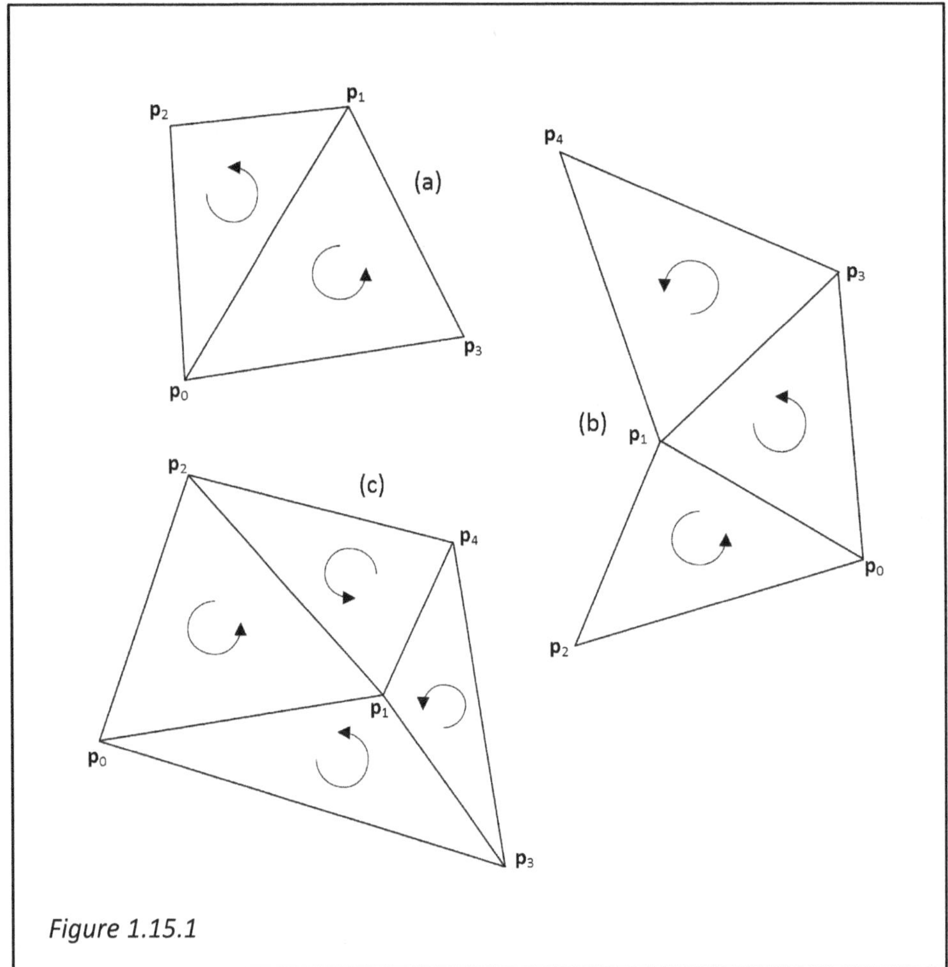

Figure 1.15.1

Now think of a triangle. It has *one* **face** ($\mathbf{p_0p_1p_2}$) – its only **2-simplex** – *three* **sides** or **1-simplexes** ($\mathbf{p_0p_1}$, $\mathbf{p_1p_2}$, $\mathbf{p_2p_0}$), and *three* **vertices** or **0-simplexes** ($\mathbf{p_0}$, $\mathbf{p_1}$, $\mathbf{p_2}$). So putting the number of **faces** as F, of **sides** or **edges** as E, and **vertices** as V, we have:

$$F + V - E = 1 + 3 - 3 = 1.$$

If we chain in another triangle, so that the two have just one side in common, as in *Figure 1.15.1*(*a*), we have a kind of quadrilateral with *four* **vertices**, *five* **edges** and *two* **faces** so now:

$$F + V - E = 2 + 4 - 5 = 1.$$

Adding in another point, whether it chains as before with *two* new **edges** and *one* new **face** as in *Figure 1.15.1*(*b*), or it connects to three of the existing **vertices** by *three* new **edges** and *two* new **faces** as in *Figure 1.15.1*(*c*), we have:

$$F + V - E = 3 + 5 - 7 = 1.$$

$$F + V - E = 4 + 5 - 8 = 1.$$

Always the result of F + V − E is 1. This number is the **Euler index** or **Euler characteristic** of an **open mesh**, which means a set of connected **2-simplexes** that do not enclose a volume. I will use the terms **Euler index** and **Euler characteristic** interchangeably.

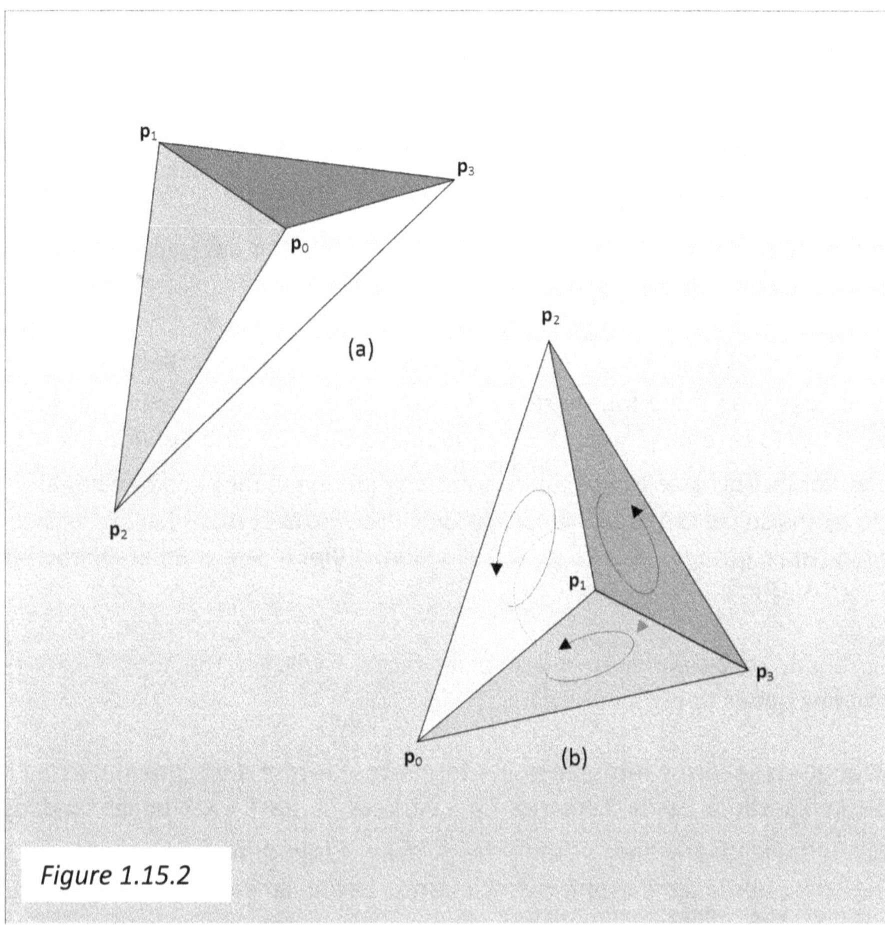

Figure 1.15.2

Now imagine that the points in the *previous* example, with (F, V, E) = (2, 4, 5) were distributed not in a plane but in three-dimensional space as in *Figure 1.15.2(a)*. We still have:

> *Two* **faces**: $\mathbf{p}_0\mathbf{p}_1\mathbf{p}_2$, $\mathbf{p}_0\mathbf{p}_1\mathbf{p}_3$
> *Four* **vertices**: $\mathbf{p}_0$, $\mathbf{p}_1$, $\mathbf{p}_2$, $\mathbf{p}_3$
> *Five* **edges**: $\mathbf{p}_0\mathbf{p}_1$, $\mathbf{p}_1\mathbf{p}_2$, $\mathbf{p}_2\mathbf{p}_0$, $\mathbf{p}_0\mathbf{p}_3$, $\mathbf{p}_1\mathbf{p}_3$.

Now suppose that we add in one more **edge** $\mathbf{p}_2\mathbf{p}_3$ so that every point or **vertex** is connected to every other, as in *Figure 1.15.2(b)*. We've added no more **vertices**, but we now have *two* more **faces**: $\mathbf{p}_3\mathbf{p}_1\mathbf{p}_2$, $\mathbf{p}_2\mathbf{p}_0\mathbf{p}_3$. So now,

$$F + V - E = 4 + 4 - 6 = 2$$

and the structure is a **closed 3-simplex** of the form of a *tetrahedron*. So the **Euler index** has gone up to 2 and the structure *does* contain a volume.

The idea holds in the plane too, but with a difference. Suppose we have a **strict 1-chain** as defined in Section 1.13 with *three* **vertices** $\mathbf{p}_0$, $\mathbf{p}_1$, $\mathbf{p}_2$ and *two* **edges** $\mathbf{p}_0\mathbf{p}_1$ and $\mathbf{p}_1\mathbf{p}_2$ and *zero* **faces**. Then:

$$F + V - E = 0 + 3 - 2 = 1$$

and this is an **open strict 1-chain**. Now connect $\mathbf{p}_0$ and $\mathbf{p}_2$ and we have the triangle as before with:

One **face**: $\mathbf{p}_0\mathbf{p}_1\mathbf{p}_2$
Three **vertices**: $\mathbf{p}_0$, $\mathbf{p}_1$, $\mathbf{p}_2$
Three **edges**: $\mathbf{p}_0\mathbf{p}_1$, $\mathbf{p}_1\mathbf{p}_2$, $\mathbf{p}_2\mathbf{p}_0$

and *the same* **Euler characteristic** $F + V - E = 1 + 3 - 3 = 1$. So in this case, the **Euler characteristic** *fails to distinguish the* **open** *from the* **closed** *form*.

We can extend the concept of a **face** to include **quadrilateral faces** or **quads** if we have two triangular **faces sharing** one **edge**, and if the four **vertices** involved lie in the same plane – or are **coplanar**. This simply gives the (F, V, E) case of *Figure 1.15.1*(*a*), equal to (2, 4, 5) with **Euler index** = 1, the standard result for an **open mesh**. This extension enables us to eliminate one **face** and one **edge** – the diagonal – each time we do it, and so we can regard a *cube*, as we would expect to, as having *six* **faces**, with *twelve* **edges** and *eight* **vertices**, giving (F, V, E) = (6, 8, 12) so with **Euler index** or **Euler characteristic** of 2, as is correct for a **closed mesh**.

When referring strictly to **2-simplexes** or triangular **faces**, I will use the term **facets** to mean these specifically, excluding **quads** or higher polygons.

Euler indices can be defined for more complex forms too. *Figure 1.15.3*(*a*) shows a triangular tube structure, with three sections. Here there are 4 × 3 **vertices**, 4 × 3 + 3 × 3 **edges**, and, assuming there are no "end-caps" closing off the ends of the tube, just 3 × 3 (**quad**) **faces**, so that $F + V - E = 9 + 12 - 21 = 0$. "Capping" the ends is done simply by adding two triangular **faces** at the ends (no new **vertices** or **edges**, so the diagram looks the same), so now $F + V - E = 11 + 12 - 21 = 2$, showing that this is now a **closed mesh** analogous to our earlier tetrahedral example. But if instead we had *joined* the end triangles together, to form a structure in which the space inside the tube is still connected but the tube itself forms a closed ring like a triangular *tyre* as in *Figure 1.15.3*(*b*), then there would be no more **faces**, but the **vertices** and the **edges** at the *ends* would now be one and the same – a process mathematicians call **identification** – and so we would have the same number of **faces** as the *open* tube, but with three less **vertices** and three less **edges**, so that:

$$F + V - E = (3 \times 3) + (3 \times 3) - (3 \times 3 + 3 \times 3) = 0.$$

Figure 1.15.3

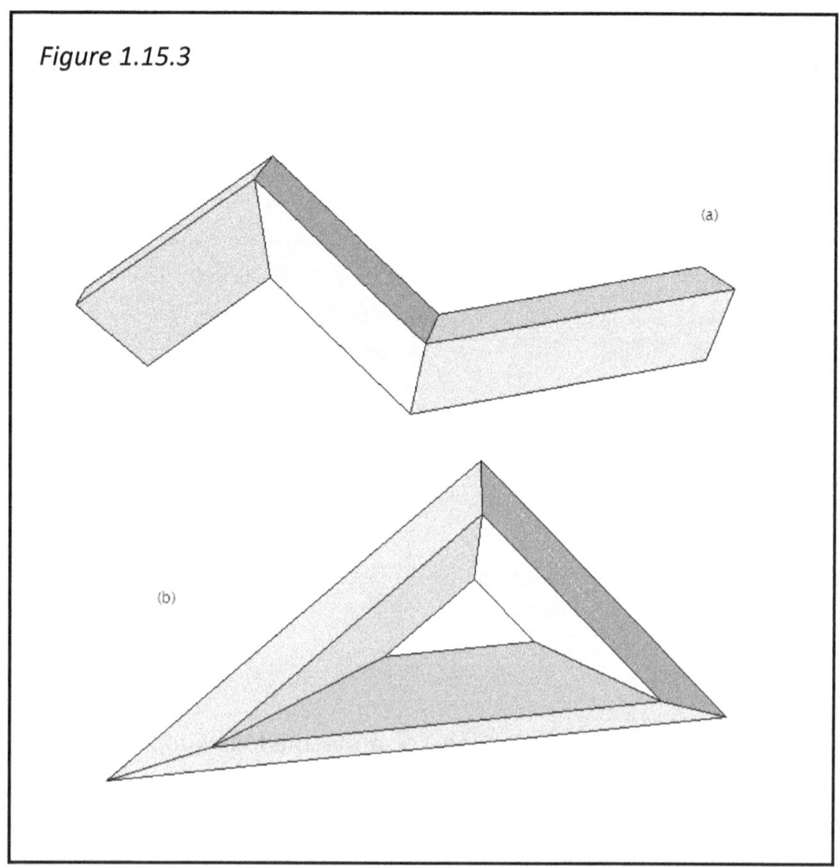

So the closed *tyre* shape has the same **Euler index** as the *open tube*, exactly by analogy with the closing of the **open strict 1-chain** into a triangle. But the *capped* tube has the same **Euler index** as the tetrahedron, or indeed of any simple (i.e. not forming any kind of ring structure) **closed mesh**.

The **Euler index** or **characteristic** is the simplest example of a **topological invariant** – a number which expresses an aspect of the **topology** of a shape. These numbers can be well-defined both for **simplexes** and for their continuum analogues, **manifolds**. Their relevance to the present discussion is quite simple. An **open mesh**, with an **Euler index** of 1, *has a* **boundary**. A (**consistently orientated**) **closed mesh**, with an **Euler index** of 2, *hasn't*. We can see this easily for the "tetrahedral" **mesh** of four **vertices**, with its **orientations** shown as in *Figure 1.15.2(b)* where:

$$\mathbf{b}(\mathbf{p}_0\mathbf{p}_1\mathbf{p}_2) = \mathbf{p}_1\mathbf{p}_2 - \mathbf{p}_0\mathbf{p}_2 + \mathbf{p}_0\mathbf{p}_1$$

$$\mathbf{b}(\mathbf{p}_0\mathbf{p}_3\mathbf{p}_1) = \mathbf{p}_3\mathbf{p}_1 - \mathbf{p}_0\mathbf{p}_1 + \mathbf{p}_0\mathbf{p}_3$$

$$\mathbf{b}(\mathbf{p}_0\mathbf{p}_2\mathbf{p}_3) = \mathbf{p}_2\mathbf{p}_3 - \mathbf{p}_0\mathbf{p}_3 + \mathbf{p}_0\mathbf{p}_2$$

and $\mathbf{b}(\mathbf{p}_2\mathbf{p}_1\mathbf{p}_3) = \mathbf{p}_1\mathbf{p}_3 - \mathbf{p}_2\mathbf{p}_3 + \mathbf{p}_2\mathbf{p}_1$

so summing all these:

$$\mathbf{p}_1\mathbf{p}_2 - \mathbf{p}_0\mathbf{p}_2 + \mathbf{p}_0\mathbf{p}_1 + \mathbf{p}_3\mathbf{p}_1 - \mathbf{p}_0\mathbf{p}_1 + \mathbf{p}_0\mathbf{p}_3 + \mathbf{p}_2\mathbf{p}_3 - \mathbf{p}_0\mathbf{p}_3 + \mathbf{p}_0\mathbf{p}_2 + \mathbf{p}_1\mathbf{p}_3 - \mathbf{p}_2\mathbf{p}_3 + \mathbf{p}_2\mathbf{p}_1$$

$$= \mathbf{p}_1\mathbf{p}_2 + \mathbf{p}_3\mathbf{p}_1 + \mathbf{p}_1\mathbf{p}_3 + \mathbf{p}_2\mathbf{p}_1 = \mathbf{p}_1\mathbf{p}_2 + \mathbf{p}_3\mathbf{p}_1 - \mathbf{p}_3\mathbf{p}_1 - \mathbf{p}_1\mathbf{p}_2 = \mathbf{0}.$$

The explanation is that $\mathbf{p}_0\mathbf{p}_1\mathbf{p}_2\mathbf{p}_3$ is a **3-simplex**, which we may call a **tet** from its *tetrahedral* prototype, and so the sum of the **boundaries** of the **faces** constitute the **boundary** of the **boundary** of the **tet**, since

$$\begin{aligned}\mathbf{b}(\mathbf{p}_0\mathbf{p}_1\mathbf{p}_2\mathbf{p}_3) &= \mathbf{p}_1\mathbf{p}_2\mathbf{p}_3 - \mathbf{p}_0\mathbf{p}_2\mathbf{p}_3 + \mathbf{p}_0\mathbf{p}_1\mathbf{p}_3 - \mathbf{p}_0\mathbf{p}_1\mathbf{p}_2 \\ &= -(\mathbf{p}_0\mathbf{p}_1\mathbf{p}_2 + \mathbf{p}_0\mathbf{p}_3\mathbf{p}_1 + \mathbf{p}_0\mathbf{p}_2\mathbf{p}_3 + \mathbf{p}_2\mathbf{p}_1\mathbf{p}_3)\end{aligned}$$

and so $\mathbf{bb}(\mathbf{p}_0\mathbf{p}_1\mathbf{p}_2\mathbf{p}_3) = \mathbf{0}$.

1.X Exercises

1. From the axioms of a **field** given in Section 1.5, prove the following:

    a. If $a + b = c$, then $b = -a + c$.

    b. If $(ab = c)$ and $a \neq 0$, then $b = (1/a) \times c$. [Allendoerfer and Oakley p.73]

2. Prove $\sqrt{2}$ is **irrational** i.e. it cannot be expressed as a ratio of two whole numbers or **integers** p/q. Hint: first prove that if x^2 is divisible by 2, then x is divisible by 2, by writing an explicit formula for **odd** and **even** numbers.

3. From the logical constructs of Section 1.4, we may define **implication** between two **predicates** as $P(x) \rightarrow Q(x) := !P(x) \vee Q(x)$. Construct the "**truth table**" for this definition – the enumeration of $P(x) \rightarrow Q(x)$ for all possible combinations of values of 0 and 1 for $P(x)$ and $Q(x)$ – and show that it gives a reasonable model of the idea of "if $P(x)$, then $Q(x)$" and show that:

    a. $[(P(x) \rightarrow Q(x)) \wedge (Q(x) \rightarrow P(x))] \equiv [P(x) \equiv Q(x)]$.

4. Show that $[P(x) \rightarrow Q(x)] \rightarrow [Q(x) \rightarrow P(x)]$ is false, but $[P(x) \rightarrow Q(x)] \rightarrow [!Q(x) \rightarrow !P(x)]$ is true. The first is the *converse*, the second the *contrapositive* of $P(x) \rightarrow Q(x)$.

5. Form the full "multiplication table" for the equilateral triangle **group** of Section 1.5. This is the grid with the six basic operations $\{e, p, q, r, s, t\}$ across the top and down the left, and the 36 "products" of the form $p * q$ as defined in that Section.

6. Imagine we had only the positive **integers** and zero, obeying just the axioms **I**, **II**, **III**, **V**, **VI**, **VII**, **VIII**, **X** and **XI** of a **field** as given in Section 1.5. We have no negative numbers. Now add an axiom:

    a. For any m and n, there exists a unique number u such that $m + u = n$.

The addition of this axiom makes these numbers into a **commutative ring with unity**. We still have no reciprocals. Define the operation "−" by: $u = n - m \equiv m + u = n$. Define $-n := 0 - n$. Now show that

a) $m + (-m) = 0$

b) $x + u = 0 \rightarrow u = -x$

c) $-(-n) = n$

d) $(-m) \times (-n) = m \times n$.

7. Now suppose we have *all* the **integers** as constructed above. Define a new kind of number from a pair of **integers** as (m, n) such that

 a. For any m and n, $m \neq 0$, there exists a unique number u such that $m \times u = n$.

 b. Define(m, n) such that $m \times (n, m) = n$.

 Show that

 a. $m \times (1, m) = 1$.

 b. $(m, n) = 0 \equiv m = 0$

 c. $(m, n) = (p, q) \equiv mq = np$,

 d. $(m, q) + (n, q) = (m + n, q)$

 e. $(m, n) + (p, q) := (mq + np, nq)$

 This gives us the **rational numbers**. (Both this and the preceding question follow S. Feferman, *The Number Systems*, Addsion-Wesley 1964.)

8. Because $x^2 > 0$ both for $x > 0$ and for $x < 0$, the **quadratic** $ax^2 + bx + c = 0$ has no solution if the **discriminant** $b^2 - 4ac < 0$. But just suppose we had a "number" j that was the square root of −1 or $j^2 = -1$. Show that with this we *could* solve $0.25x^2 + 3x + 25 = 0$.

9. The **transpose** of a **matrix A** is the **matrix** $\mathbf{A}^\mathrm{T}$ with the rows and columns exchanged, so $(\mathbf{A}^\mathrm{T})_{ij} = (\mathbf{A})_{ji}$. Show that $(\mathbf{AB})^\mathrm{T} = \mathbf{B}^\mathrm{T}\mathbf{A}^\mathrm{T}$. An **orthogonal matrix A** is one that obeys $\mathbf{A}^\mathrm{T} = \mathbf{A}^{-1}$. Show that the **matrix** $\begin{pmatrix} x & y \\ -y & x \end{pmatrix}$ is **orthogonal** if $x^2 + y^2 = 1$.

10. Using the **sharp** and **flat operators** introduced in Section 1.12, develop the entire set of Western "keys", from C major through G major, D major up to F# major (the 'circle of fifths'), and similarly the 'circle of fourths' giving F major, E♭ major through to G♭ major. What are the corresponding *minor* keys?

11. Start with the simple circuit of *Figure 1.14.1*. Now add in **resistors** R_{12} between $\mathbf{p}_1$ and $\mathbf{p}_2$, R_{30} between $\mathbf{p}_3$ and $\mathbf{p}_0$, and the crossed **resistors** R_{13} and R_{20} running diagonally. Calculate the currents through each **resistor**. Confirm that if R_{13} and R_{20} become arbitrarily large, the circuit's behaviour reverts to that of a basic **series** circuit with the three elements R_{12}, R_{23} and R_{30}. Can you construct a parallel analogue of this crossed resistors case?

12. There are five regular solids, the tetrahedron with four triangular faces, the cube with six square faces, the octahedron with eight triangular faces, the dodecahedron with twelve pentagonal faces, and the icosahedron with twenty triangular faces. Confirm that each has an **Euler characteristic** of 2. Interchanging faces and vertices establishes a duality between four of these in pairs with the last one being self-dual. Establish this.

Chapter 2

THE CALCULUS

2. 1 Introduction

The **calculus** is the gateway to higher mathematics. Its discovery in the seventeenth century revolutionised the application of mathematics to science. The key concept of the calculus is the **differential operator,**[28] which can be applied to any expression or equation to yield a new expression or equation which I will call the **linear derivative** of the original. I use these terms in a slightly more general way than is usual, but the conventional usage follows easily from this one.

An **operator** is something that transforms one mathematical entity into another, most commonly into another of the same kind.[29] The **differential operator** transforms **original variables** into new **derived variables**, **original functions** and **expressions** into **derived functions** and **expressions**, and **original equations** into **derived equations**. I will refer to all these derived forms as **derivatives**. I have also introduced a formal usage of the word **original** to refer to the entities from which the **derived** forms arise.

Applying the **differential operator** to any entity to generate its **derivative** from its **original** form is called **differentiating** the **original**. The verb is thus to **differentiate**.

Simplest are the **derivative variables**, which I will write as **d***x* and **d***y* for the variables derived from the **originals** *x* and *y*. These expressions, **d***x* and **d***y*, the result of applying the **differential operator** to *x* and *y* respectively, are new variables, new numbers, called the **differentials** of the original variables *x* and *y*. They are independent separate numbers that can be given their own values. They do *not* represent the **product** of a new variable **d** with either of the old variables. Each **differential d***x* or **d***y* is a *single new* number, represented by two letters to show its association with the original variable from which it derives. We could have written them as $x^{/}$ or $y^{/}$, a notation which is used, and Isaac Newton wrote his derivatives with a single dot over the original variable name, although these forms define a slightly different concept of **derivative**, as we shall see presently. The advantage of the **d** notation is that it allows us to operate consistently on entities from whole equations, as in **d**($y = x^2$) through to individual variables (as **d***x*) in a way which $x^{/}$ would not, at least not so clearly.

---

[28] The **differential operator** I'm introducing here is more formally called the **exterior derivative**.

[29] The term **operator** is also used to denote a simple **function** like +, × or the **Boolean operators** of Section 1.4.

The crux of the idea of the **differential operator** is *linearity*. The new **derivative** form of a **function**, **expression** or **equation** is strictly *linear* in the new variables, the **differentials** of the old variables.

The linearity is achieved by a special axiom controlling how the **differential operator** works on multiplication.

There are only two axioms needed to define how the **differential operator** applies to **expressions** and **equations**. Writing the operator itself as **d** and with *a* and *b* constant values, and *x* and *y* variable, the operator works linearly on an additive expression according to Axiom I below, and on a multiplication according to Axiom II:

Axiom I:	$\mathbf{d}(a.x + b.y) = a.\mathbf{d}x + b.\mathbf{d}y$	**Linearity**
Axiom II:	$\mathbf{d}(x \times y) = \mathbf{d}x \times y + x \times \mathbf{d}y$	**The Leibnitz Rule**

Applying just these two axioms over and over on all the terms and factors of an **expression** or **equation** always leads eventually to a *linear* form in the individual **differentials**. In effect, the **operator** cascades down through the parts of the **expression** until a strictly linear form spills out.

You may notice that we could have chosen to formulate the first axiom using a trivial "axiom 0":

Axiom 0: The **differential** of a **constant** is identically zero.

and then, when applied to an **expression** like $x^2 + y$, the **differential operator** might be defined by two further axioms:

Axiom I:	$\mathbf{d}(x + y) = \mathbf{d}x + \mathbf{d}y$	
Axiom II:	$\mathbf{d}(x \times y) = \mathbf{d}x \times y + x \times \mathbf{d}y$	**The Leibnitz Rule**

So from the "axiom 0", we have $\mathbf{d}(4) \equiv 0$, $\mathbf{d}(-27.253) \equiv 0$, and using the standard convention that in algebra letters from the start of the alphabet denote **constants**, as we've already seen in expressions like $ax^2 + bx + c$, we have $\mathbf{d}a \equiv 0$, $\mathbf{d}b \equiv 0$ and $\mathbf{d}c \equiv 0$. I use the "≡" symbol here to stress that these are not **equations**, but **identities**: **d***a always* has the value 0.

Using this approach we can recover our original axiom I, using axiom II and axiom 0. By axiom II, we have for the product of a **constant** and a **variable** as in *ax*:

$$\mathbf{d}(ax) = \mathbf{d}(a \times x) = \mathbf{d}a \times x + a \times \mathbf{d}x$$

but since $\mathbf{d}a = 0$ by our "axiom 0" this reduces to just:

$$\mathbf{d}(ax) = a \times \mathbf{d}x.$$

So we can reformulate Axiom I in the more usual form with which I introduced the **differential operator**:

Axiom I:	$\mathbf{d}(ax + by) = a.\mathbf{d}x + b.\mathbf{d}y.$	**Linearity**

This axiom simply says that **d** is a **linear operator**. Axiom II, giving the action of **d** on a product "×" is sometimes called the **derivative property**, sometimes the **Leibnitz property**.[30]

Again: applying just the two axioms I and II over and over on all the terms and factors of an **expression** (or through an **equation**) always leads eventually to an expression that is *linear* in the individual **differentials**. What this means is that however complicated the original expression, the result of applying the **differential operator d** always takes a form such as:

$$g(x, y).\mathbf{d}x + h(x, y).\mathbf{d}y$$

if there were just two **variables** x and y, or like:

$$g(x, y, z).\mathbf{d}x + h(x, y, z).\mathbf{d}y + p(x, y, z).\mathbf{d}z$$

if there are three: x, y and z. Such an expression is called a **linear combination** in the **differentials d**x, **d**y and **d**z.

In effect, the **differential operator** replaces the former complexity of the relationship between the **original variables** x, y and z by simple linear algebra in the **differentials.**

The resulting **expression** is formally called the **exterior derivative** of the original **expression**, and the form it takes, which is always that of a **linear combination** in the individual **differentials**, is called a **differential form**. The general definition of a **differential form** at this lowest level is an **expression** which can be written as:

$$f_1(x_1, x_2, ..x_n).\mathbf{d}x_1 + f_2(x_1, x_2, ..x_n).\mathbf{d}x_2 + \ldots + f_n(x_1, x_2, ..x_n).\mathbf{d}x_n$$

or in summation notation, $\Sigma_j f_j(x_1, x_2, ..x_n).\mathbf{d}x_j$, where only one **differential** appears in each term, and as a separate factor. In Chapter 6 I will extend the concept of a **differential form** by forming the **exterior derivative** of a **differential form** itself, but for this Chapter we will only need this first level of **differential form**, also called a **1-form**.[31]

Let's look at an example. Suppose we have $z = \phi(x, y) = ax^2 + 4xy + c$. We can apply the **differential operator** *across this equation* to obtain this sequence:

$$\mathbf{d}z = \mathbf{d}\phi(x, y) = \mathbf{d}(ax^2 + 4xy + c) = \mathbf{d}(ax^2) + \mathbf{d}(4xy) + \mathbf{d}c$$

using Axiom I in its simple form. So using Axiom 0 and Axiom I in its usual form, this gives:

---

[30] Gottfried Leibniz or Leibnitz (1646-1716) was one of the founders of the **calculus.**

[31] To be consistent with this concept, the ordinary **functions** and **expressions** that have been used throughout up to now are sometimes called **0-forms.** Another term for them is **scalar functions.**

$$\mathbf{d}z = a.\mathbf{d}(x^2) + 4.\mathbf{d}(xy) + 0 = a.(x.\mathbf{d}x + x.\mathbf{d}x) + 4.(y.\mathbf{d}x + x.\mathbf{d}y)$$

where the second equation comes from using the **derivative property**, Axiom II, twice, and expanding $x^2 = x \times x$ on the first term. So, collecting terms:

$$\mathbf{d}z = (2ax + 4y).\mathbf{d}x + 4x.\mathbf{d}y.$$

This is in the form $\mathbf{d}z = \phi_x(x, y).\mathbf{d}x + \phi_y(x, y).\mathbf{d}y$, with a **linear combination** in $\mathbf{d}x$ and $\mathbf{d}y$, or a **differential form** or **1-form** on the right.

When expressed like this, with a single variable on the left and a **differential form** on the right with each **differential** appearing just once in the **1-form**, the functions $\phi_x(x, y)$ and $\phi_y(x, y)$ appearing as the coefficients of the **differentials**, are called the **partial derivatives** of z with respect to x and y respectively. This is the more common use of the term "**derivative**", but for this text I will emphasize the importance of the **exterior derivative**, and so when I refer simply to a **derivative** without a qualifying adjective, I will mean the **exterior derivative**.

Applying the **differential operator** to any entity is called **differentiating** the **original**. The verb is thus to **differentiate**. I will also widely use **derivative** as an adjective to describe the result of **differentiating**.

I will also refer to the entity **differentiated** as the **original** entity.

So why complicate things by doubling up the number of variables from those we had originally? The immediate usefulness of the new variables is precisely that they abstract the *linear* part of the relationship between the old variables. Suppose we have the equation describing a simple **parabola**:

$$y = x^2.$$

Applying the **differential operator** to this equation gives:

$$\mathbf{d}y = \mathbf{d}(x^2) = \mathbf{d}(x.x) = \mathbf{d}x.x + x.\mathbf{d}x = 2x.\mathbf{d}x.$$

Now consider the graph of the parabola as shown in *Figure 2.1.1*.

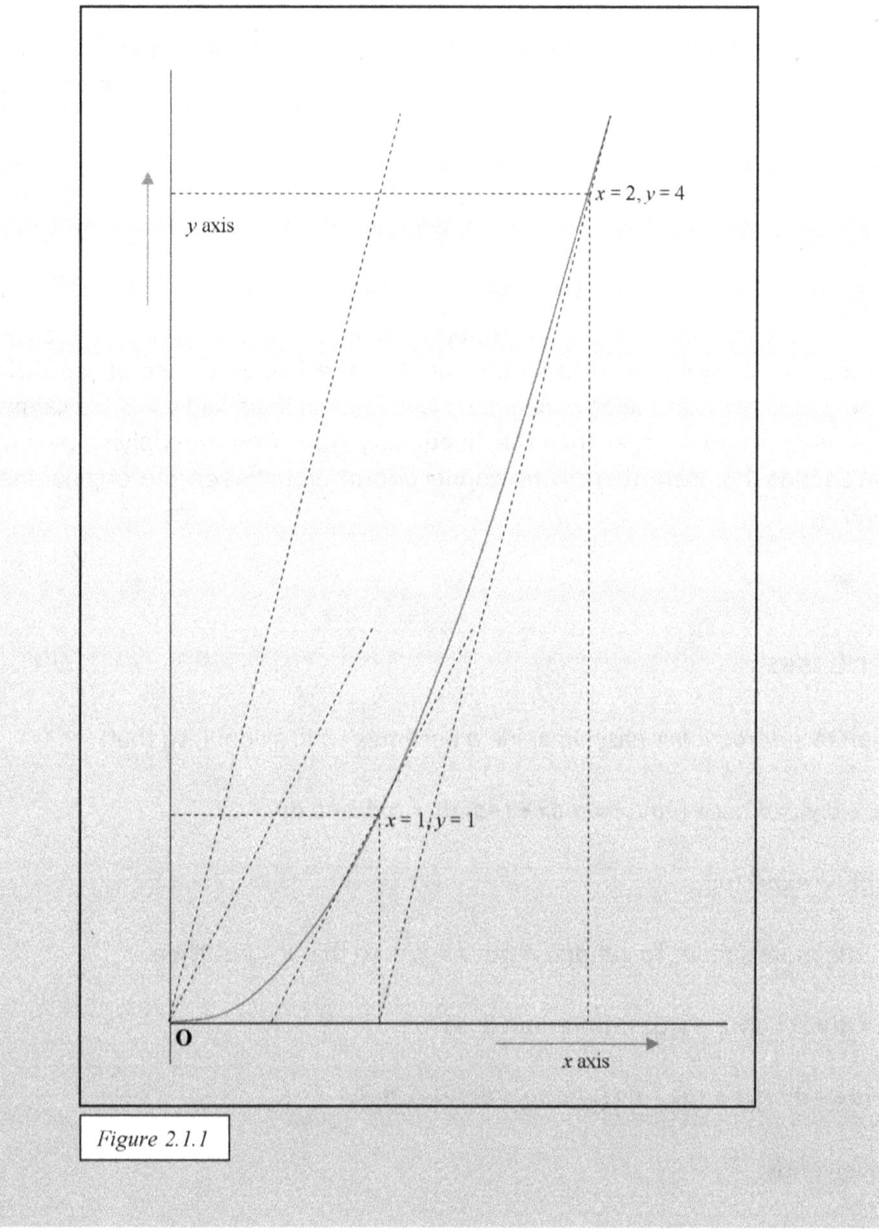

Figure 2.1.1

If we now look at the point where $x = 1$, so that $y = x^2 = 1$ also, what **d**y = 2x.**d**x = 2.**d**x shows is that at this point on the parabola the y variable is changing twice as fast as the x. The graph in *Figure 2.1.1* shows this: the equation **d**y = 2.**d**x is the equation of the straight line through the **origin** parallel to the **tangent** to the parabola at the point $x = 1$, $y = 1$. If we move to the point $x = 2$, where $y = x^2 = 4$, we will have **d**y = 2x.**d**x = 4.**d**x, so at this point on the parabola, the y value is changing four times as fast as the x, and the equation in **differentials** represents a line through the **origin** parallel to the **tangent** at $x = 2$, $y = 4$, again as shown in *Figure 2.1.1*.

Note how at each point on the original graph, there is defined a new equation in the **differentials**, which is always **linear** in them. It's as if we "freeze" or "lock" the **original** variables to see the relationship of the **derivative** ones (the **differentials**) at each point on the original graph. So although the **derivative equation** is in *four* variables – x, y, $\mathbf{d}x$ and $\mathbf{d}y$ – instead of the original *two* x and y, we look at the **derivative** form separately at each (x, y) point of the **original**. We can think of the **differential operator** as having generated a *family* of new equations in the **differentials**, one at each point on the original graph.

It is amazing what this apparently trivial concept opens up. I'll begin by working out some examples using the operator, and so start to build up an armoury of easy ready formulae to deal with various algebraic forms.

First, a word of warning: *don't* apply the **differential operator** across an inequality. Applied to an equation the **derivative equation** will also be true. But the independence of the **differentials** means that there is no guarantee that the new *inequality* will hold. So if we had $y > x^2$ we cannot infer that $\mathbf{d}y > 2x.\mathbf{d}x$. This is a more serious danger than the "**inequality trap**" from multiplying by a negative number referred to in Section 1.5. Here there is no connection *at all* between the original inequality and the **derivative** one.

2.2 Further Cases

The application to *subtraction* is easy. Imagine b becomes $-b$ in axiom I, so that:

$$\mathbf{d}(a.x - b.y) = \mathbf{d}(a.x + (-b).y) = a.\mathbf{d}x + (-b).\mathbf{d}y = a.\mathbf{d}x - b.\mathbf{d}y$$

much as might be expected.

Division is a little more subtle. To get $\mathbf{d}(x/y)$ put $z = x/y$, so that $x = y.z$. Then

$$\mathbf{d}(x) = \mathbf{d}(y.z) = \mathbf{d}y.z + y.\mathbf{d}z \text{ from axiom II, so}$$

$$\mathbf{d}z = (\mathbf{d}x - \mathbf{d}y.z)/y = (\mathbf{d}x - \mathbf{d}y.(x/y))/y = \mathbf{d}x/y - x.\mathbf{d}y/y^2$$

or, more conveniently,

$$\mathbf{d}(x/y) = (y.\mathbf{d}x - x.\mathbf{d}y)/y^2.$$

We can use this technique for the square root too. Putting $x = \sqrt{x}.\sqrt{x}$,

$$\mathbf{d}x = \sqrt{x}.\mathbf{d}\sqrt{x} + \sqrt{x}.\mathbf{d}\sqrt{x} = 2\sqrt{x}.\mathbf{d}\sqrt{x} \text{ from axiom II,}$$

$$\mathbf{d}\sqrt{x} = \tfrac{1}{2}\mathbf{d}x/\sqrt{x}.$$

This can handle the cube root too. For $x = {}_3\sqrt{x}.{}_3\sqrt{x}.{}_3\sqrt{x}$, so

$$\begin{aligned}\mathbf{d}x &= \mathbf{d}(\sqrt[3]{x}).\sqrt[3]{x}.\sqrt[3]{x} + \sqrt[3]{x}.\mathbf{d}(\sqrt[3]{x}.\sqrt[3]{x}) \\ &= \mathbf{d}(\sqrt[3]{x}).\sqrt[3]{x}.\sqrt[3]{x} + \sqrt[3]{x}.(\mathbf{d}(\sqrt[3]{x}).\sqrt[3]{x} + \sqrt[3]{x}.\mathbf{d}(\sqrt[3]{x})) \\ &= 3(\sqrt[3]{x})^2.\mathbf{d}(\sqrt[3]{x})\end{aligned}$$

So $\quad \mathbf{d}(\sqrt[3]{x}) = \tfrac{1}{3}\,\mathbf{d}x/(\sqrt[3]{x})^2.$

A particularly useful result is that for the general power of x, x^n. This can be demonstrated easily by the technique known as **induction** introduced in Section 1.10, where a result is established for $n = 1$ or $n = 2$, and then it is shown that if it holds for n, it must also hold for $n + 1$. We already have the results for $n = 1$ and $n = 2$ as $\mathbf{d}(x^1) = \mathbf{d}x$, and $\mathbf{d}(x^2) = 2.x.\mathbf{d}x$. The $n = 1$ result is special, the definition of the **differential**. The general result follows the $n = 2$ form, and is:

$$\mathbf{d}(x^n) = n.x^{n-1}.\mathbf{d}x.$$

We can show this by **induction** like this. Given the result as above for n, we can then put for $n + 1$:

$$\begin{aligned}\mathbf{d}(x^{n+1}) &= \mathbf{d}(x^n.x) = \mathbf{d}(x^n).x + x^n.\mathbf{d}x = n.x^{n-1}\mathbf{d}x.x + x^n.\mathbf{d}x \\ &= n.x^{n-1}.x\,\mathbf{d}x + x^n.\mathbf{d}x = n.x^n\,\mathbf{d}x + x^n.\mathbf{d}x = (n+1).x^n\mathbf{d}x.\end{aligned}$$

If this looks one level out, put $m = n + 1$, so $n = m - 1$, and rewrite the result as:

$$\mathbf{d}(x^m) = m.x^{m-1}.\mathbf{d}x.$$

In the next chapter, we will introduce a formalism that shows that roots can be written as *fractional* **exponents**, so that $\sqrt[3]{x}$ can be written as $x^{1/3}$ and $\sqrt{x}$ as $x^{1/2}$. The same formalism puts the reciprocal of a power like $1/x^n$ as a *negative* **exponent** x^{-n}. This leads to the formula above extending to the square and cube roots we developed earlier. So the cube root formula is just

$$\mathbf{d}(\sqrt[3]{x}) = \mathbf{d}(x^{1/3}) = \tfrac{1}{3}.x^{(1/3-1)}.\mathbf{d}x = \tfrac{1}{3}.x^{-2/3}\,\mathbf{d}x$$

because $x^{-2/3}$ (x to the *minus two-thirds*) here is the same thing as $1/(\sqrt[3]{x})^2$.

Another way of showing this is to put $y = \sqrt[3]{x}$, so that $y^3 = x$, when we will have:

$$\mathbf{d}(y^3) = 3y^2.\mathbf{d}y = \mathbf{d}x$$

so $\quad \mathbf{d}y = \mathbf{d}(\sqrt[3]{x}) = \mathbf{d}x/(3y^2) = \tfrac{1}{3}(1/(\sqrt[3]{x})^2).\mathbf{d}x = \tfrac{1}{3}(1/x^{2/3})\mathbf{d}x = \tfrac{1}{3}.x^{-2/3}.\mathbf{d}x.$

If you've followed the ideas so far, you should be comfortable with evaluating the **derivatives** of some reasonably complicated forms such as:

$$\mathbf{d}(y = 4.x^2 + 3x + 17) \text{ gives } \mathbf{d}y = (4.2x + 3).\mathbf{d}x = (8x + 3)\mathbf{d}x$$

where I use the shorthand notation $\mathbf{d}(y = \phi(x))$ for $(\mathbf{d}y = \mathbf{d}\phi(x))$, to indicate that applying **d** to both sides of an **equation** will still give an **equation**, so we can think of **d** as applied across the **equation** as a whole. I also re-use Axiom 0, that the derivative of a **constant** is zero, so that, for example, $\mathbf{d}(17) = 0$.

The division formula enables us to tackle:

$$\mathbf{d}((7.x + 4)/(x^2 + x)) = [(x^2 + x).\mathbf{d}(7.x + 4) - (7.x + 4).\mathbf{d}(x^2 + x)]/(x^2 + x)^2$$
$$= [(x^2 + x).7\mathbf{d}x - (7.x + 4).(2x + 1)\mathbf{d}x]/(x^2 + x)^2$$
$$= (7x^2 + 7x - 14x^2 - 7x - 8x - 4)\mathbf{d}x/(x^4 + 2x^3 + x^2)$$
$$= [(-7x^2 - 8x - 4) /(x^4 + 2x^3 + x^2)]\ \mathbf{d}x.$$

There need not be any variable actually standing on its own in an equation. For example, the equation:

$x^2 + y = 4x + 7z^2$ can be **differentiated** directly to give:

$$2x\mathbf{d}x + \mathbf{d}y = 4\mathbf{d}x + 14z\mathbf{d}z$$

or

$$(2x - 4)\mathbf{d}x + \mathbf{d}y - 14z\mathbf{d}z = 0.$$

Applying the **operator** to a general equation like this is sometimes called **implicit differentiation.**

If the explicit form of the **expression** or **equation** is not known, we can apply **functional notation** as described in Section 1.1. Suppose we are given y as a **function** of x in the general form $y = f(x)$. Then we know that whatever the *actual* form of f, the (**exterior**) **derivative** must be **linear** in the **differentials.** How do we write it? The most elegant way is to use the beautiful letter ∂, which seems to have no name, although it is commonly spoken as "dee" just like the operator itself, and I seem to have a very faint memory of its being sometimes called "del". Using ∂, we write the **derivative equation** as:

$$\mathbf{d}y = \partial f.\mathbf{d}x = \partial f\,(x).\mathbf{d}x$$

where $\partial f\,(x)$ is a new **function** of x, called the **ordinary derivative** of f. So for example, if the **original function** is $f\,(x) = x^2$, then $\partial f\,(x) = 2x$, because if $y = f\,(x) = x^2$, then $\mathbf{d}y = 2x.\mathbf{d}x$. Another common notation is $\mathbf{d}y = f\,'\mathbf{d}x$, using a superscript tick or "dash" against f. The commonest of all however, is the pseudo-division form:

$$\mathbf{d}y = (\mathrm{d}f(x)/\mathrm{d}x).\mathbf{d}x$$

which I will discuss in the next section. If there are two **arguments** to the function, say $z = f\,(x, y)$, the ∂ notation lends itself to the elegant form:

$$\mathbf{d}z = \partial_x f.\mathbf{d}x + \partial_y f.\mathbf{d}y$$

which is simply a statement that the (**exterior**) **derivative** must be **linear** in the **differentials** of the **arguments.** Here $\partial_x f$ and $\partial_y f$ constitute a general notation for the **partial derivatives** appearing as the *coefficients* of each **differential.**

These are new **functions** $\partial_x f\,(x, y)$ and $\partial_y f\,(x, y)$. You should note that in the conventional terminology, the word **derivative** *without a qualifying adjective* is *only used for these* **functions,** *not* for the **exterior derivative** as I prefer to do in this text. We can without confusion use this same notation for the **ordinary derivative** above as well, so that:

$$\partial f(x) \equiv \partial_x f(x)$$

and I will often write the **ordinary derivative** this way too. The various $\partial_x f(x, y)$, $\partial_y f(x, y)$ forms that appear where there are multiple **arguments** are called **partial derivatives**, as stated above, and since these comprehend the **ordinary derivative** as a special case, I will generally avoid the latter term.

These functions can be **differentiated** again, so for example:

$$\mathbf{d}(\partial_x f(x, y)) = \partial_x\partial_x f(x, y).\mathbf{d}x + \partial_y\partial_x f(x, y).\mathbf{d}y$$

and it is convenient to write this as: $\partial_x^2 f(x, y).\mathbf{d}x + \partial_y\partial_x f(x, y).\mathbf{d}y$, so $\partial_x^2 \equiv \partial_x\partial_x$, another useful shorthand.

If the underlying variables themselves are functions of other variables, say x and y depend on u and v, we can proceed by substituting for $\mathbf{d}x$ and $\mathbf{d}y$ using the equations for x and y in terms of u and v, which might be $x = g(u, v)$, $y = h(u, v)$ and which will have their own **derivatives:**

$$\mathbf{d}x = \partial_u g.\mathbf{d}u + \partial_v g.\mathbf{d}v$$

and

$$\mathbf{d}y = \partial_u h.\mathbf{d}u + \partial_v h.\mathbf{d}v.$$

If we now substitute these into $\mathbf{d}z = \partial_x f.\mathbf{d}x + \partial_y f.\mathbf{d}y$ we get:

$$\mathbf{d}z = \partial_x f.(\partial_u g.\mathbf{d}u + \partial_v g.\mathbf{d}v) + \partial_y f.(\partial_u h.\mathbf{d}u + \partial_v h).\mathbf{d}v$$

$$= (\partial_x f.\partial_u g + \partial_y f.\partial_u h).\mathbf{d}u + (\partial_x f.\partial_v g + \partial_y f.\partial_v h).\mathbf{d}v$$

a result known as the **chain rule**.

Care needs to be exercised in using this rule as we must evaluate $\partial_x f$ as $\partial_x f(x(u, v), y(u, v))$, but $\partial_u g$ as $\partial_u g(u, v)$. In other words, we must choose *the same* (u, v) point to evaluate the expression through both levels. As so often, an example may help. Suppose we have:

$$z = 3x^2 + xy$$
$$x = u + 2v$$
$$y = v^3$$

then[32]

$$\mathbf{d}z = 6x.\mathbf{d}x + y.\mathbf{d}x + x.\mathbf{d}y = (6x + y).\mathbf{d}x + x.\mathbf{d}y$$
$$\mathbf{d}x = \mathbf{d}u + 2.\mathbf{d}v$$
$$\mathbf{d}y = 3v^2.\mathbf{d}v.$$

Then $\mathbf{d}z = (6x + y).\mathbf{d}x + x.\mathbf{d}y = (6x(u, v) + y(u, v)).\mathbf{d}x + x(u, v).\mathbf{d}y$

[32] Here I use the functional notation $x = x(u, v)$ to indicate that x is a **function** of u and v. This notation is widely disparaged because "x" is being used for both a **variable** and a **function**, but even computer software can be made "intelligent" enough to see the distinction by means of *context-sensitive parsing*, and the suggestive power of the association is very helpful.

$= (6x(u, v) + y(u, v)).(\mathbf{d}u + 2.\mathbf{d}v) + x(u, v).(3v^2.\mathbf{d}v)$

$= (6(u + 2v) + (v^3)).(\mathbf{d}u + 2.\mathbf{d}v) + (u + 2v).(3v^2.\mathbf{d}v)$

$= (6u + 12v + v^3).\mathbf{d}u + (12u + 24v + 2v^3 + 3uv^2 + 6v^3).\mathbf{d}v$

$= (6u + 12v + v^3).\mathbf{d}u + (12u + 24v + 8v^3 + 3uv^2).\mathbf{d}v.$

There's no new maths here. We just need to be consistent in substituting at both the **original** level and at the **derivative** level, respectively for *z*, *x* and *y*, and for **d***z*, **d***x* and **d***y*.

2.3 The Infinitesimal Curse

Nowhere have I suggested that the **differentials** have to be small numbers, let alone *infinitely* small. This is in strong contrast to the conventional presentation, which treats **d***x* and **d***y* as being tiny changes in the **original variables** *x* and *y*. I have been at pains to emphasize that the **differential operator** merely creates a **linear differential form** from the original expression or equation.

Nevertheless, the *infinitesimal* idea played a huge role in the founding of the calculus, and we need to look at why it came to have the significance it did.

Let's go back to our parabola $y = x^2$. Imagine a point on this curve (x, y) obeying the relationship $y = x^2$ and imagine another point also on the curve at the *x* value $x + \mathbf{d}x$ where for the moment we *do* assume that **d***x* is a very small value. Call the corresponding *y* value y^*, so that $(x + \mathbf{d}x, y^*)$ is also on the curve, and we will have $y^* = (x + \mathbf{d}x)^2$. So

$$y^* = (x + \mathbf{d}x)^2 = x^2 + 2x.\mathbf{d}x + (\mathbf{d}x)^2.$$

Now if **d***x* is very small, say 0.0001, then $(\mathbf{d}x)^2$ will be much smaller again – 0.0000001 – and we can regard this term as negligible, so that

$y^* = x^2 + 2x.\mathbf{d}x$ to a very good approximation. But $y = x^2$, so this becomes:

$$y^* = y + 2x.\mathbf{d}x$$

or

$$y^* - y = 2x.\mathbf{d}x.$$

If we write the small change in *y* corresponding to **d***x* and appearing here as $y^* - y$ as **d***y* or:

$$\mathbf{d}y := y^* - y$$

using the symbol ":=" to mean "is defined as", we can write:

$$\mathbf{d}y = 2x.\mathbf{d}x$$

so the equation in **differentials** *approximates* the change in the values of the **original** variables very accurately for a small change.

In the early years of the calculus, this **infinitesimal** property was seen as the cardinal defining attribute of the calculus, and the subject actually came to be called the **infinitesimal calculus.** I simply do not see it this way. I have to admit that I am suspicious about any beliefs in infinities or infinitesimals in maths, and have a lot of sympathy with Bishop Berkeley who lambasted the calculus in the early eighteenth century on the basis that its practitioners took **differentials** to be significant or negligible simply as the whim suited them! It seems to me that the cardinal property is the **linearity** of the **derivative** forms, and this can be established purely algebraically without any recourse to infinities or infinitesimals with all their attendant philosophical difficulties. In brief, my thesis is this: *don't resort to infinities or infinitesimals if there is any other interpretation available.*

My picture is that the **differential operator** defines for any **original equation** a family of **linear** equations, which describe for two variables a family of *lines* through the **origin**, and for three variables a family of *planes* through the **origin** – equations in $\mathbf{d}x$, $\mathbf{d}y$ and $\mathbf{d}z$ – with one unique member of the family for every point on the **original** line if there are two variables, or every point on the **original** *surface* if there are three variables. For more than three variables, the geometrical imagery gets difficult, as we have to think of *hypersurfaces* and *hyperplanes*, but the algebra carries on without a murmur.

A major objection to the infinitesimal concept is that it led to the idea that **differentials** couldn't stand on their own in an equation and needed to be "inflated back up" to "proper-size" variables by always using them in a *ratio* form. This in turn led to the "pseudo-division" form for the **coefficients** of the **differentials** in any **linear derivative**, and indeed the normal presentation of the calculus uses this concept almost exclusively, so you have to know it. The idea is that we have a functional form like

$$w = f(x, y, z)$$

with a **derivative** – in my terminology:

$$\mathbf{d}w = \partial_x f.\mathbf{d}x + \partial_y f.\mathbf{d}y + \partial_z f.\mathbf{d}z.$$

Now set $\mathbf{d}y = 0$ and $\mathbf{d}z = 0$, and divide through by $\mathbf{d}x$ to give:

$$(\mathbf{d}w/\mathbf{d}x)_{y,z} = \partial_x f$$

where the subscripts indicate that y and z are being "held constant" – i.e. treated as constants, so $\mathbf{d}y = 0$ and $\mathbf{d}z = 0$. This is commonly written using the ∂ symbol as:

$$(\partial w/\partial x)_{y,z}$$

and this is the standard notation for a **partial derivative**. The **partial** bit refers to the fact that in this case there are other variables which are "held constant". If we have a relationship between only two variables like $y = f(x)$, so that $\mathbf{d}y = \partial_x f.\mathbf{d}x$ unambiguously, we dispense with the subscripts and use just "d" instead of ∂ to give the **ordinary derivative**:

$\mathrm{d}f(x)/\mathrm{d}x$.

I think this pairwise approach gives quite a wrong picture of what is going on. There isn't normally some paired connection between variables taken two by two, and an equation in **differentials** may often take the more general form:

$$\partial_x f.\mathbf{d}x + \partial_y f.\mathbf{d}y + \partial_z f.\mathbf{d}z = 0$$

where no individual variable is singled out as having a coefficient of unity (1). We don't treat variables in normal linear equations two by two in this pairwise fashion, and we shouldn't do so with equations in **differentials**.

The auxiliary concept of "holding the other variables constant" permeates the traditional presentation like a canker. It is not there in the actual algebra, and is a quite unnecessary notion.

At this point I should stress again that in the literature the term **derivative** without a qualifying adjective is *only* used for the **ordinary** and **partial derivatives** defined above. Because this usage is so established, I will still use it, referring to forms like $\partial_x f$ as **ordinary** or **partial derivatives** as appropriate, but for brevity I will also use:

OD for the **ordinary derivative** $\partial f(x)$, and
PD for the **partial derivatives** such as $\partial_y f(x, y, z)$.

2.4 Inversion of the Differential Operator

A central question in the calculus is whether we can obtain the **original** form from the **derivative** form. This procedure, the **inverse** of **differentiating**, is commonly called **integrating** or "**indefinite integration**", but the term "**integration**" is also used for a more significant operation called "**definite integration**". The two concepts are quite distinct and I am going to reserve the term **integration** for the concept introduced in Section 2.6. So I will call **inverse differentiation** simply that or else **anti-differentiation**. This suggests that the **original** form from which a **derivative** is obtained may be called the **anti-derivative**. All these terms are rather clumsy but as it happens, we will not need to use any of them very much.

An obvious possibility is to try and find an **inverse operator** to the **differential operator** which we might label something like $\mathbf{d}^{-1}$. Unfortunately, *there is no such inverse* **operator**. The **original** form can only be divined by inspection, although there are some rules. This exercise is the huge field of "Methods of Integration".

It is very important to realize two things: there simply may not *be* an **original** form at all, and even when there is it will not be unique. These two points need some explanation.

First, let's consider some **differential forms**. A **differential form** is a quite specific thing, a **linear** expression or **linear combination**[33] in **differentials**. Examples are:

[33] See the Glossary for a brief definition.

$14yz.\mathbf{d}x + x^3y.\mathbf{d}y - 22.3x.\mathbf{d}z$

$\partial_x f.\mathbf{d}x + \partial_y f.\mathbf{d}y + \partial_z f.\mathbf{d}z$

$2xy.\mathbf{d}x + x^2.\mathbf{d}y$

$(x^4 + 2xy - 3pq).\mathbf{d}s + (12p - u^3).\mathbf{d}p - (3p - xu).\mathbf{d}q$

where in the last I've tried to move away from just x and y.

Only the middle two of these are the **(exterior) derivatives** of an **original expression**. The other two are simply **differential forms** that could not owe their origin to the application of the **differential operator** to a single **original** expression. The reason is simply that they do not fit the format of the second one. For example, in the first, the coefficient of $\mathbf{d}x$ ($14yz$) would have to be $\partial_x g$ and that of $\mathbf{d}y$ (x^3y) would have to be $\partial_y g$ for some function $g(x, y, z)$, and they just don't match that prescription. We can tell that they don't because we can test for this prescription using a result known as **Young's theorem**, which is:

$$\partial_y\partial_x g(x, y) = \partial_x\partial_y g(x, y) \quad \text{always.}$$

Applying **Young's theorem** to the expressions $14yz$ =? $\partial_x g(x, y, z)$ and x^3y =? $\partial_y g(x, y, z)$ we should have:

$$\partial_y(14yz) = 14z = \partial_x(x^3y) = 3x^2y$$

and clearly we don't. (To see these are the ∂_y and ∂_x **partial derivatives** of $14yz$ and $3x^2y$ respectively, imagine applying **d** to the form $14yz$ to give $14z.\mathbf{d}y + 14y.\mathbf{d}z$ for example. In this expression $14z$ is the ∂_y part of the **(exterior) derivative** of an **original** expression $\partial_x g(x, y, z) = 14yz$).

$2xy.\mathbf{d}x + x^2.\mathbf{d}y$, however, does obey this rule, with $\partial_y(2xy) = 2x$, which *does* equal $\partial_x(x^2) = 2x$, and so this one *is* the **derivative** of an **original** form x^2y:

$$\mathbf{d}(x^2y) = 2xy.\mathbf{d}x + x^2.\mathbf{d}y.$$

A **differential form** that *is* the **(exterior) derivative** of an **original** expression is called **exact.**

The same **derivative** can also come from *more than one* **original**, because "constant" terms drop out. So both $x^2 + 3x + 2$ and $x^2 + 3x - 220$ give rise to the same **(exterior) derivative** $(2x + 3).\mathbf{d}x$. This "unknown constant" or "lost constant" is called the **constant of integration.**[34]

Inversion of **differentiation** is in general nasty, but one result we can get straight away, and this will always be useful. From $\mathbf{d}(x^n) = nx^{n-1}\mathbf{d}x$, putting $m = n - 1$, and so $n = m + 1$, we obtain:

$$(1/(m+1)).\mathbf{d}(x^{m+1}) = x^m.\mathbf{d}x$$

[34] From the "**indefinite integration**" terminology mentioned at the start of this Section, which I would prefer to avoid.

or $\quad x^m.\mathbf{d}x = \mathbf{d}(x^{m+1}/(m+1))$

which enables us to invert the terms of any **polynomial** in one variable. So, for example:

$$(ax^2 + bx + c).\mathbf{d}x = ⅓a.\mathbf{d}(x^3) + ½b.\mathbf{d}(x^2) + c.\mathbf{d}x + k$$

$$= \mathbf{d}(⅓a.x^3 + ½b.x^2 + c.x + k)$$

where k is the **constant of integration**. So this is the general form that *any* **original** expression must take to have the (**exterior**) **derivative** $(ax^2 + bx + c).\mathbf{d}x$.

You may ask whether there is a more general way in which the axiom II of the **d operator** can be used, and indeed there is. Putting it as:

$$\mathbf{d}(u(x).v(x)) = u(x).\mathbf{d}v(x) + \mathbf{d}u(x).v(x) = u(x).\partial_x v(x).\mathbf{d}x + \partial_x u(x).v(x).\mathbf{d}x$$

we have the result:

$$u(x).\partial_x v(x).\mathbf{d}x = \mathbf{d}(u(x).v(x)) - \partial_x u(x).v(x).\mathbf{d}x$$

which seems to have got us no further, as we've merely swapped the roles of u and v. But by adroit choice of u and v, this can be surprisingly useful, and we will use it in Section 2.7 to derive **Taylor's expansion**. This use of the $u-v$ swap using axiom II is called **integration by parts.**[35]

2.5 Applications: Minima and Maxima

One simple application of this theory is in finding points on a curve or surface where one of the variables attains a **minimum** or a **maximum**. *Figure 2.5.1* illustrates the principle. Since the equation in **differentials** has the same inclination as the **tangent** has to the **original** at the point where we evaluate the **derivative**, we can locate points where **minima** or **maxima** occur because they must have a **tangent** which is *horizontal* at that point (assuming we are plotting the variable we want to maximize or minimize on the vertical axis).

[35] Again using the "**indefinite integration**" terminology.

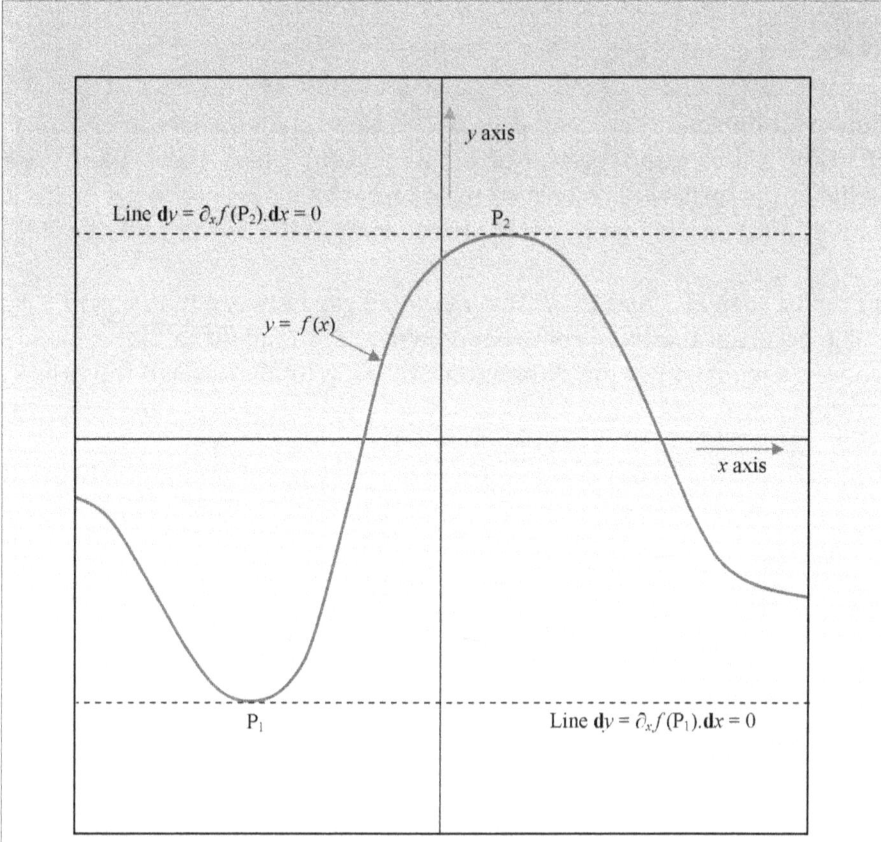

Figure 2.5.1

A **derivative** of the form **d**$y = \partial_x f(x)$.**d**x will represent a *horizontal* line through the **origin** at any point (x, y) where $\partial_x f(x) = 0$, for then whatever the value of **d**x, **d**y will always have the value 0, as can be seen in *Figure 2.5.1*. So if we take the general form of the **derivative** for arbitrary x and y and solve for $\partial_x f(x) = 0$, we will find the value of x at which the **tangent** is horizontal and so at which y has a **maximum** or **minimum**.

It's easy to see in an example. I'll take one from economics. A firm aims to maximize its *gain* G, which is the difference between its *costs* C and its *returns* R. R is equal to the *price* p at which it sells its product times the *quantity* q of the product sold. But its *costs* are a more complicated function of the quantity q. So:

$$\mathrm{G} = \mathrm{R} - \mathrm{C}$$
$$\mathrm{R} = p.q$$
$$\mathrm{C} = \mathrm{C}(q).$$

We evaluate:

$$\mathbf{d}\mathrm{G} = \mathbf{d}\mathrm{R} - \mathbf{d}\mathrm{C} = \mathbf{d}(p.q) - \partial_q\mathrm{C}(q).\mathbf{d}q = \mathbf{d}p.q + p.\mathbf{d}q - \partial_q\mathrm{C}(q).\mathbf{d}q$$

but we can assume that **d**$p = 0$ – the price is fixed by external market conditions – so that

$$\mathbf{dG} = \mathbf{dR} - \mathbf{dC} = p.\mathbf{d}q - \partial_q C(q).\mathbf{d}q = (p - \partial_q C(q)).\mathbf{d}q$$

and for a **maximum** (or **minimum**) we want **dG** = 0. This clearly holds where $(p - \partial_q C(q)) = 0$ or where $p = \partial_q C(q)$. So the firm will choose the level of production (q) where the market price p equals the **ordinary derivative** of its costs with respect to the quantity of goods produced q. Economists call this **ordinary derivative** $\partial_q C(q)$ the "marginal cost of production".

The logic of this can be seen in *Figure 2.5.2*. The maximum gap between the straight R line proportional to q and the C line occurs at the (q, p) combination where the **tangent** to C(q) is parallel to R (and so, incidentally, where the equation in the **differentials** $\mathbf{dC} = \partial_q C(q).\mathbf{d}q$ through the **origin** coincides with R).[36]

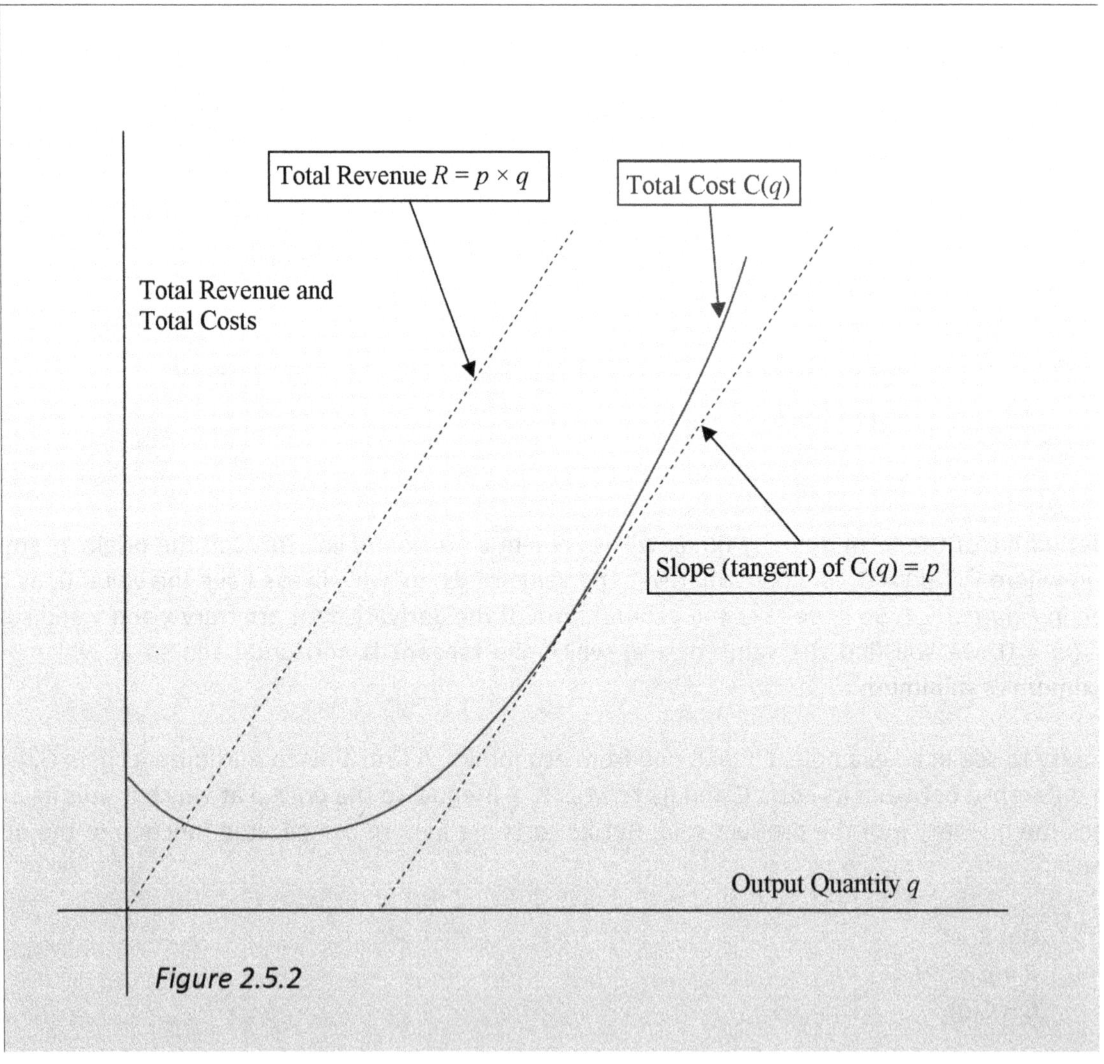

Figure 2.5.2

[36] Should you feel that this is a rather unlikely graph shape for total production costs, I have given a strong critique of this model myself in my companion book *Money and Banking*.

For another example, I'll go to physics. This is an interesting one because the basic equation that we're maximizing is subject to a **constraint condition** and our theory handles this very neatly. A physical system has a number of accessible **microstates** – basically *quantum states* – determined by its internal energy E and the number of particles in the system N. This number of **microstates** is given by a function, which can often be calculated directly from quantum mechanics theory, written $\Omega(N, E)$, at least in British usage.[37] When two such systems are brought into thermal contact, they can exchange energy but the sum of their energies $E = E_1 + E_2$ must be constant – energy cannot be created or destroyed. They will exchange energy until the number of accessible states becomes a **maximum**. The number of accessible **microstates** of the two systems together can, for any system that is large in atomic terms, be estimated very accurately by the product:

$$\Omega_1(N_1, E_1).\Omega_2(N_2, E_2)$$

and so we need to *maximize* this subject to $E = E_1 + E_2$ being a constant. Apply **d**:

$$\mathbf{d}\,(\Omega_1(N_1, E_1).\Omega_2(N_2, E_2)) = (\mathbf{d}\Omega_1(N_1, E_1)).\ \Omega_2(N_2, E_2) + \Omega_1(N_1, E_1).(\mathbf{d}\Omega_2(N_2, E_2))$$

$$= \partial_{E1}\Omega_1(N_1, E_1).\mathbf{d}E_1.\ \Omega_2(N_2, E_2) + \Omega_1(N_1, E_1).\partial_{E2}\Omega_2(N_2, E_2).\mathbf{d}E_2$$

where we assume that $\mathbf{d}N_1 = \mathbf{d}N_2 = 0$ because the numbers of particles in the two systems cannot change. For a **maximum** or **minimum** we want this **differential** to be zero, but $\mathbf{d}E = \mathbf{d}E_1 + \mathbf{d}E_2 = 0$, so $\mathbf{d}E_2 = -\mathbf{d}E_1$ so we need

$$\partial_{E1}\Omega_1(N_1, E_1).\Omega_2(N_2, E_2).\mathbf{d}E_1 - \Omega_1(N_1, E_1).\partial_{E2}\Omega_2(N_2, E_2).\mathbf{d}E_1 = 0$$

or: $$\partial_{E1}\Omega_1(N_1, E_1).\Omega_2(N_2, E_2) = \Omega_1(N_1, E_1).\partial_{E2}\Omega_2(N_2, E_2)$$

Dividing through by $\Omega_1(N_1, E_1).\Omega_2(N_2, E_2)$, we get:

$$\partial_{E1}\Omega_1(N_1, E_1) / \Omega_1 = \partial_{E2}\Omega_2(N_2, E_2) / \Omega_2.$$

Anticipating a result from the next chapter, this can be expressed in terms of the **natural logarithm function**, which here I'll write as **log**(*x*):

$$\partial_{E1}(\mathbf{log}\ \Omega_1(N_1, E_1)) = \partial_{E2}(\mathbf{log}\ \Omega_2(N_2, E_2)).$$

This is one of the most important results in physics. The quantity **log Ω** is called the **entropy** and written S and so our result can be written:

$$\partial_{E1}S_1 = \partial_{E2}S_2$$

and this states that the condition of equilibrium of two systems brought into thermal contact is that the **partial derivatives** of their **entropies** with respect to their **energies** are equal. These **partial derivatives** correspond to their **temperatures**. This is the formal theoretical definition of the concept of

[37] The model given here is a slightly simplified conflation from F. Mandl *Statistical Physics*, Wiley 1988 and the superb C.Kittel and H. Kroemer *Thermal Physics* Freeman 1980.

temperature. (The conventional definition of **temperature** corresponds to T = $1/(k(\partial_E S))$ – it's proportional to the *reciprocal* of the **derivative**, the constant *k* being Boltzmann's constant).

2.6 A First Look at Integration

Figure 2.6.1 shows a selection of points or **0-simplexes** in the *x*-*y* plane. As we saw in Section 1.13, we are quite at liberty to *sign* **0-simplexes** independently of the signs of their defining **coordinates**, and indeed the **boundary operator b** will not give the right results unless **simplexes** *are* signed. So here I've marked the points $-\mathbf{p}_1(x_1, y_1)$, $+\mathbf{p}_2(x_2, y_2)$, $-\mathbf{p}_3(x_3, y_3)$, $+\mathbf{p}_4(x_4, y_4)$, $-\mathbf{p}_5(x_5, y_5)$, $+\mathbf{p}_6(x_6, y_6)$. We can define a **scalar function** – which is simply a function that returns an ordinary real number like 27.32 or 0.00467 – of these points by applying a function *f*(*x*, *y*) to each point in turn and adding the results *taking account of the signs of the points*. Using the symbol "|" to indicate the application of the function to the set of points, we get:

$$f(x, y) \,|\, (-\mathbf{p}_1(x_1, y_1) + \mathbf{p}_2(x_2, y_2) - \mathbf{p}_3(x_3, y_3) + \mathbf{p}_4(x_4, y_4) - \mathbf{p}_5(x_5, y_5) + \mathbf{p}_6(x_6, y_6))$$

$$= -f(x_1, y_1) + f(x_2, y_2) - f(x_3, y_3) + f(x_4, y_4) - f(x_5, y_5) + f(x_6, y_6).$$

All very simple and straightforward. I'll emphasize that the signs come from the signing of the points as **simplexes**, not from the signs of the **coordinates** x_i, y_i so that $-\mathbf{p}_3(x_3, y_3)$ still has its negative sign as a **simplex** even if $x_3 = 4$ and $y_3 = 22.3$, and $+\mathbf{p}_6(x_6, y_6)$ still counts as a *positive* **simplex** even if $x_6 = -5$ and $y_6 = -0.674$.

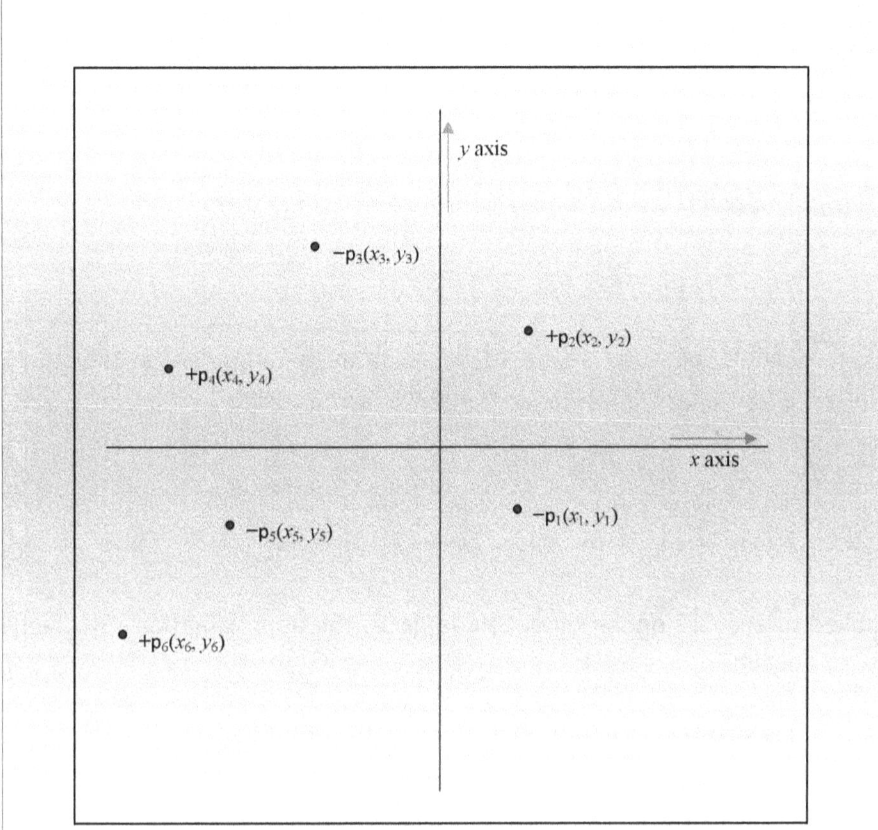

Figure 2.6.1

Let's now extend this to a set of **1-simplexes** or **line segments**. Now $\mathbf{p_1p_2}$ is one **line segment**, $\mathbf{p_3p_4}$ is another and $\mathbf{p_5p_6}$ is a third. By the definition of the **boundary operator**, $\mathbf{b}(\mathbf{p_1p_2}) = \mathbf{p_2} - \mathbf{p_1}$ and likewise our other two segments give:

$$\mathbf{b}(\mathbf{p_3p_4}) = \mathbf{p_4} - \mathbf{p_3} \text{ and } \mathbf{b}(\mathbf{p_5p_6}) = \mathbf{p_6} - \mathbf{p_5}$$

so giving us back our original points with their original signs, which is of course why I chose those signs in the first place.

Now we *define* the **operator** "$|$" for a **differential form** acting as a function on a **1-simplex** or **line segment** by using the definition we've already established for the **scalar function** on **0-simplexes** in this way:

$$\mathbf{d}\, f(x, y) \,|\, (\mathbf{p_1p_2}) = f(x, y) \,|\, \mathbf{b}(\mathbf{p_1p_2})$$

a form of relationship which is called an **adjoint**. One operator on one member of the pair – here **b** on the **line segment** $\mathbf{p_1p_2}$ – is equivalent to the other operator on the other element – here **d** on $f(x, y)$. Indeed, it is common to refer to the operators as **adjoint operators**. If the two are actually *the same* operator, then it is called **self-adjoint**. The present **adjoint** switch can be written most compactly if we write the **line segment** as λ, when we get the very simple form:

$$\mathbf{d}f \,|\, \lambda = f \,|\, \mathbf{b}\lambda.$$

By our expansion for the **scalar function** we get:

$$\mathbf{d}f(x, y) \,|\, (\mathbf{p_1p_2}) = f(x, y) \,|\, \mathbf{b}(\mathbf{p_1p_2}) = f(x, y) \,|\, (-\mathbf{p_1}(x_1, y_1) + \mathbf{p_2}(x_2, y_2))$$

$$= f(x_2, y_2) - f(x_1, y_1).$$

This definition is reasonable, because we already know from the argument at the start of Section 2.3 that for a short line segment, if we set $\mathbf{d}x = x_2 - x_1$ and $\mathbf{d}y = y_2 - y_1$, then

$$\mathbf{d}f(x, y) = \partial_x f(x, y).\mathbf{d}x + \partial_y f(x, y).\mathbf{d}y$$

$$= \partial_x f(x, y).(x_2 - x_1) + \partial_y f(x, y).(y_2 - y_1) \approx f(x_2, y_2) - f(x_1, y_1).$$

$$\uparrow$$

The product implied by the "|" operation simply replaces the approximation ("≈", highlighted by the arrow) by the *exact* equality.

If we have a **chain** of **line segments** like our original three, the definition easily extends:

$$\mathbf{d}f(x, y) \,|\, (\mathbf{p_1p_2} + \mathbf{p_3p_4} + \mathbf{p_5p_6})$$

$$= \mathbf{d}f(x, y) \,|\, (\mathbf{p_1p_2}) + \mathbf{d}f(x, y) \,|\, (\mathbf{p_3p_4}) + \mathbf{d}f(x, y) \,|\, (\mathbf{p_5p_6})$$

$$= f(x, y) \,|\, \mathbf{b}(\mathbf{p_1p_2}) + f(x, y) \,|\, \mathbf{b}(\mathbf{p_3p_4}) + f(x, y) \,|\, \mathbf{b}(\mathbf{p_5p_6})$$

$$= f(x, y) \,|\, (-\mathbf{p_1}(x_1, y_1) + \mathbf{p_2}(x_2, y_2) - \mathbf{p_3}(x_3, y_3) + \mathbf{p_4}(x_4, y_4) - \mathbf{p_5}(x_5, y_5) + \mathbf{p_6}(x_6, y_6))$$

$$= f(x_2, y_2) - f(x_1, y_1) + f(x_4, y_4) - f(x_3, y_3) + f(x_6, y_6) - f(x_5, y_5)$$

in line with our original definition of the **scalar function**.

Applying a **differential form** to a **simplex** using the "|" **operator** – for now only **line segment simplexes** will be considered – is called **definite integration**, and the result is called an **integral**. Since there is no $\mathbf{d}^{-1}$ **operator**, we are free now to call "|" the **integral operator**.

A most interesting result follows if the **chain** is a **strict chain**, with $\mathbf{p_2} = \mathbf{p_3}$ and $\mathbf{p_4} = \mathbf{p_5}$, for then we obtain:

$$\mathbf{d}f(x, y) \,|\, (\mathbf{p_1p_2} + \mathbf{p_2p_4} + \mathbf{p_4p_6})$$

$$= f(x_2, y_2) - f(x_1, y_1) + f(x_4, y_4) - f(x_2, y_2) + f(x_6, y_6) - f(x_4, y_4)$$

$= f(x_6, y_6) - f(x_1, y_1)$.

In other words, *all the intermediate terms drop out*. This is the essence of the concept of a **line integral**. The value of the **integral** of an **exact differential** is determined *entirely by the end points of the* **chain** if all the intermediate points form a **strict chain** having the leading element of each constituent **line segment** equal to the trailing element of the previous one.

This argument is commonly introduced as a *theorem*, rather than as a *definition* as I do here, called the **fundamental theorem of the calculus**, which was proved originally by Newton's teacher Barrow in the seventeenth century (in 1667). In that form it depends on a rather long-winded prior definition of the "$|$" **operator** involving the wretched concept of *infinitesimals*. Using the notation $[a, b]$ to indicate a **line segment** on the x **axis** with end points $[a]$ and $[b]$, so that:

$$[a, b] = \mathbf{p}_1(a)\mathbf{p}_2(b)$$

and $$\mathbf{b}[a, b] = \mathbf{b}(\mathbf{p}_1\mathbf{p}_2) = \mathbf{p}_2(b) - \mathbf{p}_1(a) = [b] - [a]$$

we obtain the simplest form of Barrow's result:

$$\mathbf{d}f(x) \,|\, ([a, b]) = f(x) \,|\, (\mathbf{b}[a, b]) = f(x) \,|\, (-[a] + [b]) = f(b) - f(a)$$

restricting the equation to a single **line segment** lying in the x axis. This is the form in which Barrow discovered the result and this is how it is always presented.

2.7 Taylor's Expansion[38]

We now have enough machinery to derive **Taylor's Expansion**, named after Brook Taylor (1685–1731), which is a kind of generalization of Barrow's theorem widely used in physics. We start with the form $\partial_x f(x).(x - b)$ and use axiom II to obtain its **derivative**:

$$\mathbf{d}(\partial_x f(x).(x - b)) = \partial_x f(x).\mathbf{d}(x - b) + \mathbf{d}(\partial_x f(x)).(x - b) = \partial_x f(x).\mathbf{d}x + \partial_x^{\,2}(x).(x - b).\mathbf{d}x$$

But $\partial_x f(x).\mathbf{d}x = \mathbf{d}(f(x))$, so if we apply this expression to a line segment $[a, b]$ using the $|$ **operator**, we obtain:

$$\begin{aligned}\mathbf{d}(\partial_x f(x).(x - b)) \,|\, [a, b] &= \partial_x f(x).\mathbf{d}x \,|\, [a, b] + \partial_x^{\,2}(x).(x - b).\mathbf{d}x \,|\, [a, b] \\ &= \mathbf{d}(f(x)) \,|\, [a, b] + \partial_x^{\,2}(x).(x - b).\mathbf{d}x \,|\, [a, b].\end{aligned}$$

Now, expanding the left hand side (LHS), this evaluates to

$$\mathbf{d}(\partial_x f(x).(x - b)) \,|\, [a, b] = \partial_x f(b).(b - b) - \partial_x f(a).(a - b) \;=\; -\,\partial_x f(a).(a - b)$$

since $b - b = 0$, and this equals on the other side (the RHS or right hand side) of the equation

[38] This section is quite demanding and may be omitted on a first reading.

$$\mathbf{d}(f(x))\,\Big|\,[a,b] + \partial_x^{\,2}(x).(x-b).\mathbf{dx}\,\Big|\,[a,b] \qquad = f(b) - f(a) + \partial_x^{\,2}(x).(x-b).\mathbf{dx}\,\Big|\,[a,b]$$

and we can write the whole equation as

$$\begin{aligned} f(b) &= f(a) - \partial_x f(a).(a-b) - \partial_x^{\,2}(x).(x-b).\mathbf{dx}\,\Big|\,[a,b] \\ &= f(a) + \partial_x f(a).(b-a) - \partial_x^{\,2}(x).(x-b).\mathbf{dx}\,\Big|\,[a,b]. \end{aligned}$$

Now we repeat the **integration by parts**[39] on the final term, putting the **differential form** part as $u(x).\mathbf{d}v(x)$ with $u(x) = \partial_x^{\,2}(x)$ and $\mathbf{d}v(x) = (x-b).\mathbf{dx}$ to give

$$\begin{aligned} &\mathbf{d}u(x) = \partial_x^{\,3}(x).\mathbf{dx} \\ &v(x) = \tfrac{1}{2}(x-b)^2 \end{aligned}$$

and use these to expand that term as:

$$\begin{aligned} \partial_x^{\,2}(x).(x-b).\mathbf{dx}\,\Big|\,[a,b] &= [\mathbf{d}(u(x).v(x)) - v(x).\mathbf{d}(u(x))]\,\Big|\,[a,b] \\ &= \mathbf{d}(\tfrac{1}{2}\partial_x^{\,2}(x).(x-b)^2)\,\Big|\,[a,b] - \tfrac{1}{2}(x-b)^2.\partial_x^{\,3}(x).\mathbf{dx}\,\Big|\,[a,b] \\ &= \tfrac{1}{2}\partial_x^{\,2}(b).(b-b)^2 - \tfrac{1}{2}\partial_x^{\,2}(a).(a-b)^2 - \tfrac{1}{2}(x-b)^2.\partial_x^{\,3}(x).\mathbf{dx}\,\Big|\,[a,b] \\ &= -\,\tfrac{1}{2}\partial_x^{\,2}(a).(b-a)^2 - \tfrac{1}{2}(x-b)^2.\partial_x^{\,3}(x).\mathbf{dx}\,\Big|\,[a,b] \end{aligned}$$

using $(a-b)^2 = (b-a)^2$ and so, putting this in the expression for $f(b)$ above, we have now:

$$f(b) = f(a) + \partial_x f(a).(b-a) + \tfrac{1}{2}\partial_x^{\,2}(a).(b-a)^2 + \tfrac{1}{2}(x-b)^2.\partial_x^{\,3}(x).\mathbf{dx}\,\Big|\,[a,b].$$

We can continue this process *ad infinitum* to obtain the expansion, using summation notation:

$$f(b) = f(a) + \textstyle\sum_{i=1\text{ to }n} (1/i!)\,\partial_x^{\,i}(a).(b-a)^i + R_n(b,a) \qquad (2.7.1)$$

where the remainder term $R_n(b, a) = (1/n!)\,(b-x)^n.\partial_x^{\,n+1}(x).\mathbf{dx}\,\Big|\,[a,b]$, and where $n!$ is defined as:

$$n! \equiv n \times (n-1) \times (n-2) \times \ldots \times 2 \times 1$$

which is called the **factorial function** of the **integer** n. This **infinite series** expansion for $f(b)$ in terms of its **ordinary derivatives** at a is the **Taylor Expansion**. You may spot that the $(b-x)$ factor in the remainder looks like it should have been $(x-b)$. The swap ensures that the remainder is always *positive*. Above, for example, we had:

$$R_1(b,a) = -\,\partial_x^{\,2}(x).(x-b).\mathbf{dx}\,\Big|\,[a,b]$$

But this equals $+\,\partial_x^{\,2}(x).(b-x).\mathbf{dx}\,\Big|\,[a,b]$. The even terms are unaffected since $(x-b)^{2n} = (b-x)^{2n}$.

[39] As defined at the end of Section 2.4.

2.8 Is there a Second Differential?

The theory in which the **differential** is an ordinary number – an ordinary **variable** that can be assigned a value – is called the **differential calculus**. It is not the only interpretation of the calculus.

In the **differential calculus,** the possibility of applying the **differential operator** twice is quite reasonable. Suppose we have a function $w = g(x, y)$ with **derivative**:

$$\mathbf{d}w = \partial_x g.\mathbf{d}x + \partial_y g.\mathbf{d}y$$

we could apply the **differential operator** to the whole expression again to obtain:

$$\mathbf{dd}w = \mathbf{d}(\partial_x g).\mathbf{d}x + \partial_x g.\mathbf{dd}x + \mathbf{d}(\partial_y g).\mathbf{d}y + \partial_y g.\mathbf{dd}y$$

using axiom II on each of the two terms, and this will expand to:

$$\mathbf{dd}w = (\partial_x\partial_x g\ \mathbf{d}x + \partial_y\partial_x g\ \mathbf{d}y).\mathbf{d}x + \partial_x g.\mathbf{dd}x + (\partial_x\partial_y g\ \mathbf{d}x + \partial_y\partial_y g\ \mathbf{d}y).\mathbf{d}y + \partial_y g.\mathbf{dd}y$$

or, writing **dd** as $\mathbf{d}^2$,

$$\mathbf{d}^2 w = \partial_x\partial_x g\ \mathbf{d}x.\mathbf{d}x + \partial_y\partial_x g\ \mathbf{d}y.\mathbf{d}x + \partial_x g.\mathbf{d}^2 x + \partial_x\partial_y g\ \mathbf{d}x.\mathbf{d}y + \partial_y\partial_y g\ \mathbf{d}y.\mathbf{d}y + \partial_y g.\mathbf{d}^2 y$$

and historically this led to the common notation for the **second derivatives**:

$$\mathrm{d}^2w/(\mathrm{d}x)^2 = \partial_x\partial_x g,\quad \mathrm{d}^2w/\mathrm{d}y\mathrm{d}x = \partial_y\partial_x g,\quad \mathrm{d}^2w/\mathrm{d}x\mathrm{d}y = \partial_x\partial_y g \text{ and } \mathrm{d}^2w/(\mathrm{d}y)^2 = \partial_y\partial_y g.$$

Élie Cartan (1869–1951), however, suggested a quite different interpretation of the **differential** based on a generalisation of our "$|$" **integral operator**. If we wish the formula

$$\mathbf{d}f \,|\, \lambda = f \,|\, \mathbf{b}\lambda$$

to apply in any number of dimensions, the properties of the **differential operator d** must reflect those of the **boundary operator b** much more closely. To see this, we will need to look at the extension of this definition of **integration** to higher dimensions, which I will discuss in Chapter 6. For the moment, I will just draw attention to the fact that, as discussed in Section 1.13, **bb** $\equiv 0$ – a boundary doesn't itself have a boundary. If **d** is to play the part of an **adjoint** to **b** *generally*, this property must come over to **d** too. Put simply, we would expect properties like the following (where I use arbitrary Greek letters to indicate higher level forms yet to be introduced):

$$\mathbf{dd}\omega \,|\, \Sigma = \mathbf{d}\omega \,|\, \mathbf{b}\Sigma = \omega \,|\, \mathbf{bb}\Sigma \text{ which must equal } 0$$

and this suggests we need **dd** = 0 too. We can examine the possible consequences of this requirement that **dd** = 0 already.

In our example above this would mean that all the **dd** or $\mathbf{d}^2$ terms should drop out, leaving just:

$$0 = \partial_x\partial_x g\ \mathbf{d}x.\mathbf{d}x + \partial_y\partial_x g\ \mathbf{d}y.\mathbf{d}x + 0 + \partial_x\partial_y g\ \mathbf{d}x.\mathbf{d}y + \partial_y\partial_y g\ \mathbf{d}y.\mathbf{d}y + 0.$$

But consider also the simpler case $y = f(x)$. Here we'd have $\mathbf{d}y = \partial_x f.\mathbf{d}x$ and on using the **differential operator** again:

$$\mathbf{dd}y = \partial_x\partial_x f\ \mathbf{d}x.\mathbf{d}x + \partial_x f.\mathbf{dd}x$$

and here $\mathbf{dd}y = 0$ and $\mathbf{dd}x = 0$ would require $\partial_x\partial_x f\ \mathbf{d}x.\mathbf{d}x = 0$, but, since $\partial_x\partial_x f$ need not be zero, this implies $\mathbf{d}x.\mathbf{d}x = 0$. Putting that result – in g rather than f – into the preceding equation would suggest we should have:

$$\partial_y\partial_x g\ \mathbf{d}y.\mathbf{d}x + \partial_x\partial_y g\ \mathbf{d}x.\mathbf{d}y = 0$$

or since by **Young's theorem** $\partial_y\partial_x g = \partial_x\partial_y g$,

$$\partial_y\partial_x g\ (\mathbf{d}y.\mathbf{d}x + \mathbf{d}x.\mathbf{d}y) = 0$$

clearly suggesting

$$\mathbf{d}y.\mathbf{d}x + \mathbf{d}x.\mathbf{d}y = 0$$

a result that is obviously in conflict with the idea that the **differentials** are ordinary numbers. This approach has, however, been so successful that it is best to work consistently with it. It gives the right answers in so many subtle cases and has become the foundation of the subject of **differential geometry**.

So when you use the **differential operator** on anything that already contains **differentials**, add in these axioms:

Axiom III:	**dd** $\equiv 0$ *always*. Applying **d** to any **differential** gives zero
Axiom IV:	$\mathbf{d}y.\mathbf{d}x = -\,\mathbf{d}x.\mathbf{d}y$
Axiom V:	$\mathbf{d}x.\mathbf{d}x = 0$.

Axiom V isn't strictly needed: from axiom IV we could get $\mathbf{d}x.\mathbf{d}x = -\,\mathbf{d}x.\mathbf{d}x$ and if anything equals its own negative, it must be zero. Also note that in this interpretation the product $\mathbf{d}x.\mathbf{d}y$ is normally written as:

$$\mathbf{d}x \wedge \mathbf{d}y$$

where the **operator** "$\wedge$" is called the **alternating product** operator. This is *not* the same as the **Boolean operator** "$\wedge$", but is a distinct use of the same symbol. Mathematicians were doing this long before the limitations of word processors.

It's important to realize that *both* the simple interpretation of **differentials** as ordinary numbers, as used through most of this Chapter, *and* the interpretation touched on here, that will be developed later in Chapter 6, are valid. The approach introduced tentatively in this Section works better for **integration**, but for many applications the simple model remains quite adequate. Nothing in the underlying formalism forbids either interpretation. Nothing in mathematics is set in stone.

2.X Exercises

1. Consider replacing the **Leibnitz rule** with the definition $\mathbf{d}(xy) = x^2.\mathbf{d}y + y^2.\mathbf{d}x$. Evaluate the rule for $\mathbf{d}(x/y)$ under this definition. Comment on your result. Could an acceptable "alternative calculus" be achieved by using the rule $\mathbf{d}(xy) = x.\mathbf{d}y - y.\mathbf{d}x$? If not, why not?

2. Evaluate the **integral** $\mathbf{d}(x^2y) \,|\, \mathbf{p}_0\ \mathbf{p}_1$ where $\mathbf{p}_0 = (1, 2)$ and $\mathbf{p}_1 = (3, -3)$. Do the same for the **integral** over the triangle $\mathbf{d}(x^2y) \,|\, (\mathbf{p}_0\mathbf{p}_1 + \mathbf{p}_1\mathbf{p}_2 + \mathbf{p}_2\mathbf{p}_0)$ where $\mathbf{p}_2 = (-4, -5)$.

3. Evaluate $(x\mathbf{d}y + y\mathbf{d}x) \,|\, \mathbf{p}_0\ \mathbf{p}_1$ where $\mathbf{p}_0 = (1, 2)$ and $\mathbf{p}_1 = (3, -3)$ as above. Evaluate the **integral** of $(x\mathbf{d}y + y\mathbf{d}x)$ over the triangle $(\mathbf{p}_0\mathbf{p}_1 + \mathbf{p}_1\mathbf{p}_2 + \mathbf{p}_2\mathbf{p}_0)$ where $\mathbf{p}_2 = (-4, -5)$, again as in the preceding question.

4. Using the formulae developed in Section 2.7, especially (2.7.1), for the **Taylor expansion**, evaluate the Taylor expansion of $f(x) = ax^2 + bx + c$ using $x = 0$ for the point "a" in (2.7.1). Do the same about $x = 1$. Do they match?

5. Evaluate $\mathbf{d}(y/x)$ subject to the condition $x^2 + y^2 = 1$, expressing the result in terms of x and $\mathbf{d}x$ and, alternatively, in terms of y and $\mathbf{d}y$. This function will turn out to be the **tangent** of an **angle**.

6. Evaluate $\mathbf{d}(x + y)$ subject to $x^2 - y^2 = 1$. Express it in terms of $\mathbf{d}\theta$ given $\mathbf{d}x = y.\mathbf{d}\theta$ and $\mathbf{d}y = x.\mathbf{d}\theta$.

7. The **average** of a function ϕ over an interval $[a, b]$ equals $(b - a)^{-1}.\{\phi(x).\mathbf{d}x \,|\, [a, b]\}$. Evaluate this for the parabola $\phi(x) = x^2$.

8. A can of oil is to be made in the form of an ordinary cylindrical can(a circular cylinder). What ratio of the radius of the can to its height entails the use of the least surface area of material? (G.B Thomas, Addison-Wesley 1960, p.124) [From $\mathbf{d}V = 0$ get a relation between $\mathbf{d}r$ and $\mathbf{d}h$, and from this minimize the area.]

9. Take the implicitly defined function $y^2 = (a^2 - x^2)$. Evaluate the **integral** $\pi(a^2 - x^2).\mathbf{d}x \,|\, [-a, a]$. This is a *volume*. Of what?

10. Demonstrate, at least to your own satisfaction, with reasonable assumptions, the first two **mean value theorems**:

    a. $\phi(b) - \phi(a) = \partial\phi(c).(b - a)$ for some c in the **interval** $[a, b]$

    b. $\phi(b) - \phi(a) = \partial\phi(a).(b - a) + ½\partial^2\phi(c).(b - a)^2$ for some c in the **interval** $[a, b]$.

11. Given $x = g(t)$ and $y = F(x)$, define $\mathbf{d}^2x = \partial^2 g(t).(\mathbf{d}t)^2 + \partial g(t).\mathbf{d}^2t$ and $\mathbf{d}^2y = \partial^2F(x).(\mathbf{d}x)^2 + \partial F(x).\mathbf{d}^2x$. Evaluate $\mathbf{d}^2y$ strictly as a function of t, $\mathbf{d}t$ and $\mathbf{d}^2t$. Writing the composite function $y = Fg(t)$, show that $\partial_t^2(Fg).(\mathbf{d}t)^2 = \partial^2F(x).(\mathbf{d}x)^2 + \partial F(x).\ \partial^2g(t).(\mathbf{d}t)^2$. (From Franklin, *Treatise on Advanced Calculus*, Wiley, 1940, quoted by G.B. Thomas, *op.cit.*, p.89.)

12. Show that $\frac{d^2y}{(dx)^2} = -\frac{d^2x}{(dy)^2} / \left(\frac{dx}{dy}\right)^3$ (Murray Spiegel, *Advanced Calculus*, Schaum's Outlines 1963, p.77.)

Chapter 3

LOGARITHMS AND EXPONENTIALS

3.1 The Natural Logarithm

In Section 2.4 I discussed the inversion of the **differential operator**. In particular, from the inversion formula $x^n\mathbf{d}x = \mathbf{d}x^{n+1}/(n + 1)$ we were able to get **original** forms for these cases:

$$x^3.\mathbf{d}x = \mathbf{d}(x^4/4),\ \ x^2.\mathbf{d}x = \mathbf{d}(x^3/3),\ \ x.\mathbf{d}x = \mathbf{d}(x^2/2),\ \text{ and trivially, } 1.\mathbf{d}x = \mathbf{d}x$$

and the division formula $\mathbf{d}(x/y) = (y.\mathbf{d}x - x.\mathbf{d}y)/y^2$ enables us to add, starting with:

$$\mathbf{d}(1/x) = (x.\mathbf{d}(1) - 1.\mathbf{d}x)/x^2 \ = -\ \mathbf{d}x\ /\ x^2$$

the series:

$$(1/x^2).\mathbf{d}x = \mathbf{d}(-(1/x)),\ \ (1/x^3).\mathbf{d}x = \mathbf{d}(-\tfrac{1}{2}(1/x^2)),\ \ (1/x^4).\mathbf{d}x = \mathbf{d}(-\tfrac{1}{3}(1/x^3)).$$

This latter series is consistent with the formula $x^n\mathbf{d}x = \mathbf{d}x^{n+1}/(n + 1)$ if we adopt the convention that

$$(1/x^n) \equiv x^{-n}$$

for these then become:

$$x^{-2}.\mathbf{d}x = \mathbf{d}(-x^{-1}),\qquad x^{-3}.\mathbf{d}x = \mathbf{d}(-\tfrac{1}{2}.x^{-2}),\qquad x^{-4}.\mathbf{d}x = \mathbf{d}(-\tfrac{1}{3}.x^{-3}).$$

But one element in this series is missing. We have no formula for $(1/x).\mathbf{d}x$. The series "steps over" this case. It turns out that there is no easy solution to this and we resort to the device of *defining* the **original** function whose **derivative** is $(1/x).\mathbf{d}x$ as a special function, given its own name and notation, and then derive its properties from the **derivative** formula.

The **original** function whose **derivative** is $(1/x).\mathbf{d}x$ is one of the most important functions in mathematics, and is given the name of the **natural logarithm**. It can be written in various ways such as **ln**(x), **log**(x) or **log**$_e(x)$, but I will choose in these notes to write it as an **operator** $\downarrow x$ because of that little "e" subscript that appeared in the third of the standard forms, which will be explained in a while, but which I can say for now will appear in my notation as a special case:

$$\mathbf{e} \downarrow x \equiv \downarrow x$$

i.e. these two notations (and all the others, actually) are equivalent and interchangeable. So first, the *defining* property of the **natural logarithm** is:

$\mathbf{d}(\downarrow x) = (1/x).\mathbf{d}x$ which we will use again and again.

The special properties of the **natural logarithm** begin to appear when we look at $\downarrow(x.y)$, the application of the **logarithm** to the simple product xy. From the defining property of the function, we know that:

$$\mathbf{d}(\downarrow(x.y)) = \mathbf{d}(x.y) \,/\, (x.y)$$

and we can expand the RHS (right-hand side) of this from our Axiom II as:

$$(\mathbf{d}x.y + x.\mathbf{d}y) \,/\, (xy) = (y/(xy)).\mathbf{d}x \;+\; (x/(xy)).\mathbf{d}y = (1/x).\mathbf{d}x + (1/y).\mathbf{d}y$$

$$= \mathbf{d}(\downarrow x) + \mathbf{d}(\downarrow y).$$

So we have the property that:

$$\mathbf{d}(\downarrow(x.y)) = \mathbf{d}(\downarrow x) + \mathbf{d}(\downarrow y)$$

or that *multiplication* is replaced by *addition*, at least in the **derivative equation**. But the general form of the **original equation** corresponding to this will be:

$$\downarrow(x.y) = \downarrow x + \downarrow y + c$$

where c is an arbitrary constant – the "**constant of integration**" that mirrors the fact that an infinity of **original equations** correspond to the same **derivative** since any constant terms drop out because for any constant $\mathbf{d}c = 0$.

To "solve" for c we need to impose an extra condition, and this we can *choose* for ourselves. A simple choice is that:

$$\downarrow(1) = 0$$

for this gives

$$\downarrow(x.1) = \downarrow x + \downarrow 1 + c = \downarrow x + 0 + c$$

but since $x.1 = x$ this is:

$\downarrow x = \downarrow x + 0 + c$ which means that, subtracting $\downarrow x$ from both sides, $c = 0$.

This gives us the key property of **logarithms**:

$$\downarrow(x.y) = \downarrow x + \downarrow y.$$

Further properties follow easily from this beginning.[40] Thus, putting $y = (1/x)$

$$\downarrow(x.(1/x)) = \downarrow(1) = 0 = \downarrow x + \downarrow(1/x)$$

So $\downarrow(1/x) = -\downarrow x$

which in turn gives $\downarrow(x/y) = \downarrow x + \downarrow(1/y) = \downarrow x - \downarrow y$. Again, we can start with:

$y = \downarrow(x^n)$ and apply **d**:

$$\mathbf{d}y = \mathbf{d}(\downarrow(x^n)) = \mathbf{d}(x^n)/\, x^n$$

from the defining property of the **logarithm**, and this will be:

$$(n\, x^{n-1}/x^n).\mathbf{d}x = (n/x).\mathbf{d}x = n.\mathbf{d}x/x = n.\mathbf{d}(\downarrow x)$$

so giving the result . . .

$$\mathbf{d}(\downarrow(x^n)) = n.\mathbf{d}(\downarrow x).$$

This **derivative** must "solve" for an **original** equation of the form:

$\downarrow(x^n) = n.\downarrow x + c$ for some constant c,

which can as before be solved by using $\downarrow(1) = 0$ through putting $x = 1$ to give:

$\downarrow(1^n) = \downarrow 1 = n.\downarrow 1 + c$ so that $0 = n.0 + c$ so again $c = 0$. Hence:

$$\downarrow(x^n) = n.\downarrow x$$

3.2 The Exponential Function

The **exponential function** is just the **inverse** of the **natural logarithm**. This simply means that whenever $y = \mathbf{log}(x)$ or in my notation $y = \downarrow x$, it must hold *by definition* that $x = \mathbf{exp}(y)$ for which I will introduce the notation $x = \uparrow y$. So *always*:

$y = \mathbf{log}(x) \equiv x = \mathbf{exp}(y)$ or in my notation $y = \downarrow x \equiv x = \uparrow y$.

This direct equivalence enables us to get the basic properties of the **exponential** out quite quickly. The first is rather surprising. For if $y = \downarrow x$, then we have $\mathbf{d}y = \mathbf{d}x/x$, but then $\mathbf{d}x = x.\mathbf{d}y$ so that:

---

[40] From here on I follow, albeit in my own style, the line of development in the standard text by G.B.Thomas *Calculus and Analytic Geometry*, Addison-Wesley 3rd Edition 1960. By the time of the first edition of the present work, this standard reference had reached its 11th edition!

If $x = \uparrow y$, then $\mathbf{d}x = x.\mathbf{d}y$.

To see the extraordinary significance of this, swap "x" and "y", so we see things in the more familiar way of writing a function and its argument:

If $y = \uparrow x$, then $\mathbf{d}y = y.\mathbf{d}x = \uparrow x.\mathbf{d}x$ or directly, $\mathbf{d}\uparrow x = \uparrow x.\mathbf{d}x$.

The meaning of this equation is that:

The **exponential function** *is its own* **derivative**.

Next, the "*multiplication* goes to *addition*" rule appears now reversed: the **exponential** of a *sum* is the *product* of the **exponentials** of its parts or:

$\uparrow(x + y) = \uparrow x.\uparrow y$

which we get by setting $p = \uparrow x$ and $q = \uparrow y$, so that $x = \downarrow p$ and $y = \downarrow q$, when we have:

$x + y = \downarrow p + \downarrow q = \downarrow(pq)$

but then by the **inverse** definition:

$pq = \uparrow(x + y)$, but $pq = \uparrow x.\uparrow y$ so confirming our result.

Again, if $y = \uparrow(-x)$, then it must be that $-x = \downarrow y$ or that $x = -\downarrow y$, but we know from Section 3.1 that this equals $\downarrow(1/y)$. But if $x = \downarrow(1/y)$ we must have the **inverse** $(1/y) = \uparrow x$ or $y = (1/(\uparrow x))$ and this gives the important result:

$\uparrow(-x) = (1/(\uparrow x))$

from which in the next section we will derive the relationship we've touched on a few times before, that:

$a^{-n} = (1/a^n)$.

To do this, we'll first need to establish that the **exponential function** is a special case of the familiar operation of raising something to a power. But before going into that, *Figure 3.2.1* shows the appearance of our **functions**. Note how they are mirror images of each other, reflected about the 45° diagonal line $y = x$.

On this graph, both functions are drawn with x the **argument** and y the result:

$y = \downarrow x$, or $y = \mathbf{ln}(x)$
$y = \uparrow x$, or $y = \mathbf{exp}(x)$.

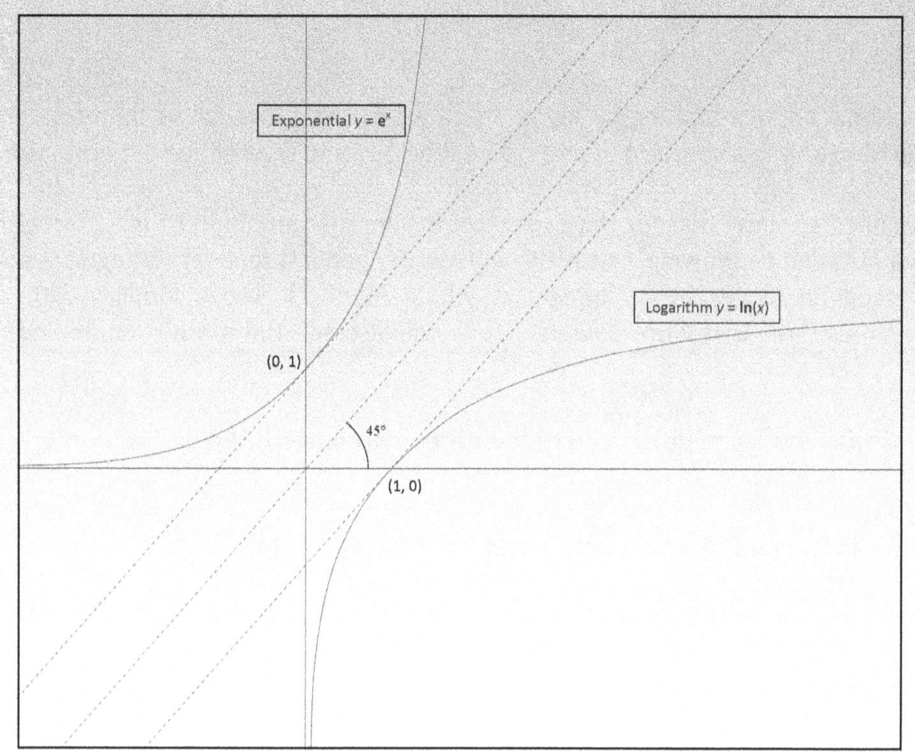

Figure 3.2.1

3.3 Generalized Exponentiation

If $y = \downarrow x$ and so $x = \uparrow y$, then $x = \uparrow(\downarrow x)$ and $y = \downarrow(\uparrow y)$. Applying a function and its **inverse** in immediate succession returns the original value.

We can now *define* the generalization of the **exponential operator** ↑ and show that it subsumes the action of raising to an integer power as a special case. This is the definition:

$$a \uparrow u = \uparrow(u.\downarrow(a))$$

where a is a constant, or applying ↓ to both sides:

$$\downarrow(a \uparrow u) = \downarrow\uparrow(u.\downarrow(a)) \;= u.\downarrow(a).$$

Again, we can define an **inverse** function on the basis that if $x = a \uparrow u$, then $u = \mathbf{log}_a x$, for which I will introduce the notation $u = a \downarrow x$ so that:

$$x = a \uparrow u \;\equiv\; u = a \downarrow x$$

which we can evaluate easily:

$$\downarrow x = \downarrow(a \uparrow u) = \downarrow\uparrow(u.\downarrow(a)) \;= u.\downarrow(a)$$

So $\quad u = \downarrow x / (\downarrow a)$

or in conventional notation u = **log** x / **log** a. These **generalized logarithms** therefore differ from the **natural logarithm** only in a constant of proportionality ($\downarrow a$) and so we have their properties already.

But we now need to show that for the **exponential** side, the definition of $a \uparrow u$ corresponds to the conventional notation a^u showing that this is indeed the general form of raising to a power (general because u need no longer be an integer, a whole number). Those familiar with programming languages such as Basic that write powers as *a^u* will also see the motivation behind my choice of notation.

Firstly, $a^0 = a \uparrow 0 = \uparrow(0.\downarrow a) = \uparrow(0) = 1$, from the **inverse** property $\downarrow(1) = 0$. Also $a^1 = a \uparrow 1 = \uparrow(\downarrow a) = a$. Next,

$$a^u.a^v = (a \uparrow u).(a \uparrow v) = \uparrow(u.\downarrow(a)).\uparrow(v.\downarrow(a)) = \uparrow(u.\downarrow(a) + v.\downarrow(a))$$

$$= \uparrow((u + v).\downarrow a) = a\uparrow(u + v)$$

so $\quad a^u.a^v = a^{(u+v)}$.

Putting $v = -u$ in this equation gives $a^u.a^{-u} = a^{u-u} = a^0 = 1$, so that:

$$a^{-u} = 1 / (a^u),$$ a result I've been promising for some time.

Then, defining $b = a^u = \uparrow(u.\downarrow(a))$, and raising *this* to the power v:

$$b^v = (a^u)^v = \uparrow(v.\downarrow(b)) = \uparrow(v.\downarrow(a^u)) = \uparrow(v.\downarrow(\uparrow(u.\downarrow(a)))) = \uparrow(v.u.\downarrow(a)) = a \uparrow(v.u)$$

so $\quad (a^u)^v = a^{u.v}$

and this gives our second long-awaited result, for if $u = 1 / v$,

$$(a^{1/v})^v = a^{v/v} = a^1 = a.$$

So $\quad a^{1/v} = {}_v\sqrt{a}$

or that *reciprocal powers correspond to roots*. Finally,

$$(ab)^u = \uparrow(u.\downarrow(ab)) = \uparrow(u.(\downarrow a + \downarrow b)) = \uparrow(u.\downarrow a + u.\downarrow b)) = \uparrow(u.\downarrow a).\uparrow(u.\downarrow b)$$

by $\uparrow(x + y) = \uparrow x.\uparrow y$, so this gives:

$$(ab)^u = a^u.b^u.$$

You may or may not like my suggested notation. A significant motivation for me was that forms like **log**(x) leave it uncertain whether we are dealing with **natural logarithms** or with logarithms to a standard base like 10, which happens to be the most common standard. So **log(**x**)** might mean $\mathbf{log}_e(x)$

or $\mathbf{log}_{10}(x)$, and surely $\downarrow x$ *and* $10\downarrow x$ should be more clear. A common attempt to resolve this problem, much used in mathematics texts, is to write $\mathbf{log}_e(x)$, the **natural logarithm**, as $\mathbf{ln}(x)$, but it seems simpler just to make the base itself explicit.

But the best way to look at my notation is to borrow an idea from Quantum Mechanics, and regard the **exponential operator** ↑ and the **logarithmic operator** ↓ as **raising** and **lowering operators**, ↑ **raising** *addition* to *multiplication*, and ↓ **lowering** *multiplication* to *addition*:

$$\uparrow(x + y) = \uparrow x.\uparrow y$$

$$\downarrow(x.y) = \downarrow x + \downarrow y$$

where the bracketing assumes that ↑ and ↓ are **high precedence operators** which will, by default, operate before *multiplication* or *addition* at the same level of bracketing (remember BODMAS from school?). Working on an arbitrary base a:

$$a\uparrow(x + y) = a\uparrow x.a\uparrow y$$

$$a\downarrow(x.y) = a\downarrow x + a\downarrow y.$$

This also suggests a spoken form for "↑" and "↓": just call "↑" "up" and "↓" "down".

This brings me to the issue of the number **e**, the base of **natural logarithms**, and the base too of what it would seem reasonable to call **natural exponentials**, but before I do, I'll issue a word of warning about my notation. *Be very careful with forms like* $x.\downarrow a$ which must be carefully distinguished from $x\downarrow a$ and similarly with $x.\uparrow a$ which must be distinguished from $x\uparrow a$. *If necessary, write* $x.(\downarrow a)$ *or* $x.(\uparrow a)$, which are actually clear even without the stop, as $x(\downarrow a)$ and $x(\uparrow a)$. This is a slight disadvantage in using the same symbol both as a **binary operator** sitting between its two **operands** and as a *function*, one solution to which appears in the next section. I assume that both my ↓ and ↑ operators will be of **high precedence** so that

$x\uparrow a.b$ will be evaluated as $(x\uparrow a).b$ and *not* as $x\uparrow(a.b)$

a usage which is already well established for **exponentiation** in computer languages, but is obviously a novel suggestion for the **logarithmic operator** ↓, since such an operator has not been suggested before! The convention of writing the argument of an **exponential** in *superscript* has one big advantage: *superscripting is self-bracketing and so superscript arguments require no parentheses.* For this reason, I did think of introducing a notation like:

a_{x+y} for $a\downarrow(x + y)$ or $\mathbf{log}_a(x + y)$

but I decided against the proliferation of superscripts and subscripts because they can be hard to read and they conflict with indicial usage that is very widespread in the theory of vectors and tensors to indicate **integer arguments** of functions, as I discussed in Section 1.1. But usage such as x^2 for *powers* is so well established and so convenient that it's definitely here to stay.

3.4 The Number **e**

"e" is just the particular number where

$\downarrow \mathbf{e} = 1$

By the **inverse** property, it therefore also obeys:

$\uparrow 1 = \mathbf{e}$.

For this reason, we can use **e** as a device to give another angle to our notation, for we can say:

$\uparrow x = \mathbf{e}\uparrow x = \mathbf{e}^x$

and this notation $\mathbf{e}^x$ is widely used in mathematics. Indeed, the form **exp**(x) is very unusual. The compactness of the $\mathbf{e}^x$ form and the automatic bracketing that the superscripting entails make this notation virtually universal. But $\mathbf{e}\uparrow x$ is quite close to the form **e**^x used in some programming languages, and opens up my suggestion of the complementary notation:

$\downarrow x = \mathbf{e}\downarrow x = \mathbf{log_e}(x) = \mathbf{ln}(x)$

as a simpler notation for the corresponding **logarithmic** form. And just as we call the **logarithm to base e** the **natural logarithm**, it would seem reasonable to call the general powers of **e** the **natural exponential**, a usage which does not seem to have caught on.
How big is **e**? G.B.Thomas gives it as:

e = 2.7 1828 1828 45 90 45 . . .

and it is an **irrational number** – it cannot be expressed as a quotient of two integers p and q in the form p/q. It is (therefore) an infinitely long decimal. In the **base e** notation, the **raising** and **lowering** rules appear as:

$\mathbf{e}\uparrow(x + y) = \mathbf{e}\uparrow x.\ \mathbf{e}\uparrow y$

$\mathbf{e}\downarrow(x.y) = \mathbf{e}\downarrow x + \mathbf{e}\downarrow y$.

3.5 Linear Differential Equations

The **exponential function** can be used to solve a large variety of physical problems, usually involving variation in *time*, described by a class of equations called **linear differential equations**.[41] The key property we will need to solve these is given by:

[41] I will explain why "**linear**" towards the end of this section. Since all the cases considered here also involve **differentiation** with respect to only one variable – time – they are also called **ordinary linear differential equations**.

$\mathbf{d}{\uparrow}(\lambda x) = {\uparrow}(\lambda x).\mathbf{d}(\lambda x) = \lambda.{\uparrow}(\lambda x).\mathbf{d}x$ in newspeak, or

$\mathbf{d}(\mathbf{e}^{\lambda x}) = \mathbf{e}^{\lambda x}.\mathbf{d}(\lambda x) = \lambda.\ \mathbf{e}^{\lambda x}.\mathbf{d}x$ in oldspeak.

The ideas are best seen by looking at an example. For this, I will return to the Electrical Circuits of Section 1.14, but now I will add in the other two kinds of line element: **capacitors** and **inductors**.

Unlike **resistors**, **capacitors** and **inductors** show a time dependence. Again, great care is needed with signs, but fortunately the rule for the two new elements follows the rule for **resistors**. They just obey **differential equations**. These are:

Capacitors: $C.\mathbf{d}v = \pm i.\mathbf{d}t$
Inductors: $L.\mathbf{d}i = \pm v.\mathbf{d}t$

where C and L are defining constants called respectively the **capacitance** and the **inductance** of their respective elements. The choice of sign is, fortunately, exactly the same as for **resistors** and depends on the same choices of **orientation** for **currents** and **voltages**. So where we would write for a **resistor**:

$v = i.R$

for **capacitors** and **inductors** we would write:

$C.\mathbf{d}v = i.\mathbf{d}t$ and $L.\mathbf{d}i = v.\mathbf{d}t$

and where for a **resistor** we would choose:

$v = -i.R$

for **capacitors** and **inductors** we would write:

$C.\mathbf{d}v = -i.\mathbf{d}t$ and $L.\mathbf{d}i = -v.\mathbf{d}t$.

But we *must* be consistent for all three.[42]

It's worth amplifying this a little. Sticking to the convention given at the end of Section 1.14 that we use $v = -i.R$, or in **indicial notation**, $v_{23} = -i_{23}.R$ here, we will have for example, labelling the plates of the **capacitor** as $\mathbf{p}_3$ for the upper below and $\mathbf{p}_4$ for the lower (as in *Figure 3.5.2*):

$C.\mathbf{d}v = -i.\mathbf{d}t \longrightarrow C.\mathbf{d}v_{34} = -i_{34}.\mathbf{d}t = -i_{01}.\mathbf{d}t$.

Here i_{01}, the current through the battery from the − pole to the +, is positive, so since, as with R, $C > 0$ *by definition*, we must have $\mathbf{d}v_{34} < 0$ as time increases. This correctly represents the accumulation of

[42] Because of its $v = \pm L.(\mathrm{d}i/\mathrm{d}t)$ form, the **inductance** term is sometimes left on the LHS as a bogus "**emf**" when the other terms are transferred in the **KVL** equation, giving the form: $E - L.(\mathrm{d}i/\mathrm{d}t) = iR + \int(i/C).\mathbf{d}t$ and this leads to the erroneous idea that the **inductance** term needs special treatment. J.R.Reitz, F.W.Milford, *Foundations of Electromagnetic Theory*, 2nd ed. Addison-Wesley 1967, p.247 footnote.

negative charge on the lower ($\mathbf{p}_4$) plate of the **capacitor**, and positive on the upper ($\mathbf{p}_3$), so that $V(\mathbf{p}_4) - V(\mathbf{p}_3) = v_{34} < 0$. Similarly if we had an **inductor**, supposing it to be placed between $\mathbf{p}_4$ and $\mathbf{p}_5$, again as in *Figure 3.5.2*:

$$L.\mathbf{d}i = -v.\mathbf{d}t \longrightarrow L.\mathbf{d}i_{45} = -v_{45}.\mathbf{d}t,$$

and with v_{45} being defined as the voltage across an element going round the circuit from the E^+ pole to the E^-, we must have $v_{45} < 0$, so, again with $L > 0$ *by definition*, we must have $\mathbf{d}i_{45} > 0$ which again correctly represents a growing positive current through the **inductor** from the moment the circuit is switched on (remember our convention is that $i_{54} = i_{10}$ is *negative*).

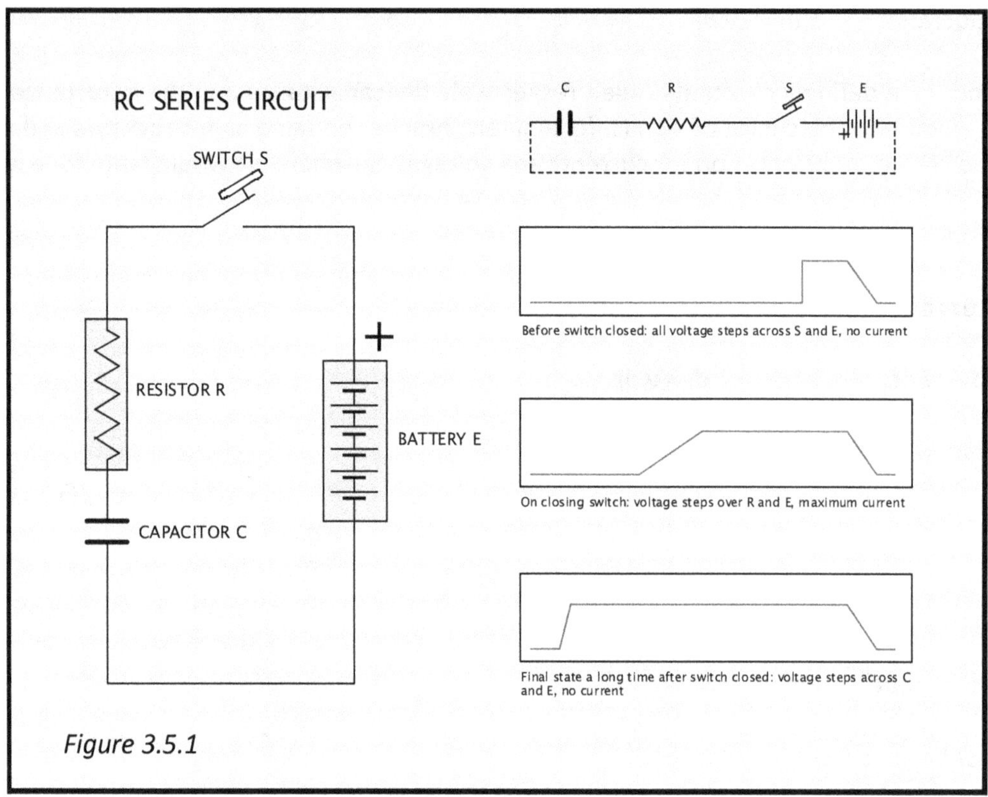
RC SERIES CIRCUIT
C
R
S
E
SWITCH S
RESISTOR R
BATTERY E
CAPACITOR C
+
Before switch closed: all voltage steps across S and E, no current
On closing switch: voltage steps over R and E, maximum current
Final state a long time after switch closed: voltage steps across C and E, no current

Figure 3.5.1

Look at the circuit in *Figure 3.5.1*. This shows an **emf** voltage E, a **resistor** and a **capacitor** in **series**. I will assume all **voltages** and the **current** – which must be the same through all three elements – are all oriented *anticlockwise* and so use the *negative* signs convention as in Section 1.14. The placing of the **emf** E, with $E > 0$ means the **voltage** *anticlockwise* through the **resistor**, v_R, must be *negative* and with that through the **capacitor** defined as v_C, we have three equations:

$$E + v_R + v_C = 0$$

$$v_R = -i.R$$

$$C.\mathbf{d}v_C = -i.\mathbf{d}t.$$

This gives us two **original** equations and one **differential** one. To solve the set, we **differentiate** the **originals** to give three **differential forms**, solve them together, and then take the answer back up.

Since $\mathbf{d}E = 0$, the first gives $\mathbf{d}v_R = -\mathbf{d}v_C$, which in the second gives $\mathbf{d}v_C = \mathbf{d}i.R$, leaving us with this and the third equation.

We can now combine these two to give:

$$C.\mathbf{d}v_C = RC.\mathbf{d}i = -i.\mathbf{d}t$$

or

$$\mathbf{d}i/i = (-1/RC).\mathbf{d}t$$

so giving:

$$\mathbf{d}(\downarrow i) = (-1/RC).\mathbf{d}t$$

or $$\downarrow i = (-1/RC).t + c = (-t/RC) + c$$

where c is the **constant of integration**. Applying $\uparrow$ throughout gives:

$$\uparrow(\downarrow i) = i = \uparrow((-t/RC) + c) = \uparrow c.\uparrow(-t/RC) = I.\uparrow(-t/RC)$$

or

$$i = I.\mathbf{e}^{(-t/RC)}$$

where I have defined $I := \mathbf{e}^c$. Hence, from $v_R = -i.R$, we get also

$v_R = -RI.\mathbf{e}^{(-t/RC)}$, and so also $v_C = -E - v_R = -E + RI.\mathbf{e}^{(-t/RC)}$.

The pattern implied by these solutions is shown on the right of the *Figure*, with the switch being thrown (i.e. the circuit being completed, shown in the second diagram or *voltage map*)[43] at $t = 0$, when, because $\mathbf{e}^0 = 1$, $i = I$, $v_R = -RI$ and $v_C = -E - v_R = -E + RI = 0$. ($E = RI$, with $E > 0$, $R > 0$, and $I > 0$.) Thereafter i and v_R gradually decline, and v_C gradually rises to equal E as the voltage difference transfers from the resistor to the capacitor with the build-up of charge on the latter.

This illustrates the solution using my emphasis on **differentials**. This isn't the way you'll see in the textbooks. The normal way in which this solution is found is to employ the additional relation that:

$$\mathbf{d}v = \partial_t v.\mathbf{d}t$$

to eliminate the **differentials** to give:

$$C.\partial_t v_C = -i$$

and using $v_C = -E + i.R$ we get:

$$C.\partial_t v_C + v_C/R + E/R = 0.$$

[43] Strictly speaking, a **potential** map.

The conventional method now makes a rather *ad hoc* move. They temporarily *ignore* the constant term E / R, obtaining the corresponding equation only in v_C and its **OD's (ordinary derivatives)**, an equation which by analogy with Section 1.7, is called the **homogeneous equation**. In this case it is:

$$C.\partial_t v_C + v_C / R = 0$$

and now *assuming* $v_C = V.\mathbf{e}^{\lambda t}$ we find much as before that:

$$RC.\lambda v_C + v_C = 0$$

and so again:

$$RC.\lambda + 1 = 0 \qquad \text{or} \qquad \lambda = -1/(RC).$$

We then proceed to the final form in a way not dissimilar to my approach.

Forms like $a.\partial_t^2 y + b.\partial_t y + c.y = 0$ are called **linear differential equations**, and since here only a single ∂_t is involved, also **ordinary differential equations** – they are equations simply in a **linear** sequence of **ordinary derivatives**. Hence the heading of this section.

I will often refer to **linear differential equations** simply as **LDE**'s.

A more complex example is where we have a **resistor**, a **capacitor** and an **inductor** all in **series**, shown in *Figure 3.5.2*. For this case, to show that the logic still works, I'll adopt the opposite sign choice, by taking the **voltages** *anticlockwise* but the *clockwise* **current**, so we have:[44]

$$E + v_R + v_C + v_L = 0$$

$$v_R = +i.R$$

$$C.\mathbf{d}v_C = +i.\mathbf{d}t$$

$$L.\mathbf{d}i = +v_L.\mathbf{d}t.$$

Here $i < 0$ because, for example, $v_R < 0$ but $R > 0$. Proceeding as before by applying **d** to the first two equations and substituting one into the other, we get:

$$\mathbf{d}i.R + \mathbf{d}v_C + \mathbf{d}v_L = 0$$

and so, using $\mathbf{d}v_C = (i / C).\mathbf{d}t$ from the **capacitor** equation,

$$\mathbf{d}i.R + (i / C).\mathbf{d}t + \mathbf{d}v_L = 0.$$

[44] Imagine just the **resistor**: $E > 0 \rightarrow v_R < 0$. Sticking to the usual assumption that $i(\mathbf{p}_-\mathbf{p}_+) > 0$ *within the* **emf**, which here is $i_{\text{anticlock}}$, we would need $v_R = -i_{\text{anticlock}}R$ to get $v_R < 0$. But this is $v_R = + i_{\text{clock}}R$.

We now want to bring down $L.\mathbf{d}i = v_L.\mathbf{d}t$, but here we have a problem, because we have no knowledge of $\mathbf{d}v_L$. We *can* proceed, however, taking a hint from the conventional approach, if we *postulate* $\uparrow(\lambda t)$ dependence throughout, assuming:

$$v_L = V_L.\uparrow(\lambda t) \qquad \text{and} \qquad i = I.\uparrow(\lambda t)$$

to give

$$\mathbf{d}(I.\uparrow(\lambda t)).R + (I.\uparrow(\lambda t) / C).\mathbf{d}t + \mathbf{d}(V_L.\uparrow(\lambda t)) = 0$$

and

$$L.\mathbf{d}(I.\uparrow(\lambda t)) = V_L.\uparrow(\lambda t).\mathbf{d}t.$$

So

$$\lambda I.R.\uparrow(\lambda t).\mathbf{d}t + (I.\uparrow(\lambda t) / C).\mathbf{d}t + \boldsymbol{\lambda} V_L.\uparrow(\lambda t).\mathbf{d}t = 0$$

and

$$\lambda LI.\uparrow(\lambda t)\ .\mathbf{d}t = V_L.\uparrow(\lambda t).\mathbf{d}t,$$

or

$$\lambda IR + I / C + \lambda V_L = 0$$

and

$$\lambda LI = V_L.$$

So $\lambda^2 LI + \lambda IR + I / C = 0.$

Note that putting $L = 0$ recovers our earlier equation for the RC circuit, $\lambda RC + 1 = 0$, so the different sign convention has made no difference to the result.

Now multiplying × $1/IL$ gives:

$$\lambda^2 + \lambda R/L + 1/LC = 0.$$

This can be solved as

$$\lambda = ½[-(R / L) \pm \surd[(R^2 / L^2) - 4/LC)] = -R / 2L \pm \surd((R^2 / 4L^2) - 1/LC)$$

by the **quadratic** equation solution demonstrated in Section 1.6.

Note that this has a solution only if the **discriminant** $(R^2 / 4L^2) - 1/LC > 0$ or $R^2 / 4L^2 > 1/LC$. In the next Chapter, I will show how the solution can be extended to the case where this is not so, but for now accept this limitation, and that putting this form back into the original equation, with arbitrary constants A and B, *and using both values of* λ, we obtain the general solution:

$$i(t) = \mathbf{e}^{(-Rt / 2L)} \times \{A.\mathbf{exp}(\surd((R^2 / 4L^2) - 1/LC).t) + B.\mathbf{exp}(-\surd((R^2 / 4L^2) - 1/LC).t)\}$$
$$i(t) = \uparrow(-Rt / 2L) \times \{A.\uparrow(\surd((R^2 / 4L^2) - 1/LC).t) + B.\uparrow(-\surd((R^2 / 4L^2) - 1/LC).t)\}$$

where the arbitrary constants A and B can be chosen to fit the initial (i.e. $t = 0$) conditions, and we adopt the convention that √ now means the *positive* root and we write the negative as −√.

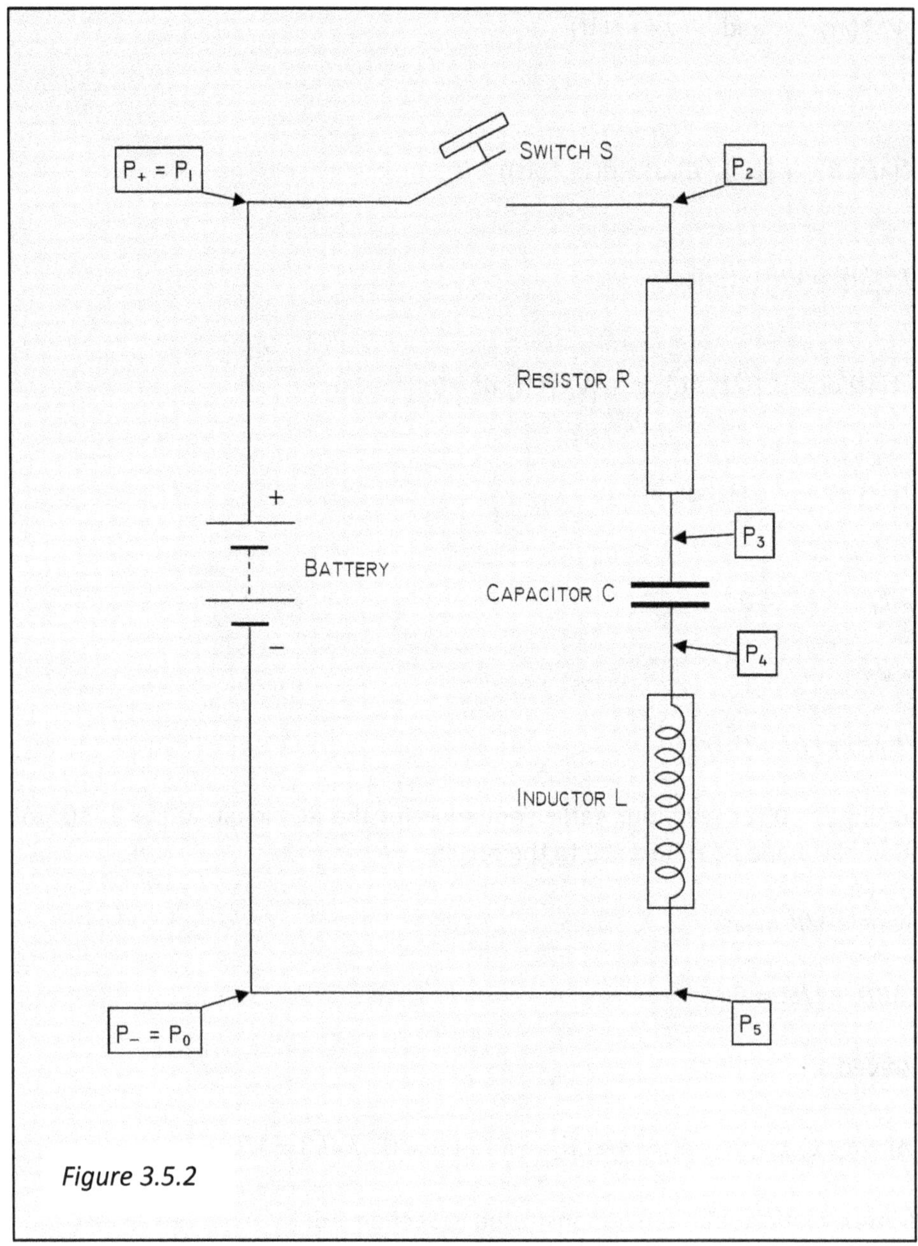

Figure 3.5.2

3.X Exercises

1. Complete the details of the derivation of the concept of **entropy** in Section 2.5.

2. Evaluate **d**$(1/\mathbf{e}^{at})$. How can this function be more easily expressed? Sketch it roughly.

3. Find **d**y for $y = ½(\mathbf{e}^x + \mathbf{e}^{-x})$ and for $y = ½(\mathbf{e}^x - \mathbf{e}^{-x})$ and for $y = (\mathbf{e}^x - \mathbf{e}^{-x})/(\mathbf{e}^x + \mathbf{e}^{-x})$. These are all **hyperbolic angle** functions that we will meet in Section 4.10.

4. Evaluate **d**y for $y = \downarrow(\downarrow x)$.

5. Prove $a\downarrow x = (b\downarrow x)/(b\downarrow a)$ and $b\downarrow a = 1/(a\downarrow b)$.

6. Prove, by comparing the areas under the curves for $(1/t)$ and $(1/\sqrt{t})$ for the **integrals** $\downarrow x = t^{-1}$.**d**$t\,|\,[1, x]$ and $t^{-½}$.**d**$t\,|\,[1, x] = (1/\sqrt{t})$.**d**$t\,|\,[1, x]$ that $\downarrow x/x \rightarrow 0$ as $x \rightarrow \infty$. (G.B Thomas, *op.cit.* p. 310.)

7. Using the formula (2.7.1) for the **Taylor expansion**, work out the first three or four terms of the **Taylor expansion** of $\mathbf{e}^x$, expanding about the point $x = 0$ (as a in (2.7.1)). Calculate the value of these terms at the point $x = 2$. This **polynomial** approximation is the normal way that $\mathbf{e}^x$ is evaluated.

8. Using the generalized **mean value theorem** from 2.X.10, we can express the **remainder term** of the **Taylor expansion** as $\partial^{n+1}f(c).(b - a)^{n+1}/(n+1)!$ where "!" indicates the **factorial function** $n! = n\times(n-1)\times(n-2) \times \ldots \times 2 \times 1$, and c lies in $[a, b]$. Using $a = 0$, show that the **remainder term** for $\mathbf{e}\uparrow x$ is $(x^{n+1}.\ \mathbf{e}\uparrow c)/(n+1)!$. Hence, taking the full **Taylor expansion** with this **remainder term**, and multiplying through by $n!$, show that **e** must be **irrational**.

9. In the circuit of *Figure 3.5.1*, imagine that the **resistor** R and the **capacitor** C are in *parallel* instead of being in series as in the *Figure*. So now instead of the same *current* through R and C we will have the same *voltage* across them. Write down the resulting circuit equations. What exactly has "gone wrong" here? Now insert a second **resistor** R_2 immediately in line with the switch S, write down the equations for this more subtle system and solve them as far as you can.

10. In the circuit of *Figure 3.5.2* above, instead of having R, C and L in **series**, set them in **parallel** and work through the full analysis analogous to that in Section 3.5 above.

11. It can be shown that: $\mathbf{lim}_{n\rightarrow\infty}\left(1 + \frac{1}{n}\right)^{nkt} = \mathbf{e}\uparrow(kt)$. See if you can get an idea of how this is proved.

Chapter 4

ANGLES

4.1 The Basic Notion of Angle

An **angle** is an ordered pair of real numbers (x, y) such that $x^2 + y^2 = 1$.

This is the sort of definition mathematicians love, because it's so obviously not an angle by any of our usual intuitive pictures, which suggest more some sort of pie-shaped thing, but it gives all the right properties. Indeed, it gives the right properties *so well* that it is surprising that this definition is not the standard one in the textbooks. I will consider the standard one later, but I will briefly say why the present definition works better. It is because this definition gives the right **topology**, which the conventional one does not. The **topological** shortcomings of the conventional definition, which treats an angle as a *single* real number, mean that its **topology** has constantly to be "corrected" by *ad hoc* assumptions.

It was from experience in programming that I came to see it this way. The definition I will develop comes naturally out of every situation in which angles appear and is much more tractable. The conventional approach – which I'll call the **single-real** or **parametric** approach – gives endless trouble when dealing with geometry, although it comes into its own in signal analysis.

I will introduce a name for my two-number system too, calling it very simply the **unit circle** approach. The **unit circle** is a well-accepted concept in mathematics: it is a circle with **radius** = 1 centred on the **origin**. A circle has quite a different **topology** to a straight line, a fact of enormous mathematical significance that is generally by-passed in conventional textbooks. By bringing it into focus at this stage, I hope I will enable readers to understand some of the oddities that certainly puzzled me when I was a student, such as how the **integrals** of certain simple algebraic functions are **inverse trigonometric functions**. The **unit circle** approach also leads smoothly to the concept of **complex numbers**, which *do* have the right **topology** for angles, and it explains why **imaginary numbers** cannot appear as part of the straight line concept of numbers, *precisely* because they conflict with its **topology**.

4.2 Addition of Angles

To get a theory out of our definition, we need an operation on angles. The operation of *addition* will be all we will need. We will want this to obey the four **Group Axioms**:

Axiom I: For two angles $\boldsymbol{\theta}$ and $\boldsymbol{\phi}$, $\boldsymbol{\theta} + \boldsymbol{\phi}$ is an angle (Closure)

Axiom II: For angles $\boldsymbol{\theta}$, $\boldsymbol{\phi}$, $\boldsymbol{\omega}$: $(\boldsymbol{\theta} + \boldsymbol{\phi}) + \boldsymbol{\omega} = \boldsymbol{\theta} + (\boldsymbol{\phi} + \boldsymbol{\omega})$ (Associativity)

Axiom III: There is a **unit** or **zero** angle **0** such that for any angle $\boldsymbol{\theta}$:

$\boldsymbol{\theta} + \mathbf{0} = \mathbf{0} + \boldsymbol{\theta} = \boldsymbol{\theta}$ (Identity Element)

Axiom IV: For any $\boldsymbol{\theta}$, there is an **inverse element** $\boldsymbol{\theta}^{-1}$ such that:

$\boldsymbol{\theta} + \boldsymbol{\theta}^{-1} = \boldsymbol{\theta}^{-1} + \boldsymbol{\theta} = \mathbf{0}$ (Inverse).

We need a formula for the sum in terms of the constituent real numbers x and y for which it will also be an idea to introduce some notation. I will commonly write:

$$\boldsymbol{\theta} = (\theta_x, \theta_y) = (\theta_c, \theta_s)$$

where the subscripts "*c*" and "*s*" come from the standard terminology that:

> *The x component of an angle is called its* ***cosine****, the y component its* ***sine****.*

$x^2 + y^2$ is a mathematical form called a **quadratic form**. It is a special case of the form $\Sigma a_{ij}x_ix_j$ which is called a **bilinear form**. A **quadratic form** is simply a **bilinear form** with $a_{ij} = 0$ wherever $i \neq j$. Because the defining condition for the **unit circle** is a **quadratic form** $x^2 + y^2 = 1$, it is reasonable to try for the more general **bilinear forms** to find the formula for angular addition in terms of components. So I will try:

$$\boldsymbol{\theta} + \boldsymbol{\phi} = (\theta_x, \theta_y) + (\varphi_x, \varphi_y)$$

$$= (a_{11}\theta_x\varphi_x + a_{12}\theta_x\varphi_y + a_{21}\theta_y\varphi_x + a_{22}\theta_y\varphi_y,\ b_{11}\theta_x\varphi_x + b_{12}\theta_x\varphi_y + b_{21}\theta_y\varphi_x + b_{22}\theta_y\varphi_y)$$

but first we'll need to choose the **unit angle**, which is the **zero** of angular addition. Some mathematician in the seventeenth century – I used to blame Euler in the eighteenth century, but the usage seems to have been established by Brook Taylor's time about 1715 – decided to choose the point where the **unit circle** intersects the x axis as the **unit angle**. The more natural choice would have been the point where the **unit circle** intersects the y axis, which would have given positive angular measure running *clockwise from North*, in agreement with navigation, and so all geographical and cartographical practice. I will briefly consider what our theory might have looked like under this more natural choice, but for the main line of development I will have to stay consistent with normal mathematical practice, even though it is in conflict with virtually everybody else's.

So $\mathbf{0} = (1, 0)$

Putting this into:

$\boldsymbol{\theta} + \mathbf{0} = \boldsymbol{\theta}$, we have

$$(\theta_x, \theta_y) + (1, 0)$$

$$= (a_{11}\theta_x + a_{12}\theta_x.0 + a_{21}\theta_y + a_{22}\theta_y.0,\ b_{11}\theta_x + b_{12}\theta_x.0 + b_{21}\theta_y + b_{22}\theta_y.0)$$

$$= (a_{11}\theta_x + a_{21}\theta_y,\ b_{11}\theta_x + b_{21}\theta_y).$$

Since this must equate to (θ_x, θ_y) for any values of θ_x and θ_y, we must have:

$a_{11} = 1$, $a_{21} = 0$ and $b_{11} = 0$, $b_{21} = 1$.

But we must also have $\mathbf{0} + \boldsymbol{\theta} = \boldsymbol{\theta}$, which means that:

$$(1, 0) + (\theta_x, \theta_y)$$

$$= (a_{11}1\theta_x + a_{12}1\theta_y + a_{21}.0.\theta_x + a_{22}.0.\theta_y,\ b_{11}1\theta_x + b_{12}1\theta_y + b_{21}.0.\theta_x + b_{22}.0.\theta_y)$$

$$= (\theta_x + a_{12}1\theta_y,\ b_{12}1\theta_y)$$

which in the same way gives $a_{12} = 0$ and $b_{12} = 1$.

So already we're down to:

$$\boldsymbol{\theta} + \boldsymbol{\phi} = (\theta_x, \theta_y) + (\varphi_x, \varphi_y) = (\theta_x\varphi_x + a_{22}\theta_y\varphi_y,\ \theta_x\varphi_y + \theta_y\varphi_x + b_{22}\theta_y\varphi_y)$$

If we now impose the *Closure* Axiom, which requires that this new product obeys $x^2 + y^2 = 1$, we have to have *for every possible set of allowed values for* θ_x, θ_y, φ_x, φ_y that:

$$\theta_x^2.\varphi_x^2 + 2a_{22}\,\theta_x\varphi_x\,\theta_y\varphi_y + a_{22}^2\theta_y^2\varphi_y^2 + \theta_x^2\varphi_y^2 + \theta_y^2\varphi_x^2 + b_{22}^2\theta_y^2\varphi_y^2$$

$$+ 2.\theta_x\varphi_y\theta_y\varphi_x + 2.b_{22}\,\theta_x\varphi_y\theta_y\varphi_y + 2.\,b_{22}\,\theta_y\varphi_x\,\theta_y\varphi_y = 1$$

or that

$$\theta_x^2.(\varphi_x^2 + \varphi_y^2) + \theta_y^2.(\varphi_x^2 + a_{22}^2\varphi_y^2 + b_{22}^2\varphi_y^2)$$

$$+ 2(a_{22} + 1)\theta_x\varphi_x\,\theta_y\varphi_y + 2b_{22}(\theta_x\varphi_y\theta_y\varphi_y + \theta_y\varphi_x\,\theta_y\varphi_y)$$

must equal 1. The last two terms involve cross-products whose value is indeterminate – i.e. we cannot evaluate them from $\theta_x^2 + \theta_y^2 = 1$ and $\varphi_x^2 + \varphi_y^2 = 1$, which we know must hold – so they must be forced out of the equation by making:

$b_{22} = 0$ and $a_{22} = -1$

and putting these values into the first two terms gives:

$$\theta_x^2.(\varphi_x^2 + \varphi_y^2) + \theta_y^2.(\varphi_x^2 + (-1)^2\varphi_y^2 + 0^2\varphi_y^2) = \theta_x^2.(\varphi_x^2 + \varphi_y^2) + \theta_y^2.(\varphi_x^2 + \varphi_y^2)$$

$= \theta_x^2.(1) + \theta_y^2.(1)$ using $\varphi_x^2 + \varphi_y^2 = 1$, and so

$= 1$ using $\theta_x^2 + \theta_y^2 = 1$.

So our **angular addition formula** is:

$$\boldsymbol{\theta} + \boldsymbol{\phi} = (\theta_x, \theta_y) + (\varphi_x, \varphi_y) = (\theta_x\varphi_x - \theta_y\varphi_y, \theta_x\varphi_y + \theta_y\varphi_x)$$

a result which, although not usually derived in this context, is one of the most famous results in mathematics. It is the formula for the *multiplication* of **complex numbers**.

Note its **logarithmic/exponential** quality: *addition* of **angles** corresponds to a *multiplication* in its components.

I will return to **complex numbers**, which are a combination of an ordinary number *r* and a **unit circle angle θ**, later. First, let's get some more results.

What of the **inverse** or **negative** of an **angle** postulated in Axiom IV? This must obey $\boldsymbol{\theta}^{-1} + \boldsymbol{\theta} = \mathbf{0}$ so in components:

$(\theta^{-1}{}_x, \theta^{-1}{}_y) + (\theta_x, \theta_y) = (1, 0)$ or

$\theta^{-1}{}_x\theta_x - \theta^{-1}{}_y\theta_y = 1,$ $\theta^{-1}{}_x\theta_y + \theta^{-1}{}_y\theta_x = 0$

from the expansion of our addition formula.

This is a pair of linear equations in two unknowns and so has a unique solution. The second equation gives:

$$\theta^{-1}{}_x / \theta_x = -\theta^{-1}{}_y / \theta_y \quad (a)$$

and the first

$$\theta^{-1}{}_x\theta_x = 1 + \theta^{-1}{}_y\theta_y . \quad (b)$$

So from (*a*)

$$\theta^{-1}{}_y = -\theta_y\,\theta^{-1}{}_x / \theta_x$$

and putting this into (*b*) gives $\theta^{-1}{}_x\theta_x = 1 - \theta_y^2\,\theta^{-1}{}_x / \theta_x$ or . . .

$\theta^{-1}{}_x(\theta_x{}^2 + \theta_y{}^2) = \theta^{-1}{}_x(1) = \theta_x$ so quite simply, $\theta^{-1}{}_x = \theta_x$ and from that $\theta^{-1}{}_y = -\theta_y$.

So in the **negative** of an **angle**, only the *y* part or the **sine** changes sign. That should be easy to remember: to form the **negative** of an angle: *only the* **sine** *changes sign*. In terms of **complex numbers** the resulting new number is called the **complex conjugate** rather than the **negative** of the **angle**.

This **negative** gives the angular subtraction formulae:

$$\boldsymbol{\theta} - \boldsymbol{\phi} = (\theta_x, \theta_y) + (\varphi_x, -\varphi_y) = (\theta_x\varphi_x + \theta_y\varphi_y, \theta_y\varphi_x - \theta_x\varphi_y)$$
$$\boldsymbol{\phi} - \boldsymbol{\theta} = (\varphi_x, \varphi_y) + (\theta_x, -\theta_y) = (\theta_x\varphi_x + \theta_y\varphi_y, \theta_x\varphi_y - \theta_y\varphi_x)$$

which also shows that **cosine(ϕ − θ)** = **cosine(θ − ϕ)**, but **sine(ϕ − θ)** = − **sine(θ − ϕ)**.[45]

I have spoken of this operation as an *addition* of **angles**. Why? Consider a simple angle like the basic right angle (going anti-clockwise from the **zero angle** (1, 0)), which will be (0, 1). "Add" this to (1, 0):

$(1, 0) + (0, 1) = (1\times0 - 0\times1, 1\times1 + 0\times0) = (0, 1)$, itself, as expected.

Add it again:

$$(0, 1) + (0, 1) = (0\times0 - 1\times1, 0\times1 + 1\times0) = (-1, 0)$$

and again:

$$(-1, 0) + (0, 1) = (-1\times0 - 0\times1, -1\times1 + 0\times0) = (0, -1)$$

and again:

$$(0, -1) + (0, 1) = (0\times0 - -1\times1, 0\times1 + -1\times0) = (+1, 0) = (1, 0)$$

So we have come full circle, and it is easy to see, as in *Figure 4.2.1*, that the four points we have passed through are equal anti-clockwise increments around the **unit circle**, taking us from East to North to West to South and back to East.

---

[45] We also get, from the addition formula, expressed in conventional notation: **cos(θ+ϕ)** = **cos(θ).cos(ϕ)** − **sin(θ).sin(ϕ)**, and **sin(θ+ϕ)** = **cos(θ).sin(ϕ)** + **sin(θ).cos(ϕ)**, and from this subtraction formula: **cos(θ−ϕ)** = **cos(θ).cos(ϕ)** + **sin(θ).sin(ϕ)** and **sin(θ−ϕ)** = **sin(θ).cos(ϕ)** − **cos(θ).sin(ϕ)**. You will find all these forms in the textbooks.

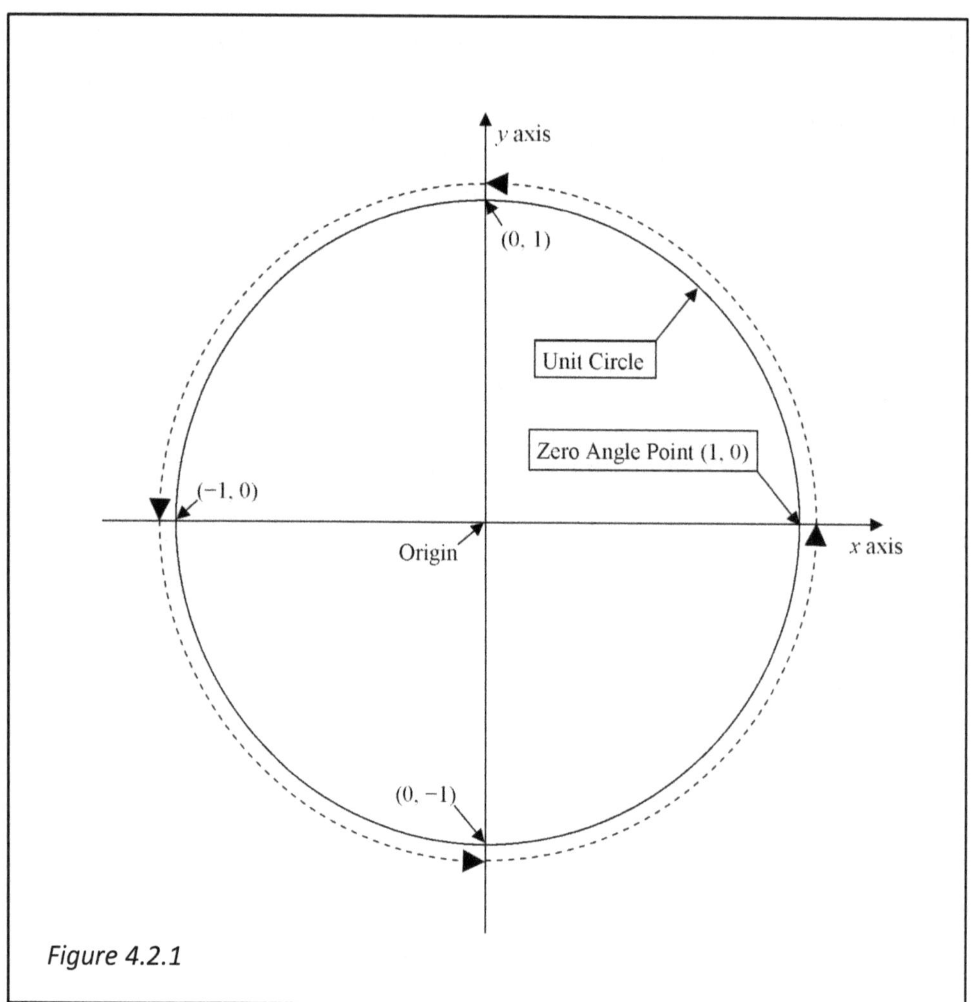

Figure 4.2.1

You can try it again with a smaller step like $\boldsymbol{\theta}_{45} = (1/\sqrt{2}, 1/\sqrt{2})$ which also obeys $x^2 + y^2 = ½ + ½ = 1$, or try $\boldsymbol{\theta}_{-45} = (1/\sqrt{2}, -1/\sqrt{2})$ which will take you round *clockwise*. The first four steps are:

$$(1, 0) + (1/\sqrt{2}, -1/\sqrt{2}) = (1 \times 1/\sqrt{2} - 0 \times (-1/\sqrt{2}), 1 \times (-1/\sqrt{2}) + 0 \times 1/\sqrt{2})$$
$$= (1/\sqrt{2}, -1/\sqrt{2})$$

$$(1/\sqrt{2}, -1/\sqrt{2}) + (1/\sqrt{2}, -1/\sqrt{2})$$
$$= ((1/\sqrt{2})(1/\sqrt{2}) - (-1/\sqrt{2})(-1/\sqrt{2}), (1/\sqrt{2})(-1/\sqrt{2}) + (-1/\sqrt{2})(1/\sqrt{2}))$$
$$= (½ - (--½), -½ + -½) = (½ - ½, -1)$$
$$= (0, -1)$$

$$(0, -1) + (1/\sqrt{2}, -1/\sqrt{2}) = (0\times(1/\sqrt{2}) - (-1)(-1/\sqrt{2}), 0\times(-1/\sqrt{2}) + (-1)(1/\sqrt{2}))$$
$$= (-1/\sqrt{2}, -1/\sqrt{2})$$

and $(-1/\sqrt{2}, -1/\sqrt{2}) + (1/\sqrt{2}, -1/\sqrt{2}) = ((-1/\sqrt{2})(1/\sqrt{2}) - (-1/\sqrt{2})(-1/\sqrt{2}),$

$$(-1/\sqrt{2})(-1/\sqrt{2}) + (-1/\sqrt{2})(1/\sqrt{2})) = (-½ -½, ½ -½)$$
$$= (-1, 0).$$

So in four *clockwise* steps we've gone from East to South-East to South to South-West to West, again all even steps.

What would have been our addition formula if we *had* chosen (0, 1) as the **zero angle**? Working through much the same logic as before, we find that *now*:

$$\boldsymbol{\theta} + \boldsymbol{\phi} = (\theta_x\varphi_y + \theta_y\varphi_x, \theta_y\varphi_y - \theta_x\varphi_x)$$

so:

$$\boldsymbol{\theta} + \mathbf{0} = (\theta_x.1 + \theta_y.0, \theta_y.1 - \theta_x.0) = (\theta_x, \theta_y) \text{ and } \ldots$$
$$\mathbf{0} + \boldsymbol{\phi} = (0.\varphi_y + 1.\varphi_x, 1.\varphi_y - 0.\varphi_x) = (\varphi_x, \varphi_y)$$

which you may find easier to follow written like that. Closure also holds:

$$(\theta_x\varphi_y + \theta_y\varphi_x)^2 + (\theta_y\varphi_y - \theta_x\varphi_x)^2 =$$
$$\theta_x^2\varphi_y^2 + \theta_y^2\varphi_x^2 + 2.\theta_x\theta_y\varphi_x\varphi_y + \theta_y^2\varphi_y^2 + \theta_x^2\varphi_x^2 - 2.\theta_x\theta_y\varphi_x\varphi_y$$
$$= \theta_x^2(\varphi_x^2 + \varphi_y^2) + \theta_y^2(\varphi_x^2 + \varphi_y^2) = \theta_x^2(1) + \theta_y^2(1) = 1$$

But the **negative** of **θ** is now $(-\theta_x, \theta_y)$ with the x component taking the minus sign. Interestingly, to be consistent with all the properties of **sines** and **cosines** in the **single-real** approach to angles, *the x value now becomes the* **sine**, *and the y value the* **cosine**, so the rule that "the **sine** changes sign" to get the **negative** still holds good.

4.3 Ordering Relations

There *is* no natural ordering of **angles**. This was Columbus's idea that on a round Earth everywhere is West if you travel far enough. As it happens, Columbus didn't, but twenty-eight years later Magellan did, although he didn't survive the experience. Similarly on the circle, from any angle you can reach any other by going clockwise *or* anticlockwise. That is not to say that we cannot establish an ordering of angles by an appropriate sleight of hand. A **simple ordering** is a relation > which obeys these axioms[46]:

Axiom I: for any two elements **a** and **b** such that **a** ≠ **b**, *either* **a** > **b** *or* **b** > **a**.

Axiom II: if **a** > **b** and **b** > **c** then **a** > **c**.

[46] S. Feferman, *The Number Systems*, Addison-Wesley 1964 p.83

and we can achieve such an ordering by use of the elementary **single-real** approach, which assigns a unique real number to each angle. The full set of **real numbers**, however, forms what mathematicians call an **ordered field**, which obeys these two extra axioms:

Axiom III: if **a** > **b** then **a** + **c** > **b** + **c** for any **c**

Axiom IV: if **a** > **b** and **c** > **0** then **a.c** > **b.c**

the last of these leading to the case mentioned in Section 1.5 which I called the **inequality trap** – that if **c** < **0** then the ordering *reverses*. These latter axioms are needed to prove that there cannot be a square root of a negative number, and it is because Axioms III and IV do *not* hold for angles that **complex numbers** involving an **angle** can circumvent this proof[47]. This was never pointed out to me as a student, which led to my feeling like Bishop Berkeley did about infinitesimals, that mathematicians make and break their own rules as they see fit.

But we can establish some very useful properties for **angles** that are akin to ordering. Firstly, we can say whether an angle **θ** can be reached from an angle **ϕ** sooner by going *clockwise* when we say it lies to the *right* of **ϕ**, or *anticlockwise* when we say it lies to the *left* of **ϕ**. This can be evaluated simply by taking:

$\boldsymbol{\delta} = \boldsymbol{\theta} - \boldsymbol{\phi}$ which implies that $\boldsymbol{\phi} + \boldsymbol{\delta} = \boldsymbol{\theta}$

so that **δ** is the angle that needs to be added to **ϕ** to obtain **θ**. If the **sine** of **δ** is *positive*, **θ** lies to the *left* of **ϕ**, if it is *negative* **θ** lies to the *right* of **ϕ**. Since, returning to the (1, 0) convention for the **zero angle**,

$$\boldsymbol{\delta} = \boldsymbol{\theta} - \boldsymbol{\phi} = (\theta_x, \theta_y) + (\varphi_x, -\varphi_y) = (\theta_x\varphi_x - \theta_y(-\varphi_y), \theta_x(-\varphi_y) + \theta_y\varphi_x),$$

so **sine(δ)** = $\theta_y\varphi_x - \theta_x\varphi_y$ and the sign of this product gives the answer.

We can also say whether an angle lies *between* two other angles. *Figure 4.3.1* shows the definition we'll use. Given two angles **a** and **b**, we say that **θ** lies *between* **a** and **b** if it lies in the **minor arc** between **a** and **b** but not if it lies in the **major arc** between them. Defining the **square distance** between two angles **θ** and **ϕ** as:

$$d^2(\boldsymbol{\theta}, \boldsymbol{\phi}) = ((\theta_x - \varphi_x)^2 + (\theta_y - \varphi_y)^2)$$

it can be shown that **θ** lies in the **minor arc** between **a** and **b** if:

$$d^2(\boldsymbol{\theta}, \mathbf{a}) + d^2(\boldsymbol{\theta}, \mathbf{b}) < d^2(\mathbf{a}, \mathbf{b}).$$

[47] Feferman, *op.cit.*, p.303

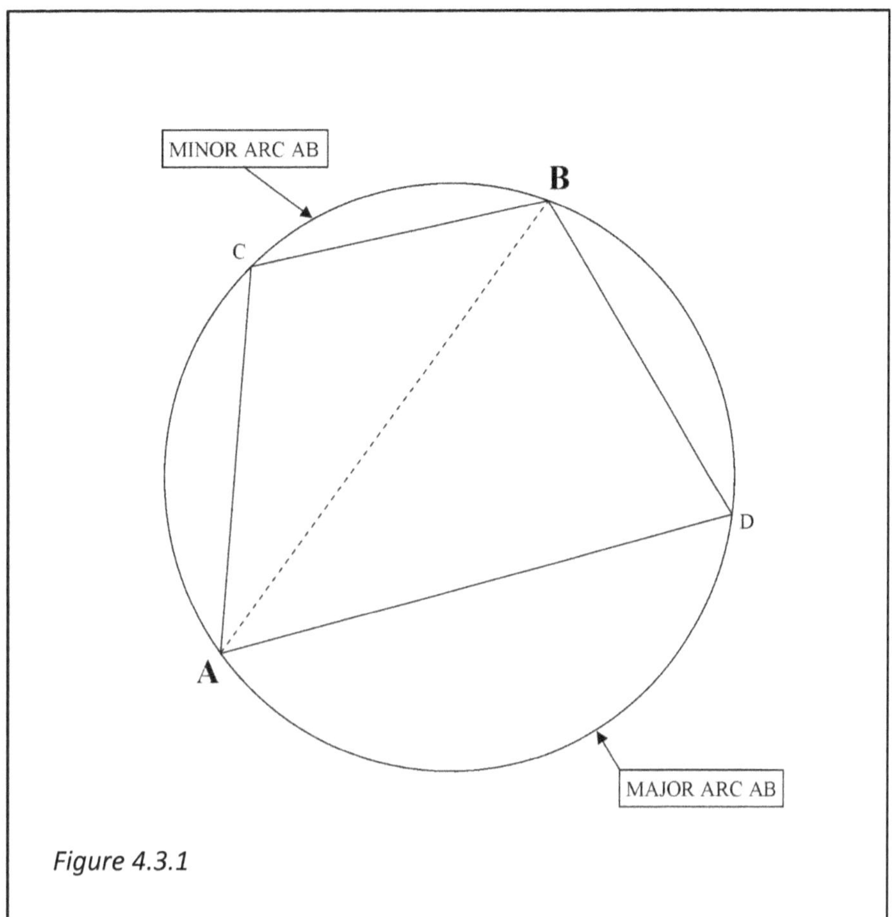

Figure 4.3.1

The proof can be sketched out thus: form the *triangle* in the *xy* plane made by the three points on the **unit circle a**, **b**, and **θ** (in the diagram **θ** is either at C or at D). Call the length of the side **b** to **θ**: *a*, that from **a** to **θ**: *b,* and that from **a** to **b**: *c* – each taking its name from the point *opposite which* it lies, with θ re-written as *c*. Then if **θ** lies in the **minor arc** between **a** and **b**, the angle *in the triangle* at **θ** must be **obtuse** – that is, greater than a right angle. The so-called **cosine rule** formula – a generalization of Pythagoras's theorem for triangles to cover non-right-angled triangles – states:

$$a^2 + b^2 - 2ab.\mathbf{cosine}(\angle \mathbf{a\theta c}) = c^2$$

where $\angle$**aθc** is the angle at **θ** *in the triangle* and if this angle is **obtuse**, **cosine(**$\angle$**aθc)** will be *negative*, so – 2*ab*.**cosine(**$\angle$**aθc)** will be *positive*, so $a^2 + b^2 < c^2$.

4.4 The Single-Real or Parametric Approach

The **single-real** approach is the traditional way of defining angles. Its development was a triumph of the early calculus in the seventeenth century, and also led to precise values for the number π which had haunted mathematics since at least the first Millennium B.C.

Although the result is quite intuitive, the theory that leads up to it is decidedly complex, but there is a deeper problem. The trouble is that **real numbers** do not have the right **topological** properties to describe angles. In particular, they *are* ordered through to infinity. To put it simply, on a straight line, as you keep adding positive numbers to your present position, you keep on going, never returning to your original position. On the circle, which is the correct **topological** context for **angles**, you *do* return to your starting point, over and over again.

In its simplest form, the **single-real** or **parametric** approach deals with this problem by artificially introducing a **discontinuity** in the circle, so that in degree measure, which is most familiar to most people, when you reach 360° you have to drop back to 0°. So you have to think of 359° and 1° as just 2° apart, not as the 358° which the actual arithmetic suggests.[48] This jump occasions endless errors in programming angles on a computer!

The preferred mathematical usage differs from the 0° to 360° version in two ways. Firstly, mathematicians prefer to regard angles as running *anticlockwise* from East through a *half circle* to just 180°, and as running *clockwise* from East through the *other half circle* to −180°. So they move the **discontinuity** from 0° to ±180°, which is reasonable, but only serves to complicate matters further, as any programme needs to keep checking which convention is in use where. Secondly, they use the actual **arc length** around the **unit circle** as their measure of angle, which, easy though it may be to visualize, is a mathematically difficult concept as we shall see below. This definition gives angular measure in **radians** rather than in *degrees*. The emphasis on this **arc length parameter** is something of a fortuitous legacy of the Ancient mystery that when you wound a string around a spindle, the number of turns a given length of string would give seemed to be a most peculiar function of the diameter of the spindle. (For centuries, if not millennia, it was thought that π must be exactly 3 and that the discrepancies were only due to what we would today call “experimental error”.)

There is a way of obtaining the **parameter** we need through a **differential form** without any need to invoke the somewhat artificial concept of **arc length**.

$$\mathbf{d}(x^2 + y^2) = 2x.\mathbf{d}x + 2y.\mathbf{d}y = 0 \text{ implies } x.\mathbf{d}x + y.\mathbf{d}y = 0$$

and since the **arc** is a one-dimensional entity, we could **parameterize** it[49] by a single appropriately chosen variable. Let’s call this variable θ and require that it obey one of two prescriptions. Either:

$$\mathbf{d}x = -y.\mathbf{d}\theta \quad \text{and} \quad \mathbf{d}y = x.\mathbf{d}\theta \qquad (4.4.1)$$

or alternatively:

$$\mathbf{d}x = y.\mathbf{d}\theta \quad \text{and} \quad \mathbf{d}y = -x.\mathbf{d}\theta \qquad (4.4.2)$$

---

[48] I will modify this rather harsh statement towards the end of this Section, but as it stands it does reflect a recurrent dilemma in programming practice.

[49] A **parameterization** of a line is the defining of a single variable such that every point on the line can be found as a **function** of that variable. So a line curving through the plane, as here, can be **parameterized** by a variable “t” if we can define **functions** $x(t)$ and $y(t)$ such that every point on the line can be obtained for some value of t as $(x(t), y(t))$.

Either of these satisfies our **derivative equation** $x.\mathbf{d}x + y.\mathbf{d}y = 0$. For example, the first gives:

$$x.\mathbf{d}x + y.\mathbf{d}y = -xy.\mathbf{d}\theta + yx.\mathbf{d}\theta = (yx - xy).\mathbf{d}\theta = 0.$$

If we now solve for $\mathbf{d}\theta$, we find:

$\mathbf{d}\theta = \mathbf{d}y / x$ using the first prescription,

and this becomes, substituting for $x = \surd(1 - y^2)$

$$\mathbf{d}\theta = \mathbf{d}y / \surd(1 - y^2). \tag{4.4.3}$$

The other prescription simply gives $\mathbf{d}\theta = -\mathbf{d}y / \surd(1 - y^2)$. The two are equivalent anyway, because the root is always $\surd \equiv \pm\surd$. So this approach gives us our correct **differential form** defining our new **parameter** θ without any of the palaver about **arc lengths**. The two prescriptions are simply the *anticlockwise* and *clockwise* **parameterization**s of the circle and do of course actually correspond to **arc lengths**.

So let's now look at the **arc length** interpretation.

The **arc length** of a curve C lying in the xy plane is defined, using the notion of **differentials** as ordinary numbers, as:

$$\surd((\mathbf{d}x)^2 + (\mathbf{d}y)^2) \,|\, \mathrm{C}$$

i.e. as the **integral** of the very non-standard looking expression in **differentials** at the left, over C. This expression can be brought into a more regular form by dividing and multiplying the **differential form** through by $(\mathbf{d}x)^2$ to give:

$$\surd(1 + (\mathbf{d}y/\mathbf{d}x)^2).\surd(\mathbf{d}x)^2 = \surd(1 + (\mathbf{d}y/\mathbf{d}x)^2).\ \mathbf{d}x = \surd(1 + (\partial_x y)^2).\mathbf{d}x$$

where I have used $\mathbf{d}y = \partial_x y(x).\mathbf{d}x$ and this at least gets it into a form consistent with the theory in Chapter 2.

This would be the natural way to proceed if we were using the (0, 1) choice of **zero angle**, for then we could evaluate around the circle clockwise through the first quarter arc starting at $x = 0$ and increasing x up to 1. For the (1, 0) choice, we need to start at $y = 0$ and go *anticlockwise* by increasing y through the first *anticlockwise* quarter arc to 1. So to do this we need to swap the roles of x and y to get:

$$\surd((\mathbf{d}x/\mathbf{d}y)^2 + 1).\surd(\mathbf{d}y)^2 = \surd((\mathbf{d}x/\mathbf{d}y)^2 + 1).\ \mathbf{d}y = \surd(1 + (\partial_y x)^2).\mathbf{d}y$$

using $\mathbf{d}x = \partial_y x(y).\mathbf{d}y$. Now from $x^2 + y^2 = 1$, we have:

$$\mathbf{d}(x^2) + \mathbf{d}(y^2) = \mathbf{d}(1) = 0$$

or $\quad 2x.\mathbf{d}x + 2y.\mathbf{d}y = 0.$

So $\mathbf{d}x/\mathbf{d}y = -y / x = -y/(\sqrt{(1 - y^2)})$, and the square of this is:

$$(\partial_y x)^2 = + y^2/(1 - y^2)$$

and so $1 + (\partial_y x)^2$ is:

$$1 + (y^2/(1 - y^2)) = (1 - y^2)/(1 - y^2) + (y^2/(1 - y^2))$$

$$= (1 - y^2 + y^2)/(1 - y^2) = 1/(1 - y^2)$$

So our **integral** becomes:

$$\sqrt{(1/(1 - y^2))}\ .\mathbf{d}y \mid C^*$$

where I write C* to indicate that having eliminated x from the **differential form**, we no longer need it in the *curve* argument either, and can work with the **pullback** of C from the curve to the simple interval in y, which I denote $C^* = [0, y_1]$ where y_1 is a number between 0 and 1 that indicates how far round the quarter circle we want to go. This **pullback** concept is a subtle point that will be dealt with in more detail in Chapter 6. For now, just accept that in eliminating x from the **differential form**, we can go ahead with just the **projection** of C onto the y axis, as in this and many simple cases, the **pullback** equals the **projection**.

So we've finally recovered the **differential form** we developed so much more easily in equation (4.4.3)!

Now we need to get this **differential form** into an **exact differential** of the form $\mathbf{d}f(y)$ in order to apply our definition in Section 2.6 that:

$$\mathbf{d}f(y) \mid C^* = \mathbf{d}f(y) \mid [0, y_1] = f(y) \mid \mathbf{b}[0, y_1] = f(y_1) - f(0).$$

This is where all the hard work usually appears in **integration** and this case is a good example. First we evaluate $(1/(1 - y^2))$ by long division. As in Section 1.9, we evaluate successive **remainders** from successive trial terms in the solution. Since we will generate an **infinite series** here and the powers will rise in steps of two, I'll use consecutive indexing rather than indexing by powers as I chose in Section 1.9. Here $N(y) = 1$, $D(y) = 1 - y^2$, so we try $Q_0 = 1$, and obtain:

$$R_1(y) = N(y) - Q_0.D(y) = 1 - 1.(1 - y^2) = + y^2$$

Now try $Q_1(y) = y^2$ to get $R_2(y) = R_1(y) - Q_1(y).D(y) = y^2 - y^2.(1 - y^2) = + y^4$

and so we carry on:

$$(try)\ Q_2(y) = y^4 \rightarrow R_3(y) = R_2(y) - Q_2(y).D(y) = y^4 - y^4.(1 - y^2) = + y^6$$

(*try*) $Q_3(y) = y^6 \rightarrow R_4(y) = R_3(y) - Q_3(y).D(y) = y^6 - y^6.(1 - y^2) = + y^8$
(*try*) $Q_4(y) = y^8 \rightarrow R_5(y) = R_4(y) - Q_4(y).D(y) = y^8 - y^8.(1 - y^2) = + y^{10}$
. . . . *etc* .

The general result is that:

$$1/(1 - y^2) = 1 + y^2 + y^4 + y^6 + y^8 + y^{10}/(1 - y^2)$$

and so on. Written this way this expansion is always valid. The **remainder term** on the right, here $y^{10}/(1 - y^2)$, is in general $y^{2n}/(1 - y^2)$. Now if the **absolute value** of y is less than 1, or $-1 < y < 1$ this remainder will become vanishingly small as y^{10} is replaced by ever higher powers or:

$$y^{2n}/(1 - y^2) \rightarrow 0 \text{ as } n \rightarrow \infty \qquad \text{if } -1 < y < 1.$$

This condition will hold for all points on the **unit circle** except where $y = \pm 1$. In this elementary treatment, I'll pass over that complication.

Now we need the square root of this. The convenient form Σy^{2i} shows that there will be no odd powers in the root either. So we need to find a form $1 + a_2y^2 + a_4y^4 + a_6y^6 + a_8y^8 + a_{10}y^{10} +$ such that:

$$(1 + a_2y^2 + a_4y^4 + a_6y^6 + a_8y^8 + a_{10}y^{10} + \ldots)(1 + a_2y^2 + a_4y^4 + a_6y^6 + a_8y^8 + a_{10}y^{10} + \ldots)$$

$$= 1 + y^2 + y^4 + y^6 + y^8 + y^{10} \ldots$$

and a table is going to be our best bet in evaluating this. *Table 4.1* shows the various products involved up to y^{10}.

Table 4.1	1	a_2y^2	a_4y^4	a_6y^6	a_8y^8	$a_{10\text{-}}y^{10}$
1	1	a_2y^2	a_4y^4	a_6y^6	a_8y^8	$a_{10}y^{10}$
a_2y^2	a_2y^2	$a_2^2y^4$	$a_2a_4y^6$	$a_2a_6y^8$	$a_2a_8y^{10}$	$a_2a_{10}y^{12}$
a_4y^4	a_4y^4	$a_2a_4y^6$	$a_4^2y^8$	$a_4a_6y^{10}$	$a_4a_8y^{12}$	$a_4a_{10}y^{14}$
a_6y^6	a_6y^6	$a_2a_6y^8$	$a_4a_6y^{10}$	$a_6^2y^{12}$	$a_6a_8y^{14}$	$a_6a_{10}y^{16}$
a_8y^8	a_8y^8	$a_2a_8y^{10}$	$a_4a_8y^{12}$	$a_6a_8y^{14}$	$a_8^2y^{16}$	$a_8a_{10}y^{18}$
$a_{10}y^{10}$	$a_{10}y^{10}$	$a_2a_{10}y^{12}$	$a_4a_{10}y^{14}$	$a_6a_{10}y^{16}$	$a_8a_{10}y^{18}$	$a_{10}^2y^{20}$

To evaluate the square root, we now equate the total coefficients of equal powers of y on both sides of our previous equation (this must hold for all values of y, and this is actually an example of the use of **linear independence** of different powers of a variable as defined in Section 1.7).

So for example we must have:

for y^2: $2.a_2 = 1$, so $a_2 = ½$

for y^4: $a_4 + a_2^2 + a_4 = 1$, so $2.a_4 + (½)^2 = 1$, or $a_4 = (1 - ¼) / 2 = 3/8$

for y^6: $2a_6 + 2a_2a_4 = 1$,

so $a_6 = ½(1 - 2a_2a_4) = ½(1 - 2.½.(3/8)) = ½(1 - (3/8)) = ½.(5/8) = 5/16$,

and so on.

The final form turns out to be:

$$\sqrt{(1/(1 - y^2))} = 1 + ½.y^2 + ((1\times3)/(2\times4)).y^4 +$$

$$+ ((1\times3\times5)/(2\times4\times6)).y^6 + ((1\times3\times5\times7)/(2\times4\times6\times8)).y^8 + \ldots$$

and this we *can* **integrate** using the formula $x^n.\mathbf{d}x = \mathbf{d}(x^{n+1}/(n+1))$ to give:

$$\mathbf{d}(\sqrt{(1/(1 - y^2))}) = \mathbf{d}(y + ½⅓.y^3 + ((1\times3)/(2\times4)).y^5/5 + ((1\times3\times5)/(2\times4\times6)).y^7/7 +$$

$$+ ((1\times3\times5\times7)/(2\times4\times6\times8)).y^9/9 + \ldots)$$

so our result is:

$$(y + ½⅓.y^3 + ((1\times3)/(2\times4)).y^5/5 + ((1\times3\times5)/(2\times4\times6)).y^7/7$$

$$+ ((1\times3\times5\times7)/(2\times4\times6\times8)).y^9/9 + \ldots) \mid [0, y_1]$$

$$= (y_1 + ½⅓.y_1^3 + ((1\times3)/(2\times4))\, y_1^5/5 + ((1\times3\times5)/(2\times4\times6))\, y_1^7/7$$

$$+ ((1\times3\times5\times7)/(2\times4\times6\times8))\, y_1^9/9 + \ldots)$$

since the powers y^i are all zero if $y = 0$.

Setting $y_1 = 1$, which represents the quarter arc point at (0, 1), and taking enough terms in this expansion gives the value $\pi/2$.

Up to this point, I've rather disparaged this **parametric** approach to **angles**. But it comes into its own when we *abandon* the idea of an artificial **discontinuity** at 0° or at 180° and simply let the single real **parameter** increase indefinitely. So we define 365° as referring to the same position on the **unit circle** as 5°, and 450° as again referring to 90°. Equally, we can say that −5° refers to the same point as 355°.

If we do this, then *every* value of the **parameter** – call it θ – still points to a position on the **unit circle**, but as θ grows indefinitely, we keep going (anticlockwise) round and round the same **unit circle** with the position (x, y) on the **unit circle** being defined by a function:

$$(x, y) = (x(\theta), y(\theta))$$

where the two component functions $x(\theta)$ and $y(\theta)$ constantly oscillate between −1 and +1 as θ increases, whilst always maintaining the relationship:

$$x^2 + y^2 = x^2(\theta) + y^2(\theta) = 1.$$

These **functions** are called the **cosine** and **sine functions**, and are commonly written $x(\theta) = \mathbf{cos}(\theta)$ and $y(\theta) = \mathbf{sin}(\theta)$. They are chosen to obey our *anticlockwise* **parametric** equations:

$$\mathbf{d}x(\theta) = -y(\theta).\mathbf{d}\theta \qquad \text{or} \qquad \mathbf{d(cos}(\theta)) = -\mathbf{sin}(\theta).\mathbf{d}\theta$$

and

$$\mathbf{d}y(\theta) = x(\theta).\mathbf{d}\theta \qquad \text{or} \qquad \mathbf{d(sin}(\theta)) = \mathbf{cos}(\theta).\mathbf{d}\theta$$

and as we will see in Section 4.6, they can be expressed very naturally by a special form of the **exponential function**. In this way, the **parametric** approach forms the foundation of the subject of Signal Analysis which we shall look at in more depth in Chapter 8.

I also give a fuller treatment of this interpretation of the **sine** and **cosine** in Section 4.9.

4.5 Complex Numbers

Any point in the *xy* plane can be defined by the combination of a point **θ** on the **unit circle** – our **angles** – and a distance from the **origin** r as shown in *Figure 4.5.1*. Such a point $(r, \boldsymbol{\theta})$ is then defined by:

$$x = r.\theta_x \quad y = r.\theta_y.$$

The *Figure* shows four such points in the plane, (r_1, θ_1), (r_2, θ_2), (r_3, θ_3) and (r_4, θ_4).

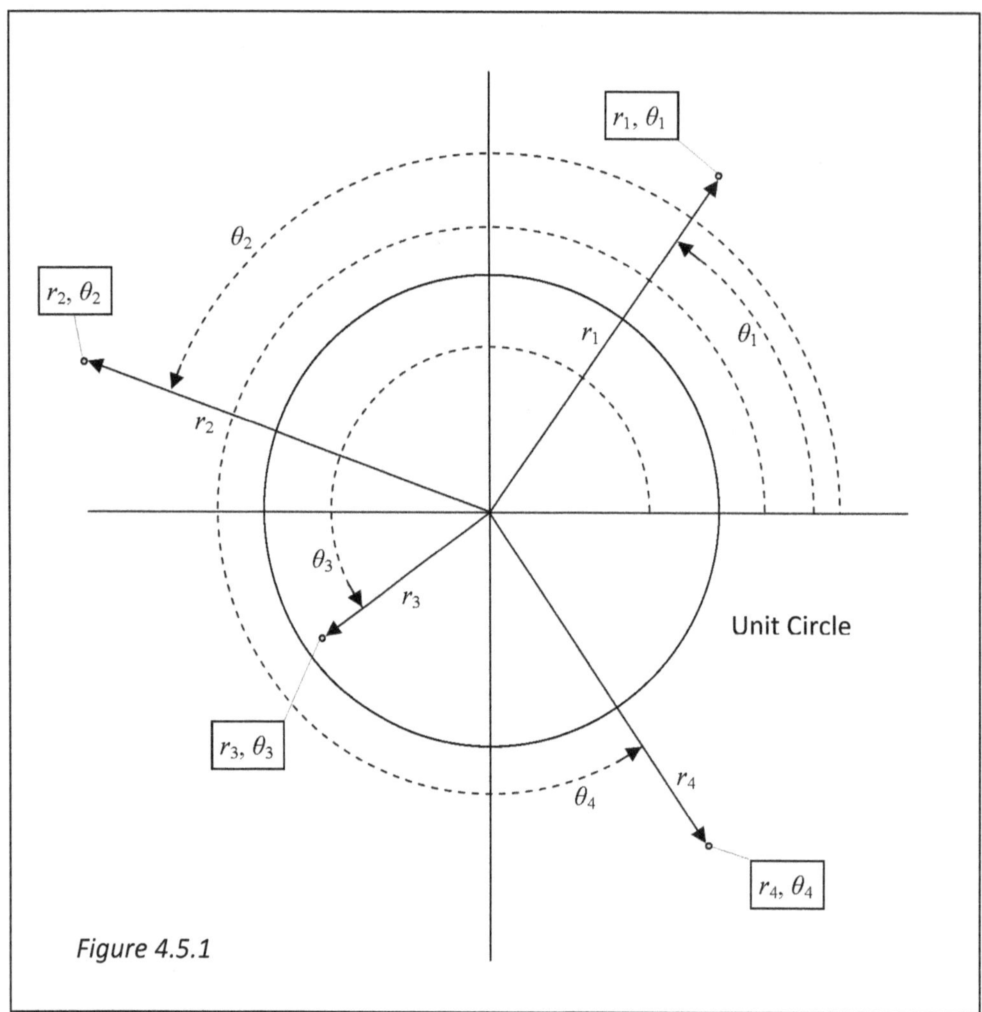

Figure 4.5.1

These two-dimensional numbers are called **complex numbers**. Their addition is defined component by component as:

$$(x, y) + (u, v) = (x + u, y + v)$$

but for multiplication, we use our **angle** concept to provide the definition:

$$(r, \boldsymbol{\theta}) \times (p, \boldsymbol{\phi}) = (r.p, \boldsymbol{\theta} + \boldsymbol{\phi})$$

which in (x, y) form will give:

$$(r, \boldsymbol{\theta}) \times (p, \boldsymbol{\phi}) = (r.p, \boldsymbol{\theta} + \boldsymbol{\phi}) = (rp(\theta_x\varphi_x - \theta_y\varphi_y), rp(\theta_x\varphi_y + \theta_y\varphi_x))$$

on the assumption of the (1, 0) **zero angle** which I will indeed assume henceforth. Substituting $x = r.\theta_x$ and $y = r.\theta_y$ and setting $u = p\varphi_x$ and $v = p\varphi_y$ we can put this in the form:

$(x, y) \times (u, v) = (xu - yv, xv + yu).$

The enormous importance of these extensions to the number system cannot be exaggerated. Their immediate significance is simpler, however. Because **complex numbers**, like our **angles**, do not admit of an **ordering** obeying the Axioms III and IV given in Section 4.3 as the axioms of an **ordered field** – which is what ordinary numbers are – this means that the conventional proof that there is no square root of a negative number breaks down. I'll give that proof in full from the axioms of an **ordered field**.[50]

$0 + 0 = 0$
$a \times (0 + 0) = a \times 0$
$(a \times 0) + (a \times 0) = a \times 0$ (Distributive Law axiom)
$a \times 0 = 0$ (Subtracting $a \times 0$ from both sides)
$b + (-b) = 0$ (Axiom of **inverse** for addition)
$a \times (b + (-b)) = a \times 0 = 0$
$a \times b + a \times (-b) = 0$ (Distributive Law and the preceding)
$a \times (-b) + a \times b = 0$ (Commutative axiom for addition)
$a \times (-b) = -(a \times b)$ (Subtract $a \times b$ from both sides)
$a > b$ (Postulate)
$a + ((-a) + (-b)) > b + ((-a) + (-b))$ (Axiom III of an **ordered field**)
$(a + -a) + (-b) > b + ((-a) + (-b))$ (Associative Law Axiom for addition)
$0 + -b > b + ((-a) + (-b))$ (Definition of additive **inverse**)
$-b > b + ((-b) + (-a))$ (Commutative Law for addition)
$-b > (b + (-b)) + (-a)$ (Associative Law for addition)
$-b > 0 + -a$ (Definition of additive **inverse**)
$-b > -a$
$-c > 0$ (Postulate)
$(-c) + c > 0 + c$ (Axiom IV of an **ordered field**)
$0 > c$ (Definition of additive **inverse**)
Define $d = -c$, so $d > 0$
$ad > bd$ (From $a > b$ and Axiom IV of an **ordered field**)
$-ad < -bd$ (From proof above that $a > b$ implies $-b > -a$)
$a(-d) < b(-d)$ (From $a \times (-b) = -(a \times b)$ shown above)
$ac < bc$

This long rigmarole shows from the axioms of an **ordered field** that:

If $a > b$ and $c < 0$ then $ac < bc$, which is our old friend the **inequality trap**.

We already have from the Axiom IV of Section 4.3 *direct* that if $a > b$ and $c > 0$, then $ac > bc$.

Now put: $b = 0$ and $a = c$ so we have:

If $a > 0$ and $c > 0$ then $ac = a^2 > 0.c = 0$.

Now we also have, putting $a = 0$ and $b = c$,

---

[50] C.B.Allendoerfer and C.O.Oakley, *Principles of Mathematics*, McGraw-Hill 1963

If $0 > b$, so $b < 0$, and $c < 0$ then $0.c = 0 < bc = b^2$ or *again* $b^2 > 0$.

So *either way*, if $a \neq 0$, $a^2 > 0$.

Because **complex numbers** don't obey the **ordering axioms**, they're not limited in this way. We can see what −1 will be in the complex plane: it is just (−1, 0) because under the definition of **complex addition**

$$(x, y) + (u, v) = (x + u, y + v)$$

we get

$$(-1, 0) + (1, 0) = (0, 0)$$

and this *is* now the **zero** (the Identity Element in Section 1.5) of **complex addition**. The **zero angle** now becomes the **unit** (again the Identity Element in Section 1.5) of **complex multiplication**, since this takes its cue from our *addition* of **angles** and so will have:

$$(x, y) \times (u, v) = (xu - yv, xv + yu)$$
$$(x, y) \times (1, 0) = (x.1 - y.0, x.0 + y.1) = (x, y)$$
$$(1, 0) \times (u, v) = (1.u - 0.v, 1.v + 0.u) = (u, v)$$

as we expect of the **unit** of multiplication. So now (0, 0) ≡ **zero** ≡ 0 and (1, 0) ≡ **unit** ≡ 1. The **additive inverse** of (1, 0) is given by the need to satisfy:

$$(x, y) + (1, 0) = (0, 0)$$

so that $x + 1 = 0$ or $x = -1$, and $y + 0 = 0$ or $y = 0$. Therefore the **additive inverse** of (1, 0) is (−1, 0) and *this* now corresponds to −1. Now look at:

$$(0, 1) \times (0, 1) = (0.0 - 1.1, 0.1 - 1.0) = (-1, 0).$$

So (0, 1) is now the *square root* of (−1, 0) ≡ −1.

In a **complex number**, the part corresponding to the **cosine** of the **angle** is called the **real part** and the part corresponding to the **sine** is called the **imaginary part**. A **complex number** of the form $(0, y)$ with the **real part** zero is called a **pure imaginary**, and it is usual to regard the numbers with zero **imaginary part** and so of the form $(x, 0)$ – sometimes called **pure reals** – as corresponding to the ordinary numbers we've been using up until now, so the **complex numbers** can then be seen as an *extension* of our ordinary **real** numbers from the **real line** – the x axis – into the *plane*.

There's a very easy way of handling **complex numbers**. Just write (x, y) as $x + j.y$ where $j = \sqrt{(-1)}$, so that $j^2 = -1$.[51] Then

---

[51] I use "*j*" as I come from an electronics background where *j* is chosen to avoid conflict with *i* for current. But in maths, *i* is generally used for √−1. I'll stick to *j* because it stands out better, and this damn' word processor won't keep capitalising it.

$$(x + jy) \times (u + jv) = xu + jyu + jxv + j^2yv = xu - yv + j(yu + xv)$$

and our product comes out automatically.

Since **complex multiplication** corresponds to **addition of angles**, so we would expect **complex division** to correspond to **subtraction of angles**, and so it does. So to *divide* by $x + jy$, we *multiply* by $x - jy$ because, if you recall, changing the sign of the **sine** gives the **negative angle**, which will now give the **divisor**. There's just one refinement: because all our **angles** were on the **unit circle**, they all in effect had $r^2 = x^2 + y^2 = 1$, where r is the **magnitude** of the **complex number** $x + jy$, so we need to divide by this to get the correct division formula. So:

$$(x, y) \div (u, v) = (x, y) \times (u, -v)/(u^2 + v^2) = (xu + yv, yu - xv)/(u^2 + v^2)$$

or

$$(x + jy) \div (u + jv) = (x + j\,y) \times (u - jv)/(u^2 + v^2) = (xu + yv + j(\,yu - xv))/(u^2 + v^2)$$

It is easy to verify that $(u - jv)/(u^2 + v^2)$ is the **multiplicative inverse** of $(u + jv)$: just multiply the two together:

$$(u + jv) \times (u - jv)/(u^2 + v^2) = (u^2 + v^2 + j(vu - uv))/(u^2 + v^2)$$

$$= (u^2 + v^2)/(u^2 + v^2) = 1$$

Complex numbers are often written with a single symbol $z = x + jy$. The **complex number** corresponding to the angular **negative** is called the **complex conjugate** and is often written either as z^* or as z with a bar over it, which I cannot do on this word processor. So $z^* = (x + jy)^* = x - jy$, and if we write the squared **magnitude** $x^2 + y^2$ as $|z|^2$ this gives the **multiplicative inverse** of z as $z^*/|z|^2$.

4.6 Calculus of Complex Numbers

In Section 4.4 I showed that the θ and $-\theta$ **parametric variables** for the **unit circle** need not be introduced as **arc lengths**. It is easier to bring them in as obeying a simple **derivative equation** such as:

$$\mathbf{d}x = -y.\mathbf{d}\theta \qquad \mathbf{d}y = x.\mathbf{d}\theta$$

Now form the **complex number** $x + jy$ and apply the **differential operator**. We get:

$$\mathbf{d}(x + jy) = \mathbf{d}x + j\,\mathbf{d}y = -y.\mathbf{d}\theta + jx.\mathbf{d}\theta = (jx - y).\mathbf{d}\theta$$

But $j.(x + jy) = (jx + j^2y) = (jx - y)$. So putting in the **parametric** dependence explicitly,

$$\mathbf{d}(x(\theta) + jy(\theta)) = (jx(\theta) - y(\theta)).\mathbf{d}\theta = j.(x(\theta) + jy(\theta))\,.\mathbf{d}\theta$$

or writing $x(\theta) + jy(\theta)$ as $z(\theta)$,

$$\mathbf{d}z(\theta) = j.z(\theta).\mathbf{d}\theta$$

But we know this form:

$$\mathbf{d}(\uparrow(ax)) = \mathbf{d}(\mathbf{e}^{ax}) = \mathbf{e}^{ax}.\mathbf{d}(ax) = a.\mathbf{e}^{ax}.\mathbf{d}x.$$

So the general form of an **original** equation whose **derivative** obeys:

$$\mathbf{d}f(x) = a.f(x).\mathbf{d}x$$

is $\quad f(x) = \uparrow(ax) + c = \mathbf{e}^{ax} + c$

remembering the **constant of integration**. So here we must have:

$$z(\theta) = \uparrow(j\theta) + c = \mathbf{e}^{j\theta} + c$$

but for $\theta = 0$, $z(\theta) = (1, 0) = 1 + j.0 = \mathbf{e}^{j0} + c = \mathbf{e}^{0} + c = 1 + c$, therefore

$$c = j.0 = 0$$

so

$$z(\theta) = x(\theta) + jy(\theta) = \uparrow(j\theta) = \mathbf{e}^{j\theta}$$

a hugely important result known as **Euler's formula**. So any point on the **unit circle** can be written as $\mathbf{e}^{j\theta}$ which means we can write any **complex number** as $r\mathbf{e}^{j\theta}$ and so we can write, for example, the multiplication rule as:

$$r\mathbf{e}^{j\theta} \times p\mathbf{e}^{j\varphi} = rp.\mathbf{e}^{j(\theta+\varphi)} = rp.(x(\theta+\varphi) + jy(\theta+\varphi)) = rp.(\theta_x\varphi_x - \theta_y\varphi_y + j(\theta_x\varphi_y + \theta_y\varphi_x))$$

so all our formulae give the same results.

There is a negative form of **Euler's formula** as well, obtained by **differentiating** the **complex conjugate** of $z(\theta)$:

$$\mathbf{d}(z^*(\theta)) = \mathbf{d}(x(\theta) - jy(\theta)) = (-jx(\theta) - y(\theta)).\mathbf{d}\theta = -j.(x(\theta) - jy(\theta)).\mathbf{d}\theta$$

or $\quad \mathbf{d}(z^*(\theta)) = -j.z^*(\theta).\mathbf{d}\theta$

leading to:

$$z^*(\theta) = \uparrow(-j\theta) = \mathbf{e}^{-j\theta}$$

The real significance of **Euler's formula**, however, comes in the analysis of **LDE**'s where the solution to the **characteristic equation** is a **complex number**. I turn to this in the next section.

4.7 LDE's and the Euler Formula

Go back to the **LDE** of Section 3.5, which we can express as an equation for a function $y(t)$:

$$a\partial_t^2 y + b\partial_t y + cy = 0$$

where on substituting $y = \uparrow\lambda t = \mathbf{e}^{\lambda t}$ we obtained:

$$(a\lambda^2 + b\lambda + c)\mathbf{e}^{\lambda t} = 0$$

with the **characteristic equation** $a\lambda^2 + b\lambda + c = 0$, with root(s):

$$\lambda = (-b \pm \sqrt{}(b^2 - 4ac)) / 2a$$

In Section 3.5, I demonstrated this for an electrical circuit equation which gave this result with:

$$a = 1, b = R/L, \ c = 1/LC.$$

Now suppose that $b^2 < 4ac$ so the **discriminant** is *negative*. Until now, that was an insurmountable problem. But with **complex numbers**, this *negative number will have a square root*. For we can now write $(j\sqrt{}|x|)^2 = j^2|x| = -|x|$, where I use the notation $|x|$ to indicate the **absolute value** of the number x, which is "the number without its sign", so $|x| = x$ if $x > 0$, and $|x| = -x$ if $x < 0$.

This means that we can write:

$$\lambda = (-b \pm j\sqrt{}|(b^2 - 4ac)|)/2a \qquad \text{when } b^2 - 4ac < 0\text{, and we still have a solution.}$$

Let's write this as $\lambda = (\alpha \pm j\omega)$ and putting this into $\mathbf{e}^{\lambda t}$ as $\mathbf{e}^{(\alpha + j\omega)t}$ we have a solution of a kind:

$$y(t) = \mathbf{e}^{\alpha t}.\mathbf{e}^{j\omega t} = \mathbf{e}^{\alpha t}.(\boldsymbol{x}(\omega t) + j\boldsymbol{y}(\omega t)) = \mathbf{e}^{\alpha t}.(\mathbf{cos}(\omega t) + j\mathbf{sin}(\omega t))$$
$$y(t) = \mathbf{e}^{\alpha t}.\mathbf{e}^{-j\omega t} = \mathbf{e}^{\alpha t}.(\boldsymbol{x}(\omega t) - j\boldsymbol{y}(\omega t)) = \mathbf{e}^{\alpha t}.(\mathbf{cos}(\omega t) - j\mathbf{sin}(\omega t))$$

where I give in and introduce the conventional notation **cos**() and **sin**() for the ***x***() and ***y***() functions giving the **cosine** and **sine** co-ordinates of points on the **unit circle** defined by the **arc length parametric variable**. I also use the **complex conjugate** form of **Euler's formula** as:

$$\mathbf{e}^{-j\theta} = \mathbf{cos}(\theta) - j.\mathbf{sin}(\theta).$$

What meaning can we give to a solution in terms of a new kind of number, a **complex number**? Firstly, $\uparrow(j\theta)$ or $\mathbf{e}^{j\theta}$ is surprisingly easy to evaluate. $\mathbf{e}^x$ has a trivial **Taylor expansion** because $\partial_x(\mathbf{e}^x) = \mathbf{e}^x$. Expanding about $a = 0$:

$$\mathbf{e}^x = \mathbf{e}^0 + \partial_x(\mathbf{e}^x)(0).(x - 0) + \tfrac{1}{2}\partial_x^2(\mathbf{e}^x)(0).(x - 0)^2 + (1/3!)\,\partial_x^3(\mathbf{e}^x)(0).(x - 0)^3 + \ldots$$

but because $\partial_x(\mathbf{e}^x) = \mathbf{e}^x$, we have $\partial_x{}^n(\mathbf{e}^x)(0) = \mathbf{e}^x(0) = \uparrow(0) = 1$ for any n, so the expansion becomes trivially:

$$\mathbf{e}^x = 1 + x + x^2/2 + x^3/3! + x^4/4! + x^5/5! + x^6/6! + \ldots$$

and for $x = j\theta$, we get the **cosine** part of $\mathbf{e}^{j\theta}$, which must be the **real** part, coming out as all the *squared terms* because $j^{2n} = (-1)^n = \pm 1$, and so:

$$\mathbf{cos}(\theta) = 1 - \theta^2/2 + \theta^4/4! - \theta^6/6! + \theta^8/8! - \theta^{10}/10! + ..$$

and the **sine** part coming out as all the *odd-powered* terms because $j^{2n+1} = (-1)^n.j = \pm j$, so:

$$\mathbf{sin}(\theta) = \theta - \theta^3/3! + \theta^5/5! - \theta^7/7! + \theta^9/9! - \theta^{11}/11! + \ldots$$

and it can be shown that both these **series** are **convergent** – successive terms diminish sufficiently fast that the sum tends to a definite **limit**.

To obtain a meaningful solution when λ in the **characteristic equation** is **complex**, we simply use **complex** values for the arbitrary constants A and B introduced in section 3.5, and fit the **real** parts of the resulting equation to the initial conditions. So in the example above, we take the two **complex conjugate solutions** together as:

$$\begin{aligned} y(t) &= A.\mathbf{e}^{\alpha t}.\mathbf{e}^{j\omega t} + B.\mathbf{e}^{\alpha t}.\mathbf{e}^{-j\omega t} \\ &= A.\mathbf{e}^{\alpha t}.(\mathbf{x}(\omega t) + j\mathbf{y}(\omega t)) + B.\mathbf{e}^{\alpha t}.(\mathbf{x}(\omega t) - j\mathbf{y}(\omega t)) \\ &= A.\mathbf{e}^{\alpha t}.(\mathbf{cos}(\omega t) + j\mathbf{sin}(\omega t)) + B.\mathbf{e}^{\alpha t}.(\mathbf{cos}(\omega t) - j\mathbf{sin}(\omega t)) \\ &= (A_x + jA_y).\mathbf{e}^{\alpha t}.(\mathbf{cos}(\omega t) + j\mathbf{sin}(\omega t)) + (B_x + jB_y).\mathbf{e}^{\alpha t}.(\mathbf{cos}(\omega t) - j\mathbf{sin}(\omega t)) \end{aligned}$$

with A and B **complex numbers**, and fit the **real** part of the solution, which will be:

$$\begin{aligned} &A_x.\mathbf{e}^{\alpha t}.\mathbf{cos}(\omega t) - A_y.\mathbf{e}^{\alpha t}.\mathbf{sin}(\omega t) + B_x.\mathbf{e}^{\alpha t}.\mathbf{cos}(\omega t) + B_y.\mathbf{e}^{\alpha t}.\mathbf{sin}(\omega t) \\ &= \mathbf{e}^{\alpha t}.((A_x + B_x).\mathbf{cos}(\omega t) - (A_y - B_y).\mathbf{sin}(\omega t)) \end{aligned}$$

to the given initial conditions.

This may not explain the philosophical problem of how we have obtained an answer in a different system of numbers from the one in which we started, but it does provide a prescription for obtaining an answer, and indeed one translated back into our original numerical system.

4.8 Further Examples of Angle Algebra

In this Section, I'll switch freely between our two interpretations of **angles**, keeping in mind that when we interpret a *sum* of two **angles** in terms of **complex numbers**, we switch to *multiplying* the numbers, and when we subtract, we *divide*, which for **angles** having **unit magnitude** ($\theta_x^2 + \theta_y^2 = 1$) corresponds to multiplication by the **complex conjugate**.

So:

$$\boldsymbol{\theta} + \boldsymbol{\phi} = (\theta_x + j\theta_y) \times (\varphi_x + j\varphi_y)$$

and

$$\boldsymbol{\theta} - \boldsymbol{\phi} = (\theta_x + j\theta_y) \times (\varphi_x - j\varphi_y).$$

Because the **complex plane** has its own **origin** (0, 0), I will amend my earlier notation. I will now write the **zero angle** (1, 0) as **Φ**. So now as **angles, Φ** = (1, 0) = 1 and **j** = (0, 1) = 0 + 1.*j* = *j*. The **angle** (−1, 0) we can provisionally write as **Ψ**, and (0, −1) as **−j**.[52]

Two **angles** are **complementary** if they add to **j**: **θ** + **ϕ** = **j**, when **ϕ** = **j** − **θ** = **j** + **θ***, where **θ*** denotes the **angle** corresponding to the **complex conjugate** $\theta_x - j\theta_y$. As stressed above, since for any **angle,** $\theta_x^2 + \theta_y^2 = 1$, we don't need to worry about the **magnitude** factor here as we did in Section 4.5. For pure **angles,** $\div (\theta_x + j\theta_y) \longrightarrow \times (\theta_x - j\theta_y)$ always. So, switching into **complex numbers** we get:

$$\boldsymbol{\phi} = j \div (\theta_x + j\theta_y) = j \times (\theta_x - j\theta_y) = \theta_y + j\theta_x.$$

So the **complementary** angle is obtained by swapping the **cos** part and the **sin** part. This is usually written something like:

$$\mathbf{cos}(90^o - \theta) = \mathbf{sin}(\theta)$$
$$\mathbf{sin}(90^o - \theta) = \mathbf{cos}(\theta).$$

Two **angles** are **supplementary** if they add to **Ψ**, so **θ** + **ϕ** = **Ψ**. so now:

$$\boldsymbol{\phi} = \boldsymbol{\Psi} - \boldsymbol{\theta} = (-1, 0) \div (\theta_x + j\theta_y) = -1 \times (\theta_x - j\theta_y) = -\theta_x + j\theta_y.$$

Here the **cosine** has changed sign to give the **angle supplementary** to **θ**. So in particular, if **θ** = **j** = 0 + *j*, its **supplementary angle** is also 0 + *j*. In other words,

$$\mathbf{j} + \mathbf{j} = 2\mathbf{j} = \boldsymbol{\Psi}.$$

This gives us an easier notation for **Ψ**: call it 2**j**. Observe also that **Ψ** + **Ψ** = (−1 + *j*.0) × (−1 + *j*.0) = −1 × −1 = 1 = **Φ**. So too half-circles return the **zero angle** or complete the circle.

Look at the triangle in *figure 4.8.1(a)*. Here the **external angles** $\boldsymbol{\theta}_{e1} + \boldsymbol{\theta}_{e2} + \boldsymbol{\theta}_{e3}$ add to **Φ** as is easily seen if we shrink the triangle itself as in *Figure 4.8.1(b)*. The three **angles** clearly add up to the complete circle, and so their sum returns to **Φ**.

---

[52] We can't write **Ψ** as **−Φ** because under our definition of subtraction between **angles**, only the **sine** changes sign in the **negative** of an **angle**.

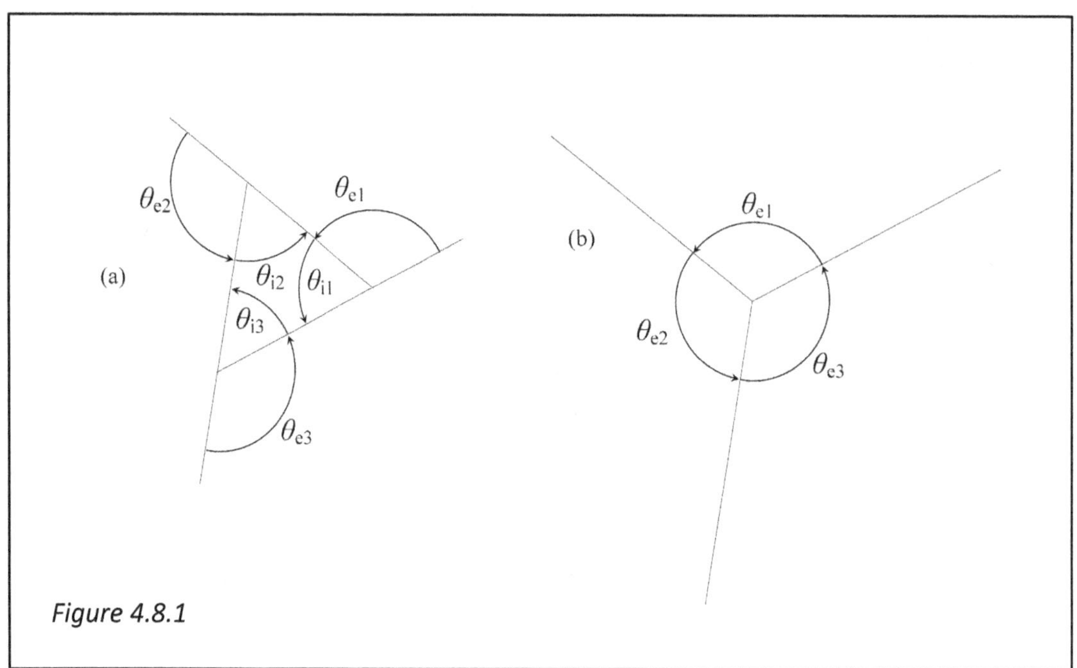

Figure 4.8.1

So $\quad \boldsymbol{\theta}_{e1} + \boldsymbol{\theta}_{e2} + \boldsymbol{\theta}_{e3} = \boldsymbol{\Phi}$.

But at each point of the triangle, the **external** and **internal angles** are **supplementary**:

$$\boldsymbol{\theta}_{e1} + \boldsymbol{\theta}_{i1} = 2\mathbf{j}, \qquad \boldsymbol{\theta}_{e2} + \boldsymbol{\theta}_{i2} = 2\mathbf{j}, \qquad \boldsymbol{\theta}_{e3} + \boldsymbol{\theta}_{i3} = 2\mathbf{j}.$$

Now add all these:

$$\boldsymbol{\theta}_{e1} + \boldsymbol{\theta}_{i1} + \boldsymbol{\theta}_{e2} + \boldsymbol{\theta}_{i2} + \boldsymbol{\theta}_{e3} + \boldsymbol{\theta}_{i3} = 2\mathbf{j} + 2\mathbf{j} + 2\mathbf{j}.$$

But $\boldsymbol{\theta}_{e1} + \boldsymbol{\theta}_{e2} + \boldsymbol{\theta}_{e3} = \boldsymbol{\Phi}$, and $2\mathbf{j} + 2\mathbf{j} = \boldsymbol{\Phi}$, so we get:

$$\boldsymbol{\Phi} + \boldsymbol{\theta}_{i1} + \boldsymbol{\theta}_{i2} + \boldsymbol{\theta}_{i3} = \boldsymbol{\Phi} + 2\mathbf{j},$$

or $\quad \boldsymbol{\theta}_{i1} + \boldsymbol{\theta}_{i2} + \boldsymbol{\theta}_{i3} = 2\mathbf{j}$.

This is the familiar result that the internal angles of a triangle add to the half-circle or 180°.

In computer graphics, this way of working with angles cuts a lot of corners, but you do need to be careful.

As another simple example, let's show that the **sin** of 30° is ½. 30° is one third of 90°, which I am writing as **j**. So we want the **angle θ** such that **θ** + **θ** + **θ** = **j**. Switching into **complex algebra**, multiplication replaces addition, so we want the angle $\theta_x + j\theta_y$ such that:

$$(\theta_x + j\theta_y) \times (\theta_x + j\theta_y) \times (\theta_x + j\theta_y) = j.$$

Multiplying out the left-hand side (LHS), we get:

$$(\theta_x^2 - \theta_y^2 + 2j.\theta_y\theta_x) \times (\theta_x + j\theta_y) = \theta_x(\theta_x^2 - \theta_y^2) - 2\theta_y^2\theta_x + 2j\theta_y\theta_x^2 + j\theta_y(\theta_x^2 - \theta_y^2) = j.$$

So, equating the **real** parts on both sides:

$$\theta_x^3 - \theta_x\theta_y^2 - 2\theta_y^2\theta_x = 0 \quad \text{or} \qquad (\text{dividing by } \theta_x)\text{: } \theta_x^2 = 3.\theta_y^2.$$

But $\theta_x^2 + \theta_y^2 = 1$, so subtracting $\theta_x^2 = 3.\theta_y^2$ from this,

$$\theta_x^2 + \theta_y^2 - \theta_x^2 = 1 - 3\theta_y^2 \quad \text{or} \qquad 4\theta_y^2 = 1, \qquad \text{so } \theta_y = ½, \qquad \text{QED!}$$

Since $\theta_x^2 = 3\theta_y^2$, this must equal ¾, so we also get $\theta_x = \sqrt{¾}$.

4.9 Hours and ε Notation

Hours can also be used as measures of **angle**, not just as measures of time. In astronomy, they are used as the standard measure corresponding to **longitude**, when they are known as **right ascension**. Astronomers measure off the night sky (well, it's the day sky as well but we can only see the objects in it at night) using **hour angles**[53] along an imaginary circle in space lying directly above the Earth's equator, and indeed this circle is actually called the **celestial equator**. Twenty-four **hours** are marked off along it going *left* or *anticlockwise* (so to the East) as seen from the Northern hemisphere, but obviously this is running to the *right* or *clockwise* (but still to the East) if you're looking from Antarctica, but then not many of us are.

Because we can't really use the position of the sun to set the zero point on this circle as the sun wanders all the way round it itself in the course of a year, a point is chosen in the constellation of Aries instead. As the sun is not itself always overhead on our equator, but wanders to the North of it in the Northern summer, and to the South of it in the Northern winter, it clearly doesn't actually pass along the **celestial equator**. Instead the sun runs along another Great Circle in the heavens inclined to the **celestial equator** at 23½° (i.e. that's the maximum separation of the two Great Circles) and called the **ecliptic**, because all eclipses inevitably lie on this circle, involving as they do the sun.

The **ecliptic**, seen as a great circle, crosses the **celestial equator** at the **spring equinox** (about 22nd of March) and the **autumn equinox** (around the 22nd of September) and this gives us our reference or zero point: the "first point of Aries" is where the two Great Circles cross in the **spring equinox**.

Figure 4.9.1 shows the general pattern, where now the **ecliptic** and the **celestial equator** are described as the *planes* in which they lie. The Earth is the round object in the middle!

---

[53] The term "**hour angles**" also has a specific technical meaning which is *not* the sense in which I'm using it here!

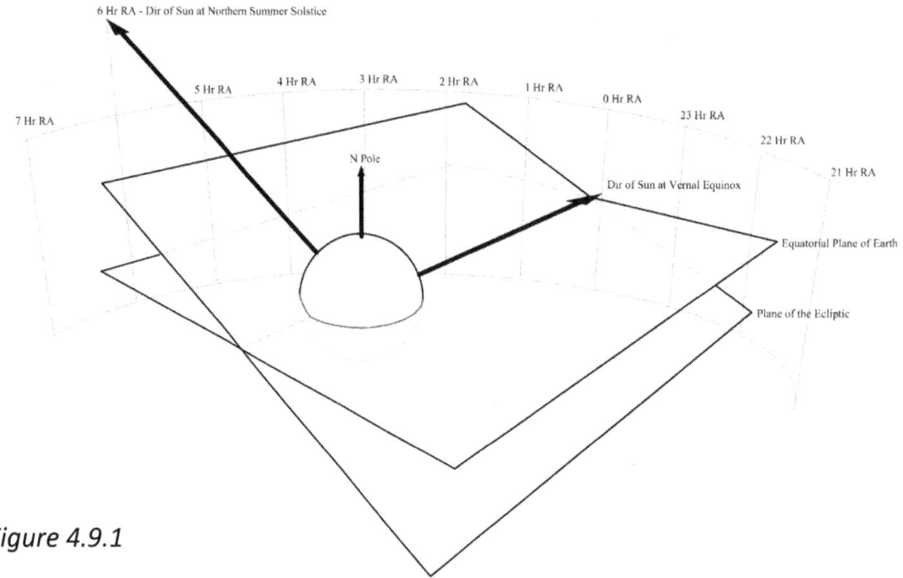

Figure 4.9.1

Because twenty-four **hours** correspond to 360°, one **hour** of **angle** corresponds to 15°. So, put another way, one degree of **angle** corresponds to four *minutes* of clock time, not to be confused with the idea that one degree of **angle** can also be equated to sixty *minutes* of **angle**, written as 60'. Each minute of **angle** is further divided into sixty seconds (60") or, originally, sixty "*second minutes*". So a minute of clock time corresponds to *fifteen* minutes of **angle**.

There is another plane that defines clock time: this is the plane containing the polar axis of the Earth and passing through (the centre of) the Sun. In the diagram this would appear as a "vertical" plane, rotating around the **right ascension** clock circle once a year. If you happen to live on or near the Greenwich **meridian**, the "zero" of **longitude** (as I do near Cambridge), you will pass through this "clock-zero" plane at almost exactly midday and midnight every twenty-four hours by *winter time* (which I'll therefore also call "true" time). But this position *won't* correspond to the zero of **right ascension**, *except at the* **spring** and **autumn equinoxes**. At the **spring** or **vernal equinox**, this plane coincides with that defined by the two arrows on the diagram labelled "N Pole" and "Dir of Sun at Vernal Equinox", and the line of intersection of all three planes is the same. So at midday at the **spring equinox** you can imagine **hours of right ascension** starting from the sun and running round to the East, 15° for each one of them, and, if you could see anything at all with the sun in your eyes, you would on this day at midday see the constellations as astronomers see them.

Summer time is a somewhat counter-intuitive concept because although we can set an equation:

$S = W + 1$ (**modulo** 12 or 24),[54]

[54] **modulo** arithmetic works in remainders; so 11 + 3 **modulo** 12 is actually the **remainder** of 14 divided by 12. All clock times are strictly **modulo** 12 or **modulo** 24.

suggesting that the time has moved around to *the right* as seen from the Northern hemisphere, our picture of the heavens has actually moved as the **inverse** to this, moving around to the *left*. So at midday *summer-time*, we're still an hour (or 15°) away from passing through the "true clock-zero" plane that passes through the Sun. You don't pass through that until 1 p.m. (and again of course at 1 a.m.). In summer we work with a "zero" 15° *right* of true North.[55]

That's why, at the summer midday, we still have an hour more of daylight to go before we do at the winter midday.

All this analysis falls apart if you live in Reykjavik, or indeed anywhere substantially removed from the defining meridian of your time zone. Reykjavik is 22° West, but because there's no other inhabited land[56] along that longitude, the Icelandics elect to use GMT. But at 22° West, the sun is still an hour and a half away in the East when it's midday in Reykjavik. So the Icelanders are on a kind of permanent summer time, with their "clock zero" plane set to 22° West. Not entirely surprisingly, they don't have summer time in Iceland.

I've expanded this question because it reveals a lot about the difficulties of human psychology in the *visualization* of mathematical ideas. Mathematically, the shift between our winter and summer clock conventions is laughably trivial, but it still causes quite a lot of difficulties in our thinking (is the sun at 5 o'clock summer time East or West of its position at 5 o'clock winter time?). Often the main difficulty people have in learning mathematical ideas is with such questions of visualization, and different people may have different ways of visualizing the same concept. It's to try to open up different perspectives on the same ideas that has been a primary motivation in developing this book.

So having used the pretty notion of **hour angles** as an introduction, I'll suggest a notation for a **single-parameter** or **parametric** measure of **angle** that is much more intuitive than the **arc length** notion. Unfortunately, I'll still have to stick with the *anticlockwise from East* approach or my readers will never be able to switch between this and conventional notation!

This is the question. Could we get a better measure of **angle** than **radians**, or even than degrees? How many degrees is 5/8th of the way round a circle? How many **radians**? We have to think a bit to figure out 225° or 3.9269908169872415480783042290994 (1.25π)! It wouldn't be quite so bad if π had been defined as the ratio of the circumference of a circle to its *radius*, for then we'd have 180° = ½π, which we could just speak as "half a circle".

It becomes particularly awkward when we get into **Fourier analysis** in Chapter 8, which builds on the theory of **periodic functions** with the property that $\phi(x) = \phi(x + T)$ for some **period** T. Obviously we'd

[55] Strictly speaking, there's a further subtlety caused by the fact that the Earth's orbit around the sun is not a true circle but an *ellipse*. This means that the actual time when the sun crosses the due North or due South **meridian** varies slightly from 12 midnight or 12 midday, *even by "true", or winter, time*. This phenomenon goes by the beautiful name of the **Equation of Time**.

[56] Part of northern Greenland is but that's not just uninhabited but seriously uninhabitable! The Cape Verde islands are 25° West. There's always Antarctica, of course.

like to develop the theory with a basic **period** of 1, but the **arc length** concept behind **radians** traps us into using a fundamental **period** of the bizarre value of 2π.[57]

So let's go the whole hog and define the obvious **parameter** for circular measure of, simply, 1. Then we'll use a function $\varepsilon:[0, 1] \rightarrow \boldsymbol{C}$ that takes values in the **unit interval** [0, 1] and maps them into the **unit circle *C***.

So we'll have:

$$\varepsilon(0) = 1 = 1 + j.0$$
$$\varepsilon(\tfrac{1}{4}) = 0 + j.1 = j.$$
$$\varepsilon(\tfrac{1}{2}) = -1$$
$$\varepsilon(\tfrac{3}{4}) = -j,$$

and

$$\varepsilon(1) = 1 = \varepsilon(0) = \varepsilon(0 + 1),$$

and indeed for any whole number or **integer** n, and any θ,

$$\varepsilon(n) = 1, \qquad \text{and} \qquad \varepsilon(\theta + n) = \varepsilon(\theta).$$

Using the **Euler formula** we can see that:

$$\varepsilon(\theta) = \mathbf{e}^{2\pi j\theta} = \mathbf{e}\uparrow(2\pi j\theta),$$

so you can think of $\varepsilon(\cdot) = \mathbf{e}\uparrow(2\pi j(\cdot))$. Note that as written here θ runs through the **unit circle** as its value goes from 0 to 1, and for now I will use θ interchangeably to mean a **parameter** that runs through ***C*** in going from 0 to 1, and as a **radian** measure, going around ***C*** as it runs from 0 to 2π. The context will usually make it clear which form I'm using, in particular the use of the ε() notation itself, but where there may be confusion I'll use θ for the [0, 2π] parameter and τ for the [0, 1] one.

> This use of a **parameter** that maps around the entire **unit circle** as its argument goes from 0 to 1, I will call the **unit angle parameterization**. The more conventional **parameterization** I'll call the **radian parameterization**.

Since ε() takes a **real** argument running from 0 to 1, and maps into a **complex** result, it can be split into two **real components**, so that we can write:

$$\varepsilon(\theta) = \varepsilon_x(\theta) + j.\varepsilon_y(\theta) = \mathbf{cos}(2\pi\theta) + j.\mathbf{sin}(2\pi\theta),$$

so again we have:

$$\varepsilon_x(\theta) = \mathbf{cos}(2\pi\theta)$$
$$\varepsilon_y(\theta) = \mathbf{sin}(2\pi\theta).$$

---

[57] It was in writing Chapter 8 that I started using the notation I'm about to introduce here; it removes a lot of 2π's!

If we're working in three-dimensional space and we're looking at a **rotation** in the *yz*-plane, the use of $\varepsilon_x(\theta)$ and $\varepsilon_y(\theta)$ might be a little confusing, so I would recommend the alternatives:

$$\varepsilon(\theta) = \varepsilon_1(\theta) + j.\varepsilon_2(\theta) = \varepsilon_c(\theta) + j.\varepsilon_s(\theta),$$

the last obviously by analogy with **cos**() and **sin**(), but remember that now, for example,

$$\varepsilon_c(\tfrac{1}{4}) = \mathbf{cos}(\pi/2) = 0$$

and

$$\varepsilon_s(\tfrac{1}{4}) = \mathbf{sin}(\pi/2) = 1,$$

so these are not mere synonyms for **cos**() and **sin**().

Now above we had:

$$\mathbf{d}x = -y.\mathbf{d}\theta \qquad \mathbf{d}y = x.\mathbf{d}\theta$$

using the **radian parameter**. Now we'll have:

$$\mathbf{d}\varepsilon_x(\theta) = \mathbf{d}(\mathbf{cos}(2\pi\theta)) = -\mathbf{sin}(2\pi\theta).\mathbf{d}(2\pi\theta) = -2\pi.\mathbf{sin}(2\pi\theta).\mathbf{d}\theta = -2\pi.\varepsilon_y(\theta).\mathbf{d}\theta,$$

and similarly

$$\mathbf{d}\varepsilon_y(\theta) = 2\pi.\varepsilon_x(\theta).\mathbf{d}\theta.$$

So the price we pay for a simpler angular measure is a more complex calculus, but surprisingly, when we get into the **Fourier analysis**, this too becomes simpler.

Recognising that $\varepsilon(\theta) = \mathbf{e}\uparrow(2\pi j\theta)$, we can treat the **domain** of the ε function in such a way as to remove that awkward "jump" from 1 back to 0 every time we get back to the end of the circle. Let's write the value of the argument of ε() as ξ instead of θ. Now when $\xi = 1$, i.e. when it's reached the end of the [0, 1] **domain** of the argument of the ε function, its value is:

$$\varepsilon(\xi = 1) = \mathbf{e}\uparrow(2\pi j1) = \mathbf{e}\uparrow(2\pi j) = (1, 0)$$ in the **complex plane**.

But (1, 0) is the **unit** of **complex multiplication** as we saw in Section 4.5: for any **complex number** $x + jy$, corresponding to any **point** (x, y) in the **plane**,

$$(1, 0) \times (x, y) \equiv (1 + j.0)(x + jy) = (1.x - 0.y) + j(0.x + 1.y) = 1.x + j1.y = x + jy.$$

Now suppose $\xi = 1 + \alpha$, where $0 < \alpha < 1$. Then

$$\varepsilon(\xi) = \mathbf{e}\uparrow(2\pi j(1+\alpha)) = \mathbf{e}\uparrow(2\pi j1).\mathbf{e}\uparrow(2\pi j\alpha)$$

by the property $\mathbf{e}^{x+y} = \mathbf{e}^x.\mathbf{e}^y$. But $\mathbf{e}\uparrow(2\pi j1)$ and $\mathbf{e}\uparrow(2\pi j\alpha)$ are both points on the **unit circle** in the **plane**, and $\mathbf{e}\uparrow(2\pi j1) = (1, 0)$, the **unit** of **complex multiplication**, so

$\mathbf{e}\uparrow(2\pi j1).\mathbf{e}\uparrow(2\pi j\alpha) = \mathbf{e}\uparrow(2\pi j\alpha)$.

Now if $\xi = 2 + \alpha$, we get

$$\varepsilon(\xi) = \mathbf{e}\uparrow(2\pi j(2+\alpha)) = \mathbf{e}\uparrow(2\pi j2).\mathbf{e}\uparrow(2\pi j\alpha) = \mathbf{e}\uparrow(2\pi j1).\mathbf{e}\uparrow(2\pi j1).\mathbf{e}\uparrow(2\pi j\alpha),$$

again using $\mathbf{e}^{x+y} = \mathbf{e}^x.\mathbf{e}^y$, and this is

$(1, 0)(1, 0).\mathbf{e}\uparrow(2\pi j\alpha) = \mathbf{e}\uparrow(2\pi j\alpha)$ again.

It's easy to see that if $\xi = n + \alpha$, for any whole number n, if we know that if $\mathbf{e}\uparrow(2\pi jn) = (1, 0)$ then

$$\mathbf{e}\uparrow(2\pi j(n + 1)) = \mathbf{e}\uparrow(2\pi jn).\mathbf{e}\uparrow(2\pi j1) = (1, 0).(1, 0) = (1, 0)$$

so the two requirements for proof by **induction** (Section 1.10) are met and we have

$\varepsilon(n + \alpha) = \varepsilon(\alpha)$ *for any n.*

So $\varepsilon()$ can be regarded as a **periodic function** over *all* $\xi > 0$ with **period** (the **interval** over which values return to the same value) of, very conveniently, just 1.

It's easy to show that $\varepsilon(\xi)$ shows this same **periodicity** for all values of $\xi < 0$ as well.

This is much better than regarding $\varepsilon()$ as being defined simply over the **interval** [0, 1]. For now we can keep on increasing the argument ξ over ever larger and larger **real numbers** and $\varepsilon(\xi)$ just keeps rotating round and round the **unit circle** anticlockwise. Equally, we can keep on *decreasing* ξ and $\varepsilon(\xi)$ just keeps on rotating around the circle clockwise.

So $\varepsilon(\xi) := \mathbf{e}\uparrow(2\pi j\xi)$ defines a function over the entire **real line mapping** onto the **unit circle** such that arbitrarily close points on the **real line map** into arbitrary close points on the **unit circle**, a property known as **continuity**.

This outwardly trivial generalization of the **single-parameter** approach will be very useful in later Chapters.

4.10 Hyperbolic Angles

The **angles** described so far in this Chapter are not the only form of **angle** used in mathematics.

A **hyperbolic angle** is an ordered pair of **reals** (x, y) such that $x^2 - y^2 = 1$.

The only difference is that "+" has gone to "−" but these new **angles** have quite different properties.

Let's suppose that these **angles** can also be expressed **parametrically** in terms of some **real** ϕ, and write them as $\eta_x(\phi) + \eta_y(\phi)$ and look for a similar *multiplicative* algebra as we found for the **circular angles** we've used until now.[58]

We'd get:

$$(\eta_x(\phi) + \eta_y(\phi)).(\eta_x(\psi) + \eta_y(\psi)) = \eta_x(\phi)\eta_x(\psi) + \eta_x(\phi)\eta_y(\psi) + \eta_y(\phi)\eta_x(\psi) + \eta_y(\phi)\eta_y(\psi),$$

so if $\phi = \psi$, we have:

$$(\eta_x(\phi) + \eta_y(\phi))^2 = \eta_x^2(\phi) + \eta_y^2(\phi) + 2\eta_x(\phi)\eta_y(\phi).$$

But we'd also like to have:

$$\eta(\phi) + \eta(-\phi) = 1 = (1, 0),$$

assuming the same angular zero as we had before. In terms of the multiplicative behaviour of the **components** this would expand to:

$$\eta_x(\phi)\eta_x(-\phi) + \eta_x(\phi)\eta_y(-\phi) + \eta_y(\phi)\eta_x(-\phi) + \eta_y(\phi)\eta_y(-\phi).$$

But we know from the definition we should have at least:

$$\eta_x^2(\phi) - \eta_y^2(\phi) = 1.$$

So if we try:

$$\eta_x(-\phi) = \eta_x(\phi)$$
$$\eta_y(-\phi) = -\eta_y(\phi)$$

exactly by analogy with $\varepsilon_x(-\phi) = \varepsilon_x(\phi)$ and $\varepsilon_y(-\phi) = -\varepsilon_y(\phi)$ which we already know to be true from **circular angles**, we'd get:

$$\eta(\phi) + \eta(-\phi) \rightarrow \eta_x(\phi)\eta_x(-\phi) + \eta_x(\phi)\eta_y(-\phi) + \eta_y(\phi)\eta_x(-\phi) + \eta_y(\phi)\eta_y(-\phi)$$

$$= \eta_x^2(\phi) - \eta_x(\phi)\eta_y(\phi) + \eta_y(\phi)\eta_x(\phi) - \eta_y^2(\phi)$$

$$= \eta_x^2(\phi) - \eta_y^2(\phi) - \eta_x(\phi)\eta_y(\phi) + \eta_y(\phi)\eta_x(\phi) = 1.$$

But then we must have $\eta_x(\phi + \psi) = \eta_x(-\phi - \psi)$ and $\eta_y(\phi + \psi) = -\eta_y(-\phi - \psi)$, but this would be consistent with the suggestion that of the four terms giving $\eta(\phi + \psi)$,

---

[58] So introducing this term (**circular angles**) for the "ordinary" **angles** we've worked with up until now when the distinction needs to be made clear. In the expansion that follows it will turn out that we can dispense with the "**imaginary**" *j* and so I do so here straight away.

$$\eta_x(\phi + \psi) = \eta_x(\phi)\eta_x(\psi) + \eta_y(\phi)\eta_y(\psi)$$
$$\eta_y(\phi + \psi) = \eta_x(\phi)\eta_y(\psi) + \eta_y(\phi)\eta_x(\psi).$$

For if $\phi \to -\phi$ and $\psi \to -\psi$, the first would give:

$$\eta_x(-\phi - \psi) = \eta_x(-\phi)\eta_x(-\psi) + \eta_y(-\phi)\eta_y(-\psi) = \eta_x(\phi)\eta_x(\psi) + (-\eta_y(\phi))(-\eta_y(\psi))$$

$$= \eta_x(\phi)\eta_x(\psi) + \eta_y(\phi)\eta_y(\psi) = \eta_x(\phi + \psi),$$

and the second would give:

$$\eta_y(-\phi - \psi) = \eta_x(-\phi)\eta_y(-\psi) + \eta_y(-\phi)\eta_x(-\psi) = \eta_x(\phi)(-\eta_y(\psi)) + -\eta_y(\phi)\eta_x(\psi)$$

$$= -\eta_x(\phi)\eta_y(\psi) - \eta_y(\phi)\eta_x(\psi) = -\eta_y(\phi + \psi).$$

So for **hyperbolic angles** it seems reasonable that we define the **addition of angles** by the rule:

$$(x, y) \mathbf{+} (p, q) = (xp + yq, xq + yp).$$

Under this rule, $\eta_x(\phi + \psi)$ captures the **symmetric** part of the result, with

$$\eta_x(-\phi - \psi) = \eta_x(\phi + \psi)$$

and $\eta_y(\phi + \psi)$ captures the **antisymmetric** part, with:

$$\eta_y(-\phi - \psi) = -\eta_y(\phi + \psi).$$

Differentiating $\eta_x^2 - \eta_y^2 = 1$, we must get:

$$2\eta_x.\mathbf{d}\eta_x - 2\eta_y.\mathbf{d}\eta_y = 0,$$

and this suggests the *symmetrical* **calculus**:

$$\mathbf{d}\eta_x = \eta_y.\mathbf{d}\phi$$
$$\mathbf{d}\eta_y = \eta_x.\mathbf{d}\phi.$$

This, taken together with the **symmetry** and **antisymmetry** rules, suggests the identification:

$$\eta_x(\phi) = \mathbf{e}^{\phi} + \mathbf{e}^{-\phi}$$
$$\eta_y(\phi) = \mathbf{e}^{\phi} - \mathbf{e}^{-\phi},$$

as these would indeed give all our expected requirements:

$$\eta_x(-\phi) = \mathbf{e}^{-\phi} + \mathbf{e}^{\phi} = \eta_x(\phi)$$
$$\eta_y(-\phi) = \mathbf{e}^{-\phi} - \mathbf{e}^{\phi} = -\eta_y(\phi)$$

and

$$\mathbf{d}\eta_x(\phi) = (\mathbf{e}^{\phi} - \mathbf{e}^{-\phi}).\mathbf{d}\phi = \eta_y(\phi).\mathbf{d}\phi$$
$$\mathbf{d}\eta_y(\phi) = (\mathbf{e}^{\phi} - -\mathbf{e}^{-\phi}).\mathbf{d}\phi = (\mathbf{e}^{\phi} + \mathbf{e}^{-\phi}).\mathbf{d}\phi = \eta_x(\phi).\mathbf{d}\phi.$$

However, this definition actually fails our defining condition $x^2 - y^2 = 1$, as it gives:

$$\eta_x^2(\phi) - \eta_y^2(\phi) = (\mathbf{e}^{\phi} + \mathbf{e}^{-\phi})(\mathbf{e}^{\phi} + \mathbf{e}^{-\phi}) - (\mathbf{e}^{\phi} - \mathbf{e}^{-\phi})(\mathbf{e}^{\phi} - \mathbf{e}^{-\phi})$$

$$= \mathbf{e}^{2\phi} + \mathbf{e}^{0} + \mathbf{e}^{0} + \mathbf{e}^{-2\phi} - (\mathbf{e}^{2\phi} - \mathbf{e}^{0} - \mathbf{e}^{0} + \mathbf{e}^{-2\phi})$$

$$= \mathbf{e}^{2\phi} + 2 + \mathbf{e}^{-2\phi} - \mathbf{e}^{2\phi} + 2 - \mathbf{e}^{-2\phi} = 4.$$

So instead we choose the identification:

$$\eta_x(\phi) = \tfrac{1}{2}(\mathbf{e}^{\phi} + \mathbf{e}^{-\phi})$$
$$\eta_y(\phi) = \tfrac{1}{2}(\mathbf{e}^{\phi} - \mathbf{e}^{-\phi}),$$

which does fit *all* our criteria. It is now clear that this can be done without any need for **imaginaries**, as indeed I anticipated at the start of this Section.

Hyperbolic angles correspond to the x and y **coordinates** of the **unit hyperbola** $x^2 - y^2 = 1$, which is drawn below in *Figure 4.10.1*. Observe that it actually has two parts or **topological components**, according to whether $x > +1$ or $x < -1$. Only the right-hand component, for $x > 1$, is shown in the *Figure*.

Mercifully, the **parameter** ϕ no longer corresponds to the awkward concept of **arc length**, and it can be shown, really quite trivially, that it corresponds to the shaded *area* in the *Figure*.

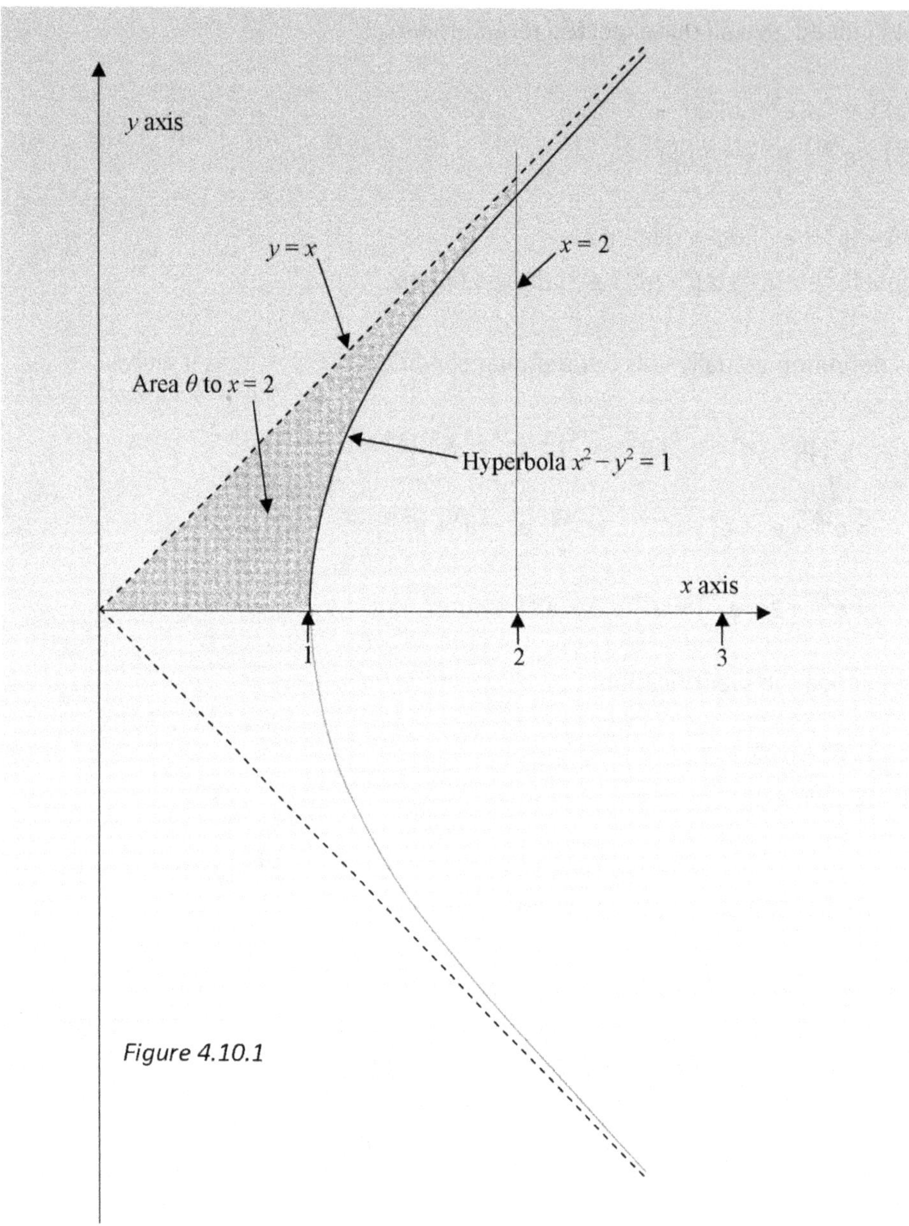

Figure 4.10.1

Also note that the arms of the **hyperbola** run off to infinity, so this geometrical object, at least in each of its **components**, is **topologically** analogous to the infinite **real line**, and indeed an **ordering relation** can easily be set up in it, simply by taking the *y* values of any two points in the *same* **component** to define the ordering:

$(x_1, y_1) < (x_2, y_2)$ if $y_1 < y_2$.

An arbitrary choice of ordering *between* the two **components** would establish an **ordering** over the whole **hyperbola**. So the proof of the non-existence of **imaginaries** discussed at length in Section 4.5 can be applied here, and that is why there are no **imaginaries** involved in **hyperbolic angles**. So there's no π either!

Hyperbolic angles are not written as $(\eta_x(\phi), \eta_y(\phi))$ in conventional textbooks. Instead a notation deriving from the ordinary notation for **circular angles** is used:

$\eta_x(\phi)$ is written as **cosh**(ϕ)
$\eta_y(\phi)$ is written as **sinh**(ϕ),

and we have the defining rules:

$\mathbf{cosh}^2(\phi) - \mathbf{sinh}^2(\phi) = 1$
$\mathbf{cosh}(\phi) = ½(\mathbf{e}^{\phi} + \mathbf{e}^{-\phi})$
$\mathbf{sinh}(\phi) = ½(\mathbf{e}^{\phi} - \mathbf{e}^{-\phi})$

with the "**h**" on the end indicating "**hyperbolic**".

I will use both notations, although the avoidance of the conventional notation is not so compelling here as we don't have those awkward $2\pi j$'s to get rid of. Also, much as suggested in the last Section, I may write:

$\eta = (\eta_c(\phi), \eta_s(\phi))$

using subscripts "c" and "s" in contexts where x and y might be confused with other uses of them.

4.X Exercises

1. Prove the **sine rule** (also known as the Law of Sines): for a triangle with sides length a, b, c and opposite angles α, β, γ, **sin**$(\alpha)/a$ = **sin**$(\beta)/b$ = **sin**$(\gamma)/c$.

2. Prove the **cosine rule** (also known as the Law of Cosines) of Section 4.3.

3. For a curve defined by the function $x^2 + Cy^2 + Gy + H = 0$, the value $e := 1 - C^2$ is called the **eccentricity**. What kinds of curve correspond to an **eccentricity** of 0, of ½, of 1, and of 2?

4. Work out the action of the 2×2 **matrix** with $a_{11} = a_{22} =$ **cos**(θ), $a_{12} =-$ **sin**(θ), $a_{21} =$ **sin**(θ) on the points (1, 0), (0, 1), (−1, 0) and (0, −1) when $\theta = \pi/2$ or 90°. Try it again for $\theta = \pi/3$ or 30°. What kind of change in the values of (x,y) are we seeing here?

5. The value y/x for an **angle** with $x^2 + y^2 = 1$ is called the **tangent** of the **angle**. Evaluate **d(tan**(θ)**)** where θ is the usual **parameterization** giving **d**$x = -y.$**d**θ, **d**$y = x.$**d**θ, and **tan**(θ) = **sin**(θ)/**cos**(θ). Put u = **tan**(θ) so that θ = **tan**$^{-1}(u)$, and so develop a formula for **d(tan**$^{-1}(u)$**)** by first expressing **d**u in terms of **d**θ. The **cotangent** is the value x/y. Evaluate **d(cot**(θ)**)** and **d(cot**$^{-1}(\theta)$**)** similarly. In these terms, what was the **integral** we were evaluating in Section 4.4?

6. Using **complex algebra**, prove the following **identities**:

   a. $\varepsilon_y(s + t) = \varepsilon_y(s).\varepsilon_x(t) + \varepsilon_x(s).\varepsilon_y(t) \equiv$ **sin**$(\theta + \varphi)$ = **sin**(θ).**cos**(φ) + **cos**(θ).**sin**(φ)

b. $\varepsilon_y(s-t) = \varepsilon_y(s).\varepsilon_x(t) - \varepsilon_x(s).\varepsilon_y(t) \equiv$ **sin**$(\theta - \varphi) =$ **sin**(θ).**cos**$(\varphi) -$ **cos**(θ).**sin**(φ)

c. $\varepsilon_x(s+t) = \varepsilon_x(s).\varepsilon_x(t) - \varepsilon_y(s).\varepsilon_y(t) \equiv$ **sin**$(\theta + \varphi) =$ **cos**(θ).**cos**$(\varphi) -$ **sin**(θ).**sin**(φ)

d. $\varepsilon_x(s-t) = \varepsilon_x(s).\varepsilon_x(t) + \varepsilon_y(s).\varepsilon_y(t) \equiv$ **cos**$(\theta - \varphi) =$ **cos**(θ).**cos**$(\varphi) +$ **sin**(θ).**sin**(φ)

e. $\varepsilon_x^2(t) = ½(1 + \varepsilon_x(2t)) \equiv$ **cos**$^2(\varphi) = ½(1 +$ **cos**$(2\varphi))$

f. $\varepsilon_y^2(t) = ½(1 - \varepsilon_x(2t)) \equiv$ **sin**$^2(\varphi) = ½(1 -$ **cos**$(2\varphi))$.

7. Again using **complex algebra** and the definition of **tangent** above, prove these **identities**:

   a. **tan**$(\theta + \varphi) = ($**tan**$(\theta) +$ **tan**$(\varphi))/(1 -$ **tan**(θ).**tan**$(\varphi))$

   b. **tan**$(\theta - \varphi) = ($**tan**$(\theta) -$ **tan**$(\varphi))/(1 +$ **tan**(θ).**tan**$(\varphi))$.

8. Any point in the **complex plane** $z = x + jy$ can be expressed as $r.\varepsilon(t) = r.\varepsilon_x(t) + rj.\varepsilon_y(t) = r.$**cos**$(\theta) + rj.$**sin**$(\theta)$. Either of these formulations is a form in **polar coordinates** (r, t) or (r, θ) where $r \in [0, \infty]$ and $\theta \in [0, 2\pi]$ and $t \in [0, 1]$. Develop **matrices** expressing **d**x and **d**y in terms of **d**r and **d**t, and for any function $\varphi(z)$, develop **matrices** expressing $\partial_r\varphi$ and $\partial_t\varphi$ in terms of $\partial_x\varphi$ and $\partial_y\varphi$. Do this also for (r, θ) rather than (r, t).

9. Show that the area highlighted in *Figure 4.10.1* equals $½yx - y(u).$**d**$x(u) \,|\, [0, \phi] = ½\eta_y(\phi).\eta_x(\phi) - \eta_y(u).$**d**$\eta_x(u) \,|\, [0, \phi] = ½\phi$ all told. The **hyperbolic** analogue of 4.X.6.f above will help.

10. The equation of a hanging cable or **catenary** is $\partial_x^2 y = (w/H)\surd(1 + (\partial_x y)^2)$, where w is the weight per unit length, and H is the horizontal tension at the lowest point. Set $p := \partial_x y$ and so show that this can be solved as **d**$\eta_y^{-1}(p) = (w/H).$**d**x. From this perform a second integration to show that $y = (H/w).\eta_x((w/H)x) + C$ for a constant C.

11. In Section 4.7 we obtained a general solution of $a\partial_t^2 y + b\partial_t y + cy = 0$ as $y(t) = A.\mathbf{e}^{\alpha t}.\mathbf{e}^{j\omega t} + B.\mathbf{e}^{\alpha t}.\mathbf{e}^{-j\omega t}$ with $\alpha = -b/2a$ and $\omega^2 = b^2 - 4ac$. Now suppose our equation is $a\partial_t^2 y + b\partial_t y + cy = k.(\mathbf{e}{\uparrow}j\Omega t + \mathbf{e}{\uparrow}(-j\Omega t))$, which is a **cos**() form. This is a **non-homogeneous** equation as opposed to the **homogeneous** form with zero on the RHS. By the **linearity** of **differentiation** the original solution to the **homogeneous** $a\partial_t^2 y + b\partial_t y + cy = 0$ will still be valid but now we have to *add* a solution to accommodate the non-zero part on the RHS. Just try an arbitrary function $y_p(t) = A.\mathbf{e}{\uparrow}(j\Omega t) + B.\mathbf{e}{\uparrow}(-j\Omega t)$, evaluate $a\partial_t^2 y_p + b\partial_t y_p + cy_p$ and equate the resulting coefficients of $\mathbf{e}{\uparrow}(j\Omega t)$ and $\mathbf{e}{\uparrow}(-j\Omega t)$ on both sides, and we have a solution $y_p(t)$. Show that if b is small, the magnitude of the oscillations represented by this solution grows very large as $\Omega \to \omega$, the phenomenon known as **resonance**. (The general solution is our original $y(t) + y_p(t)$ and $y_p(t)$ is called a **particular integral**).

Chapter 5

VECTORS AND GEOMETRIC ALGEBRA

5.1 Introduction to Vector Spaces

Vectors are most easily introduced by an example. This example is by far the most common form of **vector** and so I will give it a name: a **standard vector**. A **standard vector** is simply a finite **ordered list** of **real numbers** like [22, 37.65, 2.943, −15.87]. The "**ordered**" is important: [2, 17, 3.4] is not the same **vector** as [17, 3.4, 2]. Writing the **list** in generic **list** notation as $\{a_i\} = [a_1, a_2, \ldots a_n]$, identity of two **lists** is here defined as:

$\{a_i\} = \{b_j\}$ if and only if $a_i = b_j$ whenever $i = j$.

I use the term "**list**" rather than various less satisfactory alternatives. In computer parlance, such **ordered lists** are generally called "arrays" and "list" is reserved for a more complex kind of structure. Mathematics texts commonly use the term **ordered n-tuple**, but this is really too clumsy.

What changes the simple **ordered list** into a **standard vector** is the addition of two **operations, vector addition** and **scalar multiplication**. For a **standard vector**, these are defined:

I:	$\{a_i\} + \{b_j\} = \{a_i + b_i\}$	**(vector addition)**
II:	$\lambda\{a_i\} = \{\lambda a_i\}$	**(scalar multiplication)**

where "I" is only defined if the number of elements in both **lists** are the same, and in "II" λ is an arbitrary **real number**. The number of elements in the **ordered list** is called the **dimension** of the **vector**, so **vector addition** is only defined between **vectors** of the *same* **dimension**. The **real number** λ in **scalar multiplication** is called a **scalar** simply to distinguish it from the **vector** itself. Since it multiplies all the elements of the **vector** by the same amount, it *scales* the **vector**, hence the name.

So under these definitions, examples are:

$$\lambda.[a_1, a_2, a_3] + \mu.[b_1, b_2, b_3] = [\lambda.a_1 + \mu.b_1, \lambda.a_2 + \mu.b_2, \lambda.a_3 + \mu.b_3]$$

$$[12, 2.3, -7] + [3, -5, 4.2] = [12 + 3, 2.3 - 5, -7 + 4.2] = [15, -2.7, -3.2]$$

and

$$3 \times [2.2, -5] = [3 \times 2.2, 3 \times -5] = [6.6, -15].$$

By the Cartesian construction introduced in Section 1.2, any point in the *plane* can be defined by a **standard vector** of **dimension** two, which I will sometimes refer to as a **2-vector**, and any point in ordinary space can be defined by a **standard vector** of **dimension** three, which I may call a **3-vector**.

So **standard vectors** of these **dimensions** are the fundamental entities used to describe geometry in algebraic terms, a subject called **analytic geometry**.

Vector addition easily extends to **vector subtraction**, where the **difference** between two **vectors** is defined as $\{a_i\} - \{b_j\} = \{a_i - b_i\}$. In computer graphics, it is customary to refer to **vectors** that define *points* in space as **3-points**, and to reserve the term **3-vector** for the **differences** between **3-points**. This distinction is not normally made in mathematics, but it is best to be aware of it.

Much more rarely, **ordered lists** with an *infinite* number of elements are used, and these are thus an example of **infinite-dimensional vectors**, but we will not need these in this text. We have in fact touched on an example already, in Section 1.7: the **powers** of x, $(1, x, x^2, x^3, \ldots)$ for all values of x – i.e. as **functions** – constitute an infinite-dimensional set of **vectors**. $a_0 + a_1x^1 + a_2x^2 + \ldots$ is the general form of a **vector** in this space.

A more common kind of **infinite-dimensional vector** is obtained by defining **vector addition** and **scalar multiplication** for *continuous* **functions** on a specified **line segment** or **interval**. This is quite a different concept of **vector** to our **standard vectors**, but its vectorial nature is quite important. Writing the **function** $f(x)$ over the given **interval**, seen as a **vector**, as $[f(x)]$, the two definitions here take the form:

I: $[f(x)] + [g(x)] = [f(x) + g(x)]$
II: $\lambda[f(x)] = [\lambda.f(x)]$

and so are quite trivial.

The axiomatic formulation in Section 5.3 is set up so as to cover all these cases.

With the basic idea available, we can already introduce some other concepts that are frequently used in working with **vectors**. I'll illustrate these largely with **3-vectors**. First comes the idea of a **basis**. Using our two **operations**, any **3-vector** $[x, y, z]$ can be written as:

$$x.[1, 0, 0] + y.[0, 1, 0] + z.[0, 0, 1] = [x, 0.x, 0.x] + [0.y, y, 0.y] + [0.z, 0.z, z]$$

$$= [x, 0, 0] + [0, y, 0] + [0, 0, z] = [x + 0 + 0, 0 + y + 0, 0 + 0 + z] = [x, y, z]$$

so the three **3-vectors** [1, 0, 0], [0, 1, 0] and [0, 0, 1] can be used to express *any* **3-vector**. These three **basis vectors** are commonly written **i**, **j** and **k**, so any **3-vector** can be written in the form:

$$x.\mathbf{i} + y.\mathbf{j} + z.\mathbf{k}.$$

i, **j** and **k** are called **unit vectors** for obvious reasons that will be generalized in Section 5.4.

The **i-j-k** notation is peculiar to **3-vectors**. A more general notation for these special **basis vectors** is $\mathbf{e}_1$, $\mathbf{e}_2$ and $\mathbf{e}_3$, a form which can be used for **vectors** of any **dimension**. *The number of* **basis vectors** *always equals the* **dimension** *of the* **vectors** *in question*. The **vectors** $\{a_i\}$ where $a_i = 0$ for all i except one single one, where $a_i = 1$, is sometimes called the **natural basis** and we will later show that it is also an **orthonormal basis**, a term which will be defined in Section 5.5.

The **natural basis** is far from being the *only* **basis**. For example, we could use the three vectors [1, 2, 3], [1, 1, 1] and [5, 9, 6], when to express an arbitrary vector $[a, b, c]$ as:

$$[a, b, c] = x.[1, 2, 3] + y.[1, 1, 1] + z.[5, 9, 6]$$

we need to solve the **linear equations**:

$$x.1 + y.1 + z.5 = a$$
$$x.2 + y.1 + z.9 = b$$
$$x.3 + y.1 + z.6 = c$$

which you can see by multiplying out the right-hand side to give:

$$[x, 2x, 3x] + [y, y, y] + [5z, 9z, 6z] = [x + y + 5z, 2x + y + 9z, 3x + y + 6z]$$

and identifying the elements on both sides. The first equation gives $x = a - y - 5z$, which in the second gives

$$2a - 2y + y - 10z + 9z = b, \text{ or } 2a - z - b = y,$$

so in the third

$$3a - 3(2a - z - b) - 15z + 2a - z - b + 6z = c,$$

or

$$-3a + 3z + 3b - 15z + 2a - z - b + 6z - c = 0, \text{ so that } -7z - a + 2b - c = 0,$$

so:

$$z = (2b - a - c)/7$$
$$y = 2a - z - b = 2a - (2b - a - c)/7 - b$$

and $x = a - y - 5z = a - (2a - z - b) - 5z = -a + z - 5z + b$

$$= - a + b - (4/7)\,(2b - a - c).$$

So we get:

$$x.[1, 2, 3] + y.[1, 1, 1] + z.[5, 9, 6]$$
$$= (- a + b - (4/7)\,(2b - a - c)).[1, 2, 3] + (2a - (2b - a - c)\,/7 - b).[1, 1, 1]$$
$$+ ((2b - a - c)\,/7).[5, 9, 6]$$

$$= [- a + b - (4/7)\,(2b - a - c) + 2a - (2b - a - c)\,/7 - b + (5/7)\,(2b - a - c),$$
$$- 2a + 2b - (8/7)\,(2b - a - c) + 2a - (2b - a - c)\,/7 - b + (9/7)\,(2b - a - c),$$
$$- 3a + 3b - (12/7)\,(2b - a - c) + 2a - (2b - a - c)\,/7 - b + (6/7)\,(2b - a - c)]$$

$$= [a + (4/7)a + (1/7)a - (5/7)a, 2b - b - (9/7)\,(2b - a - c) + (9/7)\,(2b - a - c),$$
$$- a + 2b + ((-12-1+6)/7)\,(2b - a - c)]$$

$$= [a, b, -a + 2b - (2b - a - c)] = [a, b, c]$$

as required. Any three **3-vectors** which are **linearly independent** will suffice to form a **basis**. This property was introduced in Section 1.7, and here it takes the form that, writing the three **3-vectors** as $\mathbf{v}_1$, $\mathbf{v}_2$ and $\mathbf{v}_3$, then $\mathbf{v}_1$, $\mathbf{v}_2$ and $\mathbf{v}_3$ are **linearly independent** if the expression:

$$x.\mathbf{v}_1 + y.\mathbf{v}_2 + z.\mathbf{v}_3 = [0, 0, 0]$$

is true *only if* $x = 0$ *and* $y = 0$ *and* $z = 0$. The special **3-vector** [0, 0, 0] is called the **zero 3-vector** and is commonly written **0** and has the usual **zero** property that for any **vector** $\mathbf{v}$, $\mathbf{v} + \mathbf{0} = \mathbf{0} + \mathbf{v} = \mathbf{v}$. The complication emphasized in Section1.7 of the equation having to hold for all values of the **argument** x does not occur in the **vector** formulation of **linear independence**.

The significance of **linear independence** is this: if a set of **vectors** is *not* **linearly independent**, or to put it another way, *is* **linearly dependent**, then *one* **vector** *can be expressed in terms of the others*. So in the case above, where $x.\mathbf{v}_1 + y.\mathbf{v}_2 + z.\mathbf{v}_3 = \mathbf{0}$, if $x \neq 0$ and $y \neq 0$ but $z = 0$, we could say:

$$x.\mathbf{v}_1 + y.\mathbf{v}_2 = \mathbf{0} \text{ so that } \mathbf{v}_2 = -(x/y)\mathbf{v}_1.$$

So $\mathbf{v}_2$ could be expressed in terms of $\mathbf{v}_1$ or vice versa. If *none* of x, y or z were zero, we could write:

$$\mathbf{v}_2 = -(x/y)\mathbf{v}_1 - (z/y)\mathbf{v}_3.$$

From the **linear equations** above, it will be seen that **linear independence** of the three vectors [1, 2, 3], [1, 1, 1] and [5, 9, 6] corresponds to a non-zero value for the **determinant** of the **matrix**:

$$\begin{pmatrix} 1 & 2 & 3 \\ 1 & 1 & 1 \\ 5 & 9 & 6 \end{pmatrix}$$

The determinant here is in fact: 1×(1×6 − 1×9) − 2×(1×6 − 1×5) + 3×(1×9 − 1×5) = −3 − 2 + 12 = 7. In this expression you can see the determinants of the **sub-matrices** or **cofactors**:

$$\begin{pmatrix} 1 & 1 \\ 9 & 6 \end{pmatrix} \qquad \begin{pmatrix} 1 & 1 \\ 5 & 6 \end{pmatrix} \qquad \begin{pmatrix} 1 & 1 \\ 5 & 9 \end{pmatrix}$$

appearing, but you'd best refer back to Section 1.8 or look up a text on **determinants** if you want to understand this expansion! The point is simply the rule:

> To determine if n **n-vectors** are **linearly independent**, form the **matrix** whose rows are the **n-vectors** in question, and evaluate its **determinant**. If the **determinant** is non-zero, the n **n-vectors** are **linearly independent**.

So far I've talked of all **n-vectors** as the *set* of **n-vectors**. The normal mathematical term is to call this set of *all possible* **n-vectors** the **vector space** of **n-vectors**. So the subject we're developing in this chapter is the theory of **vector spaces**.

Any set of m **linearly independent vectors** in a **vector space** of **n-vectors** where $m < n$ defines a **subspace** of **dimension** m in the original **vector space**. Every **m-vector** in the **subspace** is a member of the original **n-vector space**, but *not* vice versa.

Any set of m **linearly independent m-vectors** all of which are members of the **subspace** is a **basis** for the **subspace**, so in particular, the defining set of **m-vectors** is a **basis**.

Subspaces are a very useful concept. As an example, the set of **powers** mentioned above would only rarely be used in its infinite-dimensional form, but the four **basis vectors** 1, x, x^2, x^3 define the four-dimensional **subspace** of **polynomials** comprising the **cubic polynomials**, which are widely used in computer graphics. The logic from **standard vectors** largely carries over here. So 1, x, x^2 constitute a **basis** for the **quadratic polynomials**, but from our example above, so would the three **3-vectors**:

$$\mathbf{e}_1 = 1 + 2x + 3x^2$$
$$\mathbf{e}_2 = 1 + x + x^2$$
$$\mathbf{e}_3 = 5 + 9x + 6x^2$$

so any **quadratic polynomial** could be written $a_1\mathbf{e}_1 + a_2\mathbf{e}_2 + a_3\mathbf{e}_3$.

Any set of **vectors** that form a **basis** for a **vector space** is said to **span** that **vector space**.

5.2 The Origin of Vectors: Quaternions

Standard vectors may seem a very obvious and trivial idea. But their origin lies much deeper, in an attempt to extend **complex numbers** to more than two **dimensions**. This involves perhaps the most famous anecdote in mathematical history.

Sir William Rowan Hamilton was walking beside the Royal Canal in Dublin on the 16th of October 1843 – he was then thirty-eight – thinking about this question, when the answer came to him in a flash of inspiration. Instead of using just a single **unit imaginary** j (more commonly written i by mathematicians) Hamilton proposed *three* **unit imaginaries** i, j, and k which obeyed these multiplication rules:

$$i^2 = j^2 = k^2 = -1 \qquad i.j = k \qquad j.k = i \qquad k.i = j$$

which entail $i.k = i(i.j) = (i.i)j = -j$, if associativity is to hold, and likewise $j.i = -k$ and $k.j = -j$. In other words, writing the products as we have done ($ab = c$) the **cyclic permutations** of i-j-k give positive products, and the **anti-cyclic** negative, because our **unit imaginaries** now obey an **anti-commutative** rule, that:

$$ij = -ji, \qquad ik = -ki, \qquad jk = -kj.$$

This was a novel property that was to have far-reaching consequences. Observe also that:

$$ijk = ii = -1.$$

On the strength of this insight, Hamilton introduced a kind of "super-complex" number which he called a **quaternion** and which is written in the general form:

$$\lambda + ix + jy + kz$$

where λ is the "real" part, which Hamilton called the **scalar** part, and the three **imaginary parts** *together* he called the **vector** or the **pure quaternion**. You can see what a good polemicist Hamilton was: to make sure his new **vector** part didn't suffer the fate of being called "imaginary", he called *it* "pure", implying that it was the real part that was "adulterated", from the start. It is from Hamilton's notation that the **unit vectors i, j** and **k** take their designation.

But Hamilton wasn't quite good enough as a polemicist. Through the course of the nineteenth century, **quaternions** gradually slid into the mathematical background, and a simplified **vector** system – basically our **standard vectors** of Section 5.1 – came to dominate physics, championed by the pre-eminent figures of Josiah Willard Gibbs at Yale, and Oliver Heaviside in England, who were far better at polemics than Hamilton's later supporter Peter Guthrie Tait at Edinburgh. The controversy makes entertaining reading.[59]

Quaternions remained a mathematical backwater until they started to be used again by the pioneers of computer graphics. Their significance is now beginning to be absorbed into the rapidly expanding field of **Geometric Algebra**, which we will begin to study in Section 5.8, and which is also a rediscovery of a lost nineteenth-century innovation.

5.3 Axiomatic Formulation

All **vector spaces** are covered by the **axioms** below, which define **vector addition** and **scalar multiplication** axiomatically. Here **u**, **v** and **w** are **vectors**, *a*, *b* and *c* are **scalars**.

I.	$(\mathbf{u} + \mathbf{v}) + \mathbf{w} = \mathbf{u} + (\mathbf{v} + \mathbf{w})$	(Associativity)
II.	$\mathbf{u} + \mathbf{0} = \mathbf{0} + \mathbf{u} = \mathbf{u}$	(Zero Element)
III.	$\mathbf{u} + -\mathbf{u} = -\mathbf{u} + \mathbf{u} = \mathbf{0}$	(Inverse)
IV.	$\mathbf{u} + \mathbf{v} = \mathbf{v} + \mathbf{u}$	(Commutativity)
V.	$c(\mathbf{u} + \mathbf{v}) = c\mathbf{u} + c\mathbf{v}$	
VI.	$(a + b)\mathbf{u} = a\mathbf{u} + b\mathbf{u}$	
VII.	$(ab)\mathbf{u} = a(b\mathbf{u})$	
VIII.	$1\mathbf{u} = \mathbf{u}$	(where 1 is the *number* 1)

The first four are the usual axioms for a **commutative group** as given in Section 1.5, but without the *closure* axiom. The latter four define **scalar multiplication** again without a *closure* axiom. You're just expected to assume that the results of these **operations** are themselves members of the **vector space**.

[59] C.E. Weatherburn *Elementary Vector Analysis*, Bell, London 1953.

The closure axioms are always omitted. This is just traditional, and perhaps goes to show that mathematicians are not always as rigorous as they like to think they are!

The likelihood of your needing to use these axioms is remote. Their value is more in their conceptual significance, because they show that a **vector space** is in fact an **algebra**, like the **algebra** of **real numbers** whose much more developed axioms were given in Section 1.5. Comparing this Section with that, it should be clear that a **vector space** is a much more specific thing than a **group**, but much less so than the set of **real numbers**. Both **real numbers** and **complex numbers** are special cases of a **vector space**. The main difference is that there needn't be any *multiplication* defined between **vectors**; although a kind of multiplication between the **scalars** of the **vector space** (which almost always *are* **real** or **complex numbers**) and **vectors** *is* defined, there's no *assumption* of a **vector × vector product**.

I introduced **vectors** as **lists** of **real numbers** like [3.3, 27.9, π]. This is the simplest model of a **vector** but the concept isn't limited to this model by any means. As I suggest above, we could have a **vector space** with the **basis**:

$$\mathbf{e}_0 = 1 \qquad \mathbf{e}_1 = x \qquad \mathbf{e}_2 = x^2 \qquad \mathbf{e}_3 = x^3.$$

An element of this **vector space**[60] takes the form:

$$\textstyle\sum_{i=0}^{3} a_i\, \mathbf{e}_i = a_0.\mathbf{e}_0 + a_2.\mathbf{e}_2 + a_2.\mathbf{e}_2 + a_3.\mathbf{e}_3 = a_0.1 + a_1.x + a_2.x^2 + a_3.x^3.$$

This object is the **cubic polynomial** $a_0 + a_1.x + a_2.x^2 + a_3.x^3$, and it's a **function**! We can apply it to a number like 3 as:

$$\textstyle\sum_{i=0}^{3} a_i\, \mathbf{e}_i(3) = \{ a_0 + a_1.x + a_2.x^2 + a_3.x^3\}(3) = a_0 + a_1.3 + a_2.9 + a_3.27$$

and we get a number.

This enormous conceptual "looseness" of the **vector** concept is part of its power (as with the even looser idea of a **group**), but it can be daunting to the student.

Let's go to the opposite extreme and take our **vectors** as being **lists** of **real numbers** with only *one* element in the list, like [3.3]. Then each **real number** defines a unique **vector** and each **vector** corresponds to a **real number**.

So if λ = 22 and μ = 5,

$$\lambda[3.2] + \mu.[4] = 22 \times [3.2] + 5 \times [4] = [70.4] + [20] = [90.4],$$

so effectively these **vectors** *are* the **real numbers** themselves! (Note that this depends on **scalar multiplication** and **vector addition** being defined as I did at the start of the Section 5.1, for **standard vectors**.)

---

[60] The numbering of the **basis** elements here from 0 is chosen to match the exponents of the **polynomial**.

Such a usage would be absurdly trivial, but a much less trivial case is the one developed in Chapter 4 – **complex numbers**.

A **complex number**, which is also a point in the **plane**, obviously has a **vectorial** nature. We can write it as a **vector** in the conventional bold face $\mathbf{z} = (x, y)$ where x and y are its **real components** and the **basis** is:

$$\mathbf{e}_1 = 1 \qquad\qquad \mathbf{e}_2 = j.$$

So we can write **z** as:

$$\mathbf{z} = x.\mathbf{e}_1 + y.\mathbf{e}_2 = x.1 + y.j = x + jy.$$

We can add these "**vectors**" as:

$$(x.\mathbf{e}_1 + y.\mathbf{e}_2) + (p.\mathbf{e}_1 + q.\mathbf{e}_2) = (x + p).\mathbf{e}_1 + (y + q).\mathbf{e}_2 = (x + p) + (y + q).j$$

$$= (x + p) + j(y + q).$$

We can also multiply them by a **real number scalar** a:

$$a(x.\mathbf{e}_1 + y.\mathbf{e}_2) = (ax.\mathbf{e}_1 + ay.\mathbf{e}_2) = ax.1 + j.ay = ax + jay.$$

But here we've got something else as well. We can multiply them *by each other*:

$$(x.\mathbf{e}_1 + y.\mathbf{e}_2) \times (p.\mathbf{e}_1 + q.\mathbf{e}_2) = (xp - yq).\mathbf{e}_1 + (xq + yp).\mathbf{e}_2 = (xp - yq) + j(xq + yp).$$

This is something quite outside the axioms of a **vector space** and is the simplest example of a **vector product**. It can be construed as a special case of the **Clifford product** or **geometric product** which will loom large later in this Chapter.

In computer graphics, a major distinction is made between **points** and **vectors**. Both of these look like our **standard vectors** but conceptually they're sufficiently different that computer graphics applications actually restrict what operations are applicable to which.

A **point** is a position in space described by **coordinates** $[x, y, z]$.
A **vector** is the *difference* between two **points** $[x, y, z]$ and $[u, v, w]$ and is described by its **components** $[u - x, v - y, w - z]$.

Points have **coordinates** defined by reference to **Cartesian axes** as described in Section 1.2.

Vectors don't have **coordinates**: they have **components**.

Both objects strictly obey the axioms of a **vector space** as described in this Section. But a lot of conceptual baggage now becomes attached to distinguish the two!

1. The difference between two **points** $\mathbf{p}_1$ and $\mathbf{p}_2$ is a **vector** $\mathbf{v} = \mathbf{p}_2 - \mathbf{p}_1$.
2. The sum of two **points** is commonly not defined.
3. A product may be defined between two **vectors** but *not* between two **points**.
4. Distances (or **metrics**) are defined between two **points** but not between two **vectors**.
5. **Vectors** have a **magnitude** – basically their length, but **points** generally do not.
6. The **point** $\mathbf{p}_1$ in the definition of the **vector** $\mathbf{v} = \mathbf{p}_2 - \mathbf{p}_1$.is called the **base point** of **v** and we say **base(v)** = $\mathbf{p}_1$ and refer to the **vector** as *belonging to* $\mathbf{p}_1$ (mathematicians refer to it as being an element of the **tangent space** at $\mathbf{p}_1$ – every **point p** has an attached **tangent space** which may at this stage be thought of as the set of all difference **vectors q – p**, where **q** is any other **point** of the space).

Figure 5.3.1 illustrates the differences: the fine dashed lines run from the **origin** to the **points** P_1 and P_2. At each we can define a **triad** of **basis vectors**, respectively $\{\mathbf{i}_1, \mathbf{j}_1, \mathbf{k}_1\}$ and $\{\mathbf{i}_2, \mathbf{j}_2, \mathbf{k}_2\}$. We can define **vectors** at each **point**, so at P_1 we might have $\mathbf{v}_1$ and $\mathbf{v}_2$, we can *add* **vectors** at the same **point**, e.g. $\mathbf{v}_1 + \mathbf{v}_2$, and come to that take differences, so $\mathbf{v}_2 = (\mathbf{v}_1 + \mathbf{v}_2) - \mathbf{v}_1$, and the *difference* (only) between **points** is defined as a **vector**: $\mathbf{v} = P_2 - P_1$, which would be a **vector** "belonging to" the **vector space** at P_1 with **basis** $\{\mathbf{i}_1, \mathbf{j}_1, \mathbf{k}_1\}$.

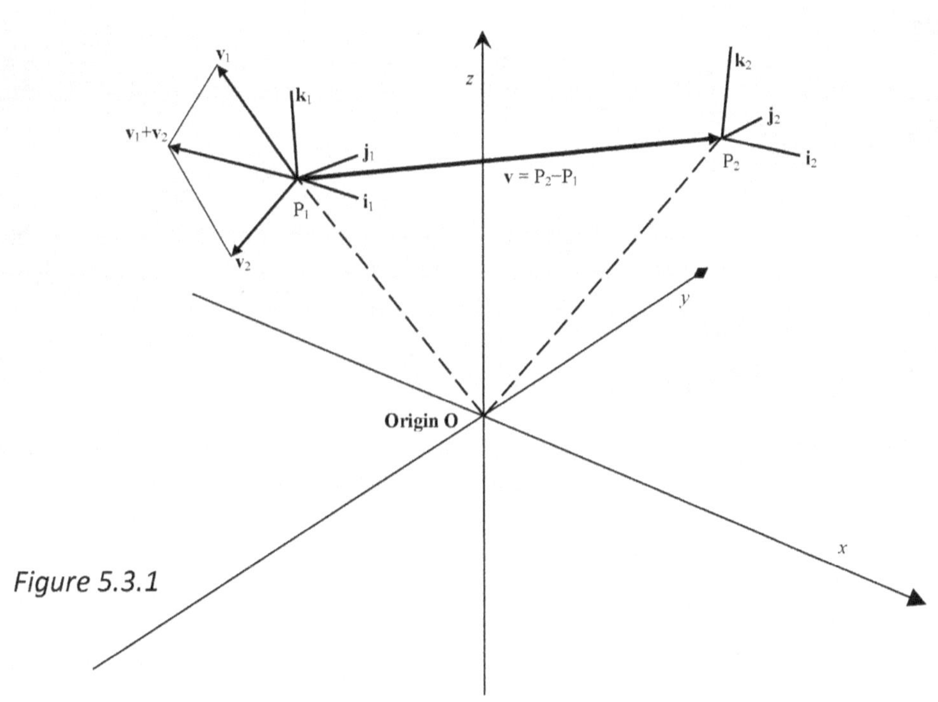

Figure 5.3.1

In my own work I find these distinctions often just get in the way – for example, the difference between two **complex numbers** (**points** in the **complex plane**) is clearly a **complex number**, not a distinct entity.

Computer graphics systems intend this distinction to be a guard against errors, to lock the programmer into "meaningful" operations. But many programmers like to exploit these "errors" as I do.

That said, the difference between **points** and **vectors** is a first taste of something very significant in more advanced work, because the set of **vectors** at a **point** (or the **tangent space** at a **point**) is the simplest illustration of the idea of a **vector bundle** which will loom large in the next volume in this series.

The distinction in nomenclature between **components** of a **vector**, and **coordinates** identifying a **point** in space is very well-established.

5.4 The Dual Space

Vector spaces, even more than the **calculus**, are a celebration of **linearity**. So the axioms above are also often referred to as the axioms of a **linear space**. **Linear spaces** and **vector spaces** are the same thing.

A **linear function** defined on a **vector space** is a **function** ϕ that has the property that:

$$\phi(\lambda\mathbf{u} + \mu\mathbf{v}) = \lambda.\phi(\mathbf{u}) + \mu.\phi(\mathbf{v}) \qquad \textbf{(Linearity)}$$

The **dual space** of a given **vector space** is the set of all **linear functions** on that space that have **scalar** or **real number** values. By the defining **linearity** property, any such **function** is completely defined by its values on a **basis** of the **vector space**. So for an arbitrary **3-vector**, we have:

$$\phi(a\mathbf{e}_1 + b\mathbf{e}_2 + c\mathbf{e}_3) = a.\phi(\mathbf{e}_1) + b.\phi(\mathbf{e}_2) + c.\phi(\mathbf{e}_3)$$

so if we know the values of $\phi(\mathbf{e}_1)$, $\phi(\mathbf{e}_2)$ and $\phi(\mathbf{e}_3)$ we can evaluate this function ϕ on *any* **3-vector**. But this opens up another possibility. For if we have a set of three **linearly independent 3-vectors** of values like:

$$[\phi_1(\mathbf{e}_1), \phi_1(\mathbf{e}_2), \phi_1(\mathbf{e}_3)] \qquad [\phi_2(\mathbf{e}_1), \phi_2(\mathbf{e}_2), \phi_2(\mathbf{e}_3)] \qquad [\phi_3(\mathbf{e}_1), \phi_3(\mathbf{e}_2), \phi_3(\mathbf{e}_3)]$$

we can set up a set of **linear equations** to solve for *any* **3-vector** of numbers $[\phi(\mathbf{e}_1), \phi(\mathbf{e}_2), \phi(\mathbf{e}_3)] = [a, b, c]$:

$$a_1.\phi_1(\mathbf{e}_1) + a_2.\phi_2(\mathbf{e}_1) + a_3.\phi_3(\mathbf{e}_1) = \phi(\mathbf{e}_1) = a$$

$a_1.\phi_1(\mathbf{e_2}) + a_2.\phi_2(\mathbf{e_2}) + a_3.\phi_3(\mathbf{e_2}) = \phi(\mathbf{e_2}) = b$
$a_1.\phi_1(\mathbf{e_3}) + a_2.\phi_2(\mathbf{e_3}) + a_3.\phi_3(\mathbf{e_3}) = \phi(\mathbf{e_3}) = c$

and so express *any* member of the **dual space** as a **linear combination** of $[\phi_1, \phi_2, \phi_3]$:

$$\phi = a_1\phi_1 + a_2\phi_2 + a_3\phi_3.$$

So the **dual space** is a **vector space** of **3-vectors** in its own right. Hence the very term "**dual space**". To check, we expand:

$$\phi(\lambda\mathbf{e_1} + \mu\mathbf{e_2} + \nu\mathbf{e_3}) = a_1\phi_1(\lambda\mathbf{e_1} + \mu\mathbf{e_2} + \nu\mathbf{e_3}) + a_2\phi_2(\lambda\mathbf{e_1} + \mu\mathbf{e_2} + \nu\mathbf{e_3}) + a_3\phi_3(\lambda\mathbf{e_1} + \mu\mathbf{e_2} + \nu\mathbf{e_3})$$

$$= \lambda(a_1\phi_1(\mathbf{e_1}) + a_2\phi_2(\mathbf{e_1}) + a_3\phi_3(\mathbf{e_1})) + \mu(a_1\phi_1(\mathbf{e_2}) + a_2\phi_2(\mathbf{e_2}) + a_3\phi_3(\mathbf{e_2}))$$

$$+ \nu(a_1\phi_1(\mathbf{e_3}) + a_2\phi_2(\mathbf{e_3}) + a_3\phi_3(\mathbf{e_3}))$$

$$= \lambda\phi(\mathbf{e_1}) + \mu\phi(\mathbf{e_2}) + \nu\phi(\mathbf{e_3}) = \lambda a + \mu b + \nu c.$$

So if we chose:

$\phi_1(\mathbf{e_1}) = 1$	$\phi_2(\mathbf{e_1}) = 2$	$\phi_3(\mathbf{e_1}) = 3$
$\phi_1(\mathbf{e_2}) = 1$	$\phi_2(\mathbf{e_2}) = 1$	$\phi_3(\mathbf{e_2}) = 1$
$\phi_1(\mathbf{e_3}) = 5$	$\phi_2(\mathbf{e_3}) = 9$	$\phi_3(\mathbf{e_3}) = 6$

then *any* member of the **dual space** defined by $\phi(\mathbf{e_1}) = a$, $\phi(\mathbf{e_2}) = b$ and $\phi(\mathbf{e_3}) = c$ could be expressed in terms of ϕ_1, ϕ_2 and ϕ_3 using our earlier solution for this set of equations,

$a_1 = - a + b - (4/7)\,(2b - a - c),$

$a_2 = 2a - (2b - a - c)\,/7 - b$

and $a_3 = (2b - a - c)\,/7.$

We can express this as:

$$\phi = (- a + b - (4/7)\,(2b - a - c))\boldsymbol{\phi}_1 + (2a - (2b - a - c)\,/7 - b)\boldsymbol{\phi}_2 + ((2b - a - c)\,/7)\boldsymbol{\phi}_3.$$

I've put the $\boldsymbol{\phi}$'s here in bold just for emphasis, although this is consistent with their vectorial nature. The defining set of **3-vectors** here would actually be:

$[\phi_1(\mathbf{e_1}), \phi_1(\mathbf{e_2}), \phi_1(\mathbf{e_3})] = [1, 1, 5]$
$[\phi_2(\mathbf{e_1}), \phi_2(\mathbf{e_2}), \phi_2(\mathbf{e_3})] = [2, 1, 9]$
$[\phi_3(\mathbf{e_1}), \phi_3(\mathbf{e_2}), \phi_3(\mathbf{e_3})] = [3, 1, 6]$

with the rows and columns **transposed** from the layout above. This switch to the **transpose** of a **matrix** – the **matrix** with the rows and columns swapped – occurs rather a lot in this area, and you need to take care which form you are using.

We now introduce a conceptual shift of a sort that mathematicians are very fond of, but which occasion endless heartaches for students. If the **dual space** is also a **vector space** of the same **dimension** as the basic **vector space**, could we somehow use the *basic space itself* to represent the **dual space**?

Suppose we introduce an **operator** $*$ between two **vectors** in the original **vector space** that is defined to be **linear** in both **arguments**:

$$\mathbf{u} * (\lambda\mathbf{v} + \mu\mathbf{w}) = \lambda(\mathbf{u} * \mathbf{v}) + \mu(\mathbf{u} * \mathbf{w})$$
$$(\theta\mathbf{u} + \nu\mathbf{v}) * \mathbf{w} = \theta(\mathbf{u} * \mathbf{w}) + \nu(\mathbf{v} * \mathbf{w}).$$

Again we can use this **linearity** to expand this **product** into terms of the **basis vectors**. So for two arbitrary **3-vectors**:

$$(a_1\mathbf{e_1} + a_2\mathbf{e_2} + a_3\mathbf{e_3}) * (b_1\mathbf{e_1} + b_2\mathbf{e_2} + b_3\mathbf{e_3})$$

$$= a_1b_1(\mathbf{e_1} * \mathbf{e_1}) + a_1b_2(\mathbf{e_1} * \mathbf{e_2}) + a_1b_3(\mathbf{e_1} * \mathbf{e_3})$$
$$+ a_2b_1(\mathbf{e_2} * \mathbf{e_1}) + a_2b_2(\mathbf{e_2} * \mathbf{e_2}) + a_2b_3(\mathbf{e_2} * \mathbf{e_3})$$
$$+ a_3b_1(\mathbf{e_3} * \mathbf{e_1}) + a_3b_2(\mathbf{e_3} * \mathbf{e_2}) + a_3b_3(\mathbf{e_3} * \mathbf{e_3})$$

and so again the result for *any* pair of **vectors** is defined once we know the result for all the **products** on the **basis vectors**, or all the $\mathbf{e}_i \times \mathbf{e}_j$ values. We're still assuming all these results are **real numbers** or **scalars**, and so any **product** of this form is called a **scalar product**.

The complete set of $\mathbf{e}_i * \mathbf{e}_j$ values is a **matrix** commonly written $\{g_{ij}\}$, and for reasons that will be apparent shortly, is called a **metric tensor**. This **matrix**, in a four-dimensional form, is the hub around which Einstein's theory of **General Relativity** revolves.

The present 3 × 3 form looks like this:

$$\begin{pmatrix} g_{11} & g_{12} & g_{13} \\ g_{21} & g_{22} & g_{23} \\ g_{31} & g_{32} & g_{33} \end{pmatrix} = \begin{pmatrix} \mathbf{e}_1 * \mathbf{e}_1 & \mathbf{e}_1 * \mathbf{e}_2 & \mathbf{e}_1 * \mathbf{e}_3 \\ \mathbf{e}_2 * \mathbf{e}_1 & \mathbf{e}_2 * \mathbf{e}_2 & \mathbf{e}_2 * \mathbf{e}_3 \\ \mathbf{e}_3 * \mathbf{e}_1 & \mathbf{e}_3 * \mathbf{e}_2 & \mathbf{e}_3 * \mathbf{e}_3 \end{pmatrix}.$$

Can we set up a **correspondence** between members of the **dual space** and members of our *basic* **vector space** using this **product**?

Well, we can do it quite simply. Given any **basis** ϕ_1, ϕ_2, ϕ_3 of the **dual space**, and a **basis** of the basic **space** $\mathbf{e_1}, \mathbf{e_2}, \mathbf{e_3}$ we need only define:

$$\mathbf{e}_i * \mathbf{e}_j \equiv \phi_i(\mathbf{e}_j)$$

and we will have the identity:

$$(a_1\mathbf{e_1} + a_2\mathbf{e_2} + a_3\mathbf{e_3}) * (b_1\mathbf{e_1} + b_2\mathbf{e_2} + b_3\mathbf{e_3})$$

$$= (a_1\phi_1 + a_2\phi_2 + a_3\phi_3)\,(b_1\mathbf{e_1} + b_2\mathbf{e_2} + b_3\mathbf{e_3}) = \phi(b_1\mathbf{e_1} + b_2\mathbf{e_2} + b_3\mathbf{e_3})$$

for any member ϕ of the **dual space**.

One particular **scalar product** is especially important. This is defined by:

$$\mathbf{e}_i * \mathbf{e}_j = \delta_{ij} := 1 \text{ if } i = j, = 0 \text{ if } i \neq j.$$

where δ_{ij} is called the **Kronecker delta** and is defined (":=") as being 1 if $i = j$, and 0 otherwise. With **3-vectors** this means that six out of the nine **basis products** are zero, or that:

$$\begin{pmatrix} g_{11} & g_{12} & g_{13} \\ g_{21} & g_{22} & g_{23} \\ g_{31} & g_{32} & g_{33} \end{pmatrix} =. \begin{pmatrix} 1 & 0 & 0 \\ 0 & 1 & 0 \\ 0 & 0 & 1 \end{pmatrix}.$$

For this $\{g_{ij}\}$ the **scalar product** reduces to:

$$(a_1\mathbf{e}_1 + a_2\mathbf{e}_2 + a_3\mathbf{e}_3) * (b_1\mathbf{e}_1 + b_2\mathbf{e}_2 + b_3\mathbf{e}_3)$$

$$= a_1b_1(\mathbf{e}_1 * \mathbf{e}_1) + a_2b_2(\mathbf{e}_2 * \mathbf{e}_2) + a_3b_3(\mathbf{e}_3 * \mathbf{e}_3) = a_1b_1 + a_2b_2 + a_3b_3.$$

This **scalar product** is so widely used that it is commonly referred to as *the* **scalar product**. I will call it a **Kronecker scalar product**. Note that it can be defined for *any* **basis**, but one particular **Kronecker scalar product** is pre-eminent for **standard vectors**. This is the unique **Kronecker scalar product** for the **natural basis**, which for **3-vectors** is: $\mathbf{e}_1 = [1, 0, 0]$, $\mathbf{e}_2 = [0, 1, 0]$, $\mathbf{e}_3 = [0, 0, 1]$, when we have:

$$[a_1, a_2, a_3] \cdot [b_1, b_2, b_3] = a_1b_1 + a_2b_2 + a_3b_3$$

replacing the $*$ with a dot "$\cdot$", from which this particular **scalar product** – *the* **scalar product** *par excellence* – takes its alternative name of the **dot product.**[61]

Note that the **dot product** is **commutative**. For any two **vectors**:

$$\mathbf{u} \cdot \mathbf{v} = \mathbf{v} \cdot \mathbf{u}$$

always. For historical reasons this **commutative** property is in this context called the **symmetric** property.

The **Kronecker scalar product** can also be expressed by the formula:

$$\mathbf{u} \cdot \mathbf{v} = \textstyle\sum_{i=1}^{3} \sum_{j=1}^{3} u_i \, \mathbf{e}_i \cdot v_j \mathbf{e}_j = \sum_{i=1}^{3} \sum_{j=1}^{3} u_i \, v_j \mathbf{e}_i \cdot \mathbf{e}_j$$

$$= \textstyle\sum_{i=1}^{3} \sum_{j=1}^{3} u_i \, v_j \delta_{ij} = \sum_{i=1}^{3} u_i \, v_i.$$

[61] The **dot product** is also sometimes called the **inner product** to distinguish it from the **alternating product** which we will encounter shortly, which is sometimes called the **outer product.**

It would be possible to have an **anti-commutative** or **anti-symmetric scalar product**, and I give this as a second example. We just choose the $\mathbf{e}_i * \mathbf{e}_j$ **matrix** that takes the form:

$$\begin{pmatrix} 0 & 1 & 1 \\ -1 & 0 & 1 \\ -1 & -1 & 0 \end{pmatrix}$$

or $\mathbf{e}_i * \mathbf{e}_j := 1$ if $i < j$, $= 0$ if $i = j$, and $= -1$ if $i > j$ where i is the *row index* and j the *column*.

Now we have:

$$(a_1\mathbf{e}_1 + a_2\mathbf{e}_2 + a_3\mathbf{e}_3) * (b_1\mathbf{e}_1 + b_2\mathbf{e}_2 + b_3\mathbf{e}_3)$$

$$= a_1b_2(\mathbf{e}_1 * \mathbf{e}_2) + a_1b_3(\mathbf{e}_1 * \mathbf{e}_3) + a_2b_1(\mathbf{e}_2 * \mathbf{e}_1)$$

$$+ a_2b_3(\mathbf{e}_2 * \mathbf{e}_3) + a_3b_1(\mathbf{e}_3 * \mathbf{e}_1) + a_3b_2(\mathbf{e}_3 * \mathbf{e}_2)$$

$$= a_1b_2 + a_1b_3 + a_2b_3 - a_2b_1 - a_3b_1 - a_3b_2$$

and swapping the arguments around

$$(b_1\mathbf{e}_1 + b_2\mathbf{e}_2 + b_3\mathbf{e}_3) * (a_1\mathbf{e}_1 + a_2\mathbf{e}_2 + a_3\mathbf{e}_3)$$

$$= b_1a_2 + b_1a_3 + b_2a_3 - b_2a_1 - b_3a_1 - b_3a_2$$

$$= - a_1b_2 - a_1b_3 - a_2b_3 + a_2b_1 + a_3b_1 + a_3b_2$$

so that $\mathbf{b} * \mathbf{a} = -\,\mathbf{a} * \mathbf{b}$. This **product** is **non-commutative** or **anti-commutative**. It could still be seen as a **scalar product**, but in practice such **non-commutative** forms are expressly excluded.

Having shown that there is a large possible number of **scalar products**, both **symmetric** (**commutative**) and **antisymmetric** (**anticommutative**), I need to make it clear that *in practice NONE of these is used except the* **dot product** *defined as the* **Kronecker scalar product** *for the* **natural basis**. Also, **scalar products** are normally *defined* as necessarily **symmetric** and the **antisymmetric** possibilities are not admitted.

So **scalar products** are usually *defined* as obeying these axioms, where α and β are **real numbers**:

Axiom I:	For all **x** and **y** in the **vector space**, $\mathbf{x} * \mathbf{y}$ is a **real number.**	Result is a **scalar**
Axiom II:	$(\alpha\mathbf{x} + \beta\mathbf{y}) * \mathbf{z} = \alpha(\mathbf{x} * \mathbf{z}) + \beta(\mathbf{y} * \mathbf{z})$ $\mathbf{x} * (\alpha\mathbf{y} + \beta\mathbf{z}) = \alpha(\mathbf{x} * \mathbf{y}) + \beta(\mathbf{x} * \mathbf{z})$	Bilinearity
AxiomIII:	$\mathbf{x} * \mathbf{y} = \mathbf{y} * \mathbf{x}$	Symmetry (Commutativity)

and *sometimes...*

Axiom IV: $\mathbf{x} * \mathbf{x} \geq 0$. Positive Definiteness

The normal **dot product** is also used to define a **magnitude** for any **vector**:[62]

$$|\mathbf{x}| = \sqrt{(\mathbf{x} \cdot \mathbf{x})}$$

and this enables us to define **unit vectors** as having a **magnitude** of 1. So in our **natural basis**, all the vectors $\mathbf{e}_i$ are **unit vectors**: $|\ \mathbf{e}_i\ | = \sqrt{(\mathbf{e}_i \cdot \mathbf{e}_i)} = (\mathbf{e}_i \cdot \mathbf{e}_i) = 1$. Also to *any* **vector x** there corresponds a **unit vector** $\underline{\mathbf{x}}$, defined by:

$$\underline{\mathbf{x}} = \mathbf{x}/|\mathbf{x}|.$$

I will introduce the notation $\underline{\mathbf{x}}$ **=: norm(x)** = $\mathbf{x}/|\mathbf{x}|$ to describe this operation of forming a **unit vector** from an arbitrary **vector.**[63] The term comes from a synonym for the **magnitude**: the **norm**, and the operation is called **normalization**.

This is especially useful with **2-vectors** as the corresponding **unit vector** is now an **angle** in the sense of Chapter 4. Thus:

$$\mathbf{x} = x\mathbf{e}_1 + y\mathbf{e}_2 = r(\theta_x\mathbf{e}_1 + \theta_y\mathbf{e}_2)$$

where $r^2 = x^2 + y^2 = \mathbf{x} \cdot \mathbf{x}$, so that $r = |\mathbf{x}|$ and so:

$$\theta_x\mathbf{e}_1 + \theta_y\mathbf{e}_2 = \mathbf{x}/r = \mathbf{x}/|\mathbf{x}|,$$

so that

$$\theta_x\mathbf{e}_1 + \theta_y\mathbf{e}_2 = \underline{\mathbf{x}}.$$

One last bit of terminology: members of the **dual space** are sometimes referred to as **covectors**.

5.5 Reciprocal Basis and Tensor Product

Given any **basis** of a **vector space**, there is one particular **basis** for the **dual space** that is particularly useful. This is the **basis** that obeys the **Kronecker** property that:

$$\varepsilon^i(\mathbf{e}_j) = \delta^i_j$$

[62] It is this concept that leads to the term **metric tensor** for the **scalar product** matrix. A **metric** is a measure.

[63] I use **norm(x)** $\rightarrow \underline{\mathbf{x}}$ here as a **vector operator** to generate a **unit vector** (hence underlined) with the same direction as the given **vector**. But the "**norm**" of a **vector** is also, as stated above, its **scalar magnitude.** I will represent this in equations as norm(**x**), not written bold so as to distinguish this **scalar** quantity. So norm(**norm**(**x**)) = 1 always.

where I introduce the special notation $\{\varepsilon^i\}$ for this **basis** of the **dual space**, and also introduce a very important convention, that *indices of items in the* **dual space** *are superscripted rather than subscripted*. The reason for this convention is that **vectors** in the two spaces transform in different ways when we switch from one **basis** to another, and we will change to a notation in which everything that transforms like the **basis** of the basic **vector space** is *subscripted*, and everything that transforms in the same way as the **basis** of the **dual space** is *superscripted*. The distinction is important.

δ^i_j is just the same **Kronecker delta** for the combination of subscripts and superscripts, and again is 1 if *$i = j$, zero otherwise.*

What would constitute a reasonable standard basis for reproducing the **functions** of the **dual space** using **vectors** in the basic **vector space** with a **scalar product**? Well, we could cheat and define *a local* **Kronecker scalar product** as:

$$\mathbf{e}_i * \mathbf{e}_j = \delta_{ij}$$

so that any ϕ in the **dual space** could be written as $\phi = a_1\mathbf{e}_1 + a_2\mathbf{e}_2 + a_3\mathbf{e}_3$ for **3-vectors**, and this would make the *same* **basis** a valid **basis** for the **dual space functions**. But we would rather avoid a multiplicity of different **scalar products** for different **bases**, so it turns out to be less confusing to use the standard **dot product** for the **natural basis** and to define the so-called **reciprocal basis** to any given **basis** as:[64]

$$\mathbf{e}^i \cdot \mathbf{e}_j = \delta^i_j.$$

To see how this works, let's look at an example. Going back to our original alternative **basis**, where:

$$\mathbf{e}_1 = [1, 2, 3], \qquad \mathbf{e}_2 = [1, 1, 1], \qquad \mathbf{e}_3 = [5, 9, 6]$$

the **reciprocal basis** to this **basis** is defined by:

$$\begin{array}{ll}
\mathbf{e}^1: & \mathbf{e}^1 \cdot \mathbf{e}_1 = (a_{11}\mathbf{e}_1 + a_{12}\mathbf{e}_2 + a_{13}\mathbf{e}_3) \cdot \mathbf{e}_1 = 1 \\
 & \mathbf{e}^1 \cdot \mathbf{e}_2 = (a_{11}\mathbf{e}_1 + a_{12}\mathbf{e}_2 + a_{13}\mathbf{e}_3) \cdot \mathbf{e}_2 = 0 \\
 & \mathbf{e}^1 \cdot \mathbf{e}_3 = (a_{11}\mathbf{e}_1 + a_{12}\mathbf{e}_2 + a_{13}\mathbf{e}_3) \cdot \mathbf{e}_3 = 0 \\
\mathbf{e}^2: & \mathbf{e}^2 \cdot \mathbf{e}_1 = (a_{21}\mathbf{e}_1 + a_{22}\mathbf{e}_2 + a_{23}\mathbf{e}_3) \cdot \mathbf{e}_1 = 0 \\
 & \mathbf{e}^2 \cdot \mathbf{e}_2 = (a_{21}\mathbf{e}_1 + a_{22}\mathbf{e}_2 + a_{23}\mathbf{e}_3) \cdot \mathbf{e}_2 = 1 \\
 & \mathbf{e}^2 \cdot \mathbf{e}_3 = (a_{21}\mathbf{e}_1 + a_{22}\mathbf{e}_2 + a_{23}\mathbf{e}_3) \cdot \mathbf{e}_3 = 0 \\
\mathbf{e}^3: & \mathbf{e}^3 \cdot \mathbf{e}_1 = (a_{31}\mathbf{e}_1 + a_{32}\mathbf{e}_2 + a_{33}\mathbf{e}_3) \cdot \mathbf{e}_1 = 0 \\
 & \mathbf{e}^3 \cdot \mathbf{e}_2 = (a_{31}\mathbf{e}_1 + a_{32}\mathbf{e}_2 + a_{33}\mathbf{e}_3) \cdot \mathbf{e}_2 = 0 \\
 & \mathbf{e}^3 \cdot \mathbf{e}_3 = (a_{31}\mathbf{e}_1 + a_{32}\mathbf{e}_2 + a_{33}\mathbf{e}_3) \cdot \mathbf{e}_3 = 1
\end{array}$$

where we write $\mathbf{e}^1 = a_{11}\mathbf{e}_1 + a_{12}\mathbf{e}_2 + a_{13}\mathbf{e}_3$, $\mathbf{e}^2 = a_{21}\mathbf{e}_1 + a_{22}\mathbf{e}_2 + a_{23}\mathbf{e}_3$ and $\mathbf{e}^3 = a_{31}\mathbf{e}_1 + a_{32}\mathbf{e}_2 + a_{33}\mathbf{e}_3$. So substituting for $\mathbf{e}_1$, $\mathbf{e}_2$ and $\mathbf{e}_3$, and expanding the **dot product**, the equation for $\mathbf{e}^1$ becomes:

$$\begin{array}{ll}
\mathbf{e}^1: & \mathbf{e}^1 \cdot \mathbf{e}_1 = a_{11} + 2a_{12} + 3a_{13} = 1 \\
 & \mathbf{e}^1 \cdot \mathbf{e}_2 = a_{11} + a_{12} + a_{13} = 0
\end{array}$$

[64] The "Big Black Book" (*Misner, Thorne and Wheeler*) calls this the **dual basis**.

$$\mathbf{e}^1 \cdot \mathbf{e}_3 = 5a_{11} + 9a_{12} + 6a_{13} = 0$$

which involves the **transpose** of the **matrix** we used in Section 5.1, so I will solve this by **Cramer's rule** using the **determinant**:

$$\begin{vmatrix} 1 & 2 & 3 \\ 1 & 1 & 1 \\ 5 & 9 & 6 \end{vmatrix}$$

$$= 1\times(6-9) - 2\times(6-5) + 3(9-5) = -3 - 2 + 12 = 7, \text{ as before}$$

and so

$$a_{11} = \begin{vmatrix} 1 & 2 & 3 \\ 0 & 1 & 1 \\ 0 & 9 & 6 \end{vmatrix} \div 7 = -3/7$$

$$a_{12} = \begin{vmatrix} 1 & 1 & 3 \\ 1 & 0 & 1 \\ 5 & 0 & 6 \end{vmatrix} \div 7 = (-1 \times (6-5))/7 = -1/7$$

where I first swap the first and second columns, changing the sign, before expanding, and

$$a_{13} = \begin{vmatrix} 1 & 2 & 1 \\ 1 & 1 & 0 \\ 5 & 9 & 0 \end{vmatrix} \div 7 = (9-5)/7 = 4/7$$

so that $\mathbf{e}^1 = -(3/7)\mathbf{e}_1 - (1/7)\mathbf{e}_2 + (4/7)\mathbf{e}_3$.

We can solve for $\mathbf{e}^2$ and $\mathbf{e}^3$ similarly. We can check the result for $\mathbf{e}^1$ by evaluating:

$\mathbf{e}^1$:

$$\mathbf{e}^1 \cdot \mathbf{e}_1 = -(3/7) + 2\times(-1/7)) + 3(4/7) = -5/7 + 12/7 = 7/7 = 1$$
$$\mathbf{e}^1 \cdot \mathbf{e}_2 = -(3/7) - 1/7 + (4/7) = 0$$
$$\mathbf{e}^1 \cdot \mathbf{e}_3 = 5\times(-(3/7)) + 9\times(-(1/7)) + 6(4/7) = -15/7 - 9/7 + 24/7 = 0$$

Note that for the **natural basis,**

$$\mathbf{e}_i \cdot \mathbf{e}_j = \delta_{ij} \qquad \textit{anyway}$$

so *the* **natural basis** *is its own* **reciprocal basis**. Any **basis** that has this property that its **basis vectors** obey the **Kronecker property** for the **dot product** is called an **orthonormal basis**. If the **basis** obeys the weaker property that:

$$\mathbf{e}_i \cdot \mathbf{e}_j = 0 \text{ if } i \neq j$$

i.e. that all off-diagonal **dot products** are zero, but the diagonal **products** are not necessarily 1, it is called an **orthogonal basis**. Indeed, any two vectors **u** and **v** for which the **dot product** $\mathbf{u} \cdot \mathbf{v} = 0$ are called **orthogonal**. *These terms are very, very important.*

The **natural basis** is not the *only* **orthonormal basis**. We can always define any new **basis** by a **transform** or **transformation matrix** which for **3-vectors** would take the form:

$$\mathbf{f}_1 = a_{11}\mathbf{e}_1 + a_{12}\mathbf{e}_2 + a_{13}\mathbf{e}_3$$
$$\mathbf{f}_2 = a_{21}\mathbf{e}_1 + a_{22}\mathbf{e}_2 + a_{23}\mathbf{e}_3$$
$$\mathbf{f}_3 = a_{31}\mathbf{e}_1 + a_{32}\mathbf{e}_2 + a_{33}\mathbf{e}_3$$

using an arbitrary $\{a_{ij}\}$ **matrix.**

What properties must the $\{a_{ij}\}$ have to make the $(\mathbf{f}_j\}$ **orthonormal**?

Now an important issue here is that the **components** of any **vector** transform *differently* to the **transformation** of the **basis** elements. To see this, I'll temporarily go back to the simplest case of **2-vectors.**

For **2-vectors**, we have just:

$$\mathbf{f}_1 = a_{11}\mathbf{e}_1 + a_{12}\mathbf{e}_2$$
$$\mathbf{f}_2 = a_{21}\mathbf{e}_1 + a_{22}\mathbf{e}_2$$

Given an arbitrary **vector x**, we have in the old **components** $\mathbf{x} = x_1\mathbf{e}_1 + x_2\mathbf{e}_2$. Let's say that in the *new* **basis**, the **components** will be (y_1, y_2): $\mathbf{x} = y_1\mathbf{f}_1 + y_2\mathbf{f}_2$.

Now, expanding the $\mathbf{f}_i$'s:

$$\mathbf{x} = y_1\mathbf{f}_1 + y_2\mathbf{f}_2 = y_1(a_{11}\mathbf{e}_1 + a_{12}\mathbf{e}_2) + y_2(a_{21}\mathbf{e}_1 + a_{22}\mathbf{e}_2)$$

$$= (y_1a_{11} + y_2a_{21})\mathbf{e}_1 + (y_1a_{12} + y_2a_{22})\mathbf{e}_2,$$

So we must have:

$$a_{11}y_1 + a_{21}y_2 = x_1$$
$$a_{12}y_1 + a_{22}y_2 = x_2.$$

Straight away, *note the* **transpose** *nature of this pair of equations*: the columns and rows have been interchanged. Now the *second* **index** of a_{ij} is indexing the equation, and the *first* the variable.

Solving these by **Cramer's rule**, we find:

$$y_1 = \begin{vmatrix} x_1 & a_{21} \\ x_2 & a_{22} \end{vmatrix} \div \begin{vmatrix} a_{11} & a_{21} \\ a_{12} & a_{22} \end{vmatrix} \qquad y_2 = \begin{vmatrix} a_{11} & x_1 \\ a_{12} & x_2 \end{vmatrix} \div \begin{vmatrix} a_{11} & a_{21} \\ a_{12} & a_{22} \end{vmatrix}$$

So

$$y_1 = (x_1a_{22} - x_2a_{21})/(a_{11}a_{22} - a_{21}a_{12}) = (x_1a_{22} - x_2a_{21})/\mathbf{det(A)}$$
$$y_2 = (x_2a_{11} - x_1a_{12})/(a_{11}a_{22} - a_{21}a_{12}) = (x_2a_{11} - x_1a_{12})/\mathbf{det(A)}$$

but from Section 1.8 this is:

$$\mathbf{y} = \begin{pmatrix} y_1 \\ y_2 \end{pmatrix} = (\mathbf{A}^{-1})^{\mathrm{T}} \times \mathbf{x} = 1/|\mathbf{A}| \times \begin{pmatrix} a_{22} & -a_{21} \\ -a_{12} & a_{11} \end{pmatrix} \times \begin{pmatrix} x_1 \\ x_2 \end{pmatrix}$$

so the **components transform** *as the* **transpose** *of the* **inverse** *of the* **basis vectors' transform**. This is a hugely important distinction, which will be central to **differential geometry** in Volume 2.

> If **basis vectors** transform like **A**, **vector components** transform like $(\mathbf{A}^{-1})^{\mathrm{T}}$.
>
> I will call **A** and $(\mathbf{A}^{-1})^{\mathrm{T}}$ **dual transforms.**

To distinguish these cases, it is customary to give things that **transform** with **A**, like the **basis vectors**, *subscript* **indices**, and to give things that **transform with** $(\mathbf{A}^{-1})^{\mathrm{T}}$, the **transpose** of the **inverse**, like the **components**, *superscript* **indices**. I'll use this convention in the rest of this Section as an exemplar, but thereafter in this book I'll return to just subscripts.

Things that transform the same way – say both by **A** – are said to transform **cogrediently**. Two things that transform in *different* ways – one according to **A** and one according to $(\mathbf{A}^{-1})^{\mathrm{T}}$ – are said to transform **contragrediently.**

Next comes a subtle twist.

> The **reciprocal basis** turns out to be one thing that **transforms** like the **components.**

To see this, let's first go back to the question above: if the $\{\mathbf{e}_i\}$ **basis** is **orthonormal**, what property does the $\{a_{ij}\}$ **matrix** need to make $\{\mathbf{f}_i\}$ also **orthonormal**? We can answer this by looking at the defining equation for the **reciprocal basis**:

$$\mathbf{f}^i \cdot \mathbf{f}_j = \delta^i_j$$

or $$(b^i_1\mathbf{e}_1 + b^i_2\mathbf{e}_2 + b^i_3\mathbf{e}_3) \cdot (a_{j1}\mathbf{e}_1 + a_{j2}\mathbf{e}_2 + a_{j3}\mathbf{e}_3) = \delta^i_j$$

where $\{b^i_j\}$ is the **matrix** defining the **reciprocal basis** from the original **basis**. Because the $\{\mathbf{e}_i\}$ **basis** is **orthonormal**, this becomes:

$$\Sigma_k\, b^i_k \times a_{jk} = \delta^i_j$$

Matrix multiplication is defined for *square* **matrices** – having the same number of rows as columns – as:

$$\mathbf{AB} = \{a_{ij}\} \times \{b_{ij}\} := \{\Sigma_k\, a_{ik} \times b_{kj}\} = \{c_{ij}\} = \mathbf{C}$$

where we sum on the *second* (the *column*) index of the *first* **argument** with the *first* (or *row*) index on the second.

The **matrix** $\{\delta_{ij}\}$ or $\{\delta^i_j\}$ is the **identity matrix** for this **multiplication** and as a **matrix** we can write it **I**. Multiplying a **matrix** by it leaves the **matrix** unchanged:

$$\{\Sigma_k\, a_{ik}\times \delta_{kj}\} = \{a_{ij}\} \qquad \text{or} \qquad \mathbf{AI} = \mathbf{IA} = \mathbf{A}$$

If two **matrices** multiply together to give the **identity matrix**, they are **inverses**. In the formula $\Sigma_k\, b^i_k\times a_{jk} = \delta^i_j$, we are summing on the *column* index of *both* **arguments**. But the **transpose** of $\{b^i_j\}$ is defined:

$$\mathbf{B}^\mathrm{T} = \{b^i_j\}^\mathrm{T} = \{b^j_i\}$$

So $\Sigma_k\, b^i_k\times a_{jk} = \delta^i_j$ is equivalent to $\mathbf{B}^\mathrm{T}\mathbf{A} = \mathbf{I}$, or $\mathbf{B}^\mathrm{T} = \mathbf{A}^{-1}$.

So the condition that $\{\mathbf{f}^i\} = \{\mathbf{f}_i\}$ or that the transformed **reciprocal basis** is the transformed *ordinary* **basis**, and so that the new **basis** is **orthonormal**, is that:

$$\mathbf{B} = \mathbf{A} \text{ or that } \mathbf{A}^\mathrm{T} = \mathbf{A}^{-1}$$

i.e. that the **transpose** of the **transform matrix** equals its **inverse**.

So **transforms** *where*

$$(\mathbf{A}^{-1})^\mathrm{T} = \mathbf{A}$$

preserve **scalar products** *and* **orthonormality**, and so are called **orthogonal transforms**.

What happens when this is *not* the case?

I'll develop the answer here by taking as an example a **transform** in the **complex plane** defined by:

$$\mathbf{e}_1 = \mathbf{1} + \mathbf{j}$$
$$\mathbf{e}_2 = -\mathbf{1},$$

where I write the original ("**natural**") **basis vectors** as **1** and **j** in bold to stress their vectorial nature. This is a good example partly because of potential pitfalls: it's easy to be side-tracked into the "wrong" interpretation here. The product that we'll need is the **real scalar product**,

$$(x + jy)\cdot(p + jq) = xp + yq,$$

not the standard **complex multiplication** $(x.\mathbf{1} + y.\mathbf{j}) \times (p.\mathbf{1} + q.\mathbf{j}) = (xp - yq)\mathbf{1} + (xq + yp)\mathbf{j}$ which in this context, as mentioned in Section 5.3, constitutes a **vector product**.

So this **transform** is defined by:

$$\mathbf{A} = \begin{pmatrix} 1 & 1 \\ -1 & 0 \end{pmatrix} \rightarrow \begin{pmatrix} \mathbf{e}_1 \\ \mathbf{e}_2 \end{pmatrix} = \begin{pmatrix} 1 & 1 \\ -1 & 0 \end{pmatrix} \begin{pmatrix} \mathbf{1} \\ \mathbf{j} \end{pmatrix}.$$

So

$$\mathbf{1} = -\mathbf{e}_2$$

$$\mathbf{j} = \mathbf{e}_2 + \mathbf{e}_1 = -\mathbf{1} + \mathbf{1} + \mathbf{j}$$

giving the **inverse matrix** $\mathbf{A}^{-1}$ as:

$$\mathbf{A}^{-1} = \begin{pmatrix} 0 & -1 \\ 1 & 1 \end{pmatrix} \qquad \rightarrow \qquad (\mathbf{A}^{-1})^{\mathrm{T}} = \begin{pmatrix} 0 & 1 \\ -1 & 1 \end{pmatrix}.$$

Clearly $\mathbf{e}_1 \cdot \mathbf{e}_2 = 1.(-1) + 1.0 = -1$, and $\mathbf{e}_1 \cdot \mathbf{e}_1 = 1.\,1 + 1.1 = 2$, so the new **basis** is *not* **orthonormal.**

This is as we would expect from the fact that here $(\mathbf{A}^{-1})^{\mathrm{T}} \neq \mathbf{A}$.

In the new **basis**, a **vector** $\mathbf{z} = x\mathbf{1} + y\mathbf{j} = x + j.y$ will transform, using $(\mathbf{A}^{-1})^{\mathrm{T}}$, into:

$$\begin{pmatrix} 0 & 1 \\ -1 & 1 \end{pmatrix} \begin{pmatrix} x \\ y \end{pmatrix} = \begin{pmatrix} y \\ -x + y \end{pmatrix},$$

which does indeed give:

$$y.\mathbf{e}_1 + (y - x).\mathbf{e}_2 = y.(\mathbf{1} + \mathbf{j}) + (y - x).(-\mathbf{1}) = (y - y + x).\mathbf{1} + y.\mathbf{j} = x\mathbf{1} + y\mathbf{j}.$$

Now in the original ("**natural**") **basis**, the **reciprocal basis** *is* the ordinary **basis**:

$\mathbf{1} \cdot \mathbf{1} = (1, 0) \cdot (1, 0) = 1$
$\mathbf{1} \cdot \mathbf{j} = (1, 0) \cdot (0, 1) = 0$
$\mathbf{j} \cdot \mathbf{1} = (0, 1) \cdot (1, 0) = 0$
$\mathbf{j} \cdot \mathbf{j} = (0, 1) \cdot (0, 1) = 1.$

So we should use the **dual transform** $(\mathbf{A}^{-1})^{\mathrm{T}}$ to get the *new* **dual basis** or **reciprocal basis**:

$$(\mathbf{A}^{-1})^{\mathrm{T}} \times \begin{pmatrix} \mathbf{1} \\ \mathbf{j} \end{pmatrix} = \begin{pmatrix} 0 & 1 \\ -1 & 1 \end{pmatrix} \begin{pmatrix} \mathbf{1} \\ \mathbf{j} \end{pmatrix} = \begin{pmatrix} \mathbf{j} \\ -\mathbf{1} + \mathbf{j} \end{pmatrix}.$$

So the new **reciprocal basis** is:

$$\boldsymbol{\varepsilon}^1 = \mathbf{j} = (0, 1), \qquad \boldsymbol{\varepsilon}^2 = -\mathbf{1} + \mathbf{j} = (-1, 1).$$

The figures in brackets here give the new **reciprocal basis** in terms of the old **basis**.

So now:

$\boldsymbol{\varepsilon}^1 \cdot \mathbf{e}_1 = (0, 1) \cdot (1, 1) = 1$
$\boldsymbol{\varepsilon}^1 \cdot \mathbf{e}_2 = (0, 1) \cdot (-1, 0) = 0$
$\boldsymbol{\varepsilon}^2 \cdot \mathbf{e}_1 = (-1, 1) \cdot (1, 1) = -1 + 1 = 0$
$\boldsymbol{\varepsilon}^2 \cdot \mathbf{e}_2 = (-1, 1) \cdot (-1, 0) = 1$

as should be, evaluating the **scalar products** in terms of the original **basis**, where and only where they are defined up until now. Hereafter we can use these $\boldsymbol{\varepsilon}^i \cdot \mathbf{e}_j$ expressions to evaluate them.

Finally, if we have a **dual space** element ϕ defined as:

$$\phi = \phi_1\mathbf{1} + \phi_2\mathbf{j},$$

we transform ϕ_1 and ϕ_2 using the *original* **transform A**:

$$\begin{pmatrix}\varphi_1^*\\ \varphi_2^*\end{pmatrix} = \mathbf{A} \times \begin{pmatrix}\varphi_1\\ \varphi_2\end{pmatrix} = \begin{pmatrix}1 & 1\\ -1 & 0\end{pmatrix}\begin{pmatrix}\varphi_1\\ \varphi_2\end{pmatrix},$$

so $\phi_1^* = \phi_1 + \phi_2$, $\phi_2^* = -\phi_1$, giving:

$$\phi = \phi_1\boldsymbol{\varepsilon}^1 + \phi_2\boldsymbol{\varepsilon}^2 = (\phi_1 + \phi_2)(\mathbf{j}) + (-\phi_1)(-\mathbf{1} + \mathbf{j}) = \phi_1\mathbf{j} + \phi_2\mathbf{j} + \phi_1\mathbf{1} - \phi_1\mathbf{j} = \phi_1\mathbf{1} + \phi_2\mathbf{j}.$$

So the **reciprocal basis** or **dual basis** transforms like the **components** of the ordinary **basis**, **contragrediently** to the ordinary **basis** elements themselves, and *its* **components** transform **cogrediently** with the ordinary **basis**. So we have **. . .**

Basis vectors transform like **A**.
Basis vector components transform like $(\mathbf{A}^{-1})^{\mathrm{T}}$.

Dual basis vectors (reciprocal basis vectors) transform like $(\mathbf{A}^{-1})^{\mathrm{T}}$.
Dual basis vector components transform like **A**.

Again, I will call **A** and $(\mathbf{A}^{-1})^{\mathrm{T}}$ **dual transforms**.

If $(\mathbf{A}^{-1})^{\mathrm{T}} = \mathbf{A}$, the transform is **orthogonal**.

This general pattern is illustrated in *Figure 5.5.1* using the foregoing example.

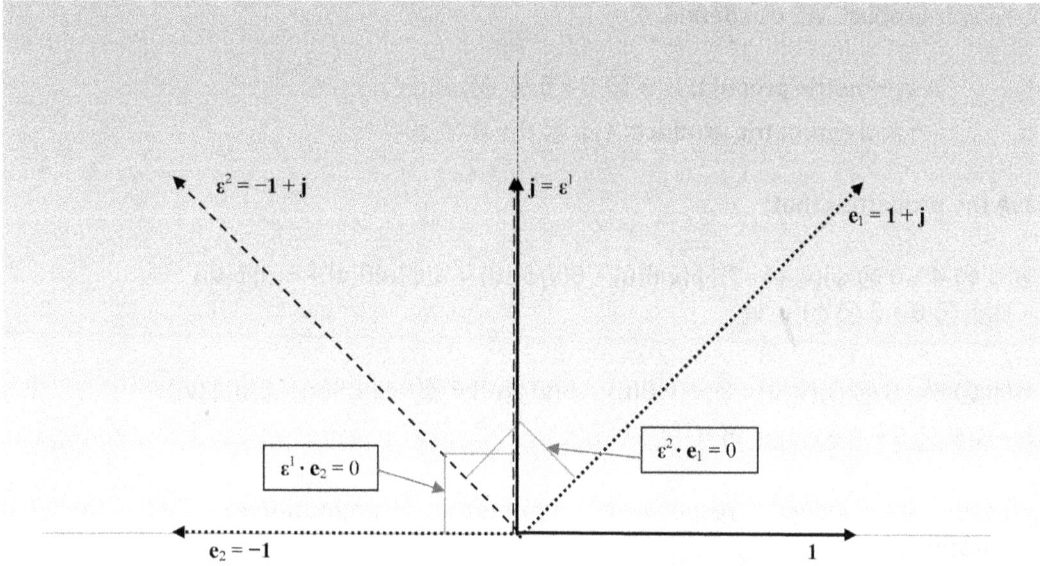

Figure 5.5.1

It is possible to define a kind of "product" of elements in the **dual space** (**covectors**) that operate on two **vectors** at once. We write these elements as:

$\phi \otimes \theta\ (\mathbf{u}, \mathbf{v}) = \phi(\mathbf{u}).\theta(\mathbf{v})$

where $\phi \otimes \theta$ is a new kind of **vector** belonging to a new **vector space** that has the **dimension** n^2 if the original **vector space** and its **dual** had the **dimension** n. This new **vector space** is called a higher **grade** space, and the "product" $\otimes$ is called the **tensor product**. The new **space** has the **basis**:

$\{\mathbf{e}^i \otimes \mathbf{e}^j\}$

which clearly has n^2 elements.

Because of the symmetry between the basic **vector space** and its **dual**, we can also regard the *basic* **vectors** as **functions** acting on members of the **dual space** and write:

$\mathbf{u} \otimes \mathbf{v}\ (\phi, \theta) = \mathbf{u}(\phi).\mathbf{v}(\theta)$

where we *define* $\mathbf{u}(\phi) := \phi(\mathbf{u})$. In other words, this **functional** form is **symmetric** or **commutative** *by definition*. This idea enables us to define a higher **grade** from the *original* **vector space** with **basis**:

$\{\mathbf{e}_i \otimes \mathbf{e}_j\}$

From any **tensor product**, we can define:

I. A **symmetric product**: $\frac{1}{2}(\phi \otimes \theta + \theta \otimes \phi)$, and . . .

II. An **antisymmetric product**: $\frac{1}{2}(\phi \otimes \theta - \theta \otimes \phi)$

which have the properties that:

$$\frac{1}{2}(\phi \otimes \theta + \theta \otimes \phi)(\mathbf{v}, \mathbf{u}) = \frac{1}{2}(\phi(\mathbf{v})\theta(\mathbf{u}) + \theta(\mathbf{v})\phi(\mathbf{u})) = \frac{1}{2}(\phi(\mathbf{u})\theta(\mathbf{v}) + \theta(\mathbf{u})\phi(\mathbf{v}))$$
$$= \frac{1}{2}(\phi \otimes \theta + \theta \otimes \phi)(\mathbf{u}, \mathbf{v})$$

and

$$\frac{1}{2}(\phi \otimes \theta - \theta \otimes \phi)(\mathbf{v}, \mathbf{u}) = \frac{1}{2}(\phi(\mathbf{v})\theta(\mathbf{u}) - \theta(\mathbf{v})\phi(\mathbf{u})) = \frac{1}{2}(-\phi(\mathbf{u})\theta(\mathbf{v}) + \theta(\mathbf{u})\phi(\mathbf{v}))$$
$$= -\frac{1}{2}(\phi \otimes \theta - \theta \otimes \phi)(\mathbf{u}, \mathbf{v})$$

hence these are indeed respectively **symmetric** (**commutative**) and **antisymmetric** (**anticommutative**).

Also:

$$\frac{1}{2}(\phi \otimes \theta + \theta \otimes \phi) + \frac{1}{2}(\phi \otimes \theta - \theta \otimes \phi) = \phi \otimes \theta + \frac{1}{2}\theta \otimes \phi - \frac{1}{2}\theta \otimes \phi = \phi \otimes \theta$$

So summed together these return the original product.

Much the more significant of these **products** is the **antisymmetric**. The **antisymmetric product**[65] is given the special symbol $\wedge$ so the example above could be written:

$$(\phi \wedge \theta)(\mathbf{u}, \mathbf{v}) = -(\phi \wedge \theta)(\mathbf{u}, \mathbf{v})$$

and since by the definition:

$$(\theta \wedge \phi)(\mathbf{u}, \mathbf{v}) = \frac{1}{2}(\theta \otimes \phi - \phi \otimes \theta)(\mathbf{u}, \mathbf{v})$$
$$= \frac{1}{2}(-(\phi \otimes \theta) + \theta \otimes \phi)(\mathbf{u}, \mathbf{v})$$
$$= -\frac{1}{2}(\phi \otimes \theta - \theta \otimes \phi)(\mathbf{u}, \mathbf{v})$$
$$= -(\phi \wedge \theta)(\mathbf{u}, \mathbf{v})$$

we also have that $\theta \wedge \phi = -\phi \wedge \theta$.

Note that this also has the consequence that:

$$\phi \wedge \phi = -\phi \wedge \phi$$

so

$$\phi \wedge \phi + \phi \wedge \phi = 2\phi \wedge \phi = 0 \qquad \text{or} \qquad \phi \wedge \phi = 0.$$

[65] This **antisymmetric product** is also called the **alternating product** or the **outer product**, the last distinguishing it from the **scalar product** which may be called the **inner product**.

5.6 Graded Algebras

The n^2 dimensional **tensor space** is the first of a series, called the second **grade** or **grade-2 space**. We could go on to define **tensors** of the form:

$$\phi \otimes \theta \otimes \omega \quad \text{or} \quad \mathbf{u} \otimes \mathbf{v} \otimes \mathbf{w}$$

which would define an n^3 dimensional space which is called the third **grade** or **grade-3 space**, and so on *ad infinitum*. Similarly, we can define **symmetric** and **antisymmetric** forms in the third **grade** by:

$$1/3!(\phi \otimes \theta \otimes \omega + \phi \otimes \omega \otimes \theta + \theta \otimes \phi \otimes \omega + \theta \otimes \omega \otimes \phi$$

$$+ \omega \otimes \phi \otimes \theta + \omega \otimes \theta \otimes \phi)$$

and

$$1/3!(\phi \otimes \theta \otimes \omega - \phi \otimes \omega \otimes \theta - \theta \otimes \phi \otimes \omega + \theta \otimes \omega \otimes \phi$$

$$+ \omega \otimes \phi \otimes \theta - \omega \otimes \theta \otimes \phi)$$

where in the latter (**antisymmetric**) case, **cyclic permutations** (123, 231, 312) get a positive sign and **anticyclic** (321, 213, 132) a negative, much as in the terms of a **determinant** discussed in Section 1.8. Note the leading 1/3! is 1/(3×2×1) = 1/6, using the 3 **factorial** symbol, as there are six **permutations**.

We will be much concerned with the **antisymmetric tensor space**, which has a finite limit to how many **grades** are possible, because of the $\phi \wedge \phi = 0$ property. The way this happens is this: each **grade** has a **basis** formed by all the possible **products** of the **basis vectors** of the underlying or **grade-1 vector space**. So if we have **3-vectors**, a **basis** for the **grade-2 tensors** is given by the possible **tensor products** of the **basis vectors** (now using the **tensor products** of the *original* space rather than of the **dual**):

$$\mathbf{e}_1 \otimes \mathbf{e}_1,\ \mathbf{e}_1 \otimes \mathbf{e}_2,\ \mathbf{e}_1 \otimes \mathbf{e}_3,\ \mathbf{e}_2 \otimes \mathbf{e}_1,\ \mathbf{e}_2 \otimes \mathbf{e}_2,\ \mathbf{e}_2 \otimes \mathbf{e}_3,\ \mathbf{e}_3 \otimes \mathbf{e}_1,\ \mathbf{e}_3 \otimes \mathbf{e}_2,\ \mathbf{e}_3 \otimes \mathbf{e}_3$$

which is the expected $3^2 = 9$ **basis vectors**. But in the **antisymmetric case**, things are much more limited, because $\mathbf{e}_1 \wedge \mathbf{e}_1 = \mathbf{0}$, $\mathbf{e}_2 \wedge \mathbf{e}_2 = \mathbf{0}$, $\mathbf{e}_3 \wedge \mathbf{e}_3 = \mathbf{0}$, and also the other six occur in **linearly dependent** pairs, because: $\mathbf{e}_2 \wedge \mathbf{e}_1 = -\mathbf{e}_1 \wedge \mathbf{e}_2$, $\mathbf{e}_3 \wedge \mathbf{e}_1 = -\mathbf{e}_1 \wedge \mathbf{e}_3$, and $\mathbf{e}_3 \wedge \mathbf{e}_2 = -\mathbf{e}_2 \wedge \mathbf{e}_3$. So for the **grade-2 algebra** restricted to the **antisymmetric tensors**, there are only *three* **basis tensors**:

$$\mathbf{e}_1 \wedge \mathbf{e}_2,\ \mathbf{e}_1 \wedge \mathbf{e}_3, \text{ and } \mathbf{e}_2 \wedge \mathbf{e}_3.$$

When we go to the **grade-3 space**, we are still further limited, because all forms like $\mathbf{e}_1 \wedge \mathbf{e}_1 \wedge \mathbf{e}_2$ with two elements the same are zero, and the six forms with all three factors different fall into just one **linearly dependent** class by alternation of their factors:

$$\mathbf{e}_1\wedge\mathbf{e}_2\wedge\mathbf{e}_3 = -\mathbf{e}_1\wedge\mathbf{e}_3\wedge\mathbf{e}_2 = \mathbf{e}_3\wedge\mathbf{e}_1\wedge\mathbf{e}_2 = -\mathbf{e}_3\wedge\mathbf{e}_2\wedge\mathbf{e}_1 = \mathbf{e}_2\wedge\mathbf{e}_3\wedge\mathbf{e}_1 = -\mathbf{e}_2\wedge\mathbf{e}_1\wedge\mathbf{e}_3.$$

In this line, I have switched two adjacent factors in each adjacent pair, which switches the sign because, for example: $\mathbf{e_3} \wedge \mathbf{e_1} = -\mathbf{e_1} \wedge \mathbf{e_3}$. So the **grade-3** space has *only one* **basis tensor**:

$$\mathbf{e_1} \wedge \mathbf{e_2} \wedge \mathbf{e_3}.$$

Beyond **grade 3**, there are *no further* **antisymmetric spaces** for **3-vectors**, since any further **antisymmetric products** of the three **basis vectors** would necessarily have two **basis vectors** the same, and so be zero.

The entire series of spaces of different **grades**, infinite for general **tensors**, but of only $n + 1$ **grades** for **antisymmetric tensors**, is called a **graded algebra**.

Note that the **antisymmetric graded algebra** follows the pattern of **Pascal's Triangle**: for **2-vectors**, there is *one* **scalar** "basis element" 1, *two* **2-vectors** $\mathbf{e_1}$ and $\mathbf{e_2}$, and *one* **grade-2 antisymmetric tensor** $\mathbf{e_1} \wedge \mathbf{e_2}$ and *nothing more*. For **3-vectors** there is *one* **scalar** "basis element" 1, *three* **3-vectors** $\mathbf{e_1}$, $\mathbf{e_2}$ and $\mathbf{e_3}$, *three* **grade-2 antisymmetric tensors** $\mathbf{e_1} \wedge \mathbf{e_2}$, $\mathbf{e_1} \wedge \mathbf{e_3}$, and $\mathbf{e_2} \wedge \mathbf{e_3}$ and *one* **grade-3 antisymmetric tensor** $\mathbf{e_1} \wedge \mathbf{e_2} \wedge \mathbf{e_3}$ and *nothing more*. These are respectively 1:2:1 and 1:3:3:1 just as in **Pascal's Triangle**.

5.7 Vectors and Angles and the Gibbs Cross Product

An **angle** can obviously be represented trivially by a **2-vector** $[\theta_x, \theta_y]$ which obeys $\theta_x^2 + \theta_y^2 = 1$. More significantly, working the other way around, any **2-vector** can be represented in the form:

$$[x, y] = [r\theta_x, r\theta_y] = r[\theta_x, \theta_y]$$

where $r^2 = x^2 + y^2$ so that $\theta_x = x/r$ and $\theta_y = y/r$ and $\theta_x^2 + \theta_y^2 = 1$.

In this form, called the **polar representation** when applied to a point in the plane, the **dot product** between two **2-vectors** takes the form:

$$[x_1, y_1] \cdot [x_2, y_2] = r_1[\theta_x, \theta_y] \cdot r_2[\varphi_x, \varphi_y] = x_1x_2 + y_1y_2 = r_1r_2(\theta_x\varphi_x + \theta_y\varphi_y).$$

Looking back at Section 4.2, we can see that this corresponds to r_1r_2 times the **cosine** part, $\theta_x\varphi_x + \theta_y\varphi_y$, of the **difference** in the **angle** parts of the two **vectors**. This is a quite general rule.

> The **dot product** *always* equals the product of the $\sqrt{(x^2 + y^2)}$ **magnitudes** of the two **vectors** times the **cosine**, $\theta_x\varphi_x + \theta_y\varphi_y$, of the **difference** in the **angle** parts of the two **vectors**.

This is true whether we are dealing with **2-vectors**, **3-vectors** or any kind of **standard vector**, and the **dot product** is indeed commonly *defined* this way.

To see that this is general, we need to transform our work into the common **plane** shared by two **vectors** in a higher dimensional space so that we have a definition of the **angle** available. Consider two **3-vectors u** and **v**, and define a **unit vector** $\mathbf{e}_1$ from **u** as:

$$\mathbf{e}_1 = (1/\sqrt{(\mathbf{u}\cdot\mathbf{u})})\mathbf{u} = \mathbf{u}/\sqrt{\mathbf{u}^2}.$$

Here I introduce the important and much-used notation $\mathbf{u}^2 := \mathbf{u}\cdot\mathbf{u}$. That this $\mathbf{e}_1$ is a **unit vector** is easy to see: $\mathbf{e}_1\cdot\mathbf{e}_1 = \mathbf{u}\cdot\mathbf{u}/(\sqrt{\mathbf{u}^2})^2 = \mathbf{u}\cdot\mathbf{u}/(\mathbf{u}^2) = \mathbf{u}^2/(\mathbf{u}^2) = 1$. Putting $r_u = \sqrt{\mathbf{u}^2}$, we have $\mathbf{u} = r_u\mathbf{e}_1$.

Now define a second **vector, orthogonal** to $\mathbf{e}_1$, by:

$$\mathbf{v}_2 = \mathbf{v} - \mathbf{u}((\mathbf{v}\cdot\mathbf{u})/(\mathbf{u}^2))$$

so that $\mathbf{v}_2\cdot\mathbf{u} = \mathbf{v}\cdot\mathbf{u} - \mathbf{u}\cdot\mathbf{u}((\mathbf{v}\cdot\mathbf{u})/(\mathbf{u}^2)) = \mathbf{v}\cdot\mathbf{u} - \mathbf{u}^2((\mathbf{v}\cdot\mathbf{u})/(\mathbf{u}^2)) = \mathbf{v}\cdot\mathbf{u} - \mathbf{v}\cdot\mathbf{u} = 0$, or $\mathbf{v}_2$ and **u** are **orthogonal**.

Next define:

$$\mathbf{e}_2 = (1/\sqrt{(\mathbf{v}_2{}^2)})\mathbf{v}_2$$

so we have $\mathbf{e}_1\cdot\mathbf{e}_1 = \mathbf{e}_2\cdot\mathbf{e}_2 = 1$ and $\mathbf{e}_1\cdot\mathbf{e}_2 = 0$, or that $\mathbf{e}_1$ and $\mathbf{e}_2$ are an **orthonormal** pair, with $\mathbf{u} = r_u\mathbf{e}_1$ and:

$$\mathbf{v} = \mathbf{v}_2 + \mathbf{u}((\mathbf{v}\cdot\mathbf{u})/(\mathbf{u}^2)) = \sqrt{(\mathbf{v}_2{}^2)}\mathbf{e}_2 + \mathbf{u}((\mathbf{v}\cdot\mathbf{u})/(\mathbf{u}^2))$$

$$= \sqrt{(\mathbf{v}_2{}^2)}\mathbf{e}_2 + r_u\mathbf{e}_1((\mathbf{v}\cdot\mathbf{u})/r_u{}^2) = \sqrt{(\mathbf{v}_2{}^2)}\mathbf{e}_2 + ((\mathbf{v}\cdot\mathbf{u})/r_u)\mathbf{e}_1.$$

> $\{\mathbf{e}_1, \mathbf{e}_2\}$ now constitute an **orthonormal basis** for the common **plane** containing **u** and **v** because they have been constructed only from **linear combinations** of **u** and **v**.

Now $\mathbf{u}\cdot\mathbf{v} = (r_u\mathbf{e}_1)\cdot(\sqrt{(\mathbf{v}_2{}^2)}\,\mathbf{e}_2 + ((\mathbf{v}\cdot\mathbf{u})/r_u)\,\mathbf{e}_1) = r_u\sqrt{(\mathbf{v}_2{}^2)}\,\mathbf{e}_1\cdot\mathbf{e}_2 + (r_u(\mathbf{v}\cdot\mathbf{u})/r_u)\mathbf{e}_1\cdot\mathbf{e}_1 = \mathbf{v}\cdot\mathbf{u}$ as expected, but also, expressed in terms of the new **orthonormal basis** $\mathbf{e}_1$, $\mathbf{e}_2$:

$$\mathbf{u} = r_u[1, 0] \qquad \text{and} \qquad \mathbf{v} = r_v[(\mathbf{v}\cdot\mathbf{u})/r_ur_v, \sqrt{(\mathbf{v}_2{}^2)}/r_v]$$

so

$$\mathbf{u}\cdot\mathbf{v} = r_ur_v(1\times(\mathbf{v}\cdot\mathbf{u})/r_ur_v + 0\times\sqrt{(\mathbf{v}_2{}^2)}/r_v)$$

which again reduces to $\mathbf{v}\cdot\mathbf{u}$ but is now clearly of the form:

$$\mathbf{u}\cdot\mathbf{v} = r_ur_v(\theta_x\varphi_x + \theta_y\varphi_y)$$

in the new **basis**. The method that I have introduced here of creating a new **orthonormal basis** from a given set of **linearly independent vectors** that will act as a **basis** for that set – i.e. that will **span** that set – is called **Gram-Schmidt orthogonalization**.

Does anything correspond to the **sine** part of the **angle difference**? Well, we can see from Section 4.2 that if **u** = $r_u[\theta_x, \theta_y]$ and **v** = $r_v[\varphi_x, \varphi_y]$, then the **sine** part would take the form:

sine(u − v) = $r_u r_v(\theta_y\varphi_x - \theta_x\varphi_y)$ or **sine(v − u)** = $r_u r_v(\theta_x\varphi_y - \theta_y\varphi_x)$

and the first thing to note about this product is that it is **antisymmetric**. If we write it provisionally in the form **u** ∧ **v** with **u** = $r_u[\theta_x, \theta_y]$ and **v** = $r_v[\varphi_x, \varphi_y]$, then:

$$\mathbf{v} \wedge \mathbf{u} = r_u r_v(\varphi_y\theta_x - \varphi_x\theta_y) = r_u r_v(-\theta_y\varphi_x + \theta_x\varphi_y) = -\,\mathbf{u} \wedge \mathbf{v}.$$

It turns out that this **alternating product** *will* be significant, and as an **antisymmetric tensor**, not as the simple **scalar** form appearing here, but with this **magnitude**.

Note that $\mathbf{v} \wedge \mathbf{u} = -\,\mathbf{u} \wedge \mathbf{v}$ implies:

$$\mathbf{u} \wedge \mathbf{u} = -\,\mathbf{u} \wedge \mathbf{u}$$

and so $\mathbf{u} \wedge \mathbf{u} + \mathbf{u} \wedge \mathbf{u} = \mathbf{0}$ or $2(\mathbf{u} \wedge \mathbf{u}) = \mathbf{0}$, so that for any **vector**:

$$\mathbf{u} \wedge \mathbf{u} = \mathbf{0}.$$

That the **antisymmetric** property returns the same result as above is easy to see: for an **orthonormal basis** of **2-vectors** $\mathbf{e}_1$ and $\mathbf{e}_2$, we can write:

$$\mathbf{u} = r_u(\theta_x\mathbf{e}_1 + \theta_y\mathbf{e}_2) \qquad \mathbf{v} = r_v(\varphi_x\mathbf{e}_1 + \varphi_y\mathbf{e}_2)$$

gives

$$\mathbf{u} \wedge \mathbf{v} = r_u r_v(\theta_x\varphi_x\mathbf{e}_1 \wedge \mathbf{e}_1 + \theta_x\varphi_y\mathbf{e}_1 \wedge \mathbf{e}_2 + \theta_y\varphi_x\mathbf{e}_2 \wedge \mathbf{e}_1 + \theta_y\varphi_y\mathbf{e}_2 \wedge \mathbf{e}_2)$$

but $\mathbf{e}_1 \wedge \mathbf{e}_1 = \mathbf{e}_2 \wedge \mathbf{e}_2 = \mathbf{0}$, and $\mathbf{e}_2 \wedge \mathbf{e}_1 = -\mathbf{e}_1 \wedge \mathbf{e}_2$, so this becomes:

$$\mathbf{u} \wedge \mathbf{v} = r_u r_v(\theta_x\varphi_y - \theta_y\varphi_x)\, \mathbf{e}_1 \wedge \mathbf{e}_2$$

which appears above as **sine(v − u)**.

In the next Section we'll develop this **antisymmetric product** as the complementary part, with the **scalar product**, of the compound **geometric product**. In that general context the product **u** ∧ **v** represents a new kind of object, with indeed a tensorial nature, which we will call a **bivector**.

But in three **dimensions**, this object can be interpreted as a new **vector**.

Here we'll work with the more familiar notation $\{\mathbf{e}_x, \mathbf{e}_y, \mathbf{e}_z\}$ rather than $\{\mathbf{e}_1, \mathbf{e}_2, \mathbf{e}_3\}$ as it's normal to talk of 3-space in terms of *x*, *y* and *z*.

Imagine that $\mathbf{u} = \mathbf{e}_x$, the **unit vector** along the *x* **axis**, and $\mathbf{v} = 2.\mathbf{e}_x + \mathbf{e}_y$. In 3-space, these two **vectors** are [1, 0, 0] and [2, 1, 0] respectively. So $r_u = 1$ and $r_v = \sqrt{(2^2 + 1^2)} = \sqrt{5}$.

So

$$\mathbf{u} \wedge \mathbf{v} = r_u r_v(\theta_x\varphi_y - \theta_y\varphi_x)\, \mathbf{e}_x \wedge \mathbf{e}_y = \sqrt{5}.(1.1 - 0.2)\, \mathbf{e}_x \wedge \mathbf{e}_y = \sqrt{5}.\mathbf{e}_x \wedge \mathbf{e}_y.$$

The hugely influential American, Josiah Willard Gibbs, suggested that in this context, rather along the lines of Hamilton's **quaternions**, we define $\mathbf{e}_x \wedge \mathbf{e}_y$ as being $\mathbf{e}_z$, and write it with a *cross* "×". So, sticking to the **cyclic permutations**:

$$\mathbf{e}_x \times \mathbf{e}_y = \mathbf{e}_z$$
$$\mathbf{e}_y \times \mathbf{e}_z = \mathbf{e}_x$$
$$\mathbf{e}_z \times \mathbf{e}_x = \mathbf{e}_y.$$

So we would define:

$$\mathbf{u} \times \mathbf{v} = r_u r_v(\theta_x\varphi_y - \theta_y\varphi_x)\, \mathbf{e}_x \times \mathbf{e}_y = \sqrt{5}.(1.1 - 0.2)\, \mathbf{e}_x \times \mathbf{e}_y = \sqrt{5}.\mathbf{e}_x \times \mathbf{e}_y = \sqrt{5}.\mathbf{e}_z.$$

But still by analogy with the **quaternions**, we now define for the **anticyclic permutations**:

$$\mathbf{e}_y \times \mathbf{e}_x = -\mathbf{e}_z$$
$$\mathbf{e}_z \times \mathbf{e}_y = -\mathbf{e}_x$$
$$\mathbf{e}_x \times \mathbf{e}_z = -\mathbf{e}_y.$$

If we now write

$$\mathbf{u} = r_u[\theta_x, \theta_y] \text{ as } r_u(\theta_x\mathbf{e}_x + \theta_y\mathbf{e}_y)$$

and

$$\mathbf{v} = r_v[\varphi_x, \varphi_y] \text{ as } r_v(\varphi_x\mathbf{e}_x + \varphi_y\mathbf{e}_y),$$

we can expand this product as a purely *algebraic* entity as:

$$\mathbf{u} \times \mathbf{v} = r_u(\theta_x\mathbf{e}_x + \theta_y\mathbf{e}_y) \times r_v(\varphi_x\mathbf{e}_x + \varphi_y\mathbf{e}_y)$$

$$= r_u r_v(\theta_x\varphi_x)\, \mathbf{e}_x \times \mathbf{e}_x + r_u r_v(\theta_x\varphi_y)\, \mathbf{e}_x \times \mathbf{e}_y + r_u r_v(\theta_y\varphi_x)\, \mathbf{e}_y \times \mathbf{e}_x + r_u r_v(\theta_y\varphi_y)\, \mathbf{e}_y \times \mathbf{e}_y.$$

As shown at the end of Section 5.5, for any **antisymmetric product**, as × now is,

$$\mathbf{u} \times \mathbf{u} = 0,$$

so the $\mathbf{e}_x \times \mathbf{e}_x$ and $\mathbf{e}_y \times \mathbf{e}_y$ terms drop out, and since $\mathbf{e}_x \times \mathbf{e}_y = \mathbf{e}_z$ and $\mathbf{e}_y \times \mathbf{e}_x = -\mathbf{e}_z$ we get:

$$\mathbf{u} \times \mathbf{v} = r_u r_v(\theta_x\varphi_y)\, \mathbf{e}_x \times \mathbf{e}_y + r_u r_v(\theta_y\varphi_x)\, \mathbf{e}_y \times \mathbf{e}_x = r_u r_v(\theta_x\varphi_y - \theta_y\varphi_x)\, \mathbf{e}_z.$$

For $\mathbf{v} \times \mathbf{u}$ we get:

$$\mathbf{v} \times \mathbf{u} = r_v(\varphi_x\mathbf{e}_x + \varphi_y\mathbf{e}_y) \times r_u(\theta_x\mathbf{e}_x + \theta_y\mathbf{e}_y)$$

$$= r_ur_v(\varphi_x\theta_y)\ \mathbf{e}_x \times \mathbf{e}_y + r_ur_v(\varphi_y\theta_x)\ \mathbf{e}_y \times \mathbf{e}_x = r_ur_v(\varphi_x\theta_y - \varphi_y\theta_x)\ \mathbf{e}_z,$$

and this indeed equals $-\ \mathbf{u} \times \mathbf{v}$, and so points *down* if $\mathbf{u} \times \mathbf{v}$ pointed *up*.

This, the **Gibbs cross product** of **vectors**, is defined only in 3-space, and the direction of the result is given intuitively by the "**right-hand rule**" illustrated in *Figure 5.7.1* below.

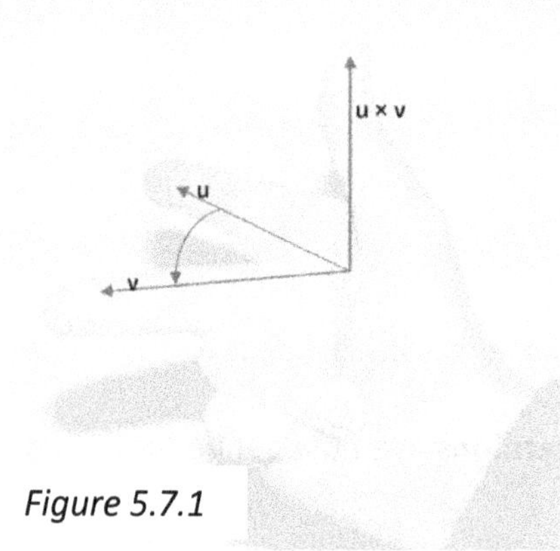

Figure 5.7.1

Note that this picture only holds good for right-handed **coordinates** (as defined in Section 1.2). But left-handed ones, with the positive *z* axis pointing *down* or *away* from the *xy* plane are commonly used in computer graphics.

The generalization to higher dimensions is described in the next Section.

5.8 The Geometric Product

The **geometric product** was introduced by William Kingdon Clifford – yes, it is "KingdoN" – who was one of quite a few mathematicians who died very young, although not as young as Galois. His dates were 1845-1879. Clifford's innovation, even more than Hamilton's **quaternions**, was largely forgotten after he died, although it cropped up in **quantum field theory** as the theory of "**Clifford Algebras**". His work was rediscovered and hugely expanded by an American, David Hestenes, from the nineteen-sixties on, and *his* work exploded onto the academic scene when three Cambridge (UK!) physicists, Stephen Gull, Anthony Lasenby and Chris Doran, discovered Hestenes's work and Doran and Lasenby published a book about it in 2003 called *Geometric Algebra for Physicists*.[66]

[66] Interestingly, Hestenes himself credits A. N. Whitehead with developing the idea that this algebra provided a *universal algebra*, in the treatise A. N. Whitehead, *A Treatise on Universal Algebra with Applications*, Cambridge University Press 1898. Whitehead is most famous for the *Principia Mathematica* written subsequesntly in collaboration with Bertrand Russell, but he isn't mentioned in the work above by his fellow Cantabrigians from a century later.

The theory of the **geometric product** is called **geometric algebra**. This was Clifford's own term, but physicists have tended to call the subject **Clifford Algebra**. The two terms mean the same thing.

The **geometric product** can be defined axiomatically, where we use *juxtaposition* of **vectors** to indicate their **geometric product** and where λ and μ are **scalars**, as:

I. For members **u**, **v** of the **graded algebra** G, **uv** is a member of G (**Closure**)

II. **u(vw) = (uv)w = uvw** (**Associativity**)

III. **u(λv + μw) = λuv + μuw** (**Bilinearity** which includes . . .
(λu + μv)w = λuw + μvw (**Distibutivity** over **Vector Addition**)

IV. For a **vector u**, **uu** = $\mathbf{u}^2$ is a **real number**. (**Modulus**).

This doesn't seem to define much at all. But note two things: we do *not* assume **closure** *within the same* **grade** and we do *not* assume **commutativity**. The **geometric product** is *not* **commutative**. Axiom I is not usually included: I have put it in for reasons that will appear very shortly. To move forward from this definition, we note that we can split the product into two parts, because:

$$\tfrac{1}{2}(\mathbf{uv} + \mathbf{vu}) + \tfrac{1}{2}(\mathbf{uv} - \mathbf{vu}) = \tfrac{1}{2}\mathbf{uv} + \tfrac{1}{2}\mathbf{vu} + \tfrac{1}{2}\mathbf{uv} - \tfrac{1}{2}\mathbf{vu} = \tfrac{1}{2}\mathbf{uv} + \tfrac{1}{2}\mathbf{uv} = \mathbf{uv}$$

and

$$(\mathbf{u} + \mathbf{v})^2 = (\mathbf{u} + \mathbf{v})(\mathbf{u} + \mathbf{v}) = \mathbf{u}^2 + \mathbf{uv} + \mathbf{vu} + \mathbf{v}^2$$

so that

$$\mathbf{uv} + \mathbf{vu} = (\mathbf{u} + \mathbf{v})^2 - \mathbf{u}^2 - \mathbf{v}^2$$

and from axiom IV, *if* **u** *and* **v** *are* ***vectors***, this must be a **real number**. *If* we assume that the *addition* in **uv** + **vu** obeys the rules of **vector addition** – which is tantamount to assuming that the result of the **geometric product** is some *new kind* of **vector** – then we can say too that **uv** + **vu** = **vu** + **uv**, because **vector addition** *is* **commutative**.[67] I have included axiom I to justify this assumption. So the ½(**uv** + **vu**) term would be identical for *both* **uv** and **vu**, and we call this the **symmetric part** of the **geometric product** and since it is a **symmetric scalar product** we will write it as **u · v**. It is indeed normally, but not necessarily, taken to be the usual **dot product**.

By contrast the part ½(**uv** – **vu**) of **uv** would, in **vu**, become ½(**vu** – **uv**) = –½(–**vu** + **uv**) = –½(**uv** – **vu**) and so *reverses sign* in going from **uv** to **vu** and this is therefore called the **antisymmetric part**.

If the **symmetric part** of **uv** is an ordinary **real number**, what is the nature of the **antisymmetric part**? If we write it as **u ∧ v**, it *might* too be just a number – a **scalar** – as suggested at the end of the last section, making the entire **geometric product** a **scalar**. But then **associativity** wouldn't hold, for if **u**, **v** and **w** are **linearly independent**, then:

$$(\mathbf{uv})\mathbf{w} = (\mathbf{u}\cdot\mathbf{v} + \mathbf{u} \wedge \mathbf{v})\mathbf{w} = \lambda\mathbf{w} = \mathbf{u}(\mathbf{vw}) = \mathbf{u}(\mathbf{v}\cdot\mathbf{w} + \mathbf{v} \wedge \mathbf{w}) = \mu\mathbf{u}$$

---

[67] I'm afraid that here we're seeing the difference between physicists' concept of axioms, and the somewhat more watertight, but still not perfect, versions preferred by mathematicians.

and then $\lambda\mathbf{w} - \mu\mathbf{u} = 0$ with $\lambda \neq 0$ and $\mu \neq 0$ which refutes **linear independence** unless the **geometric product** always evaluates to zero.

It might at a pinch be a new **vector**:

$$(\mathbf{uv})\mathbf{w} = (\mathbf{u}\cdot\mathbf{v} + \mathbf{u} \wedge \mathbf{v})\mathbf{w} = \lambda\mathbf{w} + \boldsymbol{\phi}\mathbf{w} = \mathbf{u}(\mathbf{v}\cdot\mathbf{w} + \mathbf{v} \wedge \mathbf{w}) = \mu\mathbf{u} + \mathbf{u}\boldsymbol{\theta}.$$

So this doesn't look too promising either – **associativity** blocks a lot of options!

> In practice, the result of the **alternating product** (also known as the **outer product**, which agrees with the **dot product** sometimes being called the **inner product**) is treated as an *irreducible entity*, which can be expanded in terms of the **outer products** of **basis vectors**, but not in terms of anything else.

Essentially, it corresponds to the **antisymmetric tensors** that we used to define a **graded algebra** above. If you want a physical model, a good one is developed in John Vince's excellent book *Geometric Algebra for Computer Graphics*, but he makes it clear that this is merely *a* model, not *the* model.

So if **u** and **v** are **grade-1 vectors** or **vectors** as we have used them so far, we can decompose the **geometric product** between them into the ordinary **dot product** and an **outer product** or **alternating product**:

$$\mathbf{uv} = \mathbf{u} \cdot \mathbf{v} + \mathbf{u} \wedge \mathbf{v}.$$

This decomposition is commonly used as a *definition* of the **geometric product**. As presented here, it only applies to **grade-1 vectors**. It can be extended to **products** of higher **grade vectors**, but this extension easily leads to errors, and is best avoided at least in the early stages of the subject.

Mathematicians always try to avoid going down to any particular **coordinates** as much as possible, as **coordinate**-free treatments are more general. But the rest of us need the more concrete pictures that **coordinate representations** give. **Bilinearity** means that the **geometric product** is completely defined by the way it acts on a **basis**, so that is what I explore next.

Let's start with the **geometric algebra** of **2-vectors**. I assume a **basis** $\mathbf{e}_1 = [1, 0]$, $\mathbf{e}_2 = [0, 1]$ and the **dot product** $\mathbf{e}_i \cdot \mathbf{e}_j = \delta_{ij}$. The four binary products are:

$$\mathbf{e}_1\mathbf{e}_1 = \mathbf{e}_1 \cdot \mathbf{e}_1 + \mathbf{e}_1 \wedge \mathbf{e}_1 = 1 + \mathbf{0} = 1$$
$$\mathbf{e}_2\mathbf{e}_2 = \mathbf{e}_2 \cdot \mathbf{e}_2 + \mathbf{e}_2 \wedge \mathbf{e}_2 = 1 + \mathbf{0} = 1$$
$$\mathbf{e}_1\mathbf{e}_2 = \mathbf{e}_1 \cdot \mathbf{e}_2 + \mathbf{e}_1 \wedge \mathbf{e}_2 = 0 + \mathbf{e}_1 \wedge \mathbf{e}_2 = \mathbf{e}_1 \wedge \mathbf{e}_2$$
$$\mathbf{e}_2\mathbf{e}_1 = \mathbf{e}_2 \cdot \mathbf{e}_1 + \mathbf{e}_2 \wedge \mathbf{e}_1 = 0 + \mathbf{e}_2 \wedge \mathbf{e}_1 = 0 - \mathbf{e}_1 \wedge \mathbf{e}_2 = -\mathbf{e}_1 \wedge \mathbf{e}_2.$$

The next level needs the **associative law** to evaluate all the various forms. Because $\mathbf{e}_1\mathbf{e}_1$ and $\mathbf{e}_2\mathbf{e}_2$ are **scalars** (= 1), we don't have any problems with:

$$\mathbf{e}_1\mathbf{e}_1\mathbf{e}_1 = 1.\mathbf{e}_1 = \mathbf{e}_1.1 = \mathbf{e}_1$$
$$\mathbf{e}_2\mathbf{e}_2\mathbf{e}_2 = 1.\mathbf{e}_2 = \mathbf{e}_2.1 = \mathbf{e}_2$$

$$\mathbf{e}_1\mathbf{e}_1\mathbf{e}_2 = 1.\mathbf{e}_2 = \mathbf{e}_2$$

but by the **associative law**, this must also equal $\mathbf{e}_1(\mathbf{e}_1\mathbf{e}_2) = \mathbf{e}_1(\mathbf{e}_1 \wedge \mathbf{e}_2)$, so $\mathbf{e}_1(\mathbf{e}_1 \wedge \mathbf{e}_2) = \mathbf{e}_2$. Likewise:

$$\mathbf{e}_2\mathbf{e}_2\mathbf{e}_1 = 1.\mathbf{e}_1 = \mathbf{e}_1 = \mathbf{e}_2(\mathbf{e}_2\mathbf{e}_1) = -\mathbf{e}_2(\mathbf{e}_1 \wedge \mathbf{e}_2),$$

so $\mathbf{e}_2(\mathbf{e}_1 \wedge \mathbf{e}_2) = -\mathbf{e}_1$.

Then

$$\mathbf{e}_1\mathbf{e}_2\mathbf{e}_1 = (\mathbf{e}_1 \wedge \mathbf{e}_2)\mathbf{e}_1 = \mathbf{e}_1(\mathbf{e}_2\mathbf{e}_1) = -\mathbf{e}_1(\mathbf{e}_1 \wedge \mathbf{e}_2) = -\mathbf{e}_2 \text{ from above}$$
$$\mathbf{e}_1\mathbf{e}_2\mathbf{e}_2 = (\mathbf{e}_1 \wedge \mathbf{e}_2)\mathbf{e}_2 = \mathbf{e}_1(\mathbf{e}_2\mathbf{e}_2) = 1.\mathbf{e}_1 = \mathbf{e}_1$$

and

$$\mathbf{e}_2\mathbf{e}_1\mathbf{e}_1 = -(\mathbf{e}_1 \wedge \mathbf{e}_2)\mathbf{e}_1 = \mathbf{e}_2.1 = \mathbf{e}_2$$
$$\mathbf{e}_2\mathbf{e}_1\mathbf{e}_2 = -(\mathbf{e}_1 \wedge \mathbf{e}_2)\mathbf{e}_2 = \mathbf{e}_2(\mathbf{e}_1\mathbf{e}_2) = \mathbf{e}_2(\mathbf{e}_1 \wedge \mathbf{e}_2) = -\mathbf{e}_1 \text{ from above.}$$

The unique **grade 2 vector** $\mathbf{e}_1 \wedge \mathbf{e}_2$ is the highest **grade** element in the algebra. The highest **grade** element is always the *only* **basis vector** in its **grade**, and is called the **pseudoscalar**. It is commonly written **I**. So with this convention, we can set out all these products (in ascending order of indices for the triples) as:

$$\mathbf{e}_1\mathbf{e}_1 = \mathbf{e}_2\mathbf{e}_2 = 1$$
$$\mathbf{e}_1\mathbf{e}_2 = -\mathbf{e}_2\mathbf{e}_1 = \mathbf{I}$$

$$\mathbf{e}_1\mathbf{e}_1\mathbf{e}_1 = \mathbf{e}_1$$
$$\mathbf{e}_1\mathbf{e}_1\mathbf{e}_2 = \mathbf{e}_1\mathbf{I} = \mathbf{e}_2$$
$$\mathbf{e}_1\mathbf{e}_2\mathbf{e}_1 = \mathbf{I}\mathbf{e}_1 = -\mathbf{e}_1\mathbf{I} = -\mathbf{e}_2$$
$$\mathbf{e}_1\mathbf{e}_2\mathbf{e}_2 = \mathbf{I}\mathbf{e}_2 = \mathbf{e}_1$$
$$\mathbf{e}_2\mathbf{e}_1\mathbf{e}_1 = -\mathbf{I}\mathbf{e}_1 = \mathbf{e}_2$$
$$\mathbf{e}_2\mathbf{e}_1\mathbf{e}_2 = -\mathbf{I}\mathbf{e}_2 = \mathbf{e}_2\mathbf{I} = -\mathbf{e}_1$$
$$\mathbf{e}_2\mathbf{e}_2\mathbf{e}_1 = -\mathbf{e}_2\mathbf{I} = \mathbf{e}_1$$
$$\mathbf{e}_2\mathbf{e}_2\mathbf{e}_2 = \mathbf{e}_2$$

and since these all go back to **scalars** or **grade-1 vectors**, this will suffice to give all the **2-vector** algebra.

Note the 1:2:1 pattern from the third line of **Pascal's Triangle**: *one* "scalar basis" (1), *two* **grade-1 vectors** ($\mathbf{e}_1$, $\mathbf{e}_2$), *one* **grade-2 pseudoscalar** ($\mathbf{I} = \mathbf{e}_1 \wedge \mathbf{e}_2$).

$\mathbf{I}^2 = \mathbf{e}_1\mathbf{e}_2\mathbf{e}_1\mathbf{e}_2 = -\,\mathbf{e}_1\mathbf{e}_1\mathbf{e}_2\mathbf{e}_2 = -1.1 = -1$ and this suggests a link with **complex numbers**. If we have a **2-vector** $\mathbf{x} = [x, y] = x\mathbf{e}_1 + y\mathbf{e}_2$, we can pre-multiply by $\mathbf{e}_1$ to obtain:

$$\mathbf{e}_1\mathbf{x} = x\mathbf{e}_1\mathbf{e}_1 + y\mathbf{e}_1\mathbf{e}_2 = x + y\mathbf{I},$$

and multiplying two such "**complex numbers**" together using the **geometric product** gives:

$$\mathbf{e}_1\mathbf{x}\mathbf{e}_1\mathbf{u} = (x + y\mathbf{I})(u + v\mathbf{I}) = xu + yv\mathbf{I}^2 + yu\mathbf{I} + xv\mathbf{I} = (xu - yv) + (yu + xv)\mathbf{I}$$

which exactly reproduces the form in Section 4.5 for the multiplication of **complex numbers**.

Complex numbers can also be *divided*. Is there an analogue of division here? Well, there is. Defining $\mathbf{u}^{-1}$ or $1/\mathbf{u}$ as $\mathbf{u}\mathbf{e}_1/\mathbf{u}^2$, we can evaluate:

$$\begin{aligned}\mathbf{e}_1\mathbf{x}/\mathbf{u} &= \mathbf{e}_1\mathbf{x}\mathbf{u}\mathbf{e}_1/\mathbf{u}^2 = (x + y\mathbf{I})(u\mathbf{e}_1\mathbf{e}_1 + v\mathbf{e}_2\mathbf{e}_1)/(u^2 + v^2) = (x + y\mathbf{I})(u - v\mathbf{I})/(u^2 + v^2)\\ &= [(xu - yv\mathbf{I}^2) + (yu - xv)\mathbf{I}]\,/(u^2 + v^2)\\ &= [(xu + yv) + (yu - xv)\mathbf{I}]\,/(u^2 + v^2)\end{aligned}$$

which also occurs in Section 4.5. (In this, I obtain the divisor by $\mathbf{u}^2 = (u\mathbf{e}_1 + v\mathbf{e}_2)(u\mathbf{e}_1 + v\mathbf{e}_2) = u^2\mathbf{e}_1\mathbf{e}_1 + uv\mathbf{e}_1\mathbf{e}_2 + vu\mathbf{e}_2\mathbf{e}_1 + v^2\mathbf{e}_2\mathbf{e}_2 = u^2.1 + uv\mathbf{e}_1\mathbf{e}_2 - vu\mathbf{e}_1\mathbf{e}_2 + v^2.1 = u^2 + v^2$).

Geometric Algebras are not unique. More than one algebra may be possible for the same **standard vectors** of the same **dimension** by choosing a different **symmetric scalar product** for the **symmetric** part. Although this possibility will not be developed further here, and is not even mentioned in the existing texts, I will illustrate it in passing with a simple, but exotic, example for this simplest of cases of **2-vectors**.

Suppose we take the **symmetric metric tensor** g_{ij} defined below:

$$\{g_{ij}\} = \begin{pmatrix}0 & 1\\ 1 & 0\end{pmatrix}.$$

This defines a valid **symmetric scalar product** by $\mathbf{a} * \mathbf{b} = \Sigma_i\Sigma_j a_i g_{ij} b_j$, which we can expand as:

$$\mathbf{a} * \mathbf{b} = (a_1\mathbf{e}_1 + a_2\mathbf{e}_2) * (b_1\mathbf{e}_1 + b_2\mathbf{e}_2) = a_1b_1\mathbf{e}_1{*}\mathbf{e}_1 + a_1b_2\mathbf{e}_1{*}\mathbf{e}_2 + a_2b_1\mathbf{e}_2{*}\mathbf{e}_1 + a_2b_2\mathbf{e}_2{*}\mathbf{e}_2$$

or since $\mathbf{e}_i * \mathbf{e}_j = g_{ij}$, $\mathbf{a} * \mathbf{b} = a_1b_10 + a_1b_21 + a_2b_11 + a_2b_20 = a_1b_2 + a_2b_1$. Now,

$$\mathbf{b} * \mathbf{a} = (b_1\mathbf{e}_1 + b_2\mathbf{e}_2) * (a_1\mathbf{e}_1 + a_2\mathbf{e}_2) = b_1a_2 + b_2a_1 = a_1b_2 + a_2b_1 = \mathbf{a} * \mathbf{b},$$

so this **scalar product** is **symmetric** or **commutative**, even though it isn't the **dot product**.

With the **basis** as before, $\mathbf{e}_1 = [1, 0]$ and $\mathbf{e}_2 = [0, 1]$, we have:

$$\begin{aligned}\mathbf{e}_1\mathbf{e}_1 &= \mathbf{e}_1{*}\mathbf{e}_1 + \mathbf{e}_1 \wedge \mathbf{e}_1 = 0 + 0 = 0\\ \mathbf{e}_2\mathbf{e}_2 &= 0\end{aligned}$$

but

$$\mathbf{e}_1\mathbf{e}_2 = \mathbf{e}_1{*}\mathbf{e}_2 + \mathbf{e}_1 \wedge \mathbf{e}_2 = 1 + \mathbf{e}_1 \wedge \mathbf{e}_2 = 1 + \mathbf{I}$$

and

$$\mathbf{e}_2\mathbf{e}_1 = \mathbf{e}_2{*}\mathbf{e}_1 + \mathbf{e}_2 \wedge \mathbf{e}_1 = 1 - \mathbf{e}_1 \wedge \mathbf{e}_2 = 1 - \mathbf{I}.$$

The triple products are developed analogously to the ordinary algebra, as:

$\mathbf{e}_1\mathbf{e}_1\mathbf{e}_1 = 0.\mathbf{e}_1 = \mathbf{e}_1.0 = \mathbf{0}$ and $\mathbf{e}_1\mathbf{e}_1\mathbf{e}_2 = \mathbf{e}_1(1 + \mathbf{I}) = 0.\mathbf{e}_2 = \mathbf{0}$

$\mathbf{e}_2\mathbf{e}_2\mathbf{e}_1 = 0.\mathbf{e}_1 = \mathbf{e}_2(1 - \mathbf{I}) = \mathbf{0}$ and $\mathbf{e}_2\mathbf{e}_2\mathbf{e}_2 = \mathbf{0}$

so we must have $\mathbf{e}_2\mathbf{I} = \mathbf{e}_2$ and $\mathbf{e}_1\mathbf{I} = -\mathbf{e}_1$. The other four are:

$$\mathbf{e}_1\mathbf{e}_2\mathbf{e}_1 = (1 + \mathbf{I})\mathbf{e}_1 = \mathbf{e}_1(1 - \mathbf{I})$$
$$\mathbf{e}_1\mathbf{e}_2\mathbf{e}_2 = (1 + \mathbf{I})\mathbf{e}_2 = \mathbf{e}_1.0 = \mathbf{0}$$
$$\mathbf{e}_2\mathbf{e}_1\mathbf{e}_1 = (1 - \mathbf{I})\mathbf{e}_1 = \mathbf{e}_2.0 = \mathbf{0}$$
$$\mathbf{e}_2\mathbf{e}_1\mathbf{e}_2 = (1 - \mathbf{I})\mathbf{e}_2 = \mathbf{e}_2(1 + \mathbf{I})$$

which give the added relations:

$$\mathbf{e}_1 + \mathbf{I}\mathbf{e}_1 = \mathbf{e}_1 - \mathbf{e}_1\mathbf{I}$$
$$\mathbf{I}\mathbf{e}_2 = -\mathbf{e}_2$$
$$\mathbf{I}\mathbf{e}_1 = \mathbf{e}_1$$
$$\mathbf{e}_2 - \mathbf{I}\mathbf{e}_2 = \mathbf{e}_2 + \mathbf{e}_2\mathbf{I}.$$

Are these consistent?

$\mathbf{e}_1 + \mathbf{I}\mathbf{e}_1 = \mathbf{e}_1 - \mathbf{e}_1\mathbf{I} \quad \rightarrow \quad 2\mathbf{e}_1 = \mathbf{e}_1 - (-\mathbf{e}_1) = 2\mathbf{e}_1$, and . . .

$\mathbf{e}_2 - \mathbf{I}\mathbf{e}_2 = \mathbf{e}_2 + \mathbf{e}_2\mathbf{I} \quad \rightarrow \quad \mathbf{e}_2 - (-\mathbf{e}_2) = 2\mathbf{e}_2 = \mathbf{e}_2 + \mathbf{e}_2 = 2\mathbf{e}_2$

so it seems they are!

5.9 Geometric Algebra of 3-Vectors

Moving up to **3-vectors**, we are entering the standard **vector** model of real three-dimensional space. The fourth line of **Pascal's Triangle** now gives the relevant pattern for the **graded algebra**. We have:

one **scalar** "basis element" 1,
three **3-vectors** $\mathbf{e}_1$, $\mathbf{e}_2$ and $\mathbf{e}_3$,
three **grade-2 antisymmetric tensors** $\mathbf{e}_1 \wedge \mathbf{e}_2$, $\mathbf{e}_3 \wedge \mathbf{e}_1$, and $\mathbf{e}_2 \wedge \mathbf{e}_3$ and
one **grade-3 antisymmetric tensor** $\mathbf{e}_1 \wedge \mathbf{e}_2 \wedge \mathbf{e}_3$.

Again, the **basis vectors** are:

grade-0 1
grade-1 $\mathbf{e}_1$, $\mathbf{e}_2$ and $\mathbf{e}_3$,
grade-2 $\mathbf{e}_1 \wedge \mathbf{e}_2$, $\mathbf{e}_3 \wedge \mathbf{e}_1$, and $\mathbf{e}_2 \wedge \mathbf{e}_3$
grade-3 $\mathbf{e}_1 \wedge \mathbf{e}_2 \wedge \mathbf{e}_3$.

The **grade-2 vectors** are sometimes called **bivectors**, and the **grade-3** a **trivector**, and these are all of course also **tensors**. Any **vector** (**tensor**) that is simply a single pure **outer product**, not a **linear combination** of **outer products**, such as the **bivectors** or the **trivector** here, is also known as a **blade**.

The unique **grade-3 trivector** here is now the **pseudoscalar** for *this* **graded algebra**.

Note the order 12, 23, but 31, *not* 13, for the **grade-2 bivectors**. This ensures that these are all in **cyclic permutation** order from 123: [12]31[23]12[31]23. . . Taking the order 13 would give different signs for this case for results that are otherwise directly analogous to those for the other two **bivectors**.

When $\mathbf{e}_1$, $\mathbf{e}_2$ and $\mathbf{e}_3$ are the **natural basis** with **coordinates** x, y and z, they will often be written $\mathbf{e}_x$, $\mathbf{e}_y$ and $\mathbf{e}_z$, which of course correspond to **i**, **j** and **k** in Section 5.1.

Again, the **bilinearity** axiom means that any **geometric product** can be trivially evaluated once we know the **products** on the **basis elements**. To start:

$$\mathbf{e}_1\mathbf{e}_1 = \mathbf{e}_1 \cdot \mathbf{e}_1 + \mathbf{e}_1 \wedge \mathbf{e}_1 = 1 + \mathbf{0} = 1$$

and similarly $\mathbf{e}_2\mathbf{e}_2 = \mathbf{e}_3\mathbf{e}_3 = 1$

$$\mathbf{e}_1\mathbf{e}_2 = \mathbf{e}_1 \cdot \mathbf{e}_2 + \mathbf{e}_1 \wedge \mathbf{e}_2 = 0 + \mathbf{e}_1 \wedge \mathbf{e}_2 = \mathbf{e}_1 \wedge \mathbf{e}_2$$

and

$$\mathbf{e}_3\mathbf{e}_1 = \mathbf{e}_3 \wedge \mathbf{e}_1$$
$$\mathbf{e}_2\mathbf{e}_3 = \mathbf{e}_2 \wedge \mathbf{e}_3$$
$$\mathbf{e}_1\mathbf{e}_3 = \mathbf{e}_1 \wedge \mathbf{e}_3 = -\,\mathbf{e}_3 \wedge \mathbf{e}_1$$
$$\mathbf{e}_2\mathbf{e}_1 = \mathbf{e}_2 \wedge \mathbf{e}_1 = -\,\mathbf{e}_1 \wedge \mathbf{e}_2$$
$$\mathbf{e}_1\mathbf{e}_3 = \mathbf{e}_1 \wedge \mathbf{e}_3 = -\,\mathbf{e}_3 \wedge \mathbf{e}_1.$$

The triple products get a little more difficult, and we need the **associative** axiom. Also, there are *twenty-seven* of them. First, three trivial cases are covered by:

$$\mathbf{e}_1\mathbf{e}_1\mathbf{e}_1 = \mathbf{e}_1(\mathbf{e}_1\mathbf{e}_1) = \mathbf{e}_1$$

and this covers $\mathbf{e}_2\mathbf{e}_2\mathbf{e}_2 = \mathbf{e}_2$ and $\mathbf{e}_3\mathbf{e}_3\mathbf{e}_3 = \mathbf{e}_3$ leaving twenty-four.

$$\mathbf{e}_1\mathbf{e}_2\mathbf{e}_2 = \mathbf{e}_1(\mathbf{e}_2\mathbf{e}_2) = \mathbf{e}_1.1 = \mathbf{e}_1 = (\mathbf{e}_1\mathbf{e}_2)\mathbf{e}_2 = (\mathbf{e}_1 \wedge \mathbf{e}_2)\mathbf{e}_2$$
$$\mathbf{e}_1\mathbf{e}_1\mathbf{e}_2 = \mathbf{e}_1(\mathbf{e}_1\mathbf{e}_2) = \mathbf{e}_1(\mathbf{e}_1 \wedge \mathbf{e}_2) = (\mathbf{e}_1\mathbf{e}_1)\mathbf{e}_2 = \mathbf{e}_2$$
$$\mathbf{e}_1\mathbf{e}_2\mathbf{e}_1 = \mathbf{e}_1(\mathbf{e}_2\mathbf{e}_1) = \mathbf{e}_1(\mathbf{e}_2 \wedge \mathbf{e}_1) = -\mathbf{e}_1(\mathbf{e}_1 \wedge \mathbf{e}_2)$$

$= -\mathbf{e}_2$ from the immediately preceding line,

but also $= (\mathbf{e}_1\mathbf{e}_2)\mathbf{e}_1 = (\mathbf{e}_1 \wedge \mathbf{e}_2)\mathbf{e}_1$.

This gives us all the forms $\mathbf{e}_i\mathbf{e}_j\mathbf{e}_j$, $\mathbf{e}_i\mathbf{e}_i\mathbf{e}_j$, $\mathbf{e}_i\mathbf{e}_j\mathbf{e}_i$, which have one repeating factor, of which there are eighteen, for $[ijk]$ = 122, 112, 121, 133, 113, 131, 211, 221, 212, 233, 223, 232, 311, 331, 313, 322, 332, 323:

$$\mathbf{e}_i\mathbf{e}_j\mathbf{e}_j = \mathbf{e}_i \qquad \mathbf{e}_i\mathbf{e}_i\mathbf{e}_j = \mathbf{e}_j \qquad \mathbf{e}_i\mathbf{e}_j\mathbf{e}_i = -\mathbf{e}_j.$$

This leaves us with the six which have no repeating factor:

123, 132, 312, 321, 231 and 213.

These are more subtle. We have:

$$\mathbf{e}_1\mathbf{e}_2\mathbf{e}_3 = \mathbf{e}_1(\mathbf{e}_2\mathbf{e}_3) = \mathbf{e}_1(\mathbf{e}_2\cdot\mathbf{e}_3 + \mathbf{e}_2 \wedge \mathbf{e}_3) = \mathbf{e}_1(0 + \mathbf{e}_2 \wedge \mathbf{e}_3) = \mathbf{e}_1(\mathbf{e}_2 \wedge \mathbf{e}_3)$$

and

$$\mathbf{e}_1\mathbf{e}_2\mathbf{e}_3 = (\mathbf{e}_1\mathbf{e}_2)\mathbf{e}_3 = (\mathbf{e}_1\cdot\mathbf{e}_2 + \mathbf{e}_1 \wedge \mathbf{e}_2)\mathbf{e}_3 = (0 + \mathbf{e}_1 \wedge \mathbf{e}_2)\mathbf{e}_3 = (\mathbf{e}_1 \wedge \mathbf{e}_2)\mathbf{e}_3$$

but this doesn't solve the problem. But wait:

$$\mathbf{e}_1\mathbf{e}_3\mathbf{e}_2 = \mathbf{e}_1(\mathbf{e}_3\mathbf{e}_2) = \mathbf{e}_1(\mathbf{e}_3\cdot\mathbf{e}_2 + \mathbf{e}_3 \wedge \mathbf{e}_2) = \mathbf{e}_1(0 + \mathbf{e}_3 \wedge \mathbf{e}_2) = -\mathbf{e}_1(\mathbf{e}_2 \wedge \mathbf{e}_3) = -\,\mathbf{e}_1\mathbf{e}_2\mathbf{e}_3$$

and

$$\mathbf{e}_2\mathbf{e}_1\mathbf{e}_3 = (\mathbf{e}_2\mathbf{e}_1)\mathbf{e}_3 = (\mathbf{e}_1\cdot\mathbf{e}_2 + \mathbf{e}_2 \wedge \mathbf{e}_1)\mathbf{e}_3 = (0 + \mathbf{e}_2 \wedge \mathbf{e}_1)\mathbf{e}_3 = -(\mathbf{e}_1 \wedge \mathbf{e}_2)\mathbf{e}_3 = -\,\mathbf{e}_1\mathbf{e}_2\mathbf{e}_3$$

which gives the second and last of our **permutations** above. We can repeat the logic to get the last three. I'll do it here working back from that last one (213⟶231⟶321⟶312):

$$\mathbf{e}_2\mathbf{e}_3\mathbf{e}_1 = \mathbf{e}_2(\mathbf{e}_3\mathbf{e}_1) = \mathbf{e}_2(\mathbf{e}_3\cdot\mathbf{e}_1 + \mathbf{e}_3 \wedge \mathbf{e}_1) = \mathbf{e}_2(0 + \mathbf{e}_3 \wedge \mathbf{e}_1) = -\mathbf{e}_2(\mathbf{e}_1 \wedge \mathbf{e}_3) = -\,\mathbf{e}_2\mathbf{e}_1\mathbf{e}_3 \;\; = +\mathbf{e}_1\mathbf{e}_2\mathbf{e}_3$$
$$\mathbf{e}_2\mathbf{e}_3\mathbf{e}_1 = (\mathbf{e}_2\mathbf{e}_3)\mathbf{e}_1 = (\mathbf{e}_2\cdot\mathbf{e}_3 + \mathbf{e}_2 \wedge \mathbf{e}_3)\mathbf{e}_1 = (0 + \mathbf{e}_2 \wedge \mathbf{e}_3)\mathbf{e}_1 = (\mathbf{e}_2 \wedge \mathbf{e}_3)\mathbf{e}_1$$

$$\mathbf{e}_3\mathbf{e}_2\mathbf{e}_1 = (\mathbf{e}_3\mathbf{e}_2)\mathbf{e}_1 = (\mathbf{e}_3\cdot\mathbf{e}_2 + \mathbf{e}_3 \wedge \mathbf{e}_2)\mathbf{e}_1 = (0 + \mathbf{e}_3 \wedge \mathbf{e}_2)\mathbf{e}_1 = -(\mathbf{e}_2 \wedge \mathbf{e}_3)\mathbf{e}_1 = -\,\mathbf{e}_2\mathbf{e}_3\mathbf{e}_1$$
$$\mathbf{e}_3\mathbf{e}_2\mathbf{e}_1 = \mathbf{e}_3(\mathbf{e}_2\mathbf{e}_1) = \mathbf{e}_3(\mathbf{e}_2\cdot\mathbf{e}_1 + \mathbf{e}_2 \wedge \mathbf{e}_1) = \mathbf{e}_3(0 + \mathbf{e}_2 \wedge \mathbf{e}_1) = \mathbf{e}_3(\mathbf{e}_2 \wedge \mathbf{e}_1)$$

$$\mathbf{e}_3\mathbf{e}_1\mathbf{e}_2 = \mathbf{e}_3(\mathbf{e}_1\mathbf{e}_2) = \mathbf{e}_3(\mathbf{e}_1\cdot\mathbf{e}_2 + \mathbf{e}_1 \wedge \mathbf{e}_2) = \mathbf{e}_3(0 + \mathbf{e}_1 \wedge \mathbf{e}_2) = -\mathbf{e}_3(\mathbf{e}_2 \wedge \mathbf{e}_1) = -\,\mathbf{e}_3\mathbf{e}_2\mathbf{e}_1$$

So all told:

$$\mathbf{e}_3\mathbf{e}_1\mathbf{e}_2 = -\,\mathbf{e}_3\mathbf{e}_2\mathbf{e}_1 = \mathbf{e}_2\mathbf{e}_3\mathbf{e}_1 = -\,\mathbf{e}_2\mathbf{e}_1\mathbf{e}_3 = \mathbf{e}_1\mathbf{e}_2\mathbf{e}_3 = -\,\mathbf{e}_1\mathbf{e}_3\mathbf{e}_2$$

so the $\mathbf{e}_1\mathbf{e}_2\mathbf{e}_3$ forms are fully **antisymmetric** on all factors, and so have no **symmetric** part, and therefore:

$$\mathbf{e}_1\mathbf{e}_2\mathbf{e}_3 = \mathbf{e}_1 \wedge \mathbf{e}_2 \wedge \mathbf{e}_3 = \mathbf{I},$$

writing the **pseudoscalar** here as **I** again. So we also get:

$$\mathbf{I}\mathbf{e}_2 = (\mathbf{e}_1\mathbf{e}_2\mathbf{e}_3)\mathbf{e}_2 = -\,(\mathbf{e}_1\mathbf{e}_3\mathbf{e}_2)\mathbf{e}_2 = -(\mathbf{e}_1\mathbf{e}_3)(\mathbf{e}_2\mathbf{e}_2) = -(\mathbf{e}_1\mathbf{e}_3) = \mathbf{e}_3\mathbf{e}_1$$
$$\mathbf{I}\mathbf{e}_1 = (\mathbf{e}_1\mathbf{e}_2\mathbf{e}_3)\mathbf{e}_1 = (\mathbf{e}_2\mathbf{e}_3\mathbf{e}_1)\mathbf{e}_1 = (\mathbf{e}_2\mathbf{e}_3)(\mathbf{e}_1\mathbf{e}_1) = \mathbf{e}_2\mathbf{e}_3$$
$$\mathbf{I}\mathbf{e}_3 = (\mathbf{e}_1\mathbf{e}_2\mathbf{e}_3)\mathbf{e}_3 = (\mathbf{e}_1\mathbf{e}_2)(\mathbf{e}_3\mathbf{e}_3) = \mathbf{e}_1\mathbf{e}_2$$

and

$$\mathbf{e}_1\mathbf{I} = \mathbf{e}_1(\mathbf{e}_1\mathbf{e}_2\mathbf{e}_3) = \mathbf{e}_2\mathbf{e}_3$$

$$\mathbf{e}_2\mathbf{I} = \mathbf{e}_2(\mathbf{e}_1\mathbf{e}_2\mathbf{e}_3) = -\,\mathbf{e}_2(\mathbf{e}_2\mathbf{e}_1\mathbf{e}_3) = -(\mathbf{e}_2\mathbf{e}_2)(\mathbf{e}_1\mathbf{e}_3) = \mathbf{e}_3\mathbf{e}_1$$
$$\mathbf{e}_3\mathbf{I} = \mathbf{e}_3(\mathbf{e}_1\mathbf{e}_2\mathbf{e}_3) = \mathbf{e}_3(\mathbf{e}_3\mathbf{e}_1\mathbf{e}_2) = (\mathbf{e}_3\mathbf{e}_3)(\mathbf{e}_1\mathbf{e}_2) = \mathbf{e}_1\mathbf{e}_2.$$

But also

$$\mathbf{I}^2 = (\mathbf{e}_1\mathbf{e}_2\mathbf{e}_3)(\mathbf{e}_1\mathbf{e}_2\mathbf{e}_3) = (\mathbf{e}_2\mathbf{e}_3\mathbf{e}_1)(\mathbf{e}_1\mathbf{e}_2\mathbf{e}_3) = (\mathbf{e}_2\mathbf{e}_3)(\mathbf{e}_1\mathbf{e}_1)(\mathbf{e}_2\mathbf{e}_3)$$
$$= (\mathbf{e}_2\mathbf{e}_3)(\mathbf{e}_2\mathbf{e}_3) = -(\mathbf{e}_2\mathbf{e}_3)(\mathbf{e}_3\mathbf{e}_2) = -\mathbf{e}_2(\mathbf{e}_3\mathbf{e}_3)\mathbf{e}_2 = -\,\mathbf{e}_2\mathbf{e}_2 = -\,1.$$

So

$$\mathbf{I}^2\mathbf{e}_1 = -\mathbf{e}_1 = \mathbf{I}\mathbf{e}_2\mathbf{e}_3$$
$$\mathbf{I}^2\mathbf{e}_2 = -\mathbf{e}_2 = \mathbf{I}\mathbf{e}_3\mathbf{e}_1$$
$$\mathbf{I}^2\mathbf{e}_3 = -\mathbf{e}_3 = \mathbf{I}\mathbf{e}_1\mathbf{e}_2$$

and

$$\mathbf{e}_1\mathbf{I}^2 = -\mathbf{e}_1 = \mathbf{e}_2\mathbf{e}_3\mathbf{I}$$
$$\mathbf{e}_2\mathbf{I}^2 = -\mathbf{e}_2 = \mathbf{e}_3\mathbf{e}_1\mathbf{I}$$
$$\mathbf{e}_3\mathbf{I}^2 = -\mathbf{e}_3 = \mathbf{e}_1\mathbf{e}_2\mathbf{I}.$$

These twelve relationships whereby **I** switches between the **grade-1 basis elements** and the **grade-2 basis elements** is a reflection of the symmetry within the lines of **Pascal's Triangle**, which in **differential geometry** is called **Hodge duality.**[68] I'll refer to the present more general form of it as **pseudoscalar duality**. You may wonder whether, if it works between the **bases** with three elements each, does it work between the **grade-0** and the **grade-3 bases** with *one* each? Well, of course it does, again with a sign change in one of the two directions:

$$\mathbf{I}.\mathbf{I} = \mathbf{I}^2 = -1$$
$$\mathbf{I}.1 = \mathbf{I}$$

Compare this with $\mathbf{I}\mathbf{e}_1 = \mathbf{e}_2\mathbf{e}_3$, $\mathbf{I}\mathbf{e}_2\mathbf{e}_3 = -\mathbf{e}_1$. Closing the cycle changes the sign in *both* cases.

These **basis products** enable us to evaluate arbitrary results, such as:

$$(3\mathbf{e}_1 + 4\mathbf{e}_3)(2\mathbf{e}_2\mathbf{e}_3 - 5\mathbf{e}_1\mathbf{e}_2) = 6\mathbf{e}_1\mathbf{e}_2\mathbf{e}_3 - 15\mathbf{e}_1\mathbf{e}_1\mathbf{e}_2 + 8\mathbf{e}_3\mathbf{e}_2\mathbf{e}_3 - 20\mathbf{e}_3\mathbf{e}_1\mathbf{e}_2$$

$$= 6\mathbf{I} - 15\mathbf{e}_2 - 8\mathbf{e}_2 - 20\mathbf{I} = -23\mathbf{e}_2 - 14\mathbf{I}.$$

The **trivector** is not the only "imaginary" here. The **basis bivectors** also square to -1:

$$(\mathbf{e}_1\mathbf{e}_2)(\mathbf{e}_1\mathbf{e}_2) = (\mathbf{e}_1\mathbf{e}_2)(\mathbf{e}_1 \wedge \mathbf{e}_2) = -(\mathbf{e}_1\mathbf{e}_2)(\mathbf{e}_2 \wedge \mathbf{e}_1) = -(\mathbf{e}_1\mathbf{e}_2)(\mathbf{e}_2\mathbf{e}_1)$$

---

[68] As strictly defined, **Hodge Duality** differs from this in sign as Hodge defined it in a slightly different context, and for a slightly different purpose. Hodge's definition appears in Section 7.4. Hestenes's formal definition of this **duality** is that $*\mathbf{A} := \mathbf{A}\mathbf{I}^{-1}$, although he prefers a tilde to the star. As $\mathbf{I}^2 = -1$, $\mathbf{I}^{-1} = -\mathbf{I}$ here, so $*(\mathbf{e}_1\mathbf{e}_2) = -(\mathbf{e}_1\mathbf{e}_2)\mathbf{I} = -\mathbf{e}_{12123} = +\mathbf{e}_3$ and $*\mathbf{e}_1 = -\mathbf{e}_1\mathbf{I} = -\mathbf{e}_2\mathbf{e}_3$; the **Hodge** analogues are $*\mathbf{d}x = \mathbf{d}y\mathbf{d}z$ and $*(\mathbf{d}x\mathbf{d}y) = \mathbf{d}z$.

$$= -\mathbf{e}_1(\mathbf{e}_2\mathbf{e}_2)\mathbf{e}_1 = -\,\mathbf{e}_1\mathbf{e}_1 = -1$$

and

$$(\mathbf{e}_2\mathbf{e}_3)(\mathbf{e}_2\mathbf{e}_3) = -1, \qquad (\mathbf{e}_3\mathbf{e}_1)(\mathbf{e}_3\mathbf{e}_1) = -1.$$

This has a very important consequence. For now:

$$(\mathbf{e}_2\mathbf{e}_3)(\mathbf{e}_3\mathbf{e}_1) = -\mathbf{e}_1\mathbf{e}_2$$
$$(\mathbf{e}_3\mathbf{e}_1)(\mathbf{e}_1\mathbf{e}_2) = -\mathbf{e}_2\mathbf{e}_3$$
$$(\mathbf{e}_1\mathbf{e}_2)(\mathbf{e}_2\mathbf{e}_3) = -\mathbf{e}_3\mathbf{e}_1$$

so if we identify $i \equiv \mathbf{e}_2\mathbf{e}_3$, $j \equiv \mathbf{e}_3\mathbf{e}_1$, and $k = \mathbf{e}_1\mathbf{e}_2$, we have $ij = -k$, $jk = -i$ and $ki = -j$, which are the **quaternion** formulae, *but for a sign change*. But,

$$(\mathbf{e}_3\mathbf{e}_2)(\mathbf{e}_1\mathbf{e}_3) = -\,-(\mathbf{e}_2\mathbf{e}_3)(\mathbf{e}_3\mathbf{e}_1) = +\mathbf{e}_2\mathbf{e}_1$$
$$(\mathbf{e}_1\mathbf{e}_3)(\mathbf{e}_2\mathbf{e}_1) = -\,-(\mathbf{e}_3\mathbf{e}_1)(\mathbf{e}_1\mathbf{e}_2) = +\mathbf{e}_3\mathbf{e}_2$$
$$(\mathbf{e}_2\mathbf{e}_1)(\mathbf{e}_3\mathbf{e}_2) = -\,-(\mathbf{e}_1\mathbf{e}_2)(\mathbf{e}_2\mathbf{e}_3) = +\mathbf{e}_1\mathbf{e}_3$$

so if we identify $i \equiv \mathbf{e}_3\mathbf{e}_2$, $j \equiv \mathbf{e}_1\mathbf{e}_3$, and $k = \mathbf{e}_2\mathbf{e}_1$, we *do* have $ij = k$, $jk = i$ and $ki = j$, which *are* the **quaternion** formulae.

So **quaternions** correspond to a *left-handed* **3-vector** algebra under the **geometric product**. In particular, the **geometric product** of two **3-vectors m** and **n** gives a **quaternion**:

$$\mathbf{m} = m_1\mathbf{e}_1 + m_2\mathbf{e}_2 + m_3\mathbf{e}_3$$
$$\mathbf{n} = n_1\mathbf{e}_1 + n_2\mathbf{e}_2 + n_3\mathbf{e}_3$$

gives

$$\mathbf{mn} = \mathbf{m}\cdot\mathbf{n} + (m_3n_1 - m_1n_3)\mathbf{e}_3\mathbf{e}_1 + (m_2n_3 - m_3n_2)\mathbf{e}_2\mathbf{e}_3 + (m_1n_2 - m_1n_2)\mathbf{e}_1\mathbf{e}_2$$

and this corresponds to the **quaternion**:

$$\mathbf{m}\cdot\mathbf{n} - i(m_3n_1 - m_1n_3) - j(m_2n_3 - m_3n_2) - k(m_1n_2 - m_1n_2)$$

and

$$\mathbf{nm} \equiv \mathbf{m}\cdot\mathbf{n} + i(m_3n_1 - m_1n_3) + j(m_2n_3 - m_3n_2) + k(m_1n_2 - m_1n_2).$$

As another, and telling, example, we can evaluate the **geometric product** of a **vector a** and a **bivector B**, where I use the common convention of capitalizing the symbols for **vectors** of higher **grade**. I will also adopt John Vince's excellent convention of writing $\mathbf{e}_1\mathbf{e}_2\mathbf{e}_3$ as $\mathbf{e}_{123}$, and remembering that e.g. $\mathbf{e}_1\mathbf{e}_3\mathbf{e}_1 = -\mathbf{e}_3$ and that $\mathbf{e}_{312} = -\mathbf{e}_{132} = \mathbf{e}_{123}$,

$$\mathbf{aB} = (a\mathbf{e}_1 + b\mathbf{e}_2 + c\mathbf{e}_3)(x\mathbf{e}_2\mathbf{e}_3 + y\mathbf{e}_3\mathbf{e}_1 + z\mathbf{e}_1\mathbf{e}_2)$$
$$= ax\mathbf{e}_{123} - ay\mathbf{e}_3 + az\mathbf{e}_2 + bx\mathbf{e}_3 + by\mathbf{e}_{231} - bz\mathbf{e}_1 - cx\mathbf{e}_2 + cy\mathbf{e}_1 + cz\mathbf{e}_{312}$$
$$= (cy - bz)\mathbf{e}_1 + (az - cx)\mathbf{e}_2 + (bx - ay)\mathbf{e}_3 + (ax + by + cz)\mathbf{e}_{123}$$

and

$$\mathbf{Ba} = (x\mathbf{e}_2\mathbf{e}_3 + y\mathbf{e}_3\mathbf{e}_1 + z\mathbf{e}_1\mathbf{e}_2)(a\mathbf{e}_1 + b\mathbf{e}_2 + c\mathbf{e}_3)$$
$$= ax\mathbf{e}_{231} + ay\mathbf{e}_3 - az\mathbf{e}_2 - bx\mathbf{e}_3 + by\mathbf{e}_{312} + bz\mathbf{e}_1 + cx\mathbf{e}_2 - cy\mathbf{e}_1 + cz\mathbf{e}_{123}$$
$$= (bz - cy)\mathbf{e}_1 + (cx - az)\mathbf{e}_2 + (ay - bx)\mathbf{e}_3 + (ax + by + cz)\mathbf{e}_{123}.$$

So here the "mono"-**vector** part is **anticommutative (antisymmetric)**, while the **trivector** part is **commutative (symmetric)**. We can write these as:

$\langle\mathbf{aB}\rangle_1 = -\langle\mathbf{Ba}\rangle_1$ and $\langle\mathbf{aB}\rangle_3 = \langle\mathbf{Ba}\rangle_3$ with: $\mathbf{aB} = \langle\mathbf{aB}\rangle_1 + \langle\mathbf{aB}\rangle_3$.

This is the most troublesome aspect of **geometric algebra**. For although the **geometric product** always gives a *lower* **grade** part and a *higher* **grade** part, which one is the **symmetric** and which the **antisymmetric** *varies according to the* **grades** *of the factors.*

Doran and Lasenby write $\langle\mathbf{aB}\rangle_1$ as $\mathbf{a} \cdot \mathbf{B}$ and $\langle\mathbf{aB}\rangle_3$ as $\mathbf{a} \wedge \mathbf{B}$, as "reasonable" extensions of the **dot product** and **alternating product** notations, but you need to be careful with this. For the **trivector** part, $\mathbf{a} \wedge \mathbf{B}$ is already well-defined, as, in this case:

$$\mathbf{a} \wedge \mathbf{B} = (a\mathbf{e}_1 + b\mathbf{e}_2 + c\mathbf{e}_3) \wedge (x\mathbf{e}_2\mathbf{e}_3 + y\mathbf{e}_3\mathbf{e}_1 + z\mathbf{e}_1\mathbf{e}_2) = ax\mathbf{e}_{123} + by\mathbf{e}_{231} + cz\mathbf{e}_{312}$$
$$= (ax + by + cz)\mathbf{e}_{123}$$
$$= \mathbf{B} \wedge \mathbf{a}.$$

So $\langle\mathbf{aB}\rangle_3$ actually *is* $\mathbf{a} \wedge \mathbf{B}$ and does equal $\mathbf{B} \wedge \mathbf{a}$. This apparent perversion of the **alternating product**, where it seems to cease to be "**alternating**" between a **vector** and a **bivector,** is quite general. If **B** were a **blade** of the form $\mathbf{b} \wedge \mathbf{c}$, we would have:

$$\mathbf{a} \wedge \mathbf{B} = \mathbf{a} \wedge \mathbf{b} \wedge \mathbf{c} = -\mathbf{b} \wedge \mathbf{a} \wedge \mathbf{c} = +\mathbf{b} \wedge \mathbf{c} \wedge \mathbf{a} = \mathbf{B} \wedge \mathbf{a}.$$

I am tempted to advise against writing $\langle\mathbf{aB}\rangle_1$ as $\mathbf{a} \cdot \mathbf{B}$ at least, but the usage is established, and you should just remain aware that it *isn't* **commutative** or **symmetric**.

We can find the standard form for the **outer product** between two ordinary **vectors** using the formulae we now have. Remembering that $\mathbf{e}_1 \wedge \mathbf{e}_1 = \mathbf{e}_2 \wedge \mathbf{e}_2 = \mathbf{e}_3 \wedge \mathbf{e}_3 = 0$, we have:

$$(a_1\mathbf{e}_1 + a_2\mathbf{e}_2 + a_3\mathbf{e}_3) \wedge (b_1\mathbf{e}_1 + b_2\mathbf{e}_2 + b_3\mathbf{e}_3)$$

$$= a_1b_2\mathbf{e}_1 \wedge \mathbf{e}_2 + a_1b_3\mathbf{e}_1 \wedge \mathbf{e}_3 + a_2b_1\mathbf{e}_2 \wedge \mathbf{e}_1 + a_2b_3\mathbf{e}_2 \wedge \mathbf{e}_3 + a_3b_1\mathbf{e}_3 \wedge \mathbf{e}_1 + a_3b_2\mathbf{e}_3 \wedge \mathbf{e}_2$$

$$= a_1b_2\mathbf{e}_1 \wedge \mathbf{e}_2 - a_1b_3\mathbf{e}_3 \wedge \mathbf{e}_1 - a_2b_1\mathbf{e}_1 \wedge \mathbf{e}_2 + a_2b_3\mathbf{e}_2 \wedge \mathbf{e}_3 + a_3b_1\mathbf{e}_3 \wedge \mathbf{e}_1 - a_3b_2\mathbf{e}_2 \wedge \mathbf{e}_3$$

$$= (a_1b_2 - a_2b_1)\mathbf{e}_1\mathbf{e}_2 + (a_3b_1 - a_1b_3)\mathbf{e}_3\mathbf{e}_1 + (a_2b_3 - a_3b_2)\mathbf{e}_2\mathbf{e}_3$$

which looks very like Gibbs's **cross product** of **vectors**. The difference is that this is a **linear combination** of **bivectors**, whereas the traditional **cross product** involves ordinary **grade-1 vectors** *both* as the **arguments** *and* as the *result*. We can equate the two simply by using "**Hodge**" or **pseudoscalar duality** to convert back to **vectors**. Multiply by $-\mathbf{I}$ and since $-\mathbf{I}\mathbf{e}_2\mathbf{e}_3 = \mathbf{e}_1$, $-\mathbf{I}\mathbf{e}_3\mathbf{e}_1 = \mathbf{e}_2$, $-\mathbf{I}\mathbf{e}_1\mathbf{e}_2 = \mathbf{e}_3$, we get:

$$-\mathbf{I}.[(a_1b_2 - a_2b_1)\mathbf{e_1e_2} + (a_3b_1 - a_1b_3)\mathbf{e_3e_1} + (a_2b_3 - a_3b_2)\mathbf{e_2e_3}]$$

$$= (a_1b_2 - a_2b_1)\mathbf{e_3} + (a_3b_1 - a_1b_3)\mathbf{e_2} + (a_2b_3 - a_3b_2)\mathbf{e_1}$$

and this is the ordinary **cross product a × b**. In other words:

$$\mathbf{a} \times \mathbf{b} = -\mathbf{I}(\mathbf{a} \wedge \mathbf{b}).$$

This **cross product**, together with the **dot product**, dominated **vector algebra** throughout the twentieth century, causing a total eclipse of Clifford's more general algebra.

If $a_3 = b_3 = 0$, so the **vectors** both lie in the $\mathbf{e_1}-\mathbf{e_2}$ plane, and we put $\mathbf{a} = r_a(\phi_x\mathbf{e_1} + \phi_y\mathbf{e_2})$ and $\mathbf{b} = r_b(\theta_x\mathbf{e_1} + \theta_y\mathbf{e_2})$, we can see that:

$$\mathbf{a} \times \mathbf{b} = (a_1b_2 - a_2b_1)\mathbf{e_3} = r_ar_b(\phi_x\theta_y - \phi_y\theta_x)\mathbf{e_3} = r_ar_b.\mathbf{sine}(\mathbf{b} - \mathbf{a})\mathbf{e_3}$$

a form of the **cross product** often found in the textbooks, and discussed more fully in Section 5.7.

From the above definition, $\mathbf{e_1} \times \mathbf{e_2} = -\mathbf{I}(\mathbf{e_1} \wedge \mathbf{e_2}) = -\mathbf{Ie_1e_2} = -\mathbf{e_1e_2e_3e_1e_2} = \mathbf{e_1e_2e_3e_2e_1} = -\mathbf{e_1e_3e_1} = \mathbf{e_3}$, and this is a general pattern: *the result of the* **cross product** *always lies at right angles to the plane of the two* **arguments**. Remember the "right-handed rule" here too: if you make the right hand into a fist but with the index finger pointing forwards and the thumb upwards, then as the index finger rotates *anticlockwise* from **x** to **y**, then the thumb points in the direction of **x** × **y**, as shown in *Figure 5.7.1*.

This automatic **orthogonality** of the **cross product** is very useful for forming **right-handed triads** of **3-vectors** as a **basis**. Given two arbitrary **linearly independent 3-vectors m** and **n** with **n** to the *left* of **m** in the sense defined in Section 4.3, we can take:

$$\mathbf{e_1} = \mathbf{m}/|\mathbf{m}|$$
$$\mathbf{e_2} = \mathbf{norm}(\mathbf{n} - \mathbf{e_1}(\mathbf{e_1}\cdot\mathbf{n}))$$
$$\mathbf{e_3} = \mathbf{e_1} \times \mathbf{e_2}$$

using **Gram-Schmidt orthogonalization** for $\mathbf{e_2}$ (so $\mathbf{e_1}\cdot\mathbf{e_2} = \mathbf{e_1}\cdot\mathbf{n} - (\mathbf{e_1}\cdot\mathbf{e_1})(\mathbf{e_1}\cdot\mathbf{n}) = \mathbf{e_1}\cdot\mathbf{n} - \mathbf{e_1}\cdot\mathbf{n} = 0$) and the **cross product** for $\mathbf{e_3}$, and we get a right-handed **orthonormal basis** with $\mathbf{e_1}$ and $\mathbf{e_2}$ describing the same plane as **m** and **n**.

The three-dimensional **geometric algebra** on the **basis vectors** can be expressed compactly in the form:

$$\mathbf{e}_i\mathbf{e}_j = \delta_{ij} + \varepsilon_{ijk}\mathbf{I}\mathbf{e}_k$$

where $\varepsilon_{ijk} = +1$ if $[ijk]$ is a **cyclic permutation** of [123], $= -1$ if $[ijk]$ is an **anticyclic permutation** of [123], and $= 0$ if $[ijk]$ has any two elements the same, as in [112] or [323]. I introduced this symbol briefly in the discussion of **determinants** in Section 1.8. This **algebra** has a long history in quantum mechanics, where it appears in the form:

$$\sigma_i\sigma_j = \delta_{ij}.\mathrm{I} + j\varepsilon_{ijk}\sigma_k$$

where the σ_i are the **Pauli matrices**:

$$\begin{pmatrix}0 & 1\\1 & 0\end{pmatrix} \quad \begin{pmatrix}0 & -j\\j & 0\end{pmatrix} \quad \begin{pmatrix}1 & 0\\0 & -1\end{pmatrix}$$

and I is the **unit** 2×2 **matrix**: $\begin{pmatrix}1 & 0\\0 & 1\end{pmatrix}$.

Another famous **algebra** in quantum mechanics, again originally discovered as an algebra of **matrices**, is given by the **geometric algebra** of **4-vectors** using the **Minkowski metric tensor**, which is the **symmetric scalar product** given by either of these forms:

$$\{g_{ij}\} = \begin{pmatrix}1 & 0 & 0 & 0\\0 & 1 & 0 & 0\\0 & 0 & 1 & 0\\0 & 0 & 0 & -1\end{pmatrix} \text{ or } \begin{pmatrix}-1 & 0 & 0 & 0\\0 & -1 & 0 & 0\\0 & 0 & -1 & 0\\0 & 0 & 0 & 1\end{pmatrix}.$$

which are simply negatives of each other. This **metric tensor** is the standard one used in Einstein's **special relativity**, where the "mixed signature" – three plusses and a minus, or three minuses and a plus – separates the role of *time t* from that of the *spatial* **coordinates** *x*, *y*, and *z*.

The **geometric algebra** of **4-vectors** that results from this **metric tensor** is called the **Dirac algebra** and is fundamental to **quantum field theory**, the relativistic field formulation of **quantum mechanics**.

5.10 Rotations

Rotations are to do with *planes*. Given an **n-vector** space, and two **linearly independent vectors m** and **n**, then **m** and **n** define a plane in the space. We can use the **Gram-Schmidt orthogonalization** described in Section 5.7 to create a new **basis** $\{\mathbf{e}_1, \mathbf{e}_2, \mathbf{e}_3, \ldots\}$ for the **vector space** in which $\mathbf{e}_1 = \mathbf{m}/|\mathbf{m}|$, so is the **unit vector** parallel to **m**, with $\mathbf{m} = |\mathbf{m}|\mathbf{e}_1$, and $\mathbf{e}_1 \cdot \mathbf{e}_1 = 1$, and $\mathbf{e}_2 = \mathbf{norm}(\mathbf{n} - \mathbf{m}((\mathbf{n} \cdot \mathbf{m})/(\mathbf{m}^2)))$. So $\mathbf{m} \cdot \mathbf{e}_2 = \mathbf{e}_1 \cdot \mathbf{e}_2 = \mathbf{0}$, and $\mathbf{e}_2 \cdot \mathbf{e}_2 = 1$.[69] Now **n** lies in the plane of $\mathbf{e}_1$ and $\mathbf{e}_2$, so $\mathbf{n} = \alpha\mathbf{e}_1 + \beta\mathbf{e}_2$ and so:

1. $\mathbf{e}_1$ and $\mathbf{e}_2$ define the same plane as **m** and **n**, but are **orthonormal**: $\mathbf{e}_1 \cdot \mathbf{e}_2 = 0$.

[69] Given *any* set of **linearly independent vectors** $\{\mathbf{m}_1, \mathbf{m}_2, \mathbf{m}_3, \ldots\}$, we can use **Gram-Schmidt orthogonalization** to create an **orthonormal** set $\{\mathbf{e}_1, \mathbf{e}_2, \mathbf{e}_3, \ldots\}$ that is an **orthonormal basis** for the space **spanned** by $\{\mathbf{m}_1, \mathbf{m}_2, \mathbf{m}_3, \ldots\}$. The general formula is $\mathbf{e}_k = \mathbf{norm}(\mathbf{m}_k - \Sigma_{(i<k)}\, \mathbf{e}_i(\mathbf{m}_k \cdot \mathbf{e}_i))$.

2. The entire **basis** $\{\mathbf{e}_1, \mathbf{e}_2, \mathbf{e}_3, \ldots\}$ is **orthonormal**: $\mathbf{e}_i \cdot \mathbf{e}_j = \delta_{ij}$.
3. $\mathbf{n} = \alpha\mathbf{e}_1 + \beta\mathbf{e}_2 = |\mathbf{n}|\theta_x\mathbf{e}_1 + |\mathbf{n}|\theta_y\mathbf{e}_2$.

Then a **rotation** Θ can be defined for any **vector** $x_1\mathbf{e}_1 + x_2\mathbf{e}_2 + x_3\mathbf{e}_3 + \ldots$ as:

$$\Theta(x_1\mathbf{e}_1 + x_2\mathbf{e}_2 + x_3\mathbf{e}_3 + \ldots) = (1/|\mathbf{n}|)(\alpha x_1 - \beta x_2)\mathbf{e}_1 + (1/|\mathbf{n}|)(\beta x_1 + \alpha x_2)\mathbf{e}_2 + x_3\mathbf{e}_3 + \ldots$$

where if you're very perceptive, you may recognize the angular addition form from Sections 4.2 and 5.7. In case you cannot, the argument runs like this. Since we can write **n** and **m** as:

$$\mathbf{n} = |\mathbf{n}|(\theta_x\mathbf{e}_1 + \theta_y\mathbf{e}_2) \qquad \text{and} \qquad \mathbf{m} = |\mathbf{m}|\mathbf{e}_1,$$

so that $\boldsymbol{\theta} = [\theta_x, \theta_y]$ is the **angle** between **n** and **m**.

Then any **vector** $\mathbf{x} = [x_1, x_2]$ in the $\mathbf{e}_1\text{–}\mathbf{e}_2$ plane, can also be written as

$$\mathbf{x} = (r, \boldsymbol{\varphi}) = r[\varphi_x, \varphi_y] = r\varphi_x\mathbf{e}_1 + r\varphi_y\mathbf{e}_2.$$

Next, we can precisely define a **rotation**:

$$\Theta(\mathbf{x}) = \Theta(r, \boldsymbol{\varphi})) = (r, \boldsymbol{\phi} + \boldsymbol{\theta}) = r[\varphi_x\theta_x - \varphi_y\theta_y, \varphi_x\theta_y + \varphi_y\theta_x].$$

But

$$\theta_x = \alpha/|\mathbf{n}|, \qquad \theta_y = \beta/|\mathbf{n}|$$

so

$$\Theta(\mathbf{x}) = (r/|\mathbf{n}|)[\alpha\varphi_x - \beta\varphi_y, \beta\varphi_x + \alpha\varphi_y] = (1/|\mathbf{n}|)[\alpha r\varphi_x - \beta r\varphi_y, \beta r\varphi_x + \alpha r\varphi_y]$$

$$= (1/|\mathbf{n}|)[\alpha x_1 - \beta x_2, \beta x_1 + \alpha x_2] = (1/|\mathbf{n}|)(\alpha x_1 - \beta x_2)\mathbf{e}_1 + (1/|\mathbf{n}|)(\beta x_1 + \alpha x_2)\mathbf{e}_2.$$

So a **rotation** is uniquely defined as a **transformation** of any **vector** whereby its $\mathbf{e}_1$ and $\mathbf{e}_2$ components – the components in the **m-n** plane – are rotated through the angle $\boldsymbol{\theta}$ between **m** and **n**, *but all its other components, in* $\mathbf{e}_3$, $\mathbf{e}_4$, *etc. are unaltered.*

Strictly, this only defines a **rotation** about the **origin**, but we can easily **translate** our original **coordinates** to make an arbitrary point $\mathbf{P} = [p_1, p_2, p_3, \ldots]$ become the **origin** by redefining any **vector** $\mathbf{x} = [x_1, x_2, x_3, \ldots] \to \mathbf{T}([x_1, x_2, x_3, \ldots]) = [x_1 - p_1, x_2 - p_2, x_3 - p_3, \ldots]$. So hereafter I will restrict consideration to **rotations** about the **origin**.

That only the $\mathbf{e}_1$ and $\mathbf{e}_2$ **components** of any **vector** are changed by a **rotation** about the **origin** defined for the $\mathbf{e}_1\text{–}\mathbf{e}_2 \equiv \mathbf{m}\text{–}\mathbf{n}$ plane is the key to the notion of an *axis* of **rotation**. If we are dealing with **2-vectors**, then $\mathbf{e}_1$ and $\mathbf{e}_2$ are a complete **basis** for the whole **vector space** and *only* the **origin** [0, 0] is unchanged by the **rotation**:

$$\Theta([0, 0]) = [0, 0].$$

If we are in a space of **3-vectors**, then *any point on the* $\mathbf{e}_3$ **axis** will be unchanged by the **rotation**:

$$\Theta([0, 0, z]) = [0, 0, z].$$

This means that there is a *line* of points that are **invariant** – unchanged – by the **rotation**, and we call this **invariant line** – a **one-dimensional invariant space** – the **axis** of the **rotation**.

If we are in a **four-dimensional space** with **4-vectors**, our new **basis** will consist of *four* **4-vectors** $\mathbf{e}_1$, $\mathbf{e}_2$, $\mathbf{e}_3$ and $\mathbf{e}_4$ where $\mathbf{e}_1$ and $\mathbf{e}_2$ again **span** the same **subspace** as **m** and **n**, but now *any point in the* **two-dimensional subspace spanned** *by* $\mathbf{e}_3$ *and* $\mathbf{e}_4$ *with* $x_1 = x_2 = 0$ *will be unchanged by the* **rotation**:

$$\Theta([0, 0, z, w]) = [0, 0, z, w].$$

So now the **invariant subspace** is a *plane* defined by $x_1 = x_2 = 0$. This observation prompted a famous remark by Sir Arthur Eddington:

> *"In four dimensions, things rotate about a plane. I know that it is so, but I cannot visualize it."*[70]

Our next task is to translate this concept of **rotation** into **geometric algebra**. First we replace **m** and **n** by their corresponding **unit vectors** $\underline{\mathbf{m}} = \mathbf{m}/|\mathbf{m}|$ and $\underline{\mathbf{n}} = \mathbf{n}/|\mathbf{n}|$.

How do we begin to describe the rotation?

Let's first look at something else, a *reflection* in a plane mirror in 3-space.[71] Suppose, as we're doing throughout, that everything is based at the **origin**. So assume that the plane of the mirror passes through the **origin**, and it will have two **normal vectors** $\underline{\mathbf{n}}$ and $-\underline{\mathbf{n}}$. We can use either, so let's take $\underline{\mathbf{n}}$.

Now any **vector x** must satisfy:

$$\mathbf{x} = \underline{\mathbf{n}}^2\mathbf{x}$$

because $\underline{\mathbf{n}}^2 = 1$. But this **geometric product** is associative, so we have:

$$\mathbf{x} = \underline{\mathbf{n}}(\underline{\mathbf{n}}\mathbf{x}) = \underline{\mathbf{n}}(\underline{\mathbf{n}} \cdot \mathbf{x} + \underline{\mathbf{n}} \wedge \mathbf{x}).$$

But $(\underline{\mathbf{n}} \cdot \mathbf{x})\underline{\mathbf{n}}$ is the **projection** of **x** onto $\underline{\mathbf{n}}$, or the **component** of **x** along $\underline{\mathbf{n}}$ (remembering that $\underline{\mathbf{n}}$ is a **unit vector**) *and this is precisely the part of* **x** *that will be reflected in the mirror.*

So writing the reflection as $\Re$,

$$\Re(\mathbf{x}) = -(\underline{\mathbf{n}} \cdot \mathbf{x})\underline{\mathbf{n}} + \underline{\mathbf{n}}(\underline{\mathbf{n}} \wedge \mathbf{x}).$$

---

[70] I am still trying to track down this quotation, which I repeat above just from memory. It refers primarily to the concept of a **simple rotation** in 4-space; there are also **double rotations**. See the article "Rotations in 4-dimensional Euclidean space" in Wikipedia.

[71] The illuminating argument immediately following I take from John Vince's excellent little book *Geometric Algebra for Computer Graphics*, but the main argument following is my own from the first edition of the present text.

But **vectors** and **bivectors** anticommute, so we can $\underline{\mathbf{n}}(\underline{\mathbf{n}} \wedge \mathbf{x})$ as $-(\underline{\mathbf{n}} \wedge \mathbf{x})\underline{\mathbf{n}}$, giving:

$$\Re(\mathbf{x}) = -(\underline{\mathbf{n}} \cdot \mathbf{x} - \underline{\mathbf{n}} \wedge \mathbf{x})\underline{\mathbf{n}} = -(\underline{\mathbf{n}}\mathbf{x})\underline{\mathbf{n}} = -\underline{\mathbf{n}}\mathbf{x}\underline{\mathbf{n}}.$$

Now let's imagine we've got *two* mirrors, and let's look down at the common plane in which *both* their **normal vectors** jointly lie. This is shown in *Figure 5.10.1*:

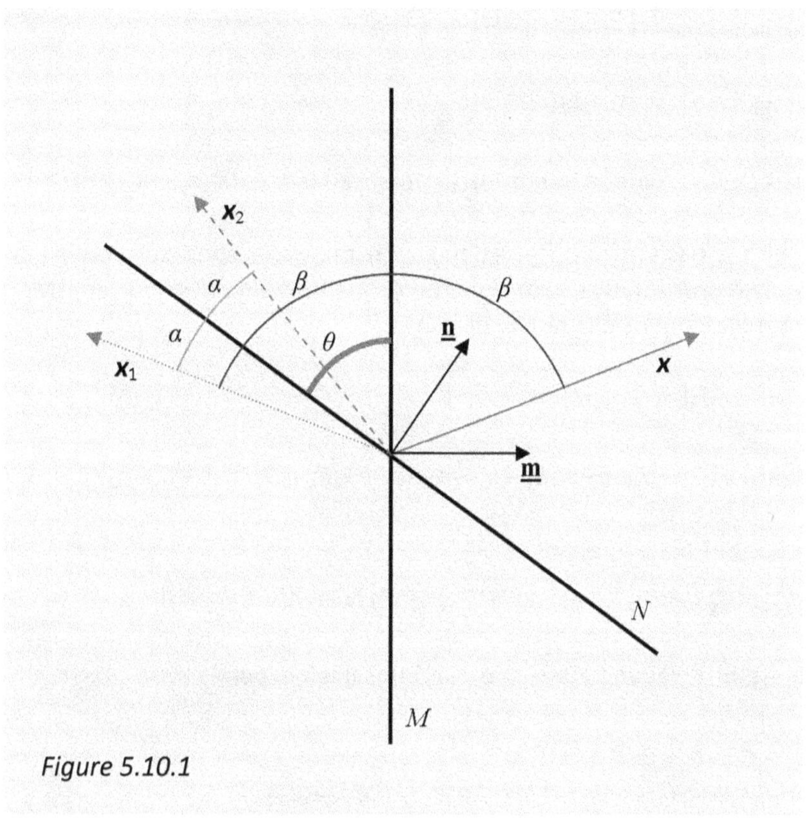

Figure 5.10.1

Here the mirrors are labelled *M* and *N* respectively, and their **normals** (**normal vectors**) are $\underline{\mathbf{m}}$ and $\underline{\mathbf{n}}$ respectively.

Now $\mathbf{x}$ is reflected about *M* into $\mathbf{x}_1$, and $\mathbf{x}_1$ is reflected about *N* into $\mathbf{x}_2$. If the angle between *M* and *N* is θ, and the angle between $\mathbf{x}$ and the plane of *M* is β, we can see from the *Figure* that the final angle between $\mathbf{x}$ and $\mathbf{x}_2$ will be $2\beta - 2\alpha$. But $\beta - \alpha = \theta$, so $2(\beta - \alpha) = 2\theta$.

But any set of **vectors** $\{\mathbf{x}, \mathbf{y}, \mathbf{z}\}$ undergoing a *double* reflection will recover their original orientation (or "handedness"), and we have from the above:

$$\mathbf{x}_1 = -\underline{\mathbf{m}}\mathbf{x}\underline{\mathbf{m}},$$

so

$$\mathbf{x}_2 = -\underline{\mathbf{n}}(\mathbf{x}_1)\underline{\mathbf{n}} = +\underline{\mathbf{n}}\underline{\mathbf{m}}(\mathbf{x})\underline{\mathbf{m}}\underline{\mathbf{n}}.$$

> So this suggests that the expression $\underline{\mathbf{n}}\mathbf{m}\mathbf{x}\underline{\mathbf{m}}\mathbf{n}$ might give a **rotation** through 2θ.

So for any **vector x** = $[x_1, x_2, x_3, \ldots]$ we'll expand this expression:

$$\Theta(\mathbf{x}) = \underline{\mathbf{n}}\mathbf{m}\mathbf{x}\underline{\mathbf{m}}\mathbf{n}.$$

We can postulate our **Gram-Schmidt basis** for the purposes of proof, and assume **x** is already given in this **basis** in which also:

$$\underline{\mathbf{n}} = \theta_x\mathbf{e}_1 + \theta_y\mathbf{e}_2 \quad \text{and} \qquad \underline{\mathbf{m}} = \mathbf{e}_1.$$

So $\underline{\mathbf{nm}} = (\theta_x\mathbf{e}_1 + \theta_y\mathbf{e}_2)\mathbf{e}_1 = \theta_x - \theta_y\mathbf{e}_1\mathbf{e}_2$ and $\underline{\mathbf{mn}} = \mathbf{e}_1(\theta_x\mathbf{e}_1 + \theta_y\mathbf{e}_2) = \theta_x + \theta_y\mathbf{e}_1\mathbf{e}_2$.

I'll start working with a **basis** of arbitrary dimension but we'll soon see, taking the fate of terms involving $\mathbf{e}_3$ and $\mathbf{e}_4$ as representative, that only the terms involving $\mathbf{e}_1$ and $\mathbf{e}_2$ matter. For the moment I'll leave in the higher elements indicated by ellipses (**. . .**).

So, expanding:

$$\Theta(\mathbf{x}) = \underline{\mathbf{n}}\mathbf{m}\mathbf{x}\underline{\mathbf{m}}\mathbf{n} = (\theta_x - \theta_y\mathbf{e}_1\mathbf{e}_2)(x_1\mathbf{e}_1 + x_2\mathbf{e}_2 + x_3\mathbf{e}_3 + x_4\mathbf{e}_4 + \ldots)(\theta_x + \theta_y\mathbf{e}_1\mathbf{e}_2)$$

$$= (\theta_x x_1\mathbf{e}_1 + \theta_x x_2\mathbf{e}_2 + \theta_x x_3\mathbf{e}_3 + \theta_x x_4\mathbf{e}_4 + \ldots$$
$$+ \theta_y x_1\mathbf{e}_2 - \theta_y x_2\mathbf{e}_1 - \theta_y x_3\mathbf{e}_1\mathbf{e}_2\mathbf{e}_3 - \theta_y x_4\mathbf{e}_1\mathbf{e}_2\mathbf{e}_4 \ldots)(\theta_x + \theta_y\mathbf{e}_1\mathbf{e}_2)$$

where I have used $\mathbf{e}_1\mathbf{e}_2\mathbf{e}_1 = -\mathbf{e}_2$ and $\mathbf{e}_1\mathbf{e}_2\mathbf{e}_2 = \mathbf{e}_1$, and this itself expands to:

$$= \theta_x^2 x_1\mathbf{e}_1 + \theta_x^2 x_2\mathbf{e}_2 + \theta_x^2 x_3\mathbf{e}_3 + \theta_x^2 x_4\mathbf{e}_4 + \ldots$$
$$+ \theta_x\theta_y x_1\mathbf{e}_2 - \theta_x\theta_y x_2\mathbf{e}_1 - \theta_x\theta_y x_3\mathbf{e}_1\mathbf{e}_2\mathbf{e}_3 - \theta_x\theta_y x_4\mathbf{e}_1\mathbf{e}_2\mathbf{e}_4 \ldots$$
$$+ \theta_x\theta_y x_1\mathbf{e}_2 - \theta_x\theta_y x_2\mathbf{e}_1 + \theta_x\theta_y x_3\mathbf{e}_3\mathbf{e}_1\mathbf{e}_2 + \theta_x\theta_y x_4\mathbf{e}_4\mathbf{e}_1\mathbf{e}_2 \ldots$$
$$- \theta_y^2 x_1\mathbf{e}_1 - \theta_y^2 x_2\mathbf{e}_2 - \theta_y^2 x_3\mathbf{e}_1\mathbf{e}_2\mathbf{e}_3\mathbf{e}_1\mathbf{e}_2 - \theta_y^2 x_4\mathbf{e}_1\mathbf{e}_2\mathbf{e}_4\mathbf{e}_1\mathbf{e}_2 \ldots$$

Now

$$\mathbf{e}_1\mathbf{e}_2\mathbf{e}_3\mathbf{e}_1\mathbf{e}_2 = -\mathbf{e}_1(\mathbf{e}_2\mathbf{e}_3\mathbf{e}_2)\mathbf{e}_1 = +\mathbf{e}_1\mathbf{e}_3\mathbf{e}_1 = -\mathbf{e}_3$$

and similarly

$$\mathbf{e}_1\mathbf{e}_2\mathbf{e}_4\mathbf{e}_1\mathbf{e}_2 = -\mathbf{e}_1\mathbf{e}_2\mathbf{e}_4\mathbf{e}_2\mathbf{e}_1 = +\mathbf{e}_1\mathbf{e}_4\mathbf{e}_1 = -\mathbf{e}_4.$$

Therefore the terms in $\mathbf{e}_3$ are $+\,\theta_x^2 x_3\mathbf{e}_3 + \theta_y^2 x_3\mathbf{e}_3 = (\theta_x^2 + \theta_y^2)x_3\mathbf{e}_3 = x_3\mathbf{e}_3$ and in $\mathbf{e}_4$ are $x_4\mathbf{e}_4$.

So indeed the terms beyond $\mathbf{e}_1$ and $\mathbf{e}_2$ are unaltered.

Then again,

$$\mathbf{e}_3\mathbf{e}_1\mathbf{e}_2 = -\mathbf{e}_1\mathbf{e}_3\mathbf{e}_2 = \mathbf{e}_1\mathbf{e}_2\mathbf{e}_3$$

and

$$\mathbf{e}_4\mathbf{e}_1\mathbf{e}_2 = -\mathbf{e}_1\mathbf{e}_4\mathbf{e}_2 = \mathbf{e}_1\mathbf{e}_2\mathbf{e}_4$$

so we have:

$$\theta_x\theta_y x_3\mathbf{e}_1\mathbf{e}_2\mathbf{e}_3 - \theta_x\theta_y x_3\mathbf{e}_1\mathbf{e}_2\mathbf{e}_3 = (\theta_x\theta_y x_3 - \theta_x\theta_y x_3)\mathbf{e}_1\mathbf{e}_2\mathbf{e}_3 = 0,$$
$$-\theta_x\theta_y x_4\mathbf{e}_1\mathbf{e}_2\mathbf{e}_4 + \theta_x\theta_y x_4\mathbf{e}_4\mathbf{e}_1\mathbf{e}_2 = (-\theta_x\theta_y x_4 + \theta_x\theta_y x_4)\mathbf{e}_1\mathbf{e}_2\mathbf{e}_4 = 0.$$

So the **trivector** parts drop out too.

In $\mathbf{e}_1$ we have:

$$\theta_x^2 x_1 - \theta_x\theta_y x_2 - \theta_x\theta_y x_2 - \theta_y^2 x_1 = \theta_x^2 x_1 - 2\theta_x\theta_y x_2 - \theta_y^2 x_1$$

and in $\mathbf{e}_2$:

$$\theta_x^2 x_2 + \theta_x\theta_y x_1 + \theta_x\theta_y x_1 - \theta_y^2 x_2 = \theta_x^2 x_2 + 2\theta_x\theta_y x_1 - \theta_y^2 x_2$$

so all told we have:

$$(\theta_x^2 x_1 - 2\theta_x\theta_y x_2 - \theta_y^2 x_1)\mathbf{e}_1 + (\theta_x^2 x_2 + 2\theta_x\theta_y x_1 - \theta_y^2 x_2)\mathbf{e}_2 + x_3\mathbf{e}_3 + x_4\mathbf{e}_4 + \ldots$$

which may not be *quite* the **rotation** we expected.

But the mystery is easily resolved. Taking a hint from the *quadratic* nature of these terms, and remembering the original motivation for using the **n̲m̲x̲m̲n̲** form in the first place, could it be that we've turned **x** through *twice* the angle?

Switching into **complex** notation to speed up the algebra, remembering that $j \equiv \mathbf{e}_1\mathbf{e}_2$,

$$(x_1 + jx_2)(\theta_x + j\theta_y)(\theta_x + j\theta_y) = ((x_1\theta_x - x_2\theta_y) + j(x_2\theta_x + x_1\theta_y))(\theta_x + j\theta_y)$$
$$= \theta_x(x_1\theta_x - x_2\theta_y) - \theta_y(x_2\theta_x + x_1\theta_y) + j(\theta_x(x_2\theta_x + x_1\theta_y) + \theta_y(x_1\theta_x - x_2\theta_y))$$
$$= x_1\theta_x^2 - x_2\theta_x\theta_y - x_2\theta_x\theta_y - x_1\theta_y^2 + j(x_2\theta_x^2 + x_1\theta_x\theta_y + x_1\theta_x\theta_y - x_2\theta_y^2)$$
$$= (\theta_x^2 x_1 - 2\theta_x\theta_y x_2 - \theta_y^2 x_1) + j(\theta_x^2 x_2 + 2\theta_x\theta_y x_1 - \theta_y^2 x_2)$$
$$= (\theta_x^2 x_1 - 2\theta_x\theta_y x_2 - \theta_y^2 x_1) + (\theta_x^2 x_2 + 2\theta_x\theta_y x_1 - \theta_y^2 x_2)\mathbf{e}_1\mathbf{e}_2$$
$$= \mathbf{e}_1\{(\theta_x^2 x_1 - 2\theta_x\theta_y x_2 - \theta_y^2 x_1)\mathbf{e}_1 + (\theta_x^2 x_2 + 2\theta_x\theta_y x_1 - \theta_y^2 x_2)\mathbf{e}_2\}.$$

So indeed $\Theta(\mathbf{x})$ = **n̲m̲x̲m̲n̲** has taken us through *twice* the angle $\boldsymbol{\theta} = [\theta_x, \theta_y]$.

There are two ways of handling this.

1. Either we accept that this attractive **rotor transform** defined using the **rotor** **<u>nm</u>** and its **conjugate** or, in **geometric algebra** terminology, its **reverse** **<u>mn</u>**, does simply give us a **rotation** through *double* the angle between **n** and **m**,
2. or we use the *bisector* of the angle instead of **n** so that we end up with the *same* angle of **rotation** after all.

The first approach we have already developed. *Figure 5.10.2* shows the construction of the bisector of **<u>m</u>** and **<u>n</u>**:

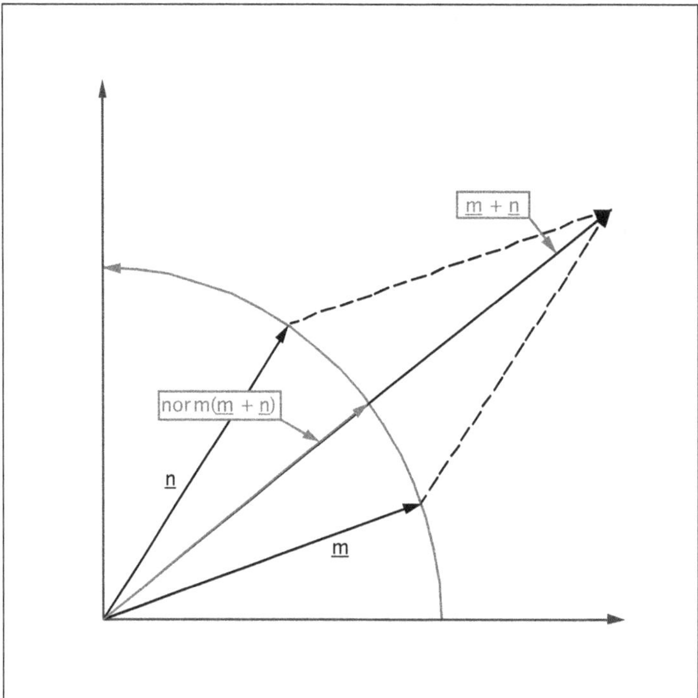

Figure 5.10.2

The *bisector* is easy to evaluate, as the **unit bisector** is given, from *Figure 5.10.2*, as simply:

$$\mathbf{norm}(\underline{\mathbf{m}} + \underline{\mathbf{n}}) = (\underline{\mathbf{m}} + \underline{\mathbf{n}})/|\underline{\mathbf{m}} + \underline{\mathbf{n}}|$$

which I will write as $\underline{\mathbf{n}}_{½}$. If we now use this as our second vector instead of just $\underline{\mathbf{n}}$, we get:

$$\Theta_{½}(\mathbf{x}) = \underline{\mathbf{n}_{½}\mathbf{m}}\ \mathbf{x}\ \underline{\mathbf{m}\mathbf{n}_{½}} = ((\underline{\mathbf{m}} + \underline{\mathbf{n}})/|\underline{\mathbf{m}} + \underline{\mathbf{n}}|)\underline{\mathbf{m}}\ \mathbf{x}\ \underline{\mathbf{m}}((\underline{\mathbf{m}} + \underline{\mathbf{n}})/|\underline{\mathbf{m}} + \underline{\mathbf{n}}|).$$

In our $\mathbf{e}_1$, $\mathbf{e}_2$ system, the **unit bisector** evaluates to:

$$(\underline{\mathbf{m}} + \underline{\mathbf{n}})/|\underline{\mathbf{m}} + \underline{\mathbf{n}}| = ((\theta_x + 1)\mathbf{e}_1 + \theta_y\mathbf{e}_2)/\sqrt{((\theta_x + 1)^2 + \theta_y^2)}$$

$$= ((\theta_x + 1)\mathbf{e}_1 + \theta_y\mathbf{e}_2)/\sqrt{(\theta_x^2 + 2\theta_x + 1 + \theta_y^2)}$$

$$= ((\theta_x + 1)\mathbf{e}_1 + \theta_y\mathbf{e}_2)/\sqrt{(2\theta_x + 2)}$$

$= ((\theta_x + 1)\mathbf{e_1} + \theta_y\mathbf{e_2})/(\sqrt{2}.\sqrt{(\theta_x + 1)})$

$= (1/\sqrt{2})(\sqrt{(\theta_x + 1)}\mathbf{e_1} + (\theta_y/\sqrt{(\theta_x + 1)})\mathbf{e_2})$.

Now we can write this as $\sqrt{½}.(\varphi_x\mathbf{e_1} + \varphi_y\mathbf{e_2})$ with $\varphi_x = \sqrt{(\theta_x + 1)}$ and $\varphi_y = \theta_y/\sqrt{(\theta_x + 1)}$.
But $\theta_y = \sqrt{(1 - \theta_x^2)}$ and $(1 + \theta_x)(1 - \theta_x) = 1 - \theta_x + \theta_x + \theta_x^2 = (1 - \theta_x^2)$.

So $\quad \varphi_y = \sqrt{((1 - \theta_x^2)/(\theta_x + 1))} = \sqrt{((1 + \theta_x)(1 - \theta_x)/(\theta_x + 1))} = \sqrt{(1 - \theta_x)}$

and this puts $(\underline{\mathbf{m}} + \underline{\mathbf{n}})/|\underline{\mathbf{m}} + \underline{\mathbf{n}}|$ in the symmetrical form:

$\sqrt{½}(\sqrt{(1 + \theta_x)}\mathbf{e_1} + \sqrt{(1 - \theta_x)}\mathbf{e_2})$

To confirm that this bisector, acting instead of $\underline{\mathbf{n}}$ in combination with $\underline{\mathbf{m}}$ *does* give a **rotation** through the full angle **θ** is rather a long-winded calculation, which I include below. First we need to evaluate the two products $\underline{\mathbf{n}_{½}\mathbf{m}}$ and $\underline{\mathbf{m}\mathbf{n}_{½}}$:

$((\underline{\mathbf{m}} + \underline{\mathbf{n}})/|\underline{\mathbf{m}} + \underline{\mathbf{n}}|)\underline{\mathbf{m}} = \sqrt{½}(\sqrt{(1 + \theta_x)}\mathbf{e_1} + \sqrt{(1 - \theta_x)}\mathbf{e_2})\mathbf{e_1}$
$\qquad = \sqrt{½}(\sqrt{(1 + \theta_x)} - \sqrt{(1 - \theta_x)}\mathbf{e_1}\mathbf{e_2})$

$\underline{\mathbf{m}}((\underline{\mathbf{m}} + \underline{\mathbf{n}})/|\underline{\mathbf{m}} + \underline{\mathbf{n}}|) = \mathbf{e_1}\sqrt{½}(\sqrt{(1 + \theta_x)}\mathbf{e_1} + \sqrt{(1 - \theta_x)}\mathbf{e_2})$
$\qquad = \sqrt{½}(\sqrt{(1 + \theta_x)} + \sqrt{(1 - \theta_x)}\mathbf{e_1}\mathbf{e_2})$

Using these expansions, the full product $\underline{\mathbf{n}_{½}\mathbf{m}}$ $(\mathbf{x})$ $\underline{\mathbf{m}\mathbf{n}_{½}}$ will be:

$\Theta_{½}(\mathbf{x}) = \sqrt{½}\sqrt{½}(\sqrt{(1 + \theta_x)} - \sqrt{(1 - \theta_x)}\mathbf{e_1}\mathbf{e_2})(x_1\mathbf{e_1} + x_2\mathbf{e_2} + x_3\mathbf{e_3} + \ldots)(\sqrt{(1 + \theta_x)} + \sqrt{(1 - \theta_x)}\mathbf{e_1}\mathbf{e_2})$

$= ½(\sqrt{(1 + \theta_x)}x_1\mathbf{e_1} + \sqrt{(1 + \theta_x)}x_2\mathbf{e_2} + \sqrt{(1 + \theta_x)}x_3\mathbf{e_3} + \ldots$

$\quad + \sqrt{(1 - \theta_x)}x_1\mathbf{e_2} - \sqrt{(1 - \theta_x)}x_2\mathbf{e_1} - \sqrt{(1 - \theta_x)}x_3\mathbf{e_1}\mathbf{e_2}\mathbf{e_3} + \ldots) (\sqrt{(1 + \theta_x)} + \sqrt{(1 - \theta_x)}\mathbf{e_1}\mathbf{e_2})$

$= ½(x_1\sqrt{(1 + \theta_x)})\sqrt{(1 + \theta_x)}\mathbf{e_1} + x_2\sqrt{(1 + \theta_x)}\sqrt{(1 + \theta_x)}\mathbf{e_2} + x_3\sqrt{(1 + \theta_x)}\sqrt{(1 + \theta_x)}\mathbf{e_3} \ldots$

$\quad + x_1\sqrt{(1 - \theta_x)}\sqrt{(1 + \theta_x)}\mathbf{e_2} - x_2\sqrt{(1 - \theta_x)}\sqrt{(1 + \theta_x)}\mathbf{e_1} - x_3\sqrt{(1 - \theta_x)}\sqrt{(1 + \theta_x)}\ \mathbf{e_1}\mathbf{e_2}\mathbf{e_3} + \ldots$

$\quad + x_1\sqrt{(1 + \theta_x)})\sqrt{(1 - \theta_x)}\mathbf{e_2} - x_2\sqrt{(1 + \theta_x)})\sqrt{(1 - \theta_x)}\mathbf{e_1} + x_3\sqrt{(1 + \theta_x)})\sqrt{(1 - \theta_x)}\ \mathbf{e_3}\mathbf{e_1}\mathbf{e_2} + \ldots$

$\quad - x_1\sqrt{(1 - \theta_x)})\sqrt{(1 - \theta_x)}\mathbf{e_1} - x_2\sqrt{(1 - \theta_x)})\sqrt{(1 - \theta_x)}\mathbf{e_2} - x_3\sqrt{(1 - \theta_x)})\sqrt{(1 - \theta_x)}\mathbf{e_1}\mathbf{e_2}\mathbf{e_3}\mathbf{e_1}\mathbf{e_2} + \ldots)$

$= ½(x_1(1 + \theta_x)\mathbf{e_1} + x_2(1 + \theta_x)\mathbf{e_2} + x_3(1 + \theta_x)\mathbf{e_3} + \ldots$

$\quad + x_1\sqrt{(1 - \theta_x^2)}\mathbf{e_2} - x_2\sqrt{(1 - \theta_x^2)}\mathbf{e_1} - x_3\sqrt{(1 - \theta_x^2)}\mathbf{e_1}\mathbf{e_2}\mathbf{e_3} + \ldots$

$$+ x_1\sqrt{(1-\theta_x^2)}\mathbf{e_2} - x_2\sqrt{(1-\theta_x^2)}\mathbf{e_1} + x_3\sqrt{(1-\theta_x^2)}\mathbf{e_1e_2e_3} + \ldots$$

$$- x_1(1-\theta_x)\mathbf{e_1} - x_2(1-\theta_x)\mathbf{e_2} + x_3(1-\theta_x)\mathbf{e_3} + \ldots)$$

but $\theta_y = \sqrt{(1-\theta_x^2)}$, so this all reduces to:

$$= ½((x_1(1+\theta_x)) - x_2\theta_y - x_2\theta_y - x_1(1-\theta_x))\mathbf{e_1}$$

$$+ (x_2(1+\theta_x) + x_1\theta_y + x_1\theta_y - x_2(1-\theta_x))\mathbf{e_2}$$

$$+ (x_3(1+\theta_x) + x_3(1-\theta_x))\mathbf{e_3} \ldots$$

$$= ½(2.x_1\theta_x - 2x_2\theta_y)\mathbf{e_1} + ½(2.x_2\theta_x + 2x_1\theta_y)\mathbf{e_2} + ½.2x_3\mathbf{e_3} \ldots$$

$$= (x_1\theta_x - x_2\theta_y)\mathbf{e_1} + (x_2\theta_x + x_1\theta_y)\mathbf{e_2} + x_3\mathbf{e_3} \ldots$$

which is our original **rotation** at the start of this Section.

The identification between a product like **nm** and a **quaternion** described in Section 5.9 now enables us to understand the use of **quaternions** as **rotors** in computer graphics, for, taking the **bisector** $\underline{\mathbf{n}}_{½}$ with $\underline{\mathbf{m}}$, we can see that:

$$\underline{\mathbf{n}}_{½}\underline{\mathbf{m}} = \underline{\mathbf{n}}_{½} \cdot \underline{\mathbf{m}} + \underline{\mathbf{n}}_{½} \wedge \underline{\mathbf{m}}$$

$$= \underline{\mathbf{n}}_{½} \cdot \underline{\mathbf{m}} - (m_3n_{(½)1} - m_1n_{(½)3})\mathbf{e_3e_1} - (m_2n_{(½)3} - m_3n_{(½)2})\mathbf{e_2e_3}$$

$$- (m_1n_{(½)2} - m_1n_{(½)2})\mathbf{e_1e_2}\,.$$

But from Section 5.7, the **dot product** of two **unit vectors** equals the **cosine** of the angle between them, so here $\underline{\mathbf{n}}_{½} \cdot \underline{\mathbf{m}} = \mathbf{cos}(\theta/2)$.

So the full **geometric product** $\underline{\mathbf{n}}_{½}\underline{\mathbf{m}}$ corresponds to the **quaternion** Q where:

$$Q \equiv \mathbf{cos}(\theta/2) + i(m_3n_{(½)1} - m_1n_{(½)3}) + j(m_2n_{(½)3} - m_3n_{(½)2}) + k(m_1n_{(½)2} - m_1n_{(½)2}).$$

Equally, again using Section 5.7, we can view this as the **scalar** plus **3-vector**:

$$\mathbf{cos}(\theta/2) + \mathbf{sin}(\theta/2)(u\mathbf{i} + v\mathbf{j} + w\mathbf{k})$$

where we have:

$u.\mathbf{sin}(\theta/2) = m_3n_{(½)1} - m_1n_{(½)3}$,
$v.\mathbf{sin}(\theta/2) = m_2n_{(½)3} - m_3n_{(½)2}$ and

$$w.\mathbf{sin}(\theta/2) = m_1 n_{(½)2} - m_1 n_{(½)2}$$

and so *ui* + **vj** + *w***k**, from the properties of the **cross product** also described in Section 5.9, is a **unit vector** lying **orthogonal** to the **mn** plane, and so in the line of the 3-space *axis* of **rotation**.

We now get our same algebra by taking a point **P** = [*x*, *y*, *z*], creating the **quaternion** *P* = *xi* + *yj* + *zk* and evaluating:

$$P^\dagger = [\mathbf{cos}(\theta/2) + \mathbf{sin}(\theta/2)(ui + vj + wk)](xi + yj + zk)[\mathbf{cos}(\theta/2) - \mathbf{sin}(\theta/2)(ui + vj + wk)]$$

using **quaternion** algebra, and then, if $P^\dagger = x^\dagger i + y^\dagger j + z^\dagger k$, then $\mathbf{P}^\dagger = [x^\dagger, y^\dagger, z^\dagger]$ is the **rotated** position of **P**. So all told:

$$P^\dagger = QPQ^{-1}$$

writing the **quaternion** with negative **imaginary** part (i.e. **pure quaternion** part) as Q^{-1}. This is legitimate because Q^{-1} is indeed the **reciprocal** of any **quaternion** *Q* of **unit magnitude** just as the **complex conjugate** is the **reciprocal** of a **complex number** of **unit magnitude** (an **angle**) in **complex algebra**.

But there's also a more direct interpretation here, because in general for two **unit vectors**:[72]

$$(\underline{\mathbf{n}} \wedge \underline{\mathbf{m}})^2 = (\underline{\mathbf{nm}} - \underline{\mathbf{n}} \cdot \underline{\mathbf{m}})^2 = \underline{\mathbf{nmnm}} - 2(\underline{\mathbf{n}} \cdot \underline{\mathbf{m}})\underline{\mathbf{nm}} + (\underline{\mathbf{n}} \cdot \underline{\mathbf{m}})^2$$

$$= (\underline{\mathbf{n}} \cdot \underline{\mathbf{m}} + \underline{\mathbf{n}} \wedge \underline{\mathbf{m}})\underline{\mathbf{nm}} - 2(\underline{\mathbf{n}} \cdot \underline{\mathbf{m}})\underline{\mathbf{nm}} + (\underline{\mathbf{n}} \cdot \underline{\mathbf{m}})^2$$

$$= -(\underline{\mathbf{m}} \wedge \underline{\mathbf{n}})\underline{\mathbf{nm}} - (\underline{\mathbf{n}} \cdot \underline{\mathbf{m}})\underline{\mathbf{nm}} + (\underline{\mathbf{n}} \cdot \underline{\mathbf{m}})^2$$

$$= -(\underline{\mathbf{mn}} - \underline{\mathbf{n}} \cdot \underline{\mathbf{m}})\underline{\mathbf{nm}} - (\underline{\mathbf{n}} \cdot \underline{\mathbf{m}})\underline{\mathbf{nm}} + (\underline{\mathbf{n}} \cdot \underline{\mathbf{m}})^2$$

$$= -\underline{\mathbf{m}}(\underline{\mathbf{nn}})\underline{\mathbf{m}} + (\underline{\mathbf{n}} \cdot \underline{\mathbf{m}})\underline{\mathbf{nm}} - (\underline{\mathbf{n}} \cdot \underline{\mathbf{m}})\underline{\mathbf{nm}} + (\underline{\mathbf{n}} \cdot \underline{\mathbf{m}})^2$$

$$= -\underline{\mathbf{mm}} + (\underline{\mathbf{n}} \cdot \underline{\mathbf{m}})^2 = -1 + (\underline{\mathbf{n}} \cdot \underline{\mathbf{m}})^2 = -(1 - \mathbf{cos}^2(\theta)) = -\mathbf{sin}^2(\theta)$$

because $\underline{\mathbf{nn}} = \underline{\mathbf{mm}} = 1$. So we can write

$$\underline{\mathbf{n}} \wedge \underline{\mathbf{m}} = \underline{\mathbf{B}}.\mathbf{sin}(\theta)$$

where $\underline{\mathbf{B}}$ is the **unit bivector** in the $\underline{\mathbf{n}}$-$\underline{\mathbf{m}}$ plane with $\underline{\mathbf{B}}^2 = -1$. Note that we also get this result if we take the *opposite* **unit bivector** $-\underline{\mathbf{B}}$, because:

$$(\underline{\mathbf{n}} \wedge \underline{\mathbf{m}})^2 = (-\underline{\mathbf{B}}.\mathbf{sin}(\theta))^2 = +\underline{\mathbf{B}}^2\mathbf{sin}^2(\theta) = -\mathbf{sin}^2(\theta).$$

Expressed in terms of $\underline{\mathbf{B}}$, using the $+\underline{\mathbf{B}}$ form:

[72] Again from John Vince!

$$\underline{\mathbf{nm}} = \underline{\mathbf{n}} \cdot \underline{\mathbf{m}} + \underline{\mathbf{n}} \wedge \underline{\mathbf{m}} = \cos(\theta) + \underline{\mathbf{B}}.\sin(\theta),$$

a form directly analogous to the **Euler's formula** of Section 4.6, and we can even write:

$$\cos(\theta) + \underline{\mathbf{B}}.\sin(\theta) = e^{\underline{\mathbf{B}}(\theta)} = e\uparrow(\underline{\mathbf{B}}.(\theta))$$

because

$$d(e\uparrow(\underline{\mathbf{B}}.(\theta))) = \underline{\mathbf{B}}.e\uparrow(\underline{\mathbf{B}}.(\theta)).d\theta = \underline{\mathbf{B}}.(\cos(\theta) + \underline{\mathbf{B}}.\sin(\theta)).d\theta = -\sin(\theta).d\theta + \underline{\mathbf{B}}.\cos(\theta).d\theta$$

which agrees with:

$$d(\cos(\theta) + \underline{\mathbf{B}}.\sin(\theta)) = -\sin(\theta).d\theta + \underline{\mathbf{B}}.\cos(\theta).d\theta.$$

For the $\underline{\mathbf{n}}_{½}$ form, we'll have:

$$\underline{\mathbf{n}}_{½}\underline{\mathbf{m}} = \underline{\mathbf{n}}_{½} \cdot \underline{\mathbf{m}} + \underline{\mathbf{n}}_{½} \wedge \underline{\mathbf{m}} = \cos(\theta/2) - \underline{\mathbf{B}}.\sin(\theta/2) = e\uparrow(-\underline{\mathbf{B}}.(\theta/2))$$

as the simplest form of **rotor** for a **rotation** through θ in the $\underline{\mathbf{n}}$-$\underline{\mathbf{m}}$ plane, where θ is the angle between $\underline{\mathbf{n}}$ and $\underline{\mathbf{m}}$, and where I've chosen the $-\underline{\mathbf{B}}$ form to be consistent with our **quaternion** representation.

So now our **rotation** is:

$$\mathbf{x}^{\dagger} = e\uparrow(-\underline{\mathbf{B}}.(\theta/2)).(\mathbf{x}).e\uparrow(\underline{\mathbf{B}}.(\theta/2)) = (\cos(\theta/2) - \underline{\mathbf{B}}.\sin(\theta/2)).(\mathbf{x}).(\cos(\theta/2) + \underline{\mathbf{B}}.\sin(\theta/2)).$$

Switching the $e\uparrow(-\underline{\mathbf{B}}.(\theta/2))$ and $e\uparrow(\underline{\mathbf{B}}.(\theta/2))$ around simply rotates **x** in the opposite sense (clockwise instead of anticlockwise) about the same axis, an effect directly equivalent to using the opposite **unit bivector**. As there's no telling which **unit bivector** we've actually chosen in the first place, this **chiral** switch always holds good!

5.11 Eigenvalues and Eigenvectors

Sometimes it is possible to find certain **vectors** which are only changed in **magnitude** by a **linear transformation**,[73] typically represented as a **matrix**. Suppose A is the **linear transformation**, so that:

$$A(\mathbf{v}) = \mathbf{u}$$

is the result of the action of A on a **vector v**. Then, because a **linear transformation** is uniquely defined by its action on the **basis vectors** of the **vector space**, this can also be defined by a **matrix A**:

$$\mathbf{u} = \mathbf{A}\mathbf{v}.$$

[73] A **linear transformation** is a **linear function** on the **vector space**, but unlike a member of the **dual space**, instead of its result being a *number*, it is another **vector**, generally in the same space as the original one.

To see this, take the example of a **3-vector**. Then we can write **v** as $v_x\mathbf{e}_1 + v_y\mathbf{e}_2 + v_z\mathbf{e}_3$, and so:

$$A(\mathbf{v}) = A(v_x\mathbf{e}_1 + v_y\mathbf{e}_2 + v_z\mathbf{e}_3) = v_xA(\mathbf{e}_1) + v_yA(\mathbf{e}_2) + v_zA(\mathbf{e}_3)$$

so if $A(\mathbf{e}_1) = A_{11}\mathbf{e}_1 + A_{12}\mathbf{e}_2 + A_{13}\mathbf{e}_3$, and similarly $A(\mathbf{e}_2) = A_{21}\mathbf{e}_1 + A_{22}\mathbf{e}_2 + A_{23}\mathbf{e}_3$, and $A(\mathbf{e}_3) = A_{31}\mathbf{e}_1 + A_{32}\mathbf{e}_2 + A_{33}\mathbf{e}_3$, then the action of A is precisely defined by the **matrix A**:

$$\begin{pmatrix} A_{11} & A_{12} & A_{13} \\ A_{21} & A_{22} & A_{23} \\ A_{31} & A_{32} & A_{33} \end{pmatrix}.$$

Now imagine that for certain **vectors** $\mathbf{v}_i$, and for certain numbers λ_i, A satisfies the equation:

$$A(\mathbf{v}_i) = \lambda_i.\mathbf{v}_i.$$

This is an **eigenvalue equation**. The values λ_i are called the **eigenvalues** (or sometimes the **characteristic values**) of A, and the **vectors** $\mathbf{v}_i$ are called the **eigenvectors** (or **characteristic vectors**) of A.

This may seem a very abstruse requirement, but in actual fact, at least for **standard vectors**, *any* **linear transformation** into the same **vector space** has exactly as many **eigenvalues** and **eigenvectors** as the **dimension** of the **vector space**. This is most easily seen for the case of **2-vectors**. In this case A corresponds to a 2×2 **matrix A** and the **eigenvalue equation** takes the form:

$$\begin{pmatrix} A_{11} & A_{12} \\ A_{21} & A_{22} \end{pmatrix} \times \begin{pmatrix} v_1 \\ v_2 \end{pmatrix} = \lambda \times \begin{pmatrix} v_1 \\ v_2 \end{pmatrix},$$

or:

$$\begin{aligned} A_{11}v_1 + A_{12}v_2 &= \lambda v_1 \\ A_{21}v_1 + A_{22}v_2 &= \lambda v_2. \end{aligned}$$

This is actually a **homogeneous** pair of **equations** since there are no **constant** terms, and from Section 1.7, it only has a solution if the **determinant** is zero:

$$\begin{vmatrix} A_{11} - \lambda & A_{12} \\ A_{21} & A_{22} - \lambda \end{vmatrix} = 0.$$

This **determinantal** equation , also known as a **characteristic equation**, is in this case the simple **quadratic**:

$$(A_{11} - \lambda)(A_{22} - \lambda) - A_{12}A_{21} = \lambda^2 - \lambda(A_{11} + A_{22}) + A_{11}A_{22} - A_{12}A_{21} = 0.$$

This always has the solution:

$$\lambda = \tfrac{1}{2}[A_{11} + A_{22} \pm \sqrt{((A_{11} + A_{22})^2 - 4.(A_{11}A_{22} - A_{12}A_{21}))}]$$

although the square root may give an **imaginary** term, so making the **eigenvalues** into **complex numbers**. As an example, take the **matrix A** to be:

$$\begin{pmatrix} 2 & 2 \\ 2 & -1 \end{pmatrix}$$

Then we will have:

$$\lambda = ½[1 \pm \sqrt{}(1 - 4(-2 - 4)] = ½[1 \pm \sqrt{}25] = ½[1 \pm 5]$$

so that $\lambda = 3$ or $\lambda = -2$. Putting these values back into the **eigenvalue equation**, we get, for $\lambda = 3$:

$$\begin{pmatrix} 2-3 & 2 \\ 2 & -1-3 \end{pmatrix} \times \begin{pmatrix} v_1 \\ v_2 \end{pmatrix} = \begin{pmatrix} -1 & 2 \\ 2 & -4 \end{pmatrix} \times \begin{pmatrix} v_1 \\ v_2 \end{pmatrix} = \begin{pmatrix} 0 \\ 0 \end{pmatrix},$$

or $-v_1 + 2v_2 = 0$ and $2v_1 - 4v_2 = 0$, both of which give $v_1 = 2v_2$. The **norm** or **magnitude** of this **vector** is $\mathbf{v} \cdot \mathbf{v} = v_1^2 + v_2^2 = (2v_2)^2 + v_2^2 = 5v_2^2$. This defines an infinite family of **vectors**. But there is a unique member of the family that has **unit magnitude**, given by $v_2 = 1/\sqrt{5}$. So we have:

$(2/\sqrt{5})\mathbf{e}_1 + (1/\sqrt{5})\mathbf{e}_2$ is the **unit eigenvector** corresponding to the **eigenvalue** of 3.

Putting $\lambda = -2$ into the original **eigenvalue equation**, we get:

$$\begin{pmatrix} 2+2 & 2 \\ 2 & -1+2 \end{pmatrix} \times \begin{pmatrix} v_1 \\ v_2 \end{pmatrix} = \begin{pmatrix} 4 & 2 \\ 2 & 1 \end{pmatrix} \times \begin{pmatrix} v_1 \\ v_2 \end{pmatrix} = \begin{pmatrix} 0 \\ 0 \end{pmatrix},$$

or $4v_1 + 2v_2 = 0$ and $2v_1 + v_2 = 0$, both of which give $v_2 = -2v_1$. The **norm** or **magnitude** of this **vector** is again $\mathbf{v} \cdot \mathbf{v} = v_1^2 + v_2^2 = v_1^2 + (-2v_1)^2 = 5v_1^2$. This again defines an infinite family of **vectors**. But there is a unique member of the family that has **unit magnitude**, given by $v_1 = 1/\sqrt{5}$. So we have:

$(1/\sqrt{5})\mathbf{e}_1 + (-2/\sqrt{5})\mathbf{e}_2$ is the **unit eigenvector** corresponding to the **eigenvalue** of -2.

In the discussion above, *don't confuse* the original **eigenvalue equation** with the **determinantal** derived **characteristic equation** which must hold for the **eigenvalue equation** to have a solution.

The set of **eigenvectors** for a **symmetric linear transformation** A, which is a **transformation** whose **matrix** is **symmetric** or equal to its own **transpose** ($\mathbf{A}^T = \mathbf{A}$), has a most important property. Take the **scalar products** of one **eigenvector** with the **linear transform** of another in opposite pairs as below:

$$\mathbf{v}_i \cdot A(\mathbf{v}_j) = \mathbf{v}_i \cdot \lambda_j \mathbf{v}_j = \lambda_j \mathbf{v}_i \cdot \mathbf{v}_j$$
$$\mathbf{v}_j \cdot A(\mathbf{v}_i) = \mathbf{v}_j \cdot \lambda_i \mathbf{v}_i = \lambda_i \mathbf{v}_i \cdot \mathbf{v}_j$$

Now, putting $\mathbf{v}_i = \Sigma_k(\mathbf{v}_i)_k\mathbf{e}_k$, with $(\mathbf{v}_i)_k$ being the k'th **coefficient** of $\mathbf{v}_i$, or the **coefficient** of $\mathbf{e}_k$ in the expansion of $\mathbf{v}_i$, the products on the left expand as:

$$\mathbf{v}_i \cdot A(\mathbf{v}_j) = \Sigma_k\Sigma_m(\mathbf{v}_i)_k\mathbf{A}_{km}(\mathbf{v}_j)_m = \Sigma_k\Sigma_m(\mathbf{v}_i)_k\mathbf{A}_{mk}(\mathbf{v}_j)_m = \Sigma_k\Sigma_m(\mathbf{v}_j)_m\mathbf{A}_{mk}(\mathbf{v}_i)_k = \mathbf{v}_j \cdot A(\mathbf{v}_i).$$

They are *the same*. So subtracting one from the other, we get:

$$\mathbf{v}_i \cdot A(\mathbf{v}_j) - \mathbf{v}_j \cdot A(\mathbf{v}_i) = 0 = \lambda_j \mathbf{v}_i \cdot \mathbf{v}_j - \lambda_i \mathbf{v}_i \cdot \mathbf{v}_j = (\lambda_j - \lambda_i)\mathbf{v}_i \cdot \mathbf{v}_j.$$

So if $\lambda_j \neq \lambda_i$, we must have $\mathbf{v}_i \cdot \mathbf{v}_j = 0$. So **eigenvectors** *of the same* **symmetric transformation** *that correspond to different* **eigenvalues** *are* **orthogonal**.

Now suppose we apply a **transformation** B of the **basis** to the **vector space**, how does it affect the **eigenvalue equation**? Well, we can write $\mathbf{v}^*_i = B\mathbf{v}_i$ or $\mathbf{v}_i = B^{-1}\mathbf{v}^*_i$, but how does the **transformation** A *itself* change?

This is quite easy normally. In **matrix** form, if $\mathbf{v} = \mathbf{Au}$, and we change $\mathbf{u}$ to $\mathbf{B}^{-1}\mathbf{u}$, then $B^{-1}(\mathbf{A})$ must give $\mathbf{B}^{-1}\mathbf{v}$, so that "$A$" is still changing the new $\mathbf{u}$ into the new $\mathbf{v}$. The change in $\mathbf{A}$ that does this is:

$$\mathbf{v} = \mathbf{B}^{-1}\mathbf{v}^* = \mathbf{Au} = \mathbf{A}(\mathbf{B}^{-1}\mathbf{u}^*).$$

So now pre-multiply by $\mathbf{B}$ to get:

$$\mathbf{Bv} = \mathbf{BB}^{-1}\mathbf{v}^* = \mathbf{v}^* = \mathbf{BA}(\mathbf{B}^{-1}\mathbf{u}^*) = (\mathbf{BAB}^{-1})\mathbf{u}^*.$$

In other words $\mathbf{BAB}^{-1}$ **transforms** the *new* $\mathbf{u}$ into the *new* $\mathbf{v}$. This **transformation** of the **matrix** of a **transformation** itself under a change of **coordinates** is called a **similarity transform**.

For the **eigenvalue equation**, then, we would get:

$$\mathbf{Av}_i = \lambda_i \mathbf{v}_i \rightarrow (\mathbf{BAB}^{-1})\mathbf{v}_i^* = \lambda_i(\mathbf{BAB}^{-1})\mathbf{v}_i^*.$$

But there's a much simpler and more interesting case here. Suppose the *new* **basis** is chosen to be the set of **eigenvectors** for A themselves. We already have the **eigenvectors** expressed in terms of the original **basis**, so we *do* have $\mathbf{B}$ as, for example:

$$\mathbf{v}_i = b_{i1}\mathbf{e}_1 + b_{i2}\mathbf{e}_2 + b_{i3}\mathbf{e}_3$$

so we merely need to work out $\mathbf{BAB}^{-1}$. But in this case we don't need to work it out. For an arbitrary **vector** $\mathbf{v} = \Sigma_k x_k \mathbf{v}_k$ expanded in terms of the **eigenvectors** of A, we get:

$$A(\mathbf{v}) = \Sigma_k(\Sigma_m(a_{mk})x_k)\mathbf{v}_k.$$

But $\quad A(\mathbf{v}) = A(\Sigma_k x_k \mathbf{v}_k) = \Sigma_k x_k A(\mathbf{v}_k) = \Sigma_k x_k \lambda_k \mathbf{v}_k.$

So identifying **coefficients** of $\mathbf{v}_k$ on both sides:

$$\Sigma_m(a_{mk}) = \lambda_k \qquad \text{or} \qquad a_{mk} = \lambda_k \delta_{mk}.$$

So the **matrix** for A becomes a *diagonal* **matrix** when expressed in terms of A's own **eigenvectors**. For our simple example, we would have:

$$\mathbf{A} = \begin{pmatrix} 3 & 0 \\ 0 & -2 \end{pmatrix}$$

giving $\mathbf{Av} = \Sigma_i(\Sigma_k a_{ik}x_k)\mathbf{v}_i = (3x_1 + 0x_2)\mathbf{v}_1 + \ (0x_1 + -2x_2)\mathbf{v}_2 = 3x_1\mathbf{v}_1 - 2x_2\mathbf{v}_2$

so if, say, $\mathbf{v} = \mathbf{v}_1$, then $x_1 = 1$ and $x_2 = 0$, and $\mathbf{Av} = \mathbf{Av}_1 = 3\mathbf{v}_1$ as expected. Does this diagonal **matrix** equal $\mathbf{BAB}^{-1}$? Well, **B** is given by:

$$\mathbf{v}_1 = (2/\sqrt{5})\mathbf{e}_1 + (1/\sqrt{5})\mathbf{e}_2$$
$$\mathbf{v}_2 = (1/\sqrt{5})\mathbf{e}_1 - (2/\sqrt{5})\mathbf{e}_2$$

with $\mathbf{det}(\mathbf{B}) = -4/5 - 1/5 = -1$, so $\mathbf{B}^{-1}$ is, from Section 1.8:

$$-1 \times \begin{pmatrix} -(2/\sqrt{5}) & -(1/\sqrt{5}) \\ -(1/\sqrt{5}) & (2/\sqrt{5}) \end{pmatrix} = \begin{pmatrix} (2/\sqrt{5}) & (1/\sqrt{5}) \\ (1/\sqrt{5}) & -(2/\sqrt{5}) \end{pmatrix}$$

So
$$(\mathbf{BB}^{-1})_{11} = (2/\sqrt{5})(2/\sqrt{5}) + (1/\sqrt{5})(1/\sqrt{5}) = 4/5 + 1/5 = 1$$
$$(\mathbf{BB}^{-1})_{12} = (2/\sqrt{5})(1/\sqrt{5}) + (1/\sqrt{5})(-2/\sqrt{5}) = 2/5 - 2/5 = 0$$
$$(\mathbf{BB}^{-1})_{21} = (1/\sqrt{5})(1/\sqrt{5}) + (-2/\sqrt{5})(1/\sqrt{5}) = 2/5 - 2/5 = 0$$
$$(\mathbf{BB}^{-1})_{22} = (1/\sqrt{5})(1/\sqrt{5}) + (-2/\sqrt{5})(-2/\sqrt{5}) = 1/5 + 4/5 = 1.$$

(So **B** is its own **inverse**.) Our original **matrix** times $\mathbf{B}^{-1}$ gives:

$$\begin{pmatrix} 2 & 2 \\ 2 & -1 \end{pmatrix} \times \begin{pmatrix} (2/\sqrt{5}) & (1/\sqrt{5}) \\ (1/\sqrt{5}) & -(2/\sqrt{5}) \end{pmatrix} = \begin{pmatrix} (6/\sqrt{5}) & (-2/\sqrt{5}) \\ (3/\sqrt{5}) & (4/\sqrt{5}) \end{pmatrix}$$

and **B** times this is:

$$\begin{pmatrix} (2/\sqrt{5}) & (1/\sqrt{5}) \\ (1/\sqrt{5}) & -(2/\sqrt{5}) \end{pmatrix} \times \begin{pmatrix} (6/\sqrt{5}) & (-2/\sqrt{5}) \\ (3/\sqrt{5}) & (4/\sqrt{5}) \end{pmatrix} = \begin{pmatrix} (15/5) & ((-4+4)/5) \\ ((6-6)/5) & ((-2-8)/5) \end{pmatrix} = \begin{pmatrix} 3 & 0 \\ 0 & -2 \end{pmatrix}.$$

So, yes, it does! The diagonal **matrix** in terms of the **eigenvectors** as **basis** does equal $\mathbf{BAB}^{-1}$.

Before leaving **standard vector spaces**, I will define one term whose continuum analogue will loom large in a later Chapter.

If a **standard n-vector** is written in an **orthogonal**, but not necessarily **orthonormal**, **basis** $\{\mathbf{v}_i\}$ in the form:

$$\mathbf{x} = x_1\mathbf{v}_1 + \ x_2\mathbf{v}_2 + x_3\mathbf{v}_3 + \ldots + x_n\mathbf{v}_n$$

then

$$\mathbf{x} \cdot \mathbf{v}_j = x_1(\mathbf{v}_1 \cdot \mathbf{v}_j) + x_2(\mathbf{v}_2 \cdot \mathbf{v}_j) + x_3(\mathbf{v}_3 \cdot \mathbf{v}_j) + \ldots + x_n(\mathbf{v}_n \cdot \mathbf{v}_j).$$

But by **orthogonality**, all these **dot products** are zero except for $\mathbf{v}_j \cdot \mathbf{v}_j$. So we have:

$\mathbf{x} \cdot \mathbf{v}_j = x_j(\mathbf{v}_j \cdot \mathbf{v}_j)$ or $x_j = (\mathbf{x} \cdot \mathbf{v}_j)/(\mathbf{v}_j \cdot \mathbf{v}_j)$.

This general formula for x_j is called the **Fourier coefficient** of **x** with respect to $\mathbf{v}_j$. The term is rare in this discrete context, but from this you should be able to recognize the same idea when we introduce it for the **integral scalar product** mentioned in Chapter 7.

5.X Exercises

1. Find the formula for the **inverse** or **reciprocal** of a **quaternion** $\lambda + xi + yj + zk$ by evaluating its product with $q^* = \lambda - xi - yj - zk$.

2. Given the operation of an element of the **dual space** ϕ on an element of the **basic space v** as $\phi(\mathbf{v})$, which we can write as $(\phi, \mathbf{v})$ for symmetry, and assuming standard 3-**vectors** of the form $\mathbf{v} = v_1\mathbf{e}_1 + v_2\mathbf{e}_2 + v_3\mathbf{e}_3$, suppose that **v** is transformed by a **matrix A**. Find the form of the **adjoint matrix** $\mathbf{A}^\dagger$ defined such that $(\mathbf{A}^\dagger\phi, \mathbf{v}) = (\phi, \mathbf{Av})$.

3. Show that the **scalar triple product** of **3-vectors** $\mathbf{x} \cdot (\mathbf{y} \times \mathbf{z})$ has the form of a **determinant**.

4. Prove that for **3-vectors** $\mathbf{x} \times (\mathbf{y} \times \mathbf{z}) = (\mathbf{x} \cdot \mathbf{z})\mathbf{y} - (\mathbf{x} \cdot \mathbf{y})\mathbf{z}$. Show that the **geometric product** gives $\mathbf{x}(\mathbf{y} \wedge \mathbf{z}) = (\mathbf{x} \cdot \mathbf{y})\mathbf{z} - (\mathbf{x} \cdot \mathbf{z})\mathbf{y} + \mathbf{xyz}$ and $(\mathbf{y} \wedge \mathbf{z})\mathbf{x} = (\mathbf{x} \cdot \mathbf{z})\mathbf{y} - (\mathbf{x} \cdot \mathbf{y})\mathbf{z} + \mathbf{xyz}$.

5. The general form for a **multivector** in the **geometric algebra** of 3-space is: $x\mathbf{e}_1 + y\mathbf{e}_2 + z\mathbf{e}_3 + \xi\mathbf{e}_{23} + \psi\mathbf{e}_{31} + \zeta\mathbf{e}_{12} + \omega\mathbf{e}_{123}$. Evaluate the general form of the **geometric product** between two such **multivectors**.

6. Imagine we have a curve in ordinary 3-space parameterized by its **arc length** s, so that the curve is described by $\mathbf{x}(s) = x(s).\mathbf{i} + y(s).\mathbf{j} + z(s).\mathbf{k}$. Show that the **vector** $\partial_s\mathbf{x}(s) = \partial_s x(s).\mathbf{i} + \partial_s y(s).\mathbf{j} + \partial_s z(s).\mathbf{k}$ is a **unit vector**. Call this **vector** $\mathbf{t}(s)$, the **tangent vector** to the curve at s.

7. Differentiating a second time, define $\kappa = |\partial_s\, \mathbf{t}(s)|$ and if $\kappa \neq 0$, call $\mathbf{p} = \kappa^{-1}.\mathbf{t}(s)$, again a **unit vector**, the **principal normal** to the curve at s. κ is the **curvature**. Finally define $\mathbf{b} = \mathbf{t} \times \mathbf{p}$ as the **binormal**, so **t**, **p** and **b** form an **orthonormal** triad of **vectors** on the curve at s. Derive the **Frenet formulae**:

   a. $\partial_s\mathbf{t}(s) = \kappa.\mathbf{p}$

   b. $\partial_s\mathbf{p}(s) = -\kappa.\mathbf{t}(s) + \tau.\mathbf{b}(s)$

   c. $\partial_s\mathbf{b}(s) = -\tau.\mathbf{p}(s)$.

   [Take $\partial_s(\mathbf{b} \cdot \mathbf{t} = 0)$ to show that $\partial_s\mathbf{b} \parallel \mathbf{p}$, and so prove (c) first. τ is defined as the **torsion**.]

8. Take a **3-vector** $xi + yj + zk$. Using simple **complex algebra**, rotate it through an **angle** θ about the **x axis** (multiply by $\varepsilon_c(\theta) + j.\varepsilon_s(\theta)$ in the yz-plane; the x value remains unchanged).

Now, similarly, rotate it through an **angle** φ about the **y axis**. Repeat the operation, starting from the original **vector**, this time taking the φ **rotation** about the **y axis** *first*, and *then* the θ **rotation** about the **x axis**. Are the results the same? If θ and φ are both quite small, how could the *difference* between the two be most easily expressed?

9. In quantum mechanics, **vector spaces** have **complex scalars**. In this context, the analogue of a **symmetric matrix** ($\mathbf{A} = \mathbf{A}^T$) is a **Hermitian matrix** defined as $\mathbf{A}^* = \mathbf{A}^T$, where $\mathbf{A}^*$ indicates the **complex conjugate**, i.e. if $a_{ij} = x_{ij} + jy_{ij}$, then $a_{ij}^* = x_{ij} - jy_{ij}$. Equally, the analogue of an **othogonal matrix** ($\mathbf{A}^{-1} = \mathbf{A}^T$) is a **unitary matrix** $\mathbf{A}^{-1} = (\mathbf{A}^T)^*$. We now have to use the **complex scalar product** $(u_x\mathbf{i} + u_y\mathbf{j} + u_z\mathbf{k}) \cdot (v_x\mathbf{i} + v_y\mathbf{j} + v_z\mathbf{k}) = u_x^*v_x + u_y^*v_y + u_z^*v_z$. By a method analogous to that used in Section 5.11 to show that the **eigenvectors** of a **symmetric matrix** are **orthogonal**, show that the **eigenvalues** of a **Hermitian matrix** are **real**.

10. Show that the **eigenvalues** of a **unitary matrix** have **unit magnitude**.

11. Repeat problem 8 using the full method of **quaternion rotors** instead of ordinary **complex algebra**.

Chapter 6

INTEGRATION

6.1 Integration

I introduced the concept of **integration** as a product between a **differential form** and a **simplex** in Section 2.6. It will be convenient to call this the **integral product**. The ramifications of this idea quite dwarf those of the **differential operator** *per se*, so that the so-called **integral calculus** turns out to be a far more complex subject than the **differential calculus**.

In Section 2.6 I gave definitions of the **integral product** for just two cases.

0) The first was between a **0-form**, which is simply a **scalar function** like $g(x, y, z)$, and a **0-simplex** or **0-chain** where the first is just a point $\mathbf{p}_0$ and the second is a **linear combination** of points like $\Sigma_k\, a_k\mathbf{p}_k$.
1) The second case extended this notion to a product between an **exact 1-form d**ϕ[74] and a **1-simplex** or **1-chain**, where a **1-simplex** is a **line segment** between two points $\mathbf{p}_0$ and $\mathbf{p}_1$, and a **1-chain** is a **linear combination** of **1-simplexes**.

So if we are in 3-space and $\mathbf{p}_0$ has **coordinates** x_0, y_0, z_0 and $\mathbf{p}_1$ has **coordinates** x_1, y_1, z_1, we have definitions for:

0) $g(x, y, z) \,|\, \mathbf{p}_0 = g(x_0, y_0, z_0)$ and $\quad g(x, y, z) \,|\, \Sigma_k\, a_k\mathbf{p}_k = \Sigma_k\, a_k.g(x_k, y_k, z_k)$,
1) $\mathbf{d}g(x, y, z) \,|\, \mathbf{p}_0\mathbf{p}_1 = g(x, y, z) \,|\, \mathbf{b}(\mathbf{p}_0\mathbf{p}_1) = g(x, y, z) \,|\, (\mathbf{p}_1 - \mathbf{p}_0) = g(x_1, y_1, z_1) - g(x_0, y_0, z_0)$.

The second case can be extended to **1-chains** as well, as:

$$\mathbf{d}g \,|\, \Sigma_k\, a_k.(\mathbf{p}_{k0}\mathbf{p}_{k1}) = g \,|\, \Sigma_k\, a_k.\mathbf{b}(\mathbf{p}_{k0}\mathbf{p}_{k1}) = g \,|\, \Sigma_k\, a_k.(\mathbf{p}_{k1} - \mathbf{p}_{k0})$$

$$= \Sigma_k\, a_k.[g(x_{k1}, y_{k1}, z_{k1}) - g(x_{k0}, y_{k0}, z_{k0})].$$

[74] An **exact form** is one which is the **exterior derivative** of an expression and so can be written **d**ϕ, as mentioned in Section 2.4.

The device used in the **1-form | 1-simplex** products is called an **adjoint switch** which *defines* the action of the **exact 1-form** on the **1-simplex** as being equal to the already defined action of the **original scalar function** or **0-form** from which the **1-form** is **derived**, acting on the **0-chain** which is the **boundary** of the **1-simplex**. Calling the **original** ϕ, and the **1-simplex** λ, the **adjoint** operation is:

$$\mathbf{d}\phi \,|\, \lambda = \phi \,|\, \mathbf{b}\lambda.$$

This Chapter will extend this concept in several ways, in particular in the following three ways:

1. To the case where the **form** is *not* **exact**,
2. From **simplexes** and **chains** to the more general geometrical structures called **manifolds**.
3. To higher-dimensional **simplexes** and **chains**, where we will need to define analogous higher-dimensional (or actually higher **grade**) **forms**.

A **manifold** will be defined in Section 6.3. Firstly, I will introduce the device of the **pullback** which enables us to extend the definition of the **integral** or **integral product** to **forms** that are not **exact**, and will also facilitate the extension to **manifolds**.

6.2 Line Integrals and the Pullback

In this Section, I will extend the simple definition of an **integral** that we have so far to cover the case where the **form** is not **exact**.[75] In doing so, I will clarify the concept of the "**pullback**", which had a brief mention in Section 4.4. The concept could be seen as the **integral** analogue of the **chain rule** introduced in Section 2.2, a major difference being that we will now need to have some way of applying this idea to the **simplex** part too.

The basic idea is quite simple, but I'll develop it in stages. First, let's revise the basic case where we have an **integral** where the **differential form** *is* **exact**, like:

$$(x^2.\mathbf{d}x + 2x.\mathbf{d}x) \,|\, [2, 4].$$

The **differential form** that is the left-hand **argument** of this **integral** is **exact** because:

$$(x^2.\mathbf{d}x + 2x.\mathbf{d}x) = \mathbf{d}(⅓.x^3 + x^2 + k)$$

where k is a **constant of integration** as defined in Section 2.4. This is easy to see by expanding the RHS, but it could also be established by using the rule given in Section 2.4:

$$x^n.\mathbf{d}x = \mathbf{d}(x^{n+1}/(n + 1)).$$

So the **integral** above is the same thing as:

$$\mathbf{d}(⅓.x^3 + x^2 + k) \,|\, [2, 4].$$

---

[75] I will sometimes refer to **forms** that are not **exact** as **inexact forms**.

In this form we can apply the **adjoint definition** of the **integral product**:

$$\mathbf{d}(⅓.x^3 + x^2 + k) \mid [2, 4] = (⅓.x^3 + x^2 + k) \mid \mathbf{b}[2, 4]$$
$$= (⅓.x^3 + x^2 + k) \mid ([4] - [2]) = (⅓.4^3 + 4^2 + k) - (⅓.2^3 + 2^2 + k)$$
$$= (⅓.64 + 16) - (⅓.8 + 4) + k - k$$
$$= 21⅓ + 16 - 2⅔ - 4 = 30⅔.$$

In evaluating **integrals, constants of integration** always cancel out and I'll ignore them in future.

Just to go over this again: [2, 4] is a **line interval** or a **1-simplex** along the real line of the x-axis from $\mathbf{p}_0 \equiv (x = 2)$ to $\mathbf{p}_1 \equiv (x = 4)$. Its **boundary** $\mathbf{b}(\mathbf{p}_0\mathbf{p}_1) = \mathbf{p}_1 - \mathbf{p}_0$ as defined in Section 1.13, and this signed pair of points may also be written [4] − [2]. To evaluate the **integration** operation "|", we invert the **differentiation** of the **differential form** $x^2.\mathbf{d}x + 2x.\mathbf{d}x$ as shown in Section 2.4 to convert the entire **exact differential** left-hand argument to the form **d**(something), which we can do easily here as:

$$\mathbf{d}(⅓.x^3 + x^2) = x^2.\mathbf{d}x + 2x.\mathbf{d}x$$

where I use the rule $x^n.\mathbf{d}x = \mathbf{d}((x^{n+1}/(n + 1))$ given in Section 2.4. Now we are in a position to apply the definition of the **integration operator** |:

$$\mathbf{d}f(x) \mid \mathbf{p}_0\mathbf{p}_1 = f(x) \mid \mathbf{b}(\mathbf{p}_0\mathbf{p}_1) = f(x) \mid (\mathbf{p}_1 - \mathbf{p}_0) = f(\mathbf{p}_1) - f(\mathbf{p}_0)$$

which here comes to $(⅓.x^3 + x^2)(4) - (⅓.x^3 + x^2)(2) = (⅓.4^3 + 4^2) - (⅓.2^3 + 2^2) = 30⅔$.

Now for the second stage of the argument. Often, for reasons which will shortly be apparent, it is advantageous to replace the variable x in the **differential form** by a different variable related to it in a defined way. So if we have x defined as a function of, say, t, in the form $x = h(t)$, we can replace the left-hand argument, the **differential form**, by substituting for x and $\mathbf{d}x$ in the way explained in Section 2.2, since $\mathbf{d}x = \partial h(t).\mathbf{d}t$, and $x = h(t)$.

Suppose, for example, $x = 3t^2$. Then by Section 2.1, $\mathbf{d}x = 3{\times}2.t.\mathbf{d}t = 6t.\mathbf{d}t$, and so:

$$x^2.\mathbf{d}x + 2x.\mathbf{d}x = (3t^2)^2.6t.\mathbf{d}t + 2(3t^2).6t.\mathbf{d}t = 9t^4.6t.\mathbf{d}t + 6t^2.6t.\mathbf{d}t = (54t^5 + 36t^3).\mathbf{d}t.$$

But if we are to get the same value for the overall **integration** (here 30⅔), *we must also alter the* **interval** [2, 4] to one defined not in x but in t as well.

This is the **pullback** problem.

The name **pullback** comes from the fact that $x = h(t)$ takes a value in t and generates a value in x. We now want to *reverse* this, "pulling back" from an **interval** defined in x to an **interval** defined in t.

In our example, we would need to solve for t as a function of x for the two end-points of the **interval** [2, 4] at $x = 2$ and $x = 4$. This means solving for t_0 and t_1 in:

$$x = 2 = 3(t_0)^2 \qquad \text{and } x = 4 = 3(t_1)^2.$$

These solve as $t_0 = \pm\sqrt{(2/3)}$ and $t_1 = \pm\sqrt{(4/3)}$. Straight away we can see the ±'s make the answer ambiguous, but if we assume the *positive* roots, we now have to evaluate the **integration**:

$$((54t^5 + 36t^3).\mathbf{d}t \,|\, [\sqrt{(2/3)}, \sqrt{(4/3)}]).$$

Much as before, we invert the **differentiation** to obtain:

$$(54t^5 + 36t^3).\mathbf{d}t = \mathbf{d}((54/6)t^6 + (36/4)t^4) = \mathbf{d}(9t^6 + 9t^4)$$

and so we obtain:

$$(\mathbf{d}(9t^6 + 9t^4) \,|\, [\sqrt{(2/3)}, \sqrt{(4/3)}]) = ((9t^6 + 9t^4) \,|\, \mathbf{b}[\sqrt{(2/3)}, \sqrt{(4/3)}])$$

$$= (9t^6 + 9t^4)(\sqrt{(4/3)}) - (9t^6 + 9t^4)(\sqrt{(2/3)})$$

$$= 9(4/3)^3 + 9(4/3)^2 - 9(2/3)^3 - 9(2/3)^2$$

because e.g. $(\sqrt{(t)})^6 = t^{½\times 6} = t^3$. This gives:

$$9\times(64/27) + 9\times(16/9) - 9\times(8/27) - 9\times(4/9) = 64/3 + 16 - 8/3 - 4$$
$$= 21⅓ + 16 - 2⅔ - 4 = 30⅔,$$

exactly as before. In this example, as it happens, because only even powers of t appear, we could have used any combination of the four solutions for $[t_0, t_1]$ and we would have obtained the same answer.

So far this may all appear to be much ado about nothing. But now we come to the third stage of the argument. Why should we introduce the complication of replacing x by t?

Suppose that instead of the **differential form** in a *single* variable we had one in *two* variables like:

$$x^2y.\mathbf{d}x + y.\mathbf{d}y.$$

Now this **form** is not **exact**: *there is no function $f(x, y)$ such that this* **form** *equals* **d**$f(x, y)$. There are two basic ways to see this. Both depend on the observation that any **differential form** in two variables that is **exact** – that *does* equal a **form d**$f(x, y)$ – must take the form:

$$\partial_x f(x, y).\mathbf{d}x + \partial_y f(x, y).\mathbf{d}y$$

and the expression for $\partial_x f(x, y)$ limits the possible forms for $\partial_y f(x, y)$ and vice versa.[76] In the example above, only a function $f(x, y)$ of the form $⅓x^3y + c$ could give rise to the $x^2y.\mathbf{d}x$ term, and only one of the form $½y^2 + d$ could give rise to the $y.\mathbf{d}y$ term, but:

$$\mathbf{d}(⅓x^3y + c) = x^2y.\mathbf{d}x + ⅓x^3\mathbf{d}y$$

$$\mathbf{d}(½y^2 + d) = 0.\mathbf{d}x + y.\mathbf{d}y = y.\mathbf{d}y.$$

So each of these forms for $f(x, y)$ would give the "wrong" result for the other term.

A more formal proof of this was introduced in Section 2.4 using **Young's theorem** which states that:

$$\partial_x\partial_y f(x, y) = \partial_y\partial_x f(x, y).$$

The idea is that we can take $\partial_x f(x, y)$ and $\partial_y f(x, y)$ from the first **differentiation** as given functions, and evaluate:

$$\mathbf{d}(\partial_x f(x, y)) = \partial_x(\partial_x f(x, y)).\mathbf{d}x + \partial_y(\partial_x f(x, y)).\mathbf{d}y$$

and $$\mathbf{d}(\partial_y f(x, y)) = \partial_x(\partial_y f(x, y)).\mathbf{d}x + \partial_y(\partial_y f(x, y)).\mathbf{d}y.$$

Then by **Young's theorem** the two **derivative functions** $\partial_y(\partial_x f(x, y))$ and $\partial_x(\partial_y f(x, y))$ will be identical.

Since any **exact differential form** must be of the form:

$$\partial_x f(x, y).\mathbf{d}x + \partial_y f(x, y).\mathbf{d}y$$

we can test if a given arbitrary **form** $A(x, y).\mathbf{d}x + B(x, y).\mathbf{d}y$ is **exact** by evaluating whether

$$\partial_y(A(x, y)) = \partial_x(B(x, y)).$$

If they are not, then $A(x, y)$ cannot correspond to a $\partial_x f(x, y)$ result for which $B(x, y)$ corresponds to the $\partial_y f(x, y)$. In our example,

$$\mathbf{d}(x^2y) = \partial_x(x^2y).\mathbf{d}x + \partial_y(x^2y).\mathbf{d}y = 2xy.\mathbf{d}x + x^2.\mathbf{d}y$$

and $$\mathbf{d}(y) = 0.\mathbf{d}x + \mathbf{d}y$$

so $\partial_y(A(x, y)) = x^2$ and $\partial_x(B(x, y)) = 0$, and clearly the two cannot be equal.

Observe how these two approaches to ascertaining whether a **differential form** is **exact** go in opposite directions: one method tries to *invert* the **differentiation** to find possible forms for any $f(x, y)$ from the **partial derivatives**, the other goes *down*, **differentiating** further with respect to the "other" variable.

[76] Here "form" *not* in bold is the ordinary English usage, **form** means a **differential form**.

This rather lengthy preamble on **exact differential forms** is to set the background to this question, which constitutes the final stage of the argument:

can we define the **integration** *when the* **differential form** *is* <u>*not*</u> **exact***?*

> The answer is that yes, we can, by defining a **parametric curve** τ: $x = x(t)$, $y = y(t)$, and evaluating the **integral pulled back** onto an **interval** in t, but *it is then dependent on the precise curve* τ *chosen in the xy plane.*

The trick is to use the **pullback** concept to switch down to a *one-dimensional* space – that of the curve **parameter**, which here is t. Taking our example of the form $\omega = x^2y.\mathbf{d}x + y.\mathbf{d}y$ (Greek letters are commonly used for **differential forms** in more advanced work), let's suppose we define a curve $x = x(t) = t^4 - 3t$, $y = y(t) = t^2 + ½.t$ from $t = 0$ to $t = 1$. This **parametric curve** has:

$$\mathbf{d}x = (4t^3 - 3).\mathbf{d}t \qquad \text{and} \qquad \mathbf{d}y = (2t + ½).\mathbf{d}t$$

so the **form** $\omega = x^2y.\mathbf{d}x + y.\mathbf{d}y$, rewritten as a function of t by substituting for x, y, $\mathbf{d}x$ and $\mathbf{d}y$ is:

$$\omega_t = (t^4 - 3t)^2(t^2 + ½.t)(4t^3 - 3).\mathbf{d}t + (t^2 + ½.t)(2t + ½).\mathbf{d}t$$

$$= ((t^4 - 3t)(t^6 + ½t^5 - 3t^3 - (3/2).t^2)(4t^3 - 3) + (2t^3 + ½t^2 + t^2 + ¼t)).\mathbf{d}t.$$

Multiplying out, this gives:

$$\omega_t = (4t^{13} + 2t^{12} - 19t^{10} - 13.5t^9 + 54t^7 + 27t^6 - 6t^5 - 27.t^4 - 11.5t^3 + 1.5.t^2 + ¼t).\mathbf{d}t.$$

Now we *can* find an **antiderivative** for this by repeated use of the $x^n.\mathbf{d}x = \mathbf{d}(x^{n+1}/(n + 1))$ formula:

$$\omega_t = \mathbf{d}((4/14)t^{14} + (2/13)t^{13} - (19/11)t^{11} - 1.35t^{10} + (54/8)t^8$$

$$+ (27/7)t^7 - t^6 - (27/5)t^5 - (23/8)t^4 + ½t^3 + (1/8)t^2) =: \mathbf{d}\phi(t).$$

In the last line I've used the definition symbol ":=" *reversed* as "=:"to indicate that I am defining $\mathbf{d}\phi(t)$ as a shorthand for the long expression in that last line. So we now have a **form** in t which *is* **exact** and this defines a valid **integral**, for which we already have the **interval** given in t as [0, 1], which is:

$$\omega_t \,|\, [0, 1] = \mathbf{d}\phi(t) \,|\, [0, 1] = \phi(t) \,|\, \mathbf{b}[0, 1] = \phi(1) - \phi(0)$$

and this equals (4/14) + (2/13) – (19/11) – 1.35 + (54/8) + (27/7) – 1 – (27/5) – (23/8) + ½ + ⅛.

This **integral** we *define* as the **line integral** of $\omega = x^2y.\mathbf{d}x + y.\mathbf{d}y$ over the **parametric curve** $\tau = (x = x(t) = t^4 - 3t, y = y(t) = t^2 + ½.t)$ from $t = 0$ to $t = 1$.

Although the value of the **line integral** of an **inexact differential form** depends on the actual curve "along which" it is **integrated**, it *doesn't* depend on the actual **parameter** of the curve. A different **parameter** specifying the *same curve in the original xy-space* will give the same result. The reason is that once we're down to the *single* **parametric** variable, we can replace the variable with another one by a function like $t = t(s)$ and $\mathbf{d}t = \partial_s t(s).\mathbf{d}s$, and as long as we correctly handle the **pullback** of the **interval** over which we're **integrating**, from an **interval** in t to an **interval** in s, we will get the same result, just as in the example with which this Section began.

As a rather trivial example, as the algebra is getting a bit complicated otherwise, suppose we used the replacement s such that $t = s^2$ and so $\mathbf{d}t = 2s.\mathbf{d}s$, then we would have:

$$\omega_s = (4s^{26} + 2s^{24} - 19s^{20} - 13.5s^{18} + 54s^{14} + 27s^{12} - 6s^{10}$$

$$- 27.s^8 - 11.5s^6 + 1.5.s^4 + ¼s^2).2s.\mathbf{d}s$$

$$= (8s^{27} + 4s^{25} - 38s^{21} - 27s^{19} + 108s^{15} + 54s^{13} - 12s^{11} - 54.s^9 - 23s^7 + 3.s^5 + ½s^3)\mathbf{d}s$$

$$= \mathbf{d}((8/28)s^{28} + (4/26)s^{26} - (38/22)s^{22} - (27/20)s^{20} + (108/16)s^{16}$$

$$+ (54/14)s^{14} - s^{12} - 5.4.s^{10} - (23/8)s^8 + ½.s^6 + ⅛s^4) =: \mathbf{d}\theta(s).$$

Since $t = 0$ at $s = 0$, and $t = 1$ at $s = 1$, the **interval** in s is still [0, 1], and we have:

$$\omega_s \,|\, [0, 1] = \mathbf{d}\theta(s) \,|\, [0, 1] = \theta(s) \,|\, \mathbf{b}[0, 1] = \theta(1) - \theta(0)$$

and this again equals (4/14) + (2/13) – (19/11) – 1.35 + (54/8) + (27/7) – 1 – (27/5) – (23/8) + ½ + ⅛.

You can see the general principle behind these changes, and why changing the *single* variable of **integration** always gives the same result as long as the **pullback** is allowed for, by looking at the change in variable as written above, and writing $s = t^{-1}(t)$ as the **inverse function** of $t = t(s)$:

$$\mathbf{d}\phi(t) \,|\, [\mathbf{p}(t = 0), \mathbf{p}(t = 1)]$$

$$= \mathbf{d}\phi(t(s)) \,|\, [\mathbf{p}(t(s = t^{-1}(0))), \mathbf{p}(t(s = t^{-1}(1)))].$$

$$= \phi(t(s)) \,|\, \mathbf{b}([\mathbf{p}(t(s = t^{-1}(0))), \mathbf{p}(t(s = t^{-1}(1)))])$$

$$= \phi(t(s)) \,|\, [\mathbf{p}(t(s = t^{-1}(1)))] - [\mathbf{p}(t(s = t^{-1}(0)))].$$

But, $t(s = t^{-1}(1)) = t(t^{-1}(1)) = 1$, and $t(s = t^{-1}(0)) = t(t^{-1}(0)) = 0$, so this equals

$$= \phi(t(s = t^{-1}(1))) - \phi(t(s = t^{-1}(0))) = \phi(t = 1) - \phi(t = 0) = \phi(1) - \phi(0).$$

We can always do this in the case of a single variable, because **differential forms** in a single variable are *always* **exact** (up to an arbitrarily close approximation, something that was foreshadowed in **Taylor's theorem** in Section 2.7).

But look at it another way. In *xy*-space, there are *two* basic **differentials**, **d***x* and **d***y*. In **pulling back** to the *t*-space, we have moved to a space in which there is only *one*, **d***t*. To anticipate again, so you will see just where we're going, when we come to switch to regarding **differential forms** as a **graded algebra**, the **pseudoscalar** for *xy*-space is **d***x***d***y*. In the *t*-space, it is **d***t*.

In any space, all **differential forms** on the **pseudoscalar** are always **exact.**

I will refer indiscriminately to ω_τ as the **pullback** of ω onto the t space and to the **interval** in t as the **pullback** of the **parametric curve** $x = x(t)$, $y = y(t)$. Again, this usage does not strictly conform to standard conventions, but is just what we need.

6.3 Manifolds

A **manifold** is the continuous analogue of a **simplex**, and like a **simplex** it has a **dimension**. The **manifold** corresponding to a **1-simplex**, a straight line in space, is a *curve* in space[77] and such a curve is called a **1-manifold**. A *surface* generalizes a **2-simplex** and may be a bounded area of a plane, or a curved surface lying in three-dimensional space: such a two-dimensional surface is called a **2-manifold**. The formal definition that I will prefer is that an **n-manifold** is defined as the **range** (see Section 1.1) of a **mapping** whose **domain** is an **n-simplex** or **chain** (i.e. sum) of **n-simplexes**. An example will make the idea clearer.

The surface of a sphere is a **2-manifold**. We can use the **arc length** variables introduced in Section 4.4 to give two **parameters** or **coordinates** for the surface, which are called **longitude** ϕ, and **latitude** θ. These two variables are said to **parameterize** the **manifold**. The number of **parameters** (or **coordinates**, the two terms being interchangeable, although **parameter** is generally preferred in this context) equals the **dimension** of the **manifold**, here two. Any point on the surface of the sphere is defined by:

[77] The "space" here is the **vector space** in which the **manifold** lies, which may be anything of dimension equal to or greater than that of the **manifold** in question. So a *curve* could lie in a *plane* as well as being in three-dimensional space, and a **line segment** or **interval** is a special case of a curve: a **1-manifold** in a 1-space.

$x = R.\mathbf{cos}(\phi).\mathbf{cos}(\theta)$
$y = R.\mathbf{sin}(\phi).\mathbf{cos}(\theta)$
$z = R.\mathbf{sin}(\theta)$.[78]

Here R is the **constant** *radius* of the sphere. The **domain** of this **mapping** is a *rectangle* in ϕ and θ, running from 0 to 2π in ϕ, which corresponds to 0° to 360°, and from $-\pi/2$ (the South Pole ≡ −90°) to $+\pi/2$ (the North Pole ≡ +90°) in θ. This rectangle is easily realized as two triangles or **2-simplexes** by dividing it by either diagonal, and this establishes that the concept fits my definition.

The *interior volume space* of the sphere is also a **manifold**, this time a **3-manifold**, the third **dimension** being given by introducing the **variable** r, in the place of the **constant** R above. Then the *cubical* **domain** that defines this **3-manifold** is given by:

$$r{:}[0, R] \times \phi{:}[0, 2\pi] \times \theta{:}[-\pi/2, +\pi/2].$$

This is more difficult to break down into **3-simplexes**, but it turns out that such a cubical volume can be represented as the sum of *six* **3-simplexes**.

Any **manifold** can also be approximated to an arbitrary degree of accuracy by a sufficiently large number of sufficiently small **simplexes** of the same **dimension**, *actually in situ*. This fact, which forms the basis of the representation of surfaces in computer graphics, is very useful in proofs, but the **mapping** idea is more useful for evaluating **integrals** on **manifolds**. An example of an *in situ* representation is shown in *Figure 6.8.1* where a circle is approximated by a large number of **2-simplexes** or triangles. Another example is the approximation of the surface of a sphere by subdivisions of the surface of an **icosahedron**, which is a regular **polyhedron** – a three-dimensional solid – with twenty identical triangular faces. Subdividing each of these into three smaller triangles gives sixty faces, and doing it again gives 180, which usually suffices to represent a sphere in computer graphics.

The definition of an **integral** over a **manifold** is now obtained by switching to its defining **simplex**:

> The **integral** of any **differential form** ω over a **manifold** Σ which is defined by a **mapping** $\Sigma = \tau(\mu)$ of a **simplex chain** μ, is itself defined as the **integral** of the **pullback** of ω by τ, or ω_τ, over the **chain** μ.

In evaluating **integrals** over **manifolds**, we use a judicious combination of both approaches, the *in situ* and the **mapping**, with a bias towards the **mapping**, so that given a definition of **integration** over **simplexes** of arbitrary **dimension**, and a way to define how we **pull back** the evaluation from the **manifold** to the appropriate **simplex**, we can obtain our result.

For **exact forms**, the "direct" definition of an **integral** and the **pullback** one coincide, as I shall now outline.

[78] These are the values in terms of **latitude**. **Spherical coordinates**, which you may see in the textbooks, have **sin**(θ) and **cos**(θ) swapped because they use the angle going down from the North Pole, which is called **colatitude**.

Suppose we have a function of *two* variables $f(x, y)$ as in Section 2.6. For that simplified case, I supposed that we have a **strict chain** of **1-simplexes** in the *xy* plane, and was able to show that if we define the **integration** on a **1-simplex** in terms of the evaluation of a **scalar function** $f(x, y)$ on a set of **0-simplexes** by:

$$\mathbf{d}f(x, y) \,|\, \mathbf{p}_0\mathbf{p}_1 := f(x, y) \,|\, \mathbf{b}(\mathbf{p}_0\mathbf{p}_1) = f(x, y) \,|\, (\mathbf{p}_1 - \mathbf{p}_0) = f(x_1, y_1) - f(x_0, y_0)$$

then for a **strict chain** of **1-simplexes**, the intermediate points drop out of the evaluation and only the end points contribute:

$$\mathbf{d}f(x, y) \,|\, \Sigma_i(\mathbf{p}_i\mathbf{p}_{i+1}) := f(x, y) \,|\, \mathbf{b}(\Sigma_i(\mathbf{p}_i\mathbf{p}_{i+1})) = f(x, y) \,|\, (\mathbf{p}_n - \mathbf{p}_0) = f(x_n, y_n) - f(x_0, y_0).$$

Now we can use the device of a **parametric variable** t to make this concept slightly more flexible. Suppose we have a *curve* in the *xy* plane defined by two functions:

$$x = x(t) \qquad \text{and} \qquad y = y(t).$$

For example, we might have:

$$x = at^3 + bt^2 + ct + d \qquad \text{and} \qquad y = pt^3 + qt^2 + rt + s$$

where I choose cubics (forms in t^3) because these are the simplest curves that give some flexibility of form, for which reason they are much used to approximate arbitrarily complex curves in computer graphics. Connecting a series of cubics together end-to-end enables one to approximate a huge variety of curve forms.

Now imagine t runs from t_0 to t_1. Then as t changes, different $x(t)$ and $y(t)$ values are selected, and these trace out a *curving path* in the *xy* plane. But the **interval** $[t_0, t_1]$ is a **1-simplex** in t. So we can extend our standard definition to that of a **line integral** along a *curve* by:

$$\mathbf{d}f(x(t), y(t)) \,|\, [t_0, t_1] = f(x(t), y(t)) \,|\, \mathbf{b}[t_0, t_1] = f(x(t), y(t)) \,|\, ([t_1] - [t_0])$$

$$= f(x(t_1), y(t_1)) - f(x(t_0), y(t_0)).$$

This is clearly equivalent to saying:

$$\mathbf{d}f(x(t), y(t)) \,|\, \mathbf{p}(t_0)\mathbf{p}(t_1) = f(x(t), y(t)) \,|\, \mathbf{b}(\mathbf{p}(t_0)\mathbf{p}(t_1)) = f(x(t), y(t)) \,|\, (\mathbf{p}(t_1) - \mathbf{p}(t_0))$$

$$= f(x(t), y(t)) \,|\, \mathbf{p}(t_1) - f(x(t), y(t)) \,|\, \mathbf{p}(t_0)$$

$$= f(x(t), y(t)) \,|\, (x(t_1), y(t_1)) - f(x(t), y(t)) \,|\, (x(t_0), y(t_0))$$

$$= f(x(t_1), y(t_1)) - f(x(t_0), y(t_0)).$$

But now we can dismiss the t and simply define this as:

$$\mathbf{d}f(x, y) \,|\, \mathbf{p}(x_0, y_0)\mathbf{p}(x_1, y_1) := f(x, y) \,|\, \mathbf{b}(\mathbf{p}(x_0, y_0)\mathbf{p}(x_1, y_1))$$

$$= f(x, y) \,|\, (\mathbf{p}(x_1, y_1) - \mathbf{p}(x_0, y_0))$$

$$= f(x_1, y_1) - f(x_0, y_0).$$

The huge difference here is that we have naturally extended the definition of **integration** from a **1-simplex** or straight **line segment** to an *arbitrary curve* or **1-manifold**.

But note another significant aspect of this:

> If the **form** in the **integral** is **exact** in *xy*-space, the value of the **line integral** is independent of the **parametric curve** over which the **line integral** is defined, as the **line integral** equals the standard form $\mathbf{d}f(x, y) \,|\, \mathbf{p}_0\mathbf{p}_1 = f(x, y) \,|\, \mathbf{b}(\mathbf{p}_0\mathbf{p}_1) = f(x, y) \,|\, \mathbf{p}_1 - f(x, y) \,|\, \mathbf{p}_0$.

If we do as we often do in computer graphics, and have a series of functions $x_i(t)$, $y_i(t)$, and a series of **line intervals** in t as $[t_0, t_1]$, $[t_2, t_3]$, $[t_4, t_5]$, . . . $[t_{n-1}, t_n]$, and have the condition analogous to the **strict chain** condition of Section 1.13 that, writing these **intervals** as $[t_{i0}, t_{i1}]$:

$$x_{i+1}(t_{(i+1)0}) = x_i(t_{i1}) \qquad \text{and} \qquad y_{i+1}(t_{(i+1)0}) = y_i(t_{i1})$$

or $\qquad \mathbf{p}_{(i+1)0} = \mathbf{p}_{i1}$

i.e. that the end point in the *xy* plane of one curve equals the start point of the next, then again the intermediate points will drop out and the **integral** depends only on the extreme end points:

$$\mathbf{d}f(x, y) \,|\, \Sigma_i\, \mathbf{p}_{i0}\mathbf{p}_{i1} = f(x, y)(\mathbf{p}_n) - f(x, y)(\mathbf{p}_0) = f(x_n, y_n) - f(x_0, y_0).$$

Note how repeatedly we have used the idea of **integration** over a **line segment** or **interval** in the **parameter** t, for which we *do* have a definition because it's a **1-simplex**, to create a definition for **integration** over a *curve* in the *xy* plane, for which we *didn't* have a definition.

Defining a **manifold** as the **range** of a **mapping** off a **simplex** or **chain** has the added advantage that the well-defined **boundary** of a **simplex** or **chain** can be used to define the **boundary** of the **manifold**:

> The **boundary** of a **manifold** M is the **range** of the **mapping** τ of the **boundary** of the underlying **simplex** or **chain** $\Sigma_k\sigma_k$: $\qquad$ if $\mathrm{M} = \tau(\Sigma_k\sigma_k)$, $\mathbf{b}\mathrm{M} := \tau(\Sigma_k\mathbf{b}\sigma_k)$.

6.4 Areas and Volumes

The third extension of our **integrals**, to higher dimensions, is defined to be consistent with the concept of higher-dimensional volumes. To clarify this point of reference, we need to look at the calculation of the area of a parallelogram and so of a triangle, and of the volume of a parallelepiped and so of a **tet**.[79]

Areas and volumes are quite subtle concepts, and methods for their calculation were one of the mainsprings in the development of the calculus. The basic idea of *rectangular area* is that any rectangle can be formed approximately from juxtaposing or **tiling** – which I put in bold to indicate this specific sense – a large number of (sufficiently small) identical squares. If we count how many little square **tiles** we can fit *inside* the rectangle, and how many will just *contain* the rectangle, we can say that the first count will be *less than* the area of the rectangle, and the second *greater than* the area. We now divide the little tiles into four smaller ones and do it again and so on *ad infinitum* and we end up with an arbitrarily precise measure of the rectangle's area. It turns out to be the **product** of the length and breadth of the sides.

Wishing to avoid this kind of argument as much as possible, I will simply *define* the area of a rectangle as that **product**. Taking that to be the case, we need Euclid I.35, which states that:

Parallelograms on the same base and between the same parallels are equal in area.

This we prove by the construction in *Figure 6.4.1*(*a*) where the common "base" is the line BC, and the two parallelograms are ABCD and EBCF. It's fairly easy to prove within the Euclidean axiom system that the triangles FDC and EAB are identical, and removing EAB from EBCF and adding FDC to it converts it into ABCD. This kind of "cut and paste" argument is somewhat intuitive, but suffices to give us a definition of the area of a parallelogram as the area of a rectangle on the same "base" (*Figure 6.4.1*(*b*)), and so as the length of one of the parallel sides times the distance between this side and its parallel twin, which distance is also called the **altitude** of the parallelogram. But of course using the other parallels as the "base" gives a different "altitude".

[79] Remember that I defined a **tet** as a shorthand term for a **3-simplex** at the end of Section 1.15.

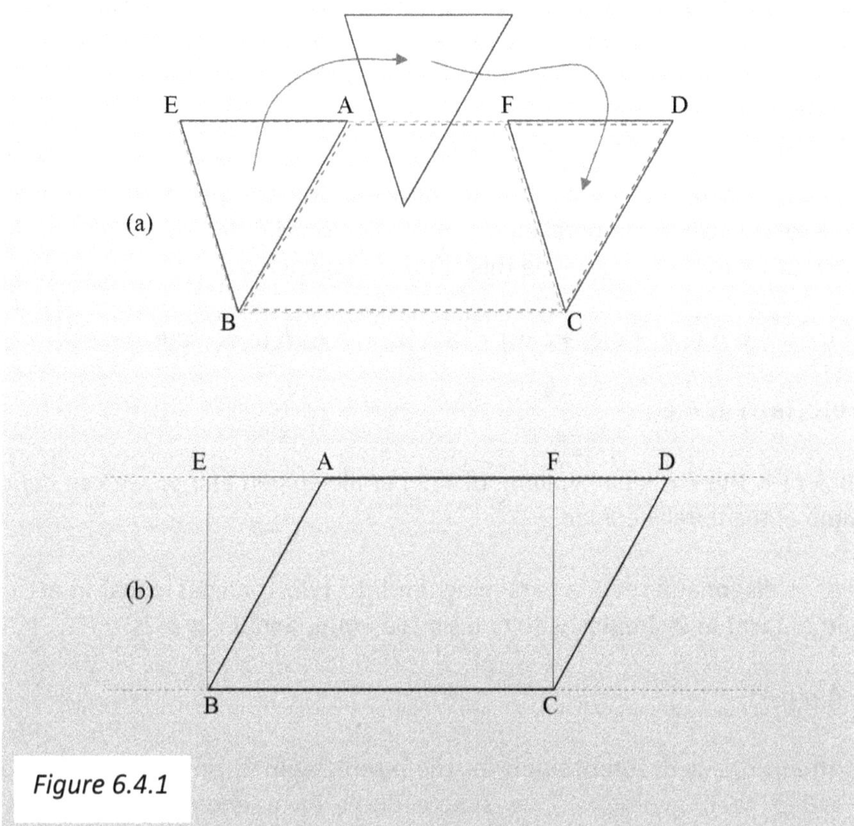

Figure 6.4.1

Any triangle can be formed by dividing a parallelogram by one of its diagonals. We need a theorem to show that either diagonal divides the parallelogram into two *identical* triangles, which apparently is Euclid I.34, and then we have that the area of a triangle is *half* the **product** of its "base" (the length of one side) and its "altitude" (the perpendicular distance to the opposite vertex).

Given any parallelogram, we can slide it (translate it) until one of its corners is at the origin, and rotate it until one of its sides aligns with the **basis vector** $\mathbf{e}_1$ in a given **coordinate** system. Alternatively, we can choose a **coordinate** system whose origin is at one corner of the parallelogram and which aligns with one of the sides. Then that side $\mathbf{a} = a_1\mathbf{e}_1$. If the other side running from this same corner (now at the origin) is $\mathbf{b} = b_1\mathbf{e}_1 + b_2\mathbf{e}_2$ then these two sides define the parallelogram, and b_2 is equal to the distance between **a** and the side parallel to it – i.e. is our "altitude". Now:

$$\mathbf{a} \wedge \mathbf{b} = a_1\mathbf{e}_1 \wedge (b_1\mathbf{e}_1 + b_2\mathbf{e}_2) = a_1b_2(\mathbf{e}_1 \wedge \mathbf{e}_2)$$

using $\mathbf{e}_1 \wedge \mathbf{e}_1 = \mathbf{0}$, so $|\mathbf{a} \wedge \mathbf{b}| = a_1b_2$ and this is the area of the parallelogram.[80] By leaving in the $\mathbf{e}_1 \wedge \mathbf{e}_2$ we can regard the **area** as a signed **bivector**, and I put this bold to indicate that henceforth I will regard this as the *definition* of the **area** of a parallelogram.

[80] I use the notation $|\mathbf{a} \wedge \mathbf{b}|$ to indicate the numerical *magnitude* of a **bivector**. Likewise, $|\mathbf{v}|$ is the magnitude of the **vector v**.

Going up to three dimensions, we can form a *parallelepiped* from three **vectors** from a common point. Call the three **vectors** **a**, **b**, and **c**, and then use a theorem analogous to Euclid I.35 which states that *parallelepipeds on the same base and between the same parallel planes are equal in volume*. The "base" now is the parallelogram defined by the vectors **a** and **b** with **bivector area** $\mathbf{a} \wedge \mathbf{b}$, and similarly choosing an "upright" parallelepiped to define the *volume* as the analogue of the representative rectangle to define the *area* for the parallelogram, we have that the volume of a parallelepiped equals the area of its "base" ($\mathbf{a} \wedge \mathbf{b}$) times the "altitude" between the **a**−**b** plane and the one parallel to it. If we translate and rotate either the parallelepiped or the **coordinates** so that **a** and **b** lie in the $\mathbf{e}_1\mathbf{e}_2$ plane, then if $\mathbf{c} = c_1\mathbf{e}_1 + c_2\mathbf{e}_2 + c_3\mathbf{e}_3$, c_3 will be this altitude and we have:

$$\mathbf{a} \wedge \mathbf{b} \wedge \mathbf{c} = \mathbf{a} \wedge \mathbf{b} \wedge (c_1\mathbf{e}_1 + c_2\mathbf{e}_2 + c_3\mathbf{e}_3) = |\mathbf{a} \wedge \mathbf{b}|\ \mathbf{e}_1 \wedge \mathbf{e}_2 \wedge (c_1\mathbf{e}_1 + c_2\mathbf{e}_2 + c_3\mathbf{e}_3)$$

$$= |\mathbf{a} \wedge \mathbf{b}|\, c_3\, (\mathbf{e}_1 \wedge \mathbf{e}_2 \wedge \mathbf{e}_3)$$

so again $|\mathbf{a} \wedge \mathbf{b} \wedge \mathbf{c}|$ is the Euclidean volume of the parallelepiped and so we can *define* $\mathbf{a} \wedge \mathbf{b} \wedge \mathbf{c}$ as the *signed* **volume** of the parallelepiped.

As argued earlier, a diagonal bisects a parallelogram into two triangles equal in area, so the signed **area** of a triangle or **facet** or **2-simplex** with two sides $\mathbf{a} = \mathbf{p}_0\mathbf{p}_1$ and $\mathbf{b} = \mathbf{p}_0\mathbf{p}_2$ is:

$$\tfrac{1}{2}(\mathbf{p}_0\mathbf{p}_1 \wedge \mathbf{p}_0\mathbf{p}_2).$$

Unfortunately, an analogous decomposition for the **parallelepiped** isn't possible. The basic reason is combinatorial rather than geometric: we simply don't have enough equal sides. Look at the **parallelepiped** in *Figure 6.4.2* below, where the analogy with the **parallelogram** is highlighted as (*a*) and the diagonals shown in (*a*) are omitted from (*b*) to give a bit of clarity:

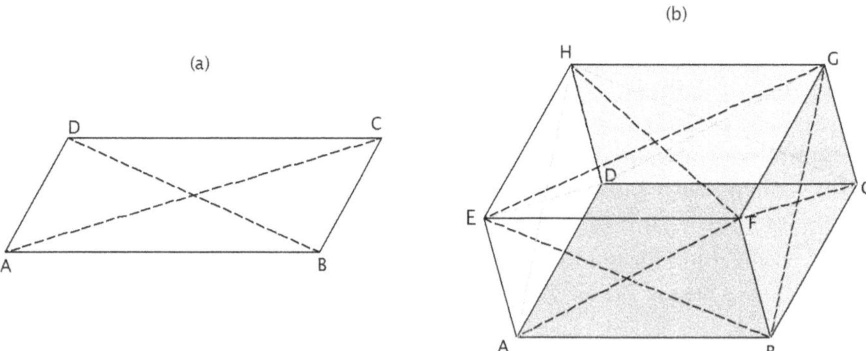

Figure 6.4.2

We have equalities {AE = BF = CG = DH}, {AB = EF = HG = DC}, {AD = BC = FG = EH} for the main sides, but the *face* diagonals drop down to just pairs {AF = DG}, {EB = HC}, {AH = BG}, {ED = FC}, {HF = DB} and {AC = EG}. For the true **parallelepiped** we can't assume e.g. that AC = BD as we can see in *6.4.2(a)*. Finally the *body* diagonals are all unique: {AG}, {BH}, {CE} and {DF}.

It turns out that the volume of a **tet** or **3-simplex** having the same lengths of its sides as a matching **parallelepiped**, as for example ABDE here, has a volume of $\frac{1}{6}$ th of that of the **parallelepiped**. So to match the decomposition of the **parallelogram** into **congruent** (equal sided) **tets** we'd need *six*. We could start with ABDE, using the sides AE, AB, AD from the first three equality sets, then work from, say, C, choosing CG, CD and CB to give CGDB, but even at this point congruence fails because the side BD in ABDE fails to match anything but HF which doesn't belong to CGDB!

I was a bit heartbroken by this discovery in the first edition of this book, but this is one place where some form of **limit** argument seems unavoidable. I sketch it in the Appendix 6.B. First, I'll give some informal definitions.

I'll use the term *n*-**volume** to indicate the analogue of **volume** of a closed figure of whatever shape in an *n*-**dimensional** space. So the *n*-**volume** of an **interval** on a line is its *length*, the *n*-**volume** of a closed shape in the plane is its *area*, 3-**volume** is ordinary **volume**, etc. Then a **cone** in *n* **dimensions** from any ($n-1$)-**dimensional** figure F to a point P *not lying in the same* ($n-1$)-**dimensional** space as F is the space of all points lying on a line joining any point of F to P. *Figure 6.4.3* shows some **cones** in 3-space.

The **vertical-distance** of P to F is the shortest distance from P to any point of F.

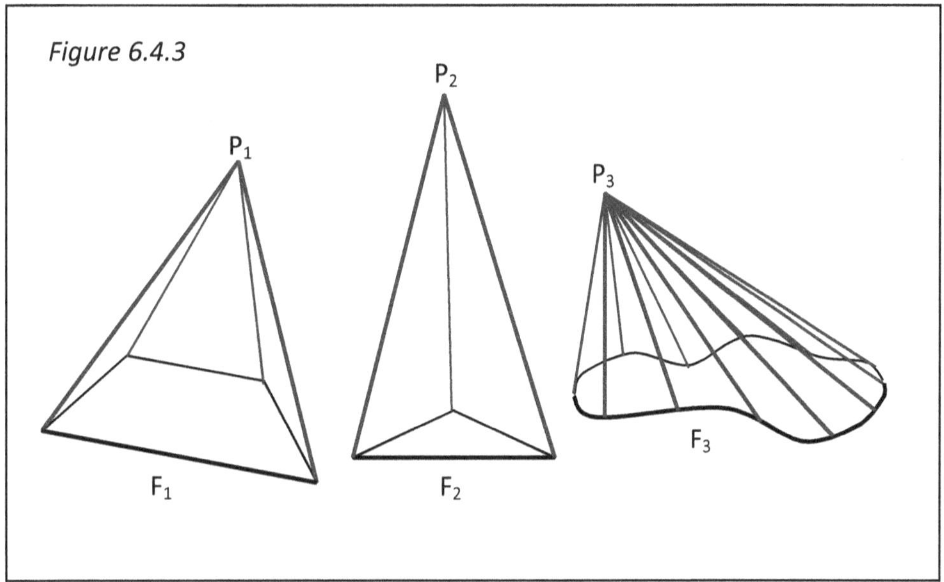

All **simplexes** are **cones**, from a **cone** in a space one **dimension** less. So from $\mathbf{p}_0$ the **cone** to $\mathbf{p}_1$ gives a **1-simplex** $\mathbf{p}_0\mathbf{p}_1$. From $\mathbf{p}_0\mathbf{p}_1$, given $\mathbf{p}_2$ not **collinear** with $\mathbf{p}_0\mathbf{p}_1$, we can define a **cone** $\mathbf{p}_0\mathbf{p}_1\mathbf{p}_2$. From $\mathbf{p}_0\mathbf{p}_1\mathbf{p}_2$, given $\mathbf{p}_3$ not **coplanar** with $\mathbf{p}_0\mathbf{p}_1\mathbf{p}_2$, we can define the **tet** $\mathbf{p}_0\mathbf{p}_1\mathbf{p}_2\mathbf{p}_3$, etc.

The result in the appendix is the 3-space version of the general result that:

> The *n*-**volume** of the **cone** from the ($n-1$)-**dimensional** figure F to a point P not in the same ($n-1$)-space as F itself is equal to $(1/n)((n-1)$-**volume**(F)) × **vertical-distance**(PF).

So the area of the triangle from the **interval** [A, B] to C is ½|B − A| × *height*([A, B], C).
Similarly the 3-**volume** of the **cone** from a triangle **simplex** $\mathbf{p}_0\mathbf{p}_1\mathbf{p}_2$ to a point $\mathbf{p}_3$ is:

$$\tfrac{1}{3} \times \mathit{area}(\mathbf{p}_0\mathbf{p}_1\mathbf{p}_2) \times \textbf{vertical-distance}(\mathbf{p}_0\mathbf{p}_1\mathbf{p}_2, \mathbf{p}_3)$$

$$= \tfrac{1}{3} \times \tfrac{1}{2} \times \mathit{length}(\mathbf{p}_0\mathbf{p}_1) \times \mathit{height}(\mathbf{p}_0\mathbf{p}_1, \mathbf{p}_2) \times \textbf{vertical-distance}(\mathbf{p}_0\mathbf{p}_1\mathbf{p}_2, \mathbf{p}_3)$$

and this is where the $\frac{1}{6}$th comes from.

Historically, the discovery of the need for **limit** arguments in evaluating the volumes of 3-dimensional objects was of enormous influence on the development of the calculus, leading to the idea (that I'm trying to challenge) that the **calculus** *essentially* depends on **limit** concepts and cannot be developed algebraically.

It also, incidentally, led to the idea that **integrals** shouldn't be signed as **volumes** are always positive, yet **integrals**, developed as algebraic objects, are *naturally* signed.

Perhaps surprisingly, despite all those rectilinear **vertical-distances**, the **volume** of a **tet** *can* in fact be evaluated by the formula below, which <u>without</u> the $\frac{1}{6}$th factor would give the volume of the analogous **parallelepiped** (as A = $\mathbf{p}_0$, B = $\mathbf{p}_1$, D = $\mathbf{p}_2$, E = $\mathbf{p}_3$ in *Figure 6.4.2*):

$$(1/6)(\mathbf{p}_0\mathbf{p}_1 \wedge \mathbf{p}_0\mathbf{p}_2 \wedge \mathbf{p}_0\mathbf{p}_3).$$

Be careful when the points are given emanating from the **origin**, because it's the **vectors** forming the **edges** or running *between* the **vertices** that matter.[81] So if in the triangle we have:

$$\mathbf{p}_0 = x_0\mathbf{e}_1 + y_0\mathbf{e}_2$$
$$\mathbf{p}_1 = x_1\mathbf{e}_1 + y_1\mathbf{e}_2$$
$$\mathbf{p}_2 = x_2\mathbf{e}_1 + y_2\mathbf{e}_2$$

then $\mathbf{p}_0\mathbf{p}_1 = (x_1 - x_0)\mathbf{e}_1 + (y_1 - y_0)\mathbf{e}_2$ and $\mathbf{p}_0\mathbf{p}_2 = (x_2 - x_0)\mathbf{e}_1 + (y_2 - y_0)\mathbf{e}_2$,

and so:

$$\begin{aligned}
\tfrac{1}{2}(\mathbf{p}_0\mathbf{p}_1 \wedge \mathbf{p}_0\mathbf{p}_2) &= \tfrac{1}{2}((x_1 - x_0)(y_2 - y_0)\, \mathbf{e}_1 \wedge \mathbf{e}_2 + (y_1 - y_0)(x_2 - x_0)\, \mathbf{e}_2 \wedge \mathbf{e}_1) \\
&= \tfrac{1}{2}((x_1 - x_0)(y_2 - y_0) - (y_1 - y_0)(x_2 - x_0))\, \mathbf{e}_1 \wedge \mathbf{e}_2 \\
&= \tfrac{1}{2}(x_1y_2 - x_1y_0 - x_0y_2 + x_0y_0 - (x_2y_1 - x_2y_0 - x_0y_1 + x_0y_0))\mathbf{e}_1 \wedge \mathbf{e}_2 \\
&= \tfrac{1}{2}(x_1y_2 - x_1y_0 - x_0y_2 - x_2y_1 + x_2y_0 + x_0y_1)\mathbf{e}_1 \wedge \mathbf{e}_2 \\
&= \tfrac{1}{2}((x_1y_2 - x_2y_1) + (x_0y_1 - x_1y_0) + (x_2y_0 - x_0y_2))\mathbf{e}_1 \wedge \mathbf{e}_2
\end{aligned}$$

and this also equals:

---

[81] It is in fact improper to treat **1-simplexes** as **vectors**. Instead we should define the **vector** *corresponding to* a **1-simplex** $\mathbf{p}_0\mathbf{p}_1$ as, say in 3-space, $(x_1 - x_0)\mathbf{e}_1 + (y_1 - y_0)\mathbf{e}_2 + (z_1 - z_0)\mathbf{e}_3$, but the intuitive picture is so compelling that it is admissible at this stage. I will return to this distinction later.

$\frac{1}{2}((\mathbf{p}_1 \wedge \mathbf{p}_2) + (\mathbf{p}_0 \wedge \mathbf{p}_1) + (\mathbf{p}_2 \wedge \mathbf{p}_0))$.

A way to remember this last formula is by the analogy:

$\mathbf{b}(\mathbf{p}_0\mathbf{p}_1\mathbf{p}_2) = \mathbf{p}_1\mathbf{p}_2 - \mathbf{p}_0\mathbf{p}_2 + \mathbf{p}_0\mathbf{p}_1$, and similarly,

$\mathbf{area}(\mathbf{p}_0\mathbf{p}_1\mathbf{p}_2) = \frac{1}{2}((\mathbf{p}_1 \wedge \mathbf{p}_2) - (\mathbf{p}_0 \wedge \mathbf{p}_2) + (\mathbf{p}_0 \wedge \mathbf{p}_1))$.

The switch to a vectorial concept of **area** and of **volume** can cause a little confusion. For example, it can be shown that the area of a parallelogram *at an angle in space* is still given by the $\mathbf{a} \wedge \mathbf{b}$ formula.

But now we will have $\mathbf{a} = (x_1 - x_0)\mathbf{e}_1 + (y_1 - y_0)\mathbf{e}_2 + (z_1 - z_0)\mathbf{e}_3$ and $\mathbf{b} = (x_2 - x_0)\mathbf{e}_1 + (y_2 - y_0)\mathbf{e}_2 + (z_1 - z_0)\mathbf{e}_3$, and so, simplifying these forms to $\mathbf{a} = a_1\mathbf{e}_1 + a_2\mathbf{e}_2 + a_3\mathbf{e}_3$ and $\mathbf{b} = b_1\mathbf{e}_1 + b_2\mathbf{e}_2 + b_3\mathbf{e}_3$, we get:

$\mathbf{a} \wedge \mathbf{b} = (a_1b_2 - a_2b_1)\mathbf{e}_1 \wedge \mathbf{e}_2 + (a_2b_3 - a_3b_2)\mathbf{e}_2 \wedge \mathbf{e}_3 + (a_3b_1 - a_1b_3)\mathbf{e}_3 \wedge \mathbf{e}_1$

which is a fully vectorial form. So to obtain a single number representing the conventional "area", regardless of sign, we must take the *magnitude* or **norm** of this **bivector**, which is:

$\sqrt{[(a_1b_2 - a_2b_1)^2 + (a_2b_3 - a_3b_2)^2 + (a_3b_1 - a_1b_3)^2]}$.

This case will actually arise in the discussion in Section 6.7.

6.5 Multiple Integrals

Multiple integrals are the simplest case of the third of the extensions of **integration** listed in Section 6.1, the extension of **integrals** to higher dimensions. A **multiple integral** is an **integral** between an **n-form** and an **n-simplex** in *n*-space. This case is simple because, at least in principle, the **form** is always **exact**. The more subtle case of an **integral** between an **n-form** and an **n-simplex** in *m*-space, where $m > n$ and so the **form** may not be **exact** I will introduce in Section 6.7.

The simplest **multiple integral** is this one:

$\mathbf{d}x.\mathbf{d}y \mid [x_0, x_1] \times_c [y_0, y_1]$

where $[x_0, x_1] \times_c [y_0, y_1]$ is not a **simplex**, but is the **Cartesian product** of two **intervals** $[x_0, x_1]$ and $[y_0, y_1]$ on different **coordinate axes**. Since $[x_0, x_1]$ is the set of all points on the *x* **axis** with *x* such that $x \geq x_0$ and $x \leq x_1$ and $[y_0, y_1]$ is the set of all points on the *y* **axis** with *y* such that $y \geq y_0$ and $y \leq y_1$, the **Cartesian product** is defined as the set of all *pairs* (x, y) – i.e. points in the *xy plane* – such that *x* is in $[x_0, x_1]$ and *y* is in $[y_0, y_1]$. So $[x_0, x_1] \times_c [y_0, y_1]$ is a *rectangle* in the *xy* plane.

This **integral** is trivially evaluated by letting $\mathbf{d}x$ and $\mathbf{d}y$ act *independently* on their respective **intervals**:

$$\mathbf{d}x.\mathbf{d}y \,|\, [x_0, x_1] \times_c [y_0, y_1] = (\mathbf{d}x \,|\, [x_0, x_1]).(\mathbf{d}y \,|\, [y_0, y_1]) = (x \,|\, \mathbf{b}[x_0, x_1]).(y \,|\, \mathbf{b}[y_0, y_1])$$

$$= (x_1 - x_0).(y_1 - y_0).$$

This does indeed equal the (unsigned) area of the $[x_0, x_1] \times_c [y_0, y_1]$ rectangle. It is perhaps worth remarking that the rectangle *can* be defined as a sum of two **2-simplexes** if we define $\mathbf{p}_0 \equiv (x_0, y_0)$, $\mathbf{p}_1 = (x_1, y_0)$, $\mathbf{p}_2 \equiv (x_0, y_1)$, $\mathbf{p}_3 = (x_1, y_1)$, then $[x_0, x_1] \times_c [y_0, y_1] = \mathbf{p}_0\mathbf{p}_1\mathbf{p}_3 + \mathbf{p}_0\mathbf{p}_3\mathbf{p}_2$.

The simplicity of this **integral** is misleading, because it is unusual that an **integral** can be treated as the product of two quite independent simple **integrals**. We will see why when we apply the same **differential form** $\mathbf{d}x\mathbf{d}y$ to a **simplex**.

To that end, consider the **integral**:

$$\mathbf{d}x\mathbf{d}y \,|\, \mathbf{p}_0\mathbf{p}_1\mathbf{p}_2$$

where the second **argument** of the **integral**, often called the **domain of integration**, is a **2-simplex** $\mathbf{p}_0\mathbf{p}_1\mathbf{p}_2$. For now, you won't know how to apply the **differential operator** to a **grade 1 differential form** (a **1-form**) to get a **grade 2 form** or, simply, a **2-form**. But it's quite simple: all that will turn out to be needed will be the Axioms III, IV and V given in Section 2.8, and to solve this problem, all you will need will be one of them, Axiom III: $\mathbf{dd}\omega = \mathbf{0}$ *always*.

$\mathbf{d}x\mathbf{d}y$ is a **2-form** because it involves a product of **differentials**. To apply our **adjoint** definition of **integration**, we need to find a **1-form** ϕ such that:

$$\mathbf{d}\phi = \mathbf{d}x\mathbf{d}y.$$

Using just what you know, arguably assisted by Axiom III, you can verify that $\phi = x.\mathbf{d}y$ satisfies this condition:

$$\mathbf{d}(x.\mathbf{d}y) = \mathbf{d}x.\mathbf{d}y + x.\mathbf{dd}y = \mathbf{d}x\mathbf{d}y,$$

because $\mathbf{dd}y = \mathbf{0}$. So from the usual **adjoint definition** extended now to **2-forms**, we have:

$$\mathbf{d}x\mathbf{d}y \,|\, \mathbf{p}_0\mathbf{p}_1\mathbf{p}_2 = \mathbf{d}(x.\mathbf{d}y) \,|\, \mathbf{p}_0\mathbf{p}_1\mathbf{p}_2 = x.\mathbf{d}y \,|\, \mathbf{b}(\mathbf{p}_0\mathbf{p}_1\mathbf{p}_2) = x.\mathbf{d}y \,|\, (\mathbf{p}_1\mathbf{p}_2 - \mathbf{p}_0\mathbf{p}_2 + \mathbf{p}_0\mathbf{p}_1)$$

$$= x.\mathbf{d}y \,|\, (\mathbf{p}_0\mathbf{p}_1 + \mathbf{p}_1\mathbf{p}_2 + \mathbf{p}_2\mathbf{p}_0).$$

In the last line I have put the **boundary** elements of the triangle in their normal order, switching the **orientation** of $\mathbf{p}_0\mathbf{p}_2$ to show this.

Now this is a simple **line integral** much like those in the Section 6.2. To evaluate it, we **parameterize** each **edge** of the three. Because the result is independent of the actual choice of **parameterization** as long as we follow the same path in *xy*-space, we can use a **pullback** onto the *same* **interval** t **:** [0, 1] for each **edge**.

Thus for $\mathbf{p}_0\mathbf{p}_1$, which is a straight line running from (x_0, y_0) to (x_1, y_1), we can choose:

$$x = x_0 + (x_1 - x_0).t$$
$$y = y_0 + (y_1 - y_0).t$$

which, calling this **mapping** T, does indeed give $T(t = 0) = (x_0, y_0)$ and $T(t = 1) = (x_1, y_1)$. The equation in y gives:

$$\mathbf{d}y = (y_1 - y_0).\mathbf{d}t$$

so the $\mathbf{p}_0\mathbf{p}_1$ part of our **integral** is now:

$$x.\mathbf{d}y \,|\, \mathbf{p}_0\mathbf{p}_1 = (x_0 + (x_1 - x_0).t)(y_1 - y_0).\mathbf{d}t \,|\, T^{-1}(\mathbf{p}_0\mathbf{p}_1)$$

$$= (x_0 + (x_1 - x_0).t)(y_1 - y_0).\mathbf{d}t \,|\, [0, 1].$$

This is a familiar form, and we can evaluate it by repeating the **adjoint** definition:

$$x_0(y_1 - y_0).\mathbf{d}t \,|\, [0, 1] + (x_1 - x_0).(y_1 - y_0).t.\mathbf{d}t \,|\, [0, 1]$$

$$= x_0(y_1 - y_0)(1 - 0) + ½(x_1 - x_0).(y_1 - y_0).\mathbf{d}(t^2) \,|\, [0, 1]$$

$$= x_0(y_1 - y_0) + ½(x_1 - x_0).(y_1 - y_0).(1^2 - 0^2)$$

$$= (x_0 + ½x_1 - ½x_0).(y_1 - y_0) = ½(x_0 + x_1).(y_1 - y_0).$$

Analogously, we get:

$$x.\mathbf{d}y \,|\, \mathbf{p}_1\mathbf{p}_2 = ½(x_1 + x_2).(y_2 - y_1)$$
$$x.\mathbf{d}y \,|\, \mathbf{p}_2\mathbf{p}_0 = ½(x_0 + x_2).(y_0 - y_2)$$

so our entire **integral** for the whole **2-simplex** comes to:

$$½\{(x_0 + x_1).(y_1 - y_0) + (x_1 + x_2).(y_2 - y_1) + (x_0 + x_2).(y_0 - y_2)\}$$

$$= ½[x_0y_1 - x_0y_0 + x_1y_1 - x_1y_0 + x_1y_2 - x_1y_1 + x_2y_2 - x_2y_1 + x_2y_0 - x_2y_2 + x_0y_0 - x_0y_2]$$

$$= ½[x_0y_1 - x_1y_0 + x_1y_2 - x_2y_1 + x_2y_0 - x_0y_2].$$

By comparison, we have for the **bivector area** of a **2-simplex**:

$$½[((x_1 - x_0)\mathbf{e}_x + (y_1 - y_0)\mathbf{e}_y) \wedge ((x_2 - x_0)\mathbf{e}_x + (y_2 - y_0)\mathbf{e}_y)]$$

$$= \tfrac{1}{2}[(x_1 - x_0)(y_2 - y_0) - (x_2 - x_0)(y_1 - y_0)]\mathbf{e}_x\mathbf{e}_y$$

$$= \tfrac{1}{2}[x_1y_2 - x_1y_0 - x_0y_2 + x_0y_0 - x_2y_1 + x_0y_1 + x_2y_0 - x_0y_0]\mathbf{e}_x\mathbf{e}_y$$

$$= \tfrac{1}{2}[x_1y_2 - x_1y_0 - x_0y_2 - x_2y_1 + x_0y_1 + x_2y_0]\mathbf{e}_x\mathbf{e}_y$$

and so the **integral** has given the value of the magnitude of the **area**.

We can now see why the first example seemed to come out so trivially. In the rectangle in the *xy* plane, there are two identical **2-simplexes**, as shown in *Figure 6.5.1*, with $\mathbf{p}_0\mathbf{p}_1\mathbf{p}_2$ having $(y_1 - y_0) = 0$, so the second term of the **exterior** or **alternating product** above is zero, and since $\mathbf{p}_0\mathbf{p}_2\mathbf{p}_3$ has the same area, we have in total:

$$|\mathbf{area}| = 2 \times \tfrac{1}{2}(x_1 - x_0)(y_2 - y_0) = (x_1 - x_0)(y_2 - y_0).$$

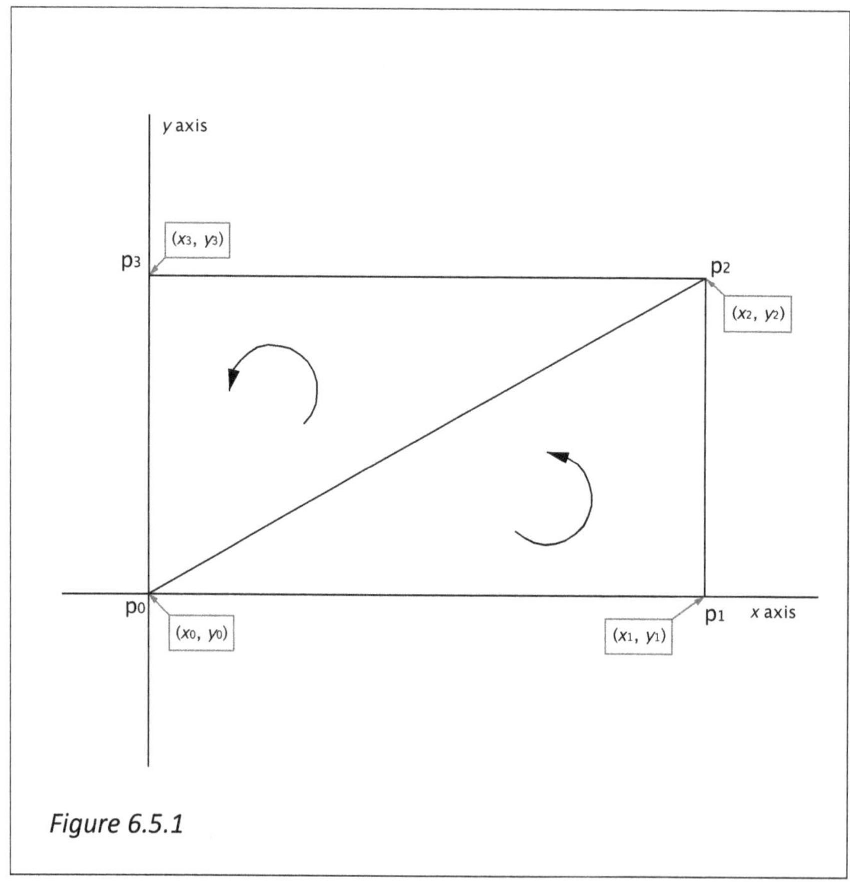

Figure 6.5.1

In this expression I have written |**area**| for the *magnitude* of the **bivector area**, which is what the **integration** gives us.

This result, together with René Descartes's (then) recent introduction of rectangular **coordinate** systems, may have contributed to early work on the **calculus** being biased towards rectangles. It will take a long time for this baleful influence to fade. **Integration** is nothing to do with rectangles *per se*, and everything to do with **simplexes**. Interestingly, the Euclidean school in classical times realized that triangles were geometrically more significant than rectangles, but the Renaissance moved away from this.

The **integration** for the case of a **3-simplex** or **tet** is a bit more arduous because of the proliferation of products. I give it in full in Appendix 6.B, but as it contains some important lessons, I'll discuss it briefly here.

The **integral** for a **tet** is:

$$\mathbf{d}x\mathbf{d}y\mathbf{d}z \mid \mathbf{p}_0\mathbf{p}_1\mathbf{p}_2\mathbf{p}_3.$$

We start its evaluation by recognizing that:

$$\mathbf{d}x\mathbf{d}y\mathbf{d}z = \mathbf{d}(x\mathbf{d}y\mathbf{d}z)$$

because the **differential operator** *acting on a* **form** *of higher* **grade** *acts, in effect, only on the* **coefficient scalar function** *part*, so that $\mathbf{d}(x\mathbf{d}y\mathbf{d}z) = \mathbf{d}(x).\mathbf{d}y\mathbf{d}z$, as any operation on the actual **differential** part gives zero by Axiom III of Section 2.8.[82]

We *can* now apply the **adjoint definition** of **integration** since we have an **exact form** $\mathbf{d}(*)$ on the left, to get:[83]

$$\mathbf{d}(x\mathbf{d}y\mathbf{d}z) \mid \mathbf{p}_0\mathbf{p}_1\mathbf{p}_2\mathbf{p}_3 = x\mathbf{d}y\mathbf{d}z \mid \mathbf{b}(\mathbf{p}_0\mathbf{p}_1\mathbf{p}_2\mathbf{p}_3)$$

$$= x\mathbf{d}y\mathbf{d}z \mid (\mathbf{p}_1\mathbf{p}_2\mathbf{p}_3 - \mathbf{p}_0\mathbf{p}_2\mathbf{p}_3 + \mathbf{p}_0\mathbf{p}_1\mathbf{p}_3 - \mathbf{p}_0\mathbf{p}_1\mathbf{p}_2).$$

Now $x\mathbf{d}y\mathbf{d}z$ is *not* **exact**. There is no **1-form** $A(x, y, z)\mathbf{d}x + B(x, y, z)\mathbf{d}y + C(x, y, z)\mathbf{d}z$ such that:

$$\mathbf{d}(A(x, y, z)\mathbf{d}x + B(x, y, z)\mathbf{d}y + C(x, y, z)\mathbf{d}z) = x\mathbf{d}y\mathbf{d}z.$$

The reason is that any **1-form** that gives $x\mathbf{d}y\mathbf{d}z$ will also throw up unwanted terms in $\mathbf{d}z\mathbf{d}x$ and $\mathbf{d}x\mathbf{d}y$. So for example, $xy\mathbf{d}z$ gives:

$$\mathbf{d}(xy\mathbf{d}z) = \mathbf{d}(xy).\mathbf{d}z = y\mathbf{d}x\mathbf{d}z + x\mathbf{d}y\mathbf{d}z.$$

Observe how, again, the **differential operator** acts only on the **scalar function** part.

So we'll need to circumvent the fact that $x\mathbf{d}y\mathbf{d}z$ is not **exact**. To do this, we create a **pullback** for each of the four **faces** $\mathbf{p}_1\mathbf{p}_2\mathbf{p}_3$, $\mathbf{p}_0\mathbf{p}_2\mathbf{p}_3$, $\mathbf{p}_0\mathbf{p}_1\mathbf{p}_3$ and $\mathbf{p}_0\mathbf{p}_1\mathbf{p}_2$, by a **mapping** from xyz space to a *two-dimensional* space which we label with **coordinates** u and v, arranging that each of the **faces maps** precisely onto the triangle in uv space described by the three points (0, 0), i.e. the **origin**, (1, 0) and (0, 1), which I'll label as $\mathbf{u}_0$, $\mathbf{u}_1$ and $\mathbf{u}_2$. (The **simplex** formed by the **origin** and the n points each of which has all its

[82] This point will be amplified in the next Section.
[83] As late as when I started on this Chapter, I had believed the **adjoint definition** could not be used at these higher **grades**!

coordinates zero except for one of them, which is 1, is often called the **standard n-simplex.**) Under this **mapping**, the *x***d***y***d***z* integral is transformed into **integrals** in *u***d***u***d***v*, *v***d***u***d***v* and **d***u***d***v*. In *uv* space, these are all **exact** because there is no third variable to throw up unwanted **2-forms** and we can *again* apply the **adjoint definition** to give, for example:

$$u\,\mathbf{d}u\,\mathbf{d}v \,\big|\, \mathbf{u}_0\mathbf{u}_1\mathbf{u}_2 = \mathbf{d}(\tfrac{1}{2}u^2.\mathbf{d}v) \,\big|\, \mathbf{u}_0\mathbf{u}_1\mathbf{u}_2 = \tfrac{1}{2}u^2.\mathbf{d}v \,\big|\, \mathbf{b}(\mathbf{u}_0\mathbf{u}_1\mathbf{u}_2)$$

$$= \tfrac{1}{2}u^2.\mathbf{d}v \,\big|\, (\mathbf{u}_1\mathbf{u}_2 - \mathbf{u}_0\mathbf{u}_2 + \mathbf{u}_0\mathbf{u}_1)$$

and this **line integral** we can evaluate by the methods introduced in Section 6.2, **mapping** to a space **parameterized** by a *single* variable *t*.

At each stage, we circumvent the **exactness** problem by **mapping** to a space in which the **grade** of the **form** we've got is the highest allowed. That this means it corresponds to the **pseudoscalar** on that space and **forms** on the **pseudoscalar** are **exact** will be a recurring principle in the rest of this Chapter.

6.6 Reinterpretation of the Differential

Differentials look awfully like **vectors**. They have the same **linearity** pattern, and there's a strong feeling when any **1-form** in, say, 3-space $\boldsymbol{R}^3$ is written as

$$\phi_x(x, y, z).\mathbf{d}x + \phi_y(x, y, z).\mathbf{d}y + \phi_z(x, y, z).\mathbf{d}z,$$

that this is an expansion in terms of a **basis** {**d***x*, **d***y*, **d***z*}.

Can we formalize this notion?

Well, the answer is obviously going to be yes, but first notice something that wasn't widely apparent in Chapter 5. The **components** of this "**vector**" or **linear combination** aren't just numbers, but *functions*.

Towards the end of Section 5.3 I gave a first intimation of this sort of concept in discussing how in computer graphics, **points** and **vectors** are commonly distinguished, and, more than that, *the total set of all* **vectors** *with a common* **base point** *constitutes a* **vector space**.

Might it be that the set of all **1-forms** written as above, as

$$\phi_x(x, y, z).\mathbf{d}x + \phi_y(x, y, z).\mathbf{d}y + \phi_z(x, y, z).\mathbf{d}z,$$

is the set of all **vectors** with the **base point** P having **coordinates** (*x*, *y*, *z*)?

The answer is: *almost*!

To see why it isn't quite, and just what it really is, let's go back to our old stamping ground of the **complex plane**.

Any point P in the **complex plane** can be identified with a **complex number z** = $(x, y) = z + jy$. But P or **z** doesn't have to be represented like that. We can also write P as:

$$r.\varepsilon(\theta) = r(\varepsilon_x(\theta) + j.\varepsilon_y(\theta)) \quad \text{where } r \in \,]0, \infty], \text{ and } \theta \in [0, 1].$$ [84]

So this implies

$$r.\varepsilon_x(\theta) = x$$
$$r.\varepsilon_y(\theta) = y.$$

This is a **transformation** of **coordinates** of the **plane** (into **polar coordinates**), so far nothing to do with **vectors**.

As a **vector space**, the **complex plane** can be thought of as having two **basis** elements, the **unit vector 1** pointing away from the **origin** to the right, and the **unit vector j** pointing away from the **origin** *upwards*. We could bring this into line with the normal notation in *three* dimensions introduced in Section 5.1, by writing **i** := **1**.

So now the **complex plane**, seen as a **vector space**, has the **basis** {**i**, **j**} with **i** clearly indicating the direction of increasing x values for *constant* y values, and **j** indicating the direction of increasing y values for *constant* x.

If we now make the distinction between **points** over the **complex plane** as a whole, and **vectors** defined at a specific **point** P defined by **coordinates** (x, y), this same **basis** {**i**, **j**} could be translated to P and serve just as well as a **basis** for the **vectors** with **base point** P.

So if we had a **point** Q with **coordinates** (ξ, ψ) we could define a **vector v** *at* P as:

$$\mathbf{v}(x, y) = \mathrm{P} - \mathrm{Q} = (\xi - x)\mathbf{i} + (\psi - y)\mathbf{j}.$$

I've put **v** as a function of x and y to indicate that different **v**'s might be defined at different **points** P themselves determined by different values of the **coordinates** (x, y).

Now what do we do when we have **polar coordinates** (r, θ)?

It would seem reasonable to replace **i** and **j** by **unit vectors** $\underline{\mathbf{r}}$ and $\underline{\boldsymbol{\theta}}$ defined as:

$\underline{\mathbf{r}}$ indicates the direction of increasing r values for *constant* θ values,

and

$\underline{\boldsymbol{\theta}}$ indicates the direction of increasing θ values for *constant* r.

This would be directly analogous to our definitions for **i** and **j** themselves.

[84] This is using the "set inclusion" operator "∈" to mean "belongs to" i.e. r is a value in the **interval**]0, 1] where the initial bracket's being turned around is taken to mean "not including 0".

> But now, at different **points** P defined by different (x, y) **coordinates**, these directions are no longer the same across the whole **complex plane**. $\underline{\mathbf{r}}$ and $\underline{\boldsymbol{\theta}}$ are now, in effect, *functions* of (x, y) *themselves*.

In Chapter 2, we saw that **partial derivatives** like ∂_r can be construed as indicating the variation in something *for other* **coordinates** (here θ) *constant*.

So we might try to establish $\underline{\mathbf{r}}$ from the expression of P itself as a **point** written vectorially as:

$$P = x\mathbf{i} + y\mathbf{j} = r.\varepsilon_x(\theta)\mathbf{i} + r.\varepsilon_y(\theta)\mathbf{j}.$$

We can **differentiate** this expression as:

$$\partial_r(x\mathbf{i} + y\mathbf{j}) = \partial_r(r.\varepsilon_x(\theta)\mathbf{i} + r.\varepsilon_y(\theta)\mathbf{j}) = \varepsilon_x(\theta)\mathbf{i} + \varepsilon_y(\theta)\mathbf{j}.$$

Because $\varepsilon_x^2(\theta) + \varepsilon_y^2(\theta) = 1$, and this matches the expression for the **scalar product** of this **vector** with itself, this *is* a **unit vector**. It's clearly pointing in the same direction as the line from the **origin** of the **complex plane** to P, which is clearly the direction of increasing r, so it looks like this is a good candidate for $\underline{\mathbf{r}}$:

$$\underline{\mathbf{r}} := \varepsilon_x(\theta)\mathbf{i} + \varepsilon_y(\theta)\mathbf{j}.$$

Note that it's independent of r itself – not quite as good as **i** and **j**, which were independent of *both x* and y, but good enough.

To define $\underline{\boldsymbol{\theta}}$, we might try the analogous:

$$\partial_\theta(x\mathbf{i} + y\mathbf{j}) = \partial_\theta(r.\varepsilon_x(\theta)\mathbf{i} + r.\varepsilon_y(\theta)\mathbf{j}) = r(\partial_\theta\, \varepsilon_x(\theta)\mathbf{i} + \partial_\theta\, \varepsilon_y(\theta)\mathbf{j})$$

and from Section 4.9, these are:

$$r(-2\pi.\varepsilon_y(\theta).\mathbf{i} + 2\pi.\varepsilon_x(\theta).\mathbf{j}) = -2\pi r.(\varepsilon_y(\theta).\mathbf{i} - \varepsilon_x(\theta).\mathbf{j}).$$

If we take the **scalar product** of this with $\underline{\mathbf{r}}$, we get:

$$\partial_\theta(x\mathbf{i} + y\mathbf{j}) \cdot \underline{\mathbf{r}} = -2\pi r.(\varepsilon_y(\theta).\mathbf{i} - \varepsilon_x(\theta).\mathbf{j}) \cdot (\varepsilon_x(\theta)\mathbf{i} + \varepsilon_y(\theta)\mathbf{j})$$

$$= -2\pi r(\varepsilon_y(\theta).\varepsilon_x(\theta) - \varepsilon_x(\theta).\varepsilon_y(\theta)) = 0.$$

So $\underline{\mathbf{r}}$ and our provisional $\underline{\boldsymbol{\theta}}$ **vector** are **orthogonal**. It's just that the $\underline{\boldsymbol{\theta}}$ **vector** isn't a **unit vector**.
That's easily remedied by taking the **norm**() operation to return the analogous **unit vector**:

$$\underline{\boldsymbol{\theta}} := -\,(\varepsilon_y(\theta).\mathbf{i} - \varepsilon_x(\theta).\mathbf{j}).$$

This clearly *does* point in the direction of increasing θ as you may convince yourself by first taking the point where $\varepsilon_x(\theta) = 1$, $\varepsilon_y(\theta) = 0$, corresponding to the (1, 0) element of the **unit circle**, where $\underline{\boldsymbol{\theta}}(1, 0) = \mathbf{j}$. But at the (0, 1) = *j* point, we have instead $\varepsilon_x(\theta) = 0$, $\varepsilon_y(\theta) = 1$, so indeed the **unit vector** $\underline{\boldsymbol{\theta}}(0, 1) = -\mathbf{i}$.

Figure 6.6.1 illustrates these possibilities, showing the different **basis vectors** at different points round the circle.

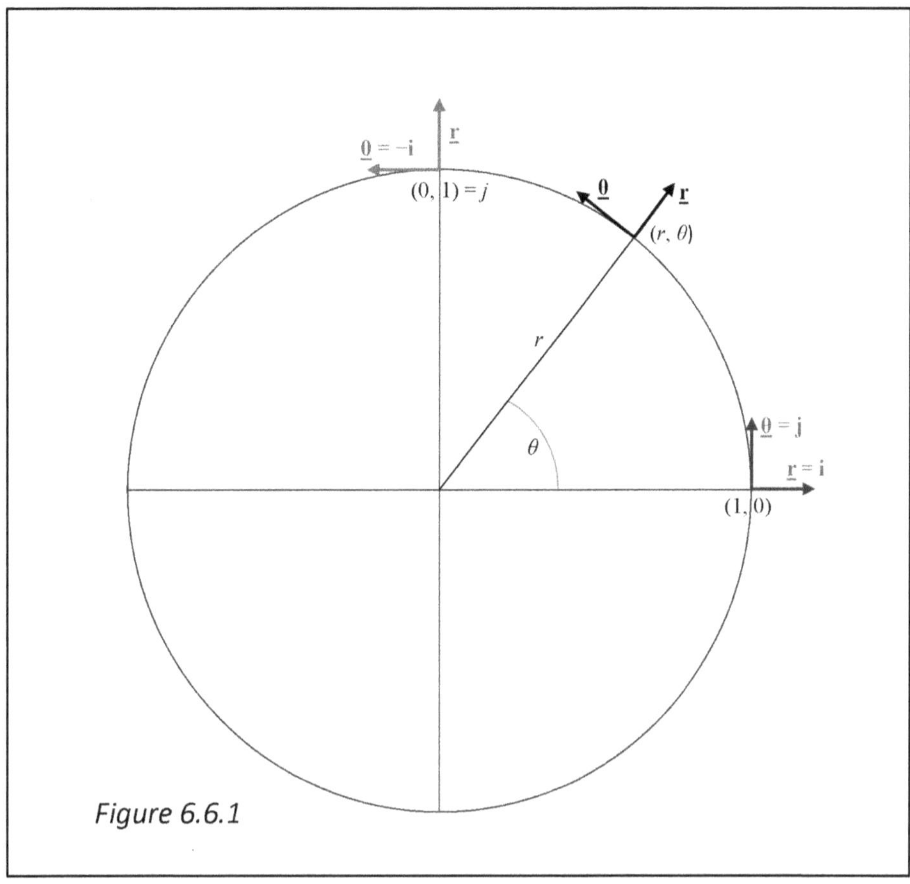

Figure 6.6.1

But let's look at this in a different way, accepting that the simply transformed **basis vectors**, although **orthogonal**, aren't **orthonormal**. Let's write these as just {**r**, **θ**}. Then:

$$\mathbf{r} = \varepsilon_x(\theta)\mathbf{i} + \varepsilon_y(\theta)\mathbf{j}$$
$$\boldsymbol{\theta} = -2\pi r.(\varepsilon_y(\theta).\mathbf{i} - \varepsilon_x(\theta).\mathbf{j}).$$

So $$\begin{pmatrix}\mathbf{r}\\ \boldsymbol{\theta}\end{pmatrix} = \begin{pmatrix}\varepsilon_x(\theta) & \varepsilon_y(\theta)\\ -2\pi r.\varepsilon_y(\theta) & 2\pi r.\varepsilon_x(\theta)\end{pmatrix}\begin{pmatrix}\mathbf{i}\\ \mathbf{j}\end{pmatrix}.$$

Now let's go back to the original equations,

$$r.\varepsilon_x(\theta) = x$$
$$r.\varepsilon_y(\theta) = y.$$

and calculate the **differentials**:

$$\mathbf{d}x = \mathbf{d}r.\varepsilon_x(\theta) + r.\mathbf{d}\varepsilon_x(\theta) = \mathbf{d}r.\varepsilon_x(\theta) - 2\pi r.\varepsilon_y(\theta).\mathbf{d}\theta$$
$$\mathbf{d}y = \mathbf{d}r.\varepsilon_y(\theta) + r.\mathbf{d}\varepsilon_y(\theta) = \mathbf{d}r.\varepsilon_y(\theta) + 2\pi r.\varepsilon_x(\theta).\mathbf{d}\theta.$$

or

$$\begin{pmatrix}\mathbf{d}x\\ \mathbf{d}y\end{pmatrix} = \begin{pmatrix}\varepsilon_x(\theta) & -2\pi r.\varepsilon_y(\theta)\\ \varepsilon_y(\theta) & 2\pi r.\ \varepsilon_x(\theta)\end{pmatrix}\begin{pmatrix}\mathbf{d}r\\ \mathbf{d}\theta\end{pmatrix}.$$

But this is the **inverse** of the **transform** from $\begin{pmatrix}\mathbf{d}x\\ \mathbf{d}y\end{pmatrix}$ to $\begin{pmatrix}\mathbf{d}r\\ \mathbf{d}\theta\end{pmatrix}$, and it's the **transpose** of the **matrix** above.

So the **transform** from {**i**, **j**} to {**r**, **θ**} is the **dual transform**, in the sense of Section 5.5, of the **transform** from {**d**x, **d**y} to {**d**r, **d**θ}.

So the space of **differential forms** at a **point** P is the **dual** of the space of **vectors** at P.
So **differentials** are **covectors.**

By the **chain rule** we also have that for any function $\phi(x, y)$:

$$\partial_r\phi(x, y) = \partial_r(x(r, \theta)).\partial_x\phi(x, y) + \partial_r(y(r, \theta)).\partial_y\phi(x, y)$$

which in this case is:

$$\partial_r\phi(x, y) = \partial_r(r.\varepsilon_x(\theta)).\partial_x\phi(x, y) + \partial_r(r.\varepsilon_y(\theta)).\partial_y\phi(x, y)$$

$$= \varepsilon_x(\theta).\partial_x\phi(x, y) + \varepsilon_y(\theta).\partial_y\phi(x, y).$$

Equally,

$$\partial_\theta\phi(x, y) = \partial_\theta(x(r, \theta)).\partial_x\phi(x, y) + \partial_\theta(y(r, \theta)).\partial_y\phi(x, y)$$

$$= \partial_\theta(r.\varepsilon_x(\theta)).\partial_x\phi(x, y) + \partial_\theta(r.\varepsilon_y(\theta)).\partial_y\phi(x, y)$$

$$= -2\pi r.\varepsilon_y(\theta).\partial_x\phi(x, y) + 2\pi r.\varepsilon_x(\theta).\partial_y\phi(x, y).$$

This suggests that there is a 1:1 correspondence between the equations:

$$\mathbf{r} = \varepsilon_x(\theta)\mathbf{i} + \varepsilon_y(\theta)\mathbf{j}$$
$$\boldsymbol{\theta} = -2\pi r.\varepsilon_y(\theta).\mathbf{i} + 2\pi r.\varepsilon_x(\theta).\mathbf{j}$$

and the formal *operator* representation:

$$\partial_r \equiv \varepsilon_x(\theta).\partial_x + \varepsilon_y(\theta).\partial_y$$
$$\partial_\theta \equiv -2\pi r.\varepsilon_y(\theta).\partial_x + 2\pi r.\varepsilon_x(\theta).\partial_y$$

which we obtain simply by omitting the $\phi(x, y)$ arbitrary operands from the foregoing equations. Since the **operator equations** just given must hold for *any* $\phi(x, y)$, we can work with them as entities in their own right.

This is the standard formalism used in all advanced work.

At any **point** P in a space (or more generally, a **manifold**) a set of **basis vectors** is defined which we may write as {**i**, **j**, **k**, . . .} or as $\{\mathbf{e}_1, \mathbf{e}_2, \mathbf{e}_3, \ldots\}$ or as $\{\partial_x, \partial_y, \partial_z, \ldots\}$ which transform under a change of **coordinates** according to a **matrix A**(P), and a set of **dual vectors**, also called **covectors** {**d***x*, **d***y*, **d***z*, . . .} which are **differentials**, which transform according to $((\mathbf{A}^{-1})^T)(P)$.

The **matrix A** is defined according to the **chain rule** for a change of **coordinates**

$$\{x^j\} \rightarrow \{\xi^j\} \text{ as } \left\{\frac{\partial x^i}{\partial \xi^j}\right\}.$$

So in our 2 × 2 case here,

$$\mathbf{A} = \begin{pmatrix} \frac{\partial x}{\partial r} & \frac{\partial y}{\partial r} \\ \frac{\partial x}{\partial \theta} & \frac{\partial y}{\partial \theta} \end{pmatrix}.$$

By the same **chain rule** the **inverse** mappings are:

$$\mathbf{d}r = \frac{\partial r}{\partial x}.\mathbf{d}x + \frac{\partial r}{\partial y}.\mathbf{d}y$$

and

$$\mathbf{d}\theta = \frac{\partial \theta}{\partial x}.\mathbf{d}x + \frac{\partial \theta}{\partial y}.\mathbf{d}y, \qquad (6.6.1)$$

$$\text{so that } ((\mathbf{A}^{-1})^T)(P) = \begin{pmatrix} \frac{\partial r}{\partial x} & \frac{\partial r}{\partial y} \\ \frac{\partial \theta}{\partial x} & \frac{\partial \theta}{\partial y} \end{pmatrix},$$

a rather neat inversion. It's easy to see that this is the required **transpose** by expanding **d***x* and **d***y* in the preceding equation:

$$\mathbf{d}r = \frac{\partial r}{\partial x}.\mathbf{d}x + \frac{\partial r}{\partial y}.\mathbf{d}y = \frac{\partial r}{\partial x}.\left(\frac{\partial x}{\partial r}.\mathbf{d}r + \frac{\partial x}{\partial \theta}.\mathbf{d}\theta\right) + \frac{\partial r}{\partial y}.\left(\frac{\partial y}{\partial r}.\mathbf{d}r + \frac{\partial y}{\partial \theta}.\mathbf{d}\theta\right)$$

$$= \left(\frac{\partial r}{\partial x}.\frac{\partial x}{\partial r} + \frac{\partial r}{\partial y}.\frac{\partial y}{\partial r}\right)\mathbf{d}r + \left(\frac{\partial r}{\partial x}.\frac{\partial x}{\partial \theta} + \frac{\partial r}{\partial y}\frac{\partial y}{\partial \theta}\right).\mathbf{d}\theta$$

$$d\theta = \frac{\partial\theta}{\partial x}.\mathbf{d}x + \frac{\partial\theta}{\partial y}.\mathbf{d}y = \frac{\partial\theta}{\partial x}.(\frac{\partial x}{\partial r}.\mathbf{d}r + \frac{\partial x}{\partial\theta}.\mathbf{d}\theta) + \frac{\partial\theta}{\partial y}.(\frac{\partial y}{\partial r}.\mathbf{d}r + \frac{\partial y}{\partial\theta}.\mathbf{d}\theta)$$

$$= (\frac{\partial\theta}{\partial x}.\frac{\partial x}{\partial r}. + \frac{\partial\theta}{\partial y}.\frac{\partial y}{\partial r})\mathbf{d}r + (\frac{\partial\theta}{\partial x}.\frac{\partial x}{\partial\theta} + \frac{\partial\theta}{\partial y}\frac{\partial y}{\partial\theta}).\mathbf{d}\theta$$

so the form that gives the **inverse $\mathbf{A}^{-1}$** must be:

$$\mathbf{A}.\mathbf{A}^{-1} = \begin{pmatrix} \frac{\partial x}{\partial r} & \frac{\partial y}{\partial r} \\ \frac{\partial x}{\partial\theta} & \frac{\partial y}{\partial\theta} \end{pmatrix}\begin{pmatrix} \frac{\partial r}{\partial x} & \frac{\partial\theta}{\partial x} \\ \frac{\partial r}{\partial y} & \frac{\partial\theta}{\partial y} \end{pmatrix} = \begin{pmatrix} 1 & 0 \\ 0 & 1 \end{pmatrix}.$$

This is indeed the **transpose** of the **matrix** above. Indeed, regarding (**d**x, **d**y) and (**d**r, **d**θ) as **covectors**, we see that the elements in these two **matrices** *must* make them **inverses** *by definition*. All the entities like $\frac{\partial x}{\partial r}$ and $\frac{\partial\theta}{\partial y}$ in equations like (6.6.1) above are *defined* to be the coefficients of the **inverse transformations** of these two **covector bases**.

So how do we define the operation of a **dual space** element like $\Sigma\phi_i.\mathbf{d}x^i$ on a **vector** like $\Sigma v^j.\mathbf{e}_j$?[85]

Well, because the **A** : $(\mathbf{A}^{-1})^T$ relationship is that of a **vector basis** to its **reciprocal** or **dual basis**, that's now all defined for us. It doesn't matter that we may have $\phi_i = \phi_i(\mathrm{P}) = \phi_i(x, y, z)$ or that $v^j = v^j(\mathrm{P}) = v^j(x, y, z)$ because everything's restricted to the **vector space** and **dual space** now at P, so the variation beyond that doesn't matter. We can treat the $\{ \phi_i \}$ and the $\{ v^j \}$ as just being *numbers* – the **vector components**.

What matters is the **reciprocal basis** defining relationship, that:

$$\mathbf{d}x^i(\mathbf{e}_j) = \delta^i_j.$$

So

$$(\phi_x.\mathbf{d}x + \phi_y.\mathbf{d}y + \phi_z.\mathbf{d}z)(v^x.\mathbf{e}_x + v^y.\mathbf{e}_y + v^z.\mathbf{e}_z)$$

must equal

$$\phi_x\, v^x.\mathbf{d}x(\mathbf{e}_x) + \phi_x\, v^y.\mathbf{d}x(\mathbf{e}_y) + \phi_x\, v^z.\mathbf{d}x(\mathbf{e}_z)$$

$$+ \phi_y\, v^x.\mathbf{d}y(\mathbf{e}_x) + \phi_y\, v^y.\mathbf{d}y(\mathbf{e}_y) + \phi_y\, v^z.\mathbf{d}y(\mathbf{e}_z)$$

$$+ \phi_z\, v^x.\mathbf{d}z(\mathbf{e}_x) + \phi_z\, v^y.\mathbf{d}z(\mathbf{e}_y) + \phi_z\, v^z.\mathbf{d}z(\mathbf{e}_z)$$

and by the **Kronecker condition** above this must reduce to:

[85] Note that now the **components** in the ordinary **vector space** get *upper indices* because they transform *contragrediently* to the **vectors** themselves, and conversely in the **dual** space where it's the **dual vectors** that get the upper indices and their **components** the lower.

$$\phi_x v^x.\mathbf{d}x(\mathbf{e}_x) + \phi_y v^y.\mathbf{d}y(\mathbf{e}_y) + \phi_z v^z.\mathbf{d}z(\mathbf{e}_z) = \phi_x v^x + \phi_y v^y + \phi_z v^z.$$

The whole thing is defined by the relation between the **basis** and **dual basis** elements.

A typical textbook statement of this definition is this one:[86]

$$\mathbf{d}f(\text{P}, \mathbf{v}) := (\partial_x f(\text{P}).\mathbf{e}_x + \partial_y f(\text{P}).\mathbf{e}_y + \partial_z f(\text{P}).\mathbf{e}_z) \cdot \mathbf{v}$$

where P = (x, y, z) is a point in *xyz*-space, $f()$ a **function** on *xyz*-space, and **v** a **3-vector**, and the dot before the **v** is the ordinary **dot product**. This defines the **differential** as a function of a **vector** at a point. I'll call this the **dual space definition** or the **covector definition** of a **differential**.

If $f()$ is one of the **coordinate functions**, say $x()$, $\partial_x x(\text{P}).\mathbf{e}_x + \partial_y x(\text{P}).\mathbf{e}_y + \partial_z x(\text{P}).\mathbf{e}_z = 1.\mathbf{e}_x + 0.\mathbf{e}_y + 0.\mathbf{e}_z = \mathbf{e}_x$ and with $\mathbf{v} = v_x\mathbf{e}_x + v_y\mathbf{e}_y + v_z\mathbf{e}_z$, we have:

$$\mathbf{d}x(\text{P}, \mathbf{v}) = \mathbf{e}_x \cdot \mathbf{v} = \mathbf{e}_x \cdot (v_x\mathbf{e}_x + v_y\mathbf{e}_y + v_z\mathbf{e}_z) = v_x.$$

The **vector** $\partial_x f(\text{P}).\mathbf{e}_x + \partial_y f(\text{P}).\mathbf{e}_y + \partial_z f(\text{P}).\mathbf{e}_z$, which can be defined similarly in any number of dimensions, is called the **gradient** of f, and looms very large in most conventional treatments of vector calculus. It is often written as **grad** f, or as ∇f.

This definition formally gives **differentials** a clear vectorial nature and places them unambiguously in the **dual space** as **covectors**. The **dual space** functionality so defined is very similar to the property where an **exact form** operates on a **1-simplex** through the **integral product**, but they are not quite the same and I would like briefly to elucidate the distinction.

I mentioned in a footnote to Section 6.4 that it is in fact somewhat improper to treat **1-simplexes** as **vectors**, however intuitively they may suggest themselves as such. Taking up this cue, **differentials** can be defined as **functions** on **1-simplexes** by the usual **integration operator**:

$$\mathbf{d}f \,|\, \mathbf{p}_0\mathbf{p}_1 = f(\mathbf{p}_1) - f(\mathbf{p}_0)$$

and this, incidentally, gives us a very simple, and quite elegant, definition of a conversion **function** from **1-simplexes** to **vectors** using the **differentials** of the **coordinate functions** (here shown in *xyz*-space), as:

$$\mathbf{vec}(\mathbf{p}_0\mathbf{p}_1) = (\mathbf{d}x \,|\, \mathbf{p}_0\mathbf{p}_1)\mathbf{e}_x + (\mathbf{d}y \,|\, \mathbf{p}_0\mathbf{p}_1)\mathbf{e}_y + (\mathbf{d}z \,|\, \mathbf{p}_0\mathbf{p}_1)\mathbf{e}_z$$

$$= (x_1 - x_0)\mathbf{e}_x + (y_1 - y_0)\mathbf{e}_y + (z_1 - z_0)\mathbf{e}_z.$$

This is in just the form in which we need it.

---

[86] Which I take from R. E. Johnson and F. L. Kiokemeister, *Calculus with Analytic Geometry*, Allyn and Bacon 1964, p.566.

Note that for any of the **coordinate differentials** applied to a **simplex**, for example **d***x*, the **integral product** and the **d***x*(P, **v**) definitions give the same answer:

$$\mathbf{d}x \,|\, \mathbf{p}_0\mathbf{p}_1 = \mathbf{d}x(\mathbf{p}_0, \mathbf{vec}(\mathbf{p}_0\mathbf{p}_1)) = (x_1 - x_0).$$

But for any *other* function, the two differ: while $\mathbf{d}f \,|\, \mathbf{p}_0\mathbf{p}_1 = f(\mathbf{p}_1) - f(\mathbf{p}_0)$ *precisely*, the action of **d***f* as a **dual space** element on the **vector** deriving from $\mathbf{p}_0\mathbf{p}_1$ only *approximates* the change in *f* between $\mathbf{p}_0$ and $\mathbf{p}_1$(note the "≈" at the end):

$$\mathbf{d}f\,(\mathrm{P}, \mathbf{v}) = \mathbf{grad}(f) \cdot \mathbf{v} = \mathbf{grad}(f) \cdot \mathbf{vec}(\mathbf{p}_0\mathbf{p}_1)$$

$$= \partial_x f(\mathrm{P}).v_x + \partial_y f\,(\mathrm{P}).v_y + \partial_z f\,(\mathrm{P}).v_z$$

$$= \partial_x f\,(\mathrm{P}).(x_1 - x_0) + \partial_y f\,(\mathrm{P}).(y_1 - y_0) + \partial_z f\,(\mathrm{P}).(z_1 - z_0) \approx f\,(\mathbf{p}_1) - f\,(\mathbf{p}_0).$$

This is the generalization of the approximation discussed in Section 2.3, and relates directly to the idea of **differentials** as *numbers*. Further discussion of this appears in Appendix 6.C. We will not need this interpretation hereafter.

So armed with a vectorial definition of **differentials**, we can go on to define an **alternating product** between them which obeys the axioms introduced in Section 2.8:

Axiom III:	**dd** ≡ **0** *always.* Applying **d** to any **differential** gives zero
Axiom IV:	$\mathbf{d}y \wedge \mathbf{d}x = -\,\mathbf{d}x \wedge \mathbf{d}y$
Axiom V:	$\mathbf{d}x \wedge \mathbf{d}x = 0$

Again, Axiom V isn't strictly needed: from axiom IV we could get $\mathbf{d}x\wedge\mathbf{d}x = -\,\mathbf{d}x\wedge\mathbf{d}x$ and if anything equals its own negative, it must be zero. I've brought down the **dd** ≡ **0** axiom here too, because we never need a **second differential** in **integration**. This enables us to extend the action of the **differential operator** to act on a **differential form** as we have used them so far – a straight **linear combination** of **differentials**, which henceforth we will call a **1-form** – to create a **linear combination** in **alternating products** of **differentials**, which is an object of the next **grade** in the algebra, and which we will call a **2-form**. Axiom III is also known as **Poincaré's Lemma**, and will be discussed more fully towards the end of this Section.

These are the axioms I usually use, which together with our original axioms I and II from Chapter 2 suffice to evaluate any **exterior derivative**. So, for example,

$$\mathbf{d}(g(x, y).\mathbf{d}x + h(x, y).\mathbf{d}y) = \mathbf{d}(g(x, y).\mathbf{d}x) + \mathbf{d}(h(x, y).\mathbf{d}y) \qquad \text{[Axiom I]}$$

$$= \mathbf{d}g(x, y).\mathbf{d}x + g(x, y).\mathbf{dd}x + \mathbf{d}h(x, y).\mathbf{d}y + h(x, y).\mathbf{dd}y \qquad \text{[Axiom II]}$$

$$= \mathbf{d}g(x, y).\mathbf{d}x + \mathbf{d}h(x, y).\mathbf{d}y \qquad \text{[Axiom III]}$$

$= (\partial_x g(x, y)\mathbf{d}x + \partial_y g(x, y)\mathbf{d}y).\mathbf{d}x + (\partial_x h(x, y)\mathbf{d}x + \partial_y h(x, y)\mathbf{d}y).\mathbf{d}y$

$= \partial_x g(x, y)\mathbf{d}x\mathbf{d}x + \partial_y g(x, y)\mathbf{d}y\mathbf{d}x + \partial_x h(x, y)\mathbf{d}x\mathbf{d}y + \partial_y h(x, y)\mathbf{d}y\mathbf{d}y$

$= \partial_y g(x, y)\mathbf{d}y\mathbf{d}x + \partial_x h(x, y)\mathbf{d}x\mathbf{d}y$ [Axiom V]

$= (\partial_x h(x, y) - \partial_y g(x, y)).\mathbf{d}x\mathbf{d}y$ [Axiom IV].

More formally, axioms IV and V could be replaced by the single extra axiom, that for any two **forms** ω and ϕ of whatever **grade**, including **grade** 0:

Axiom IVb: $\mathbf{d}(\omega \wedge \phi) = \mathbf{d}\omega \wedge \phi + (-1)^p \omega \wedge \mathbf{d}\phi$

where p is the **grade** of ω. Axiom III remains. So, for example, where $g(x, y)$ and $h(x, y)$ are **0-forms** giving $p = 0$[87]:

$\mathbf{d}(g(x, y).\mathbf{d}x + h(x, y).\mathbf{d}y)$

$= \mathbf{d}g(x, y) \wedge \mathbf{d}x + g(x, y) \wedge \mathbf{dd}x + \mathbf{d}h(x, y) \wedge \mathbf{d}y + h(x, y) \wedge \mathbf{dd}y$

$= \mathbf{d}g(x, y) \wedge \mathbf{d}x + \mathbf{d}h(x, y) \wedge \mathbf{d}y$

because $\mathbf{dd}x = \mathbf{dd}y = 0$.

Either set of axioms justifies the argument that **d** effectively operates only on the **scalar coefficient** functions (g and h above) and $\mathbf{d}(\mathbf{d}x) = \mathbf{d}(\mathbf{d}x \wedge \mathbf{d}y) = 0$ for any **differential** or product (⋀) of **differentials**.

This alternative axiom IVb is most useful in more subtle cases, as we shall see in Chapter 7.

Now, given that **alternating product**, we can refer back to the **basis vectors** of our spaces of **differentials** to see that we now must have:

In the plane:

grade 1: (2 **basis vectors**): $\mathbf{d}x$, $\mathbf{d}y$
grade 2: (1 **basis bivector** ≡ the unique **pseudoscalar**): $\mathbf{d}x \wedge \mathbf{d}y$.

In three-dimensional space:

grade 1: (3 **basis vectors**) $\mathbf{d}x$, $\mathbf{d}y$, $\mathbf{d}z$
grade 2: (3 **basis bivectors**) $\mathbf{d}x \wedge \mathbf{d}y$, $\mathbf{d}y \wedge \mathbf{d}z$, $\mathbf{d}z \wedge \mathbf{d}x$
grade 3: (1 **basis trivector** ≡ the unique **pseudoscalar**): $\mathbf{d}x \wedge \mathbf{d}y \wedge \mathbf{d}z$.

[87] We assume that the **alternating product** reduces to ordinary multiplication between **0-forms**.

These products are often written *assuming* the "$\wedge$" sign, so that, for example, henceforth, we can *define*:

$$\mathbf{d}x\mathbf{d}y := \mathbf{d}x \wedge \mathbf{d}y.$$

The definition of **d** acting on a **1-form** in *xyz*-space is:

$$\mathbf{d}(f(x, y, z)\mathbf{d}x + g(x, y, z)\mathbf{d}y + h(x, y, z)\mathbf{d}z) :=$$

$$\mathbf{d}f(x, y, z) \wedge \mathbf{d}x + \mathbf{d}g(x, y, z) \wedge \mathbf{d}y + \mathbf{d}h(x, y, z) \wedge \mathbf{d}z$$

$$= (\partial_x f(x, y, z).\mathbf{d}x + \partial_y f(x, y, z).\mathbf{d}y + \partial_z f(x, y, z).\mathbf{d}z) \wedge \mathbf{d}x + (\partial_x g(x, y, z).\mathbf{d}x + \partial_y g(x, y, z).\mathbf{d}y + \partial_z g(x, y, z).\mathbf{d}z) \wedge \mathbf{d}y + (\partial_x h(x, y, z).\mathbf{d}x + \partial_y h(x, y, z).\mathbf{d}y + \partial_z h(x, y, z).\mathbf{d}z) \wedge \mathbf{d}z$$

and by Axiom V all the symmetrical products drop out, and we get:

$$= \partial_y f(x, y, z).(\mathbf{d}y \wedge \mathbf{d}x) + \partial_z f(x, y, z).(\mathbf{d}z \wedge \mathbf{d}x) + \partial_x g(x, y, z).(\mathbf{d}x \wedge \mathbf{d}y) + \partial_z g(x, y, z).(\mathbf{d}z \wedge \mathbf{d}y) + \partial_x h(x, y, z).(\mathbf{d}x \wedge \mathbf{d}z) + \partial_y h(x, y, z).(\mathbf{d}y \wedge \mathbf{d}z)$$

and by Axiom IV:

$$= (\partial_x g(x, y, z) - \partial_y f(x, y, z)).(\mathbf{d}x \wedge \mathbf{d}y) + (\partial_y h(x, y, z) - \partial_z g(x, y, z)).(\mathbf{d}y \wedge \mathbf{d}z) + (\partial_z f(x, y, z) - \partial_x h(x, y, z)).(\mathbf{d}z \wedge \mathbf{d}x).$$

If $f(x, y, z) \equiv \partial_x\phi(x, y, z)$, $g(x, y, z) \equiv \partial_y\phi(x, y, z)$, and $h(x, y, z) \equiv \partial_z\phi(x, y, z)$, so that the original **1-form** was **exact**:

$$f.\mathbf{d}x + g.\mathbf{d}y + h.\mathbf{d}z = \partial_x\phi.\mathbf{d}x + \partial_y\phi.\mathbf{d}y + \partial_z\phi.\mathbf{d}z = \mathbf{d}\phi$$

then:

$$\mathbf{d}(f.\mathbf{d}x + g.\mathbf{d}y + h.\mathbf{d}z) = \mathbf{d}(\partial_x\phi.\mathbf{d}x + \partial_y\phi.\mathbf{d}y + \partial_z\phi.\mathbf{d}z)$$

$$= (\partial_x g - \partial_y f).\mathbf{d}x \wedge \mathbf{d}y + (\partial_y h - \partial_z g).\mathbf{d}y \wedge \mathbf{d}z + (\partial_z f - \partial_x h).\mathbf{d}z \wedge \mathbf{d}x$$

$$= (\partial_x\partial_y\phi - \partial_y\partial_x\phi).\mathbf{d}x \wedge \mathbf{d}y + (\partial_y\partial_z\phi - \partial_z\partial_y\phi).\mathbf{d}y \wedge \mathbf{d}z + (\partial_z\partial_x\phi - \partial_x\partial_z\phi).\mathbf{d}z \wedge \mathbf{d}x$$

and by **Young's theorem**,

$$= (0).\mathbf{d}x \wedge \mathbf{d}y + (0).\mathbf{d}y \wedge \mathbf{d}z + (0).\mathbf{d}z \wedge \mathbf{d}x = \mathbf{0}.$$

In other words, **d(d**ϕ**)** = **dd**ϕ = **0** in accordance with Axiom III. (We need Axiom III as an axiom in order to block the expansion of the **differentials** themselves which we might be tempted to do from Axiom II in Chapter 2. But Axiom III means these will all give results like **d(d***x*) = **0**.)

> This result is the key reason for using an **alternating product** for the higher **grade differential forms**. A fuller treatment of this question is given in Appendix 6.D.

So this interpretation sees the **differential operator** or **exterior derivative d** as:

1. Operating on a **scalar function** *f* to create a **vector** object, a **1-form**, that operates as a **function** on **1-simplexes**, and
2. Operating on **1-forms** to create **bivector** objects or **2-forms** that operate as **functions** on **2-simplexes**.

The logic can be continued. A **3-form** in *xyz*-space takes the form:

$$\phi(x, y, z).\mathbf{d}x\wedge\mathbf{d}y\wedge\mathbf{d}z$$

as there is only a single *unique* **trivector** in a **3-vector graded algebra**.

Consider again a transformation of **coordinates** in the plane where:

$$x = x(\xi, \psi)$$
$$y = y(\xi, \psi)$$

the transformation of the **differentials** at any given point is still linear:

$$\mathbf{d}x = \partial_\xi x(\xi, \psi).\mathbf{d}\xi + \partial_\psi x(\xi, \psi).\mathbf{d}\psi$$
$$\mathbf{d}y = \partial_\xi y(\xi, \psi).\mathbf{d}\xi + \partial_\psi y(\xi, \psi).\mathbf{d}\psi$$

This gives:

$$\mathbf{d}x \wedge \mathbf{d}y = \partial_\xi x(\xi, \psi)\, \partial_\psi y(\xi, \psi).(\mathbf{d}\xi \wedge \mathbf{d}\psi) + \partial_\psi x(\xi, \psi)\, \partial_\xi y(\xi, \psi).(\mathbf{d}\psi \wedge \mathbf{d}\xi)$$

$$= (\partial_\xi x(\xi, \psi)\, \partial_\psi y(\xi, \psi) - \partial_\psi x(\xi, \psi)\, \partial_\xi y(\xi, \psi)).(\mathbf{d}\xi \wedge \mathbf{d}\psi).$$

It can be shown that substituting this expansion for **d***x***d***y* in the original **integral**, and evaluating the result over the **pullback** of the original **domain of integration**, gives the same result as **integrating** directly in **d***x***d***y*. In other words, if the **functions** above:

$$x = x(\xi, \psi)$$
$$y = y(\xi, \psi)$$

are written as $(x, y) = T(\xi, \psi)$, then:

$\phi(x, y)\,\mathbf{d}x\mathbf{d}y \mid \Omega$

$= \phi(x(\xi, \psi), x(\xi, \psi))\,(\partial_\xi x(\xi, \psi)\,\partial_\psi y(\xi, \psi) - \partial_\psi x(\xi, \psi)\,\partial_\xi y(\xi, \psi)).\mathbf{d}\xi\mathbf{d}\psi \mid (T)^{-1}(\Omega)$.

A proof of this highly important result, in a slightly broader context, is outlined in the next section, together with some discussion of its significance.

The more rigorous definition of **differentials** introduced in this Section enables me to highlight an interpretation of Axiom III that will be central to subsequent work.

I'll first formally extend the definition of an **exact differential form** to **forms** of arbitrary **grade**.

A **differential form** of **grade** *n* (an **n-form**) ω is **exact** if there exists an **(*n* − 1)-form** ϕ such that:

$\omega = \mathbf{d}\phi$.

Here a **0-form** is a **scalar function** like $g(x, y, z)$.

I will at this point also introduce the term **closed**: an **n-form** ω is **closed** if **d**ω = **0**.

By Axiom III in Section 6.6, an **exact form** is always **closed**, because if ω = **d**ϕ, then **d**ω = **dd**ϕ = **0**. This interpretation of Axiom III is known as **Poincaré's Lemma**.

The *converse*, that a **closed form** is always **exact**, is more difficult to demonstrate, but it is so within certain rather abstract constraints: I will refer to it as the **Poincaré Lemma Converse**. It will prove to be a very useful guide in the rest of our analysis.

In summary, for ω an **n-form**, $n > 0$, and using the logical implication symbol "→":

- **Poincaré Lemma**: An **exact form** is **closed**: ω = **d**ϕ → **d**ω = **dd**ϕ = 0.
- **Poincaré Lemma Converse**: A **closed form** is *locally* **exact**:
  dω = 0 → ω = **d**ϕ for some **(n−1)-form** ϕ.

Also two words of warning:

- The "**Converse**" breaks down in certain singular cases, so treat it more as a *guide*.
- Some authors refer to the "**Converse**" above as the **Poincaré Lemma**, leaving the assertion that an **exact form** is **closed** as just being a trivial consequence of axiom III.

We can use the **Poincaré Lemma Converse** to give a proof of the assertion at the end of Section 6.2 that a **1-form** in a single variable is always **exact**, because if $\omega = f(x)\mathbf{d}x$, then $\mathbf{d}\omega = \partial f(x)(\mathbf{d}x \wedge \mathbf{d}x) = \mathbf{0}$, so ω is **closed**, so it must be **exact**: there must be a function $g(x)$ such that $\omega = \mathbf{d}g(x) = f(x)\mathbf{d}x$.

Now this result for a **1-form** in a single variable is all we need to show how we can *always* **integrate** an **n-form** in *n*-space, that is to say a **form** on the **pseudoscalar**. For we can always find the **antiderivative** *for any one variable* in this case, and all these *n* **antiderivatives** will all give the **pseudoscalar form** on **differentiation**. An example should make the point clear. Let's take a very simple **integral**:

$x^2y.\mathbf{d}x\mathbf{d}y \mid \mathbf{p}_0\mathbf{p}_1\mathbf{p}_2$

where $\mathbf{p}_0\mathbf{p}_1\mathbf{p}_2$ is the **standard 2-simplex** in the *xy*-plane with $\mathbf{p}_0$ the **origin** (0, 0), $\mathbf{p}_1$ = (1, 0), and $\mathbf{p}_2$ = (0, 1), as shown in *Figure 6.6.2*. The three legs of the **simplex** are defined by:

- $\mathbf{p}_0\mathbf{p}_1$: x runs from 0 to 1, $y = 0$.
- $\mathbf{p}_1\mathbf{p}_2$: $x + y = 1$, both vary, inversely to each other, from 0 to 1.
- $\mathbf{p}_2\mathbf{p}_1$: y runs from 1 to 0, $x = 0$.

Now in 2-space, **d**x**d**y is the **pseudoscalar**, so this is a **form** on the **pseudoscalar**.

With two variables, we have two possible **antiderivatives**.

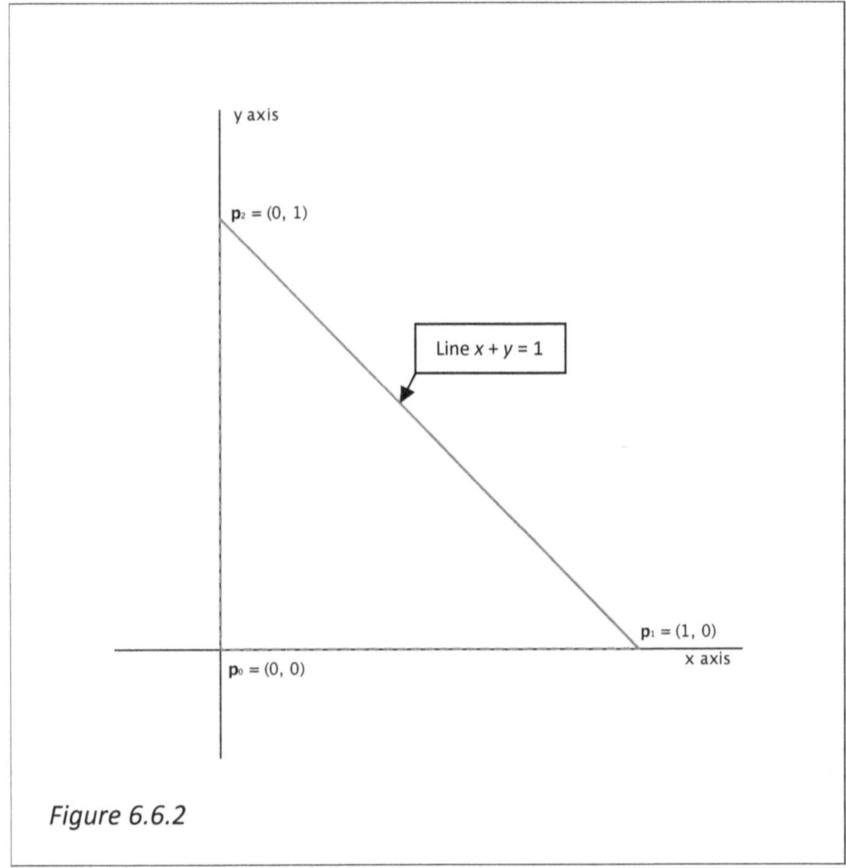

Figure 6.6.2

The **antiderivative** for x is:

$\tfrac{1}{3}x^3y.\mathbf{d}y$

because

$\mathbf{d}(\tfrac{1}{3}x^3y.\mathbf{d}y) = 3.\tfrac{1}{3}.x^2y.\mathbf{d}x\mathbf{d}y + \tfrac{1}{3}x^3.\mathbf{d}y\mathbf{d}y = x^2y.\mathbf{d}x\mathbf{d}y$

and the **antiderivative** for y is $-\frac{1}{2}x^2y^2.\mathbf{d}x$ because:

$$\mathbf{d}(-\tfrac{1}{2}x^2y^2.\mathbf{d}x) = -\tfrac{1}{2}.2.xy^2.\mathbf{d}x\mathbf{d}x - \tfrac{1}{2}.2.x^2y.\mathbf{d}y\mathbf{d}x = +\, x^2y.\mathbf{d}x\mathbf{d}y.$$

So *both* these **1-forms**, writing them as $\phi^x = \frac{1}{3}x^3y.\mathbf{d}y$ and $\phi^y = -\frac{1}{2}x^2y^2.\mathbf{d}x$, make the original **2-form exact** because:

$$\mathbf{d}\phi^x = \mathbf{d}\phi^y = x^2y.\mathbf{d}x\mathbf{d}y.$$

A remarkable theorem called **Fubini's theorem**, which I will not prove, states that we can evaluate the **integral** using *either* **antiderivative** and we get the same result barring a possible difference in sign.[88] In this very simple example:

$$x^2y.\mathbf{d}x\mathbf{d}y \,|\, \mathbf{p}_0\mathbf{p}_1\mathbf{p}_2 = \mathbf{d}(\tfrac{1}{3}x^3y.\mathbf{d}y) \,|\, \mathbf{p}_0\mathbf{p}_1\mathbf{p}_2 = \tfrac{1}{3}x^3y.\mathbf{d}y \,|\, \mathbf{b}(\mathbf{p}_0\mathbf{p}_1\mathbf{p}_2)$$

$$= \tfrac{1}{3}x^3y.\mathbf{d}y \,|\, (\mathbf{p}_0\mathbf{p}_1 + \mathbf{p}_1\mathbf{p}_2 + \mathbf{p}_2\mathbf{p}_0).$$

Along $\mathbf{p}_0\mathbf{p}_1$, $y = 0$, so $0.\mathbf{d}y \,|\, \mathbf{p}_0\mathbf{p}_1 = 0$, along $\mathbf{p}_2\mathbf{p}_0$, $x = 0$, so again $0.\mathbf{d}y \,|\, \mathbf{p}_2\mathbf{p}_0 = 0$, and along $\mathbf{p}_1\mathbf{p}_2$, we can choose to **parameterize** the **line segment** using y, with $x = 1 - y$ and y running from 0 to 1, so the **pullback** onto y gives:

$$\tfrac{1}{3}(1-y)^3y.\mathbf{d}y \,|\, [0, 1] = \tfrac{1}{3}\{y - 3y^2 + 3y^3 + y^4\}.\mathbf{d}y \,|\, [0, 1]$$

$$= \tfrac{1}{3}\mathbf{d}\{\tfrac{1}{2}y^2 - y^3 + \tfrac{3}{4}y^4 + y^5/5\} \,|\, [0, 1]$$

$$= \tfrac{1}{3}(\tfrac{1}{2} - 1 + \tfrac{3}{4} + 1/5) - \tfrac{1}{3}(0) = -1/60.$$

Doing it the other way:

$$x^2y.\mathbf{d}x\mathbf{d}y \,|\, \mathbf{p}_0\mathbf{p}_1\mathbf{p}_2 = \mathbf{d}(-\tfrac{1}{2}x^2y^2.\mathbf{d}x) \,|\, \mathbf{p}_0\mathbf{p}_1\mathbf{p}_2 = -\tfrac{1}{2}x^2y^2.\mathbf{d}x \,|\, \mathbf{b}(\mathbf{p}_0\mathbf{p}_1\mathbf{p}_2)$$

$$= -\tfrac{1}{2}x^2y^2.\mathbf{d}x \,|\, (\mathbf{p}_0\mathbf{p}_1 + \mathbf{p}_1\mathbf{p}_2 + \mathbf{p}_2\mathbf{p}_0).$$

Again, along $\mathbf{p}_0\mathbf{p}_1$ and $\mathbf{p}_2\mathbf{p}_0$ either $x = 0$ or $y = 0$ so these do not contribute, and **parameterizing** $\mathbf{p}_1\mathbf{p}_2$ with x, where $y = 1 - x$, and x runs *back* from 1 to 0, we have for the **pullback** onto x:

$$-\tfrac{1}{2}x^2(1-x)^2.\mathbf{d}x \,|\, [1, 0] = -\tfrac{1}{2}(x^2 - 2x^3 + x^4).\mathbf{d}x \,|\, [1, 0]$$

$$= -\tfrac{1}{2}\mathbf{d}(\tfrac{1}{3}x^3 - \tfrac{1}{2}x^4 + x^5/5) \,|\, [1, 0]$$

[88] Since *both* the factors in the **integral product** are **alternating** or **anticommutative**, changing the order of the **differentials** in the **differential form** *or* changing the order of the points in the **simplex** will change the sign of an **integral**, but these differences are usually disregarded.

$$= -\tfrac{1}{2}(0) + \tfrac{1}{2}(\tfrac{1}{3} - \tfrac{1}{2} + 1/5) = \tfrac{1}{2}(1/30) = 1/60.$$

This device of using the **antiderivative** for one chosen variable in the **form** and using the other variables to **parameterize** the **boundary** space is called **partial integration**. The sign difference here comes from $\mathbf{d}(-\tfrac{1}{2}x^2y^2.\mathbf{d}x) = -\,x^2y.\mathbf{d}y\mathbf{d}x = +\,x^2y.\mathbf{d}x\mathbf{d}y$ where I have explicitly taken account of $\mathbf{d}y\mathbf{d}x = -\,\mathbf{d}x\mathbf{d}y$, but in conventional presentations this is commonly ignored.

6.7 Surface Integrals

A **surface integral** is an **integral** of a **2-form** in 3-space or *xyz*-space. Evaluating these encounters exactly the same sort of problem that we had in Section 6.2 in evaluating a **1-form** in 2-space (the plane) or, indeed, in 3-space. For, in general, an arbitrary **2-form** in 3-space is *not* **exact**: that is to say, that given an arbitrary **2-form** like:

$$f(x, y, z).\mathbf{d}y\mathbf{d}z + g(x, y, z).\mathbf{d}z\mathbf{d}x + h(x, y, z).\mathbf{d}x\mathbf{d}y$$

there is usually *no* **1-form** ω such that $\mathbf{d}\omega = f(x, y, z).\mathbf{d}y\mathbf{d}z + g(x, y, z).\mathbf{d}z\mathbf{d}x + h(x, y, z).\mathbf{d}x\mathbf{d}y$. This means we cannot use the **adjoint definition** for **integration**:

$$\mathbf{d}\omega \,|\, \Sigma = \omega \,|\, \mathbf{b}\Sigma$$

because what we've got on the left of the **integral operator** "|" is *not* in the form **d**ω.

We circumvent this in a way that is directly analogous to what was done in Section 6.2: by using a **mapping** onto a space where the **pullback** of our original **2-form** *is* **exact**. The method used in this Section can lead to a proof that products of **differentials** must be **alternating**, or that $\mathbf{d}y\mathbf{d}x = -\,\mathbf{d}x\mathbf{d}y$, and this also automatically gives the correct rule for the transformation of variables in **differential forms** of higher **grade**. So although this Section appears to deal with a special case, it has huge general importance. An outline of how these proofs proceed is given in Appendix 6.C.

Just as a set of **parametric functions** of a *single* **parameter** t taking the form $x(t)$, $y(t)$, $z(t)$ can describe a *curve* in *xyz*-space, so a set of **parametric functions** of *two* **parameters** u, v of the form $x(u,v)$, $y(u,v)$, $z(u,v)$ can describe a *surface* in *xyz*-space.[89] I may refer to the particular *surface* so defined as Σ for Σurface just as I referred to the particular *curve* τ in Section 6.2, so that by analogy with that Section I may write $\Sigma = [x = x(u, v), y = y(u, v), z = z(u, v)]$. It will be reasonable to assume that the **domain** of Σ is a **chain** of **2-simplexes** or triangles in the *uv*-plane, just as in Section 6.2 we assumed the **domain** of

[89] Again I freely use the form $x = x(u, v)$ where the variable for the *value* of the **function** and the **function** itself are denoted by the same letter. As I mentioned in an earlier footnote, this "bastard" notation is much disparaged, but both human intelligence and computer programmes are quite capable of **context-sensitive parsing**, and the intuitive and associative advantages of this "bastard" notation are obvious. In much the same way, I may refer to *either* the **mapping** or to its **range** as the "surface".

the *curve* τ to be an **interval** in t-space (on the t **axis**). Most commonly, this **domain** will be a rectangle $[u_0, u_1] \times_c [v_0, v_1]$ which we can represent as the sum of just two **2-simplexes** as in Section 6.5. As in *Figure 6.5.1*, we can call these $\mathbf{p}_0\mathbf{p}_1\mathbf{p}_2 + \mathbf{p}_0\mathbf{p}_2\mathbf{p}_3$.

When I need to make it quite clear which "end" of the **mapping** we're dealing with, I'll write Σ_{uv} for the **domain** in uv-space, and the resulting "actual" surface in xyz-space I'll call Σ_{xyz}. So, describing the **function** Σ in *aggregate* as in Section 1.1, Σ **maps**:

$$\Sigma\colon \Sigma_{uv} \longrightarrow \Sigma_{xyz} \qquad \text{or} \qquad \Sigma_{xyz} = \Sigma(\Sigma_{uv}).$$

If the **2-form**, which I'll call ψ, is **exact** it must **derive** from a **1-form** like $\omega = \alpha(x, y, z).\mathbf{d}x + \beta(x, y, z).\mathbf{d}y + \gamma(x, y, z).\mathbf{d}z$, and it will then take the form, as we saw in the last Section using our extended **exterior derivative** axioms:

$$\psi = \mathbf{d}\omega = \mathbf{d}(\alpha(x, y, z).\mathbf{d}x + \beta(x, y, z).\mathbf{d}y + \gamma(x, y, z).\mathbf{d}z)$$

$$= (\partial_x\beta(x, y, z) - \partial_y\alpha(x, y, z)).(\mathbf{d}x \wedge \mathbf{d}y) + (\partial_y\gamma(x, y, z) - \partial_z\beta(x, y, z)).(\mathbf{d}y \wedge \mathbf{d}z) +$$

$$(\partial_z\alpha(x, y, z) - \partial_x\gamma(x, y, z)).(\mathbf{d}z \wedge \mathbf{d}x).$$

If this is the case then our original **integral** is well defined by the usual **adjoint** switch:

$$\psi \mid \Sigma_{xyz} = \mathbf{d}\omega \mid \Sigma_{xyz} = \omega \mid \mathbf{b}\Sigma_{xyz} = \omega \mid \Sigma(\mathbf{b}\Sigma_{uv})$$

and, for our standard rectangular **domain**, this is now a **line integral** along the four **line segments** given by:

$$\mathbf{b}\Sigma_{uv} = \mathbf{b}(\mathbf{p}_0\mathbf{p}_1\mathbf{p}_2 + \mathbf{p}_0\mathbf{p}_2\mathbf{p}_3) = \mathbf{p}_1\mathbf{p}_2 - \mathbf{p}_0\mathbf{p}_2 + \mathbf{p}_0\mathbf{p}_1 + \mathbf{p}_0\mathbf{p}_2 - \mathbf{p}_0\mathbf{p}_3 + \mathbf{p}_2\mathbf{p}_3$$

$$= \mathbf{p}_1\mathbf{p}_2 + \mathbf{p}_0\mathbf{p}_1 - \mathbf{p}_0\mathbf{p}_3 + \mathbf{p}_2\mathbf{p}_3$$

$$= \mathbf{p}_0\mathbf{p}_1 + \mathbf{p}_1\mathbf{p}_2 + \mathbf{p}_2\mathbf{p}_3 + \mathbf{p}_3\mathbf{p}_0.$$

From *Figure 6.5.1*, you can clearly see that this **chain** describes the outside of the rectangle bounding the **domain** in uv-space.

We can now evaluate the **integral** as the **line integral**:

$$\omega \mid \Sigma(\mathbf{b}\Sigma_{uv}) = \omega_{uv} \mid \mathbf{b}\Sigma_{uv} = \omega_{uv} \mid (\mathbf{p}_0\mathbf{p}_1 + \mathbf{p}_1\mathbf{p}_2 + \mathbf{p}_2\mathbf{p}_3 + \mathbf{p}_3\mathbf{p}_0)$$

wherein we evaluate ω_{uv} from ω by substituting for $\Sigma = [x = x(u, v), y = y(u, v), z = z(u, v)]$, and for:

$$\mathbf{d}x = \partial_u x(u, v)\mathbf{d}u + \partial_v x(u, v)\mathbf{d}v$$
$$\mathbf{d}y = \partial_u y(u, v)\mathbf{d}u + \partial_v y(u, v)\mathbf{d}v$$
$$\mathbf{d}z = \partial_u z(u, v)\mathbf{d}u + \partial_v z(u, v)\mathbf{d}v$$

to give:

$$\omega_{uv} = \alpha(x(u, v), y(u, v), z(u, v))(\partial_u x(u, v)\mathbf{d}u + \partial_v x(u, v)\mathbf{d}v) + \beta(x(u, v), y(u, v), z(u, v))(\partial_u y(u, v)\mathbf{d}u + \partial_v y(u, v)\mathbf{d}v) + \gamma(x(u, v), y(u, v), z(u, v))(\partial_u z(u, v)\mathbf{d}u + \partial_v z(u, v)\mathbf{d}v).$$

Obviously most **2-forms** will not conform to this and will not be **exact**, that is to say they will be **inexact**. To handle this case, we proceed by analogy to the way we defined the **line integral**, making the switch to *uv*-space *this time at the* **2-form** *level* to obtain a **2-form** in **d***u***d***v* *directly*. We use the same device of substituting for **d***x*, **d***y* and **d***z* as we did above, *but now we have to take into account the* **alternating** *nature of the products between the* **differentials**. So, for example, the original $f(x, y, z).\mathbf{d}y\mathbf{d}z$ term now expands to:

$$f(x, y, z).\mathbf{d}y\mathbf{d}z = f(x(u, v), y(u, v), z(u, v)).[(\partial_u y(u, v)\mathbf{d}u + \partial_v y(u, v)\mathbf{d}v) \wedge (\partial_u z(u, v)\mathbf{d}u + \partial_v z(u, v)\mathbf{d}v)]$$

$$= f(x(u,v), y(u, v), z(u, v))[\, \partial_u y(u, v).\partial_v z(u, v)\mathbf{d}u\mathbf{d}v + \partial_v y(u, v).\partial_u z(u, v)\mathbf{d}v\mathbf{d}u]$$

$$= f(x(u,v), y(u, v), z(u, v))[\partial_u y(u, v).\partial_v z(u, v) - \partial_v y(u, v).\partial_u z(u, v)].\mathbf{d}u\mathbf{d}v.$$

The *g* and *h* terms give similarly complex results, and all told we end up with a new **2-form** ψ_{uv} in **arguments** *u* and *v* defined in terms of **d***u***d***v*:

$$\psi_{uv} = \phi(u, v).\mathbf{d}u\mathbf{d}v.$$

So we now have a transformed **integral**:

$$\psi \,|\, \Sigma_{xyz} = \psi_{uv} \,|\, \Sigma_{uv} = \phi(u, v).\mathbf{d}u\mathbf{d}v \;|\, (\,\mathbf{p}_0\mathbf{p}_1\mathbf{p}_2 + \mathbf{p}_0\mathbf{p}_2\mathbf{p}_3).$$

This we *can* evaluate. Why?

Because on a two-dimensional **pullback** space where **d***u* ∧ **d***v* will be the **pseudoscalar**, *any* **2-form** $\phi(u, v)\mathbf{d}u\mathbf{d}v$ *will in general* be **exact** by the **Poincaré Lemma Converse**, that if a **form** ω is **closed** i.e. **d**ω = 0, it is **exact**.

In the present case, $\mathbf{d}(\phi(u, v)\mathbf{d}u\mathbf{d}v) = (\phi_u.\mathbf{d}u + \phi_v.\mathbf{d}v).\mathbf{d}u\mathbf{d}v = 0$, as this expression only involves **d***u***d***u***d***v* = **d***v***d***u***d***v* = 0, so $\phi(u, v)\mathbf{d}u\mathbf{d}v$ *will* be **exact** and we have a validly defined **integral** in the **pullback** space on **d***u***d***v*.

This gives us a general definition of a **surface integral**.

Look back at Section 6.2 to see the logic there repeating, but now one **grade** higher. In Section 6.2, an **integral** in the **1-form** $\psi = f(x, y)\mathbf{d}x + g(x, y)\mathbf{d}y$ was undefined unless ψ happened to be **exact**, but **pulling back** using a **mapping** τ onto a one-dimensional *t*-space with **pseudoscalar d***t* gave a **form** ψ_τ of the form $p(t)\mathbf{d}t$ on an **interval** in *t* which *was* **exact** because $\mathbf{d}(p(t)\mathbf{d}t) = \partial_t p(t).\mathbf{d}t\mathbf{d}t = 0$. So $\mathbf{d}\psi_\tau = 0$, so ψ_τ is **exact**.

This pattern repeats over and over again in the development of **integration**.

> Here, like the **line integrals** in Section 6.2, the resulting **surface integral** depends on the particular **parameterization** Σ chosen to define the **pullback**, and so **surface integrals** are not independent of Σ unless the original **2-form** ψ in *xyz*-space was itself **exact**.

Stokes's Theorem, to be covered in the next Section, is the statement that **surface integrals** are only independent of the actual surface Σ if the original **2-form** ψ is **exact**. If it is, then the **surface integral** can be obtained directly from the **adjoint definition** *and depends only on the* **boundary** *of* Σ:

If $\psi = \mathbf{d}\omega$, then $\psi \,|\, \boldsymbol{\Sigma}$ over *any* surface $\boldsymbol{\Sigma}$ with the same **boundary** $\mathbf{b}\boldsymbol{\Sigma}$ can be evaluated as:

$$\psi \,|\, \boldsymbol{\Sigma} = \mathbf{d}\omega \,|\, \boldsymbol{\Sigma} = \omega \,|\, \mathbf{b}\boldsymbol{\Sigma}.$$

This is exactly analogous to the **line integral** case where the **integral** depends only on the **end points**, which *are* the **boundary** of a curve, if the **1-form** is **exact**.

It is precisely in developing the direct transformation of the **2-form**:

$$\psi = f(x, y, z).\mathbf{d}y\mathbf{d}z + g(x, y, z).\mathbf{d}z\mathbf{d}x + h(x, y, z).\mathbf{d}x\mathbf{d}y$$

into

$$\psi_\tau = \phi(u, v)\mathbf{d}u\mathbf{d}v$$

that the need to use an **alternating product** between **differentials** becomes apparent. An intuitive presentation of why this should be so is given in Appendix 6.D.

6.8 Integral Theorems

In the textbooks, you will find mention of three special theorems relating to **integrals** of the kinds we have been discussing so far: **Green's Theorem**, **Stokes's Theorem** and the **Divergence Theorem**. I will discuss these briefly now.

In the present formulation **Green's Theorem** and the **Divergence Theorem** are hardly theorems at all, but are simply the statement of our **adjoint definition** of **integration** for **manifolds** in, respectively, 2-space and 3-space. **Stokes's theorem**, as was mentioned in the last Section, is the statement that the **surface integral** of an **exact 2-form** in 3-space depends only on the **boundary** of the **manifold** over which the **integral** is defined.

Therefore much of this Section will revise ideas we have already looked at, whilst adding a little detail.

All three of our **integral theorems** take much the same form, the form familiar from our original definition of **integration**, that if ω is an **exact n-form** and Σ is an **n-manifold** (where a curve in space is a **1-manifold**, a surface a **2-manifold** and so on), then:

$$\omega \mid \Sigma = \mathbf{d}\phi \mid \Sigma = \phi \mid \mathbf{b}\Sigma.$$

In three dimensional space, an **exact 2-form** has to be very specific in its arrangement. It must derive from applying the **differential operator** to a **1-form** ϕ,

and if $\omega = \mathbf{d}\phi = \mathbf{d}(f(x, y, z)\mathbf{d}x + g(x, y, z)\mathbf{d}y + h(x, y, z)\mathbf{d}z)$, then:

$$\omega = \mathbf{d}(f(x, y, z)) \wedge \mathbf{d}x + \mathbf{d}(g(x, y, z)) \wedge \mathbf{d}x + \mathbf{d}(h(x, y, z)) \wedge \mathbf{d}z$$

$$= (\partial_x f(x, y, z)\mathbf{d}x + \partial_y f(x, y, z)\mathbf{d}y + \partial_z f(x, y, z)\mathbf{d}z) \wedge \mathbf{d}x + (\partial_x g(x, y, z)\mathbf{d}x + \partial_y g(x, y, z)\mathbf{d}y + \partial_z g(x, y, z)\mathbf{d}z) \wedge \mathbf{d}y + (\partial_x h(x, y, z)\mathbf{d}x + \partial_y h(x, y, z)\mathbf{d}y + \partial_z h(x, y, z)\mathbf{d}z) \wedge \mathbf{d}z$$

$$= (\partial_x g(x, y, z) - \partial_y f(x, y, z))\mathbf{d}x \wedge \mathbf{d}y + (\partial_y h(x, y, z) - \partial_z g(x, y, z))\mathbf{d}y \wedge \mathbf{d}z + (\partial_z f(x, y, z) - \partial_x h(x, y, z))\mathbf{d}z \wedge \mathbf{d}x.$$

Bivectors that take this form are called the **curl** of the original **vector** $f(x, y, z)\mathbf{d}x + g(x, y, z)\mathbf{d}y + h(x, y, z)\mathbf{d}z$.

They are ordinarily presented as their **vector** analogues, formed from the above using their **Hodge duals**[90] to replace $\mathbf{d}x \wedge \mathbf{d}y$ with $\mathbf{d}z$, $\mathbf{d}y \wedge \mathbf{d}z$ with $\mathbf{d}x$, and $\mathbf{d}z \wedge \mathbf{d}x$ with $\mathbf{d}y$ as in Section 5.9.

So what you will see in the standard books is this form: writing $\mathbf{e}_1$ as **i**, $\mathbf{e}_2$ as **j**, and $\mathbf{e}_3$ as **k**, the standard notation for 3-space, we define:

$$\mathbf{curl}(f(x, y, z)\mathbf{i} + g(x, y, z)\mathbf{j} + h(x, y, z)\mathbf{k}) :=$$

$$(\partial_x g(x, y, z) - \partial_y f(x, y, z))\mathbf{i} + (\partial_y h(x, y, z) - \partial_z g(x, y, z))\mathbf{j} +$$

$$(\partial_z f(x, y, z) - \partial_x h(x, y, z))\mathbf{k}.$$

Note too, in particular, that the **(n−1)-forms** ϕ that give rise to **exact n-forms** ω = **d**ϕ, *are precisely those* **(n−1)-forms** *that are <u>not</u>* **exact** *themselves*. If ϕ were **exact**, then ϕ = **d**ξ for some **(n−2)-form** ξ, and so **d**ϕ = **dd**ξ = **0**.

So **2-forms** in 3-space that do *not* match the form of the **curl vector** in their **coefficient functions** will *not* be **exact** and we will *not* be able to apply our **adjoint definition** of **integration** directly.

[90] In this context this is the relevant term.

But we *can* use our **adjoint definition** to define **integration** *wherever the* **differential form** *is a* **form** *on the* **pseudoscalar** *for the space*, which any **2-form** *will* be if we **pull** it **back** onto a 2-space. Much of the rest of this Section will amplify this point.

Green's theorem is the simplest.

Our **adjoint definition** for a **1-form** ω, whose **derivative** **d**ω is therefore a **2-form**, in the *xy* plane is just:

$$\mathbf{d}\omega \,\big|\, \mathbf{p}_0\mathbf{p}_1\mathbf{p}_2 = \omega \,\big|\, (\mathbf{p}_0\mathbf{p}_1 + \mathbf{p}_1\mathbf{p}_2 + \mathbf{p}_2\mathbf{p}_0).$$

Green's Theorem extends this to an arbitrary **closed bounded region** in the *xy* plane. This can easily be done from the **simplex** definition. *Figure 6.8.1* shows a **mesh** which gives an **adaptive subdivision triangulation** of a circle. The central equilateral triangle disposes of about half the area of the circle, then smaller triangles on its sides take this up to a regular hexagon, then we go to a dodecagon, and so rapidly approach filling the complete circle, which of course we never actually do.

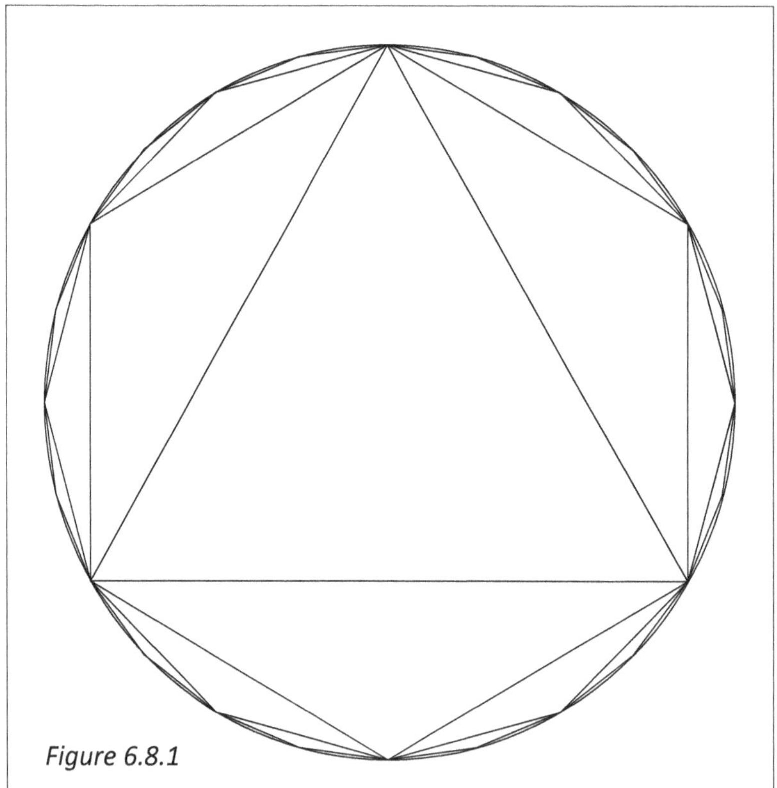

Figure 6.8.1

Now set up the sum of all these triangular **facets**, which I will label $(\mathbf{p}_0\mathbf{p}_1\mathbf{p}_2)_i$ indexed by **facet**, *not* by the points, as the same points (and **edges**) may occur in more than one **facet**. Then the **integral** for the whole circle can be approximated to an arbitrary degree of accuracy by:

$$\mathbf{d}\omega \,\big|\, \Sigma_i (\mathbf{p}_0\mathbf{p}_1\mathbf{p}_2)_i = \omega \,\big|\, \Sigma_i (\mathbf{p}_0\mathbf{p}_1 + \mathbf{p}_1\mathbf{p}_2 + \mathbf{p}_2\mathbf{p}_0)_i.$$

If the **mesh** is **consistently orientated**, then every **shared edge** appears just twice, with opposite signs. So every **shared edge** is cancelled from the sum, and we need only retain those **edges** on the *outside*, the ones that actually do approximate the circle.[91] If we label all the points around the outside, all of which actually lie on the true circle, with the index j = 1 to n, then the **line integral** taken around the outer edge:

$$\omega \,|\, \Sigma_j\, (\mathbf{p}_j\mathbf{p}_{(j+1)}) \qquad \text{where } \mathbf{p}_{(n+1)} = \mathbf{p}_1$$

gives the value of our original **integral**. If we call the circle Ω, then we have:

$$\mathbf{d}\omega \,|\, \Omega = \omega \,|\, \mathbf{b}\Omega,$$

where **b**Ω is the (intuitive) **boundary** of the circle. This is the gist of **Green's Theorem**. Note how the cancellation of **shared edges** reflects the cancellation of intermediate points in our first concept of a **line integral** itself introduced in Section 2.6.

We can *always* use this definition to evaluate an **integral** in the plane. Why? *Because any* **2-form** *φ in the xy plane is* **exact***: there is some ω for which φ =* **d***ω.*

Why again? Because any **2-form** in the xy plane must be a form on **d**x**d**y, and **d**x**d**y *is the* **pseudoscalar** *for* **differential forms** *in the plane.* A **2-form** in the plane doesn't have to be **d**x**d**y itself. It can be something like x^2y.**d**x**d**y or any $g(x, y)$.**d**x**d**y. Any such form is **closed** because:

$$\mathbf{d}(g(x, y).\mathbf{d}x\mathbf{d}y) = (\partial_x g(x, y).\mathbf{d}x + \partial_y g(x, y).\mathbf{d}y).\mathbf{d}x\mathbf{d}y = \mathbf{0}$$

since the first term gives **d**x**d**x, which must be **0**, and the second **d**y**d**x**d**y = − **d**x**d**y**d**y = **0**, and by the **Poincaré Lemma Converse**, it must therefore be **exact.**[92] In practice, what this means is that we can always find an appropriate **antiderivative** by inverting the **differential operator** *for just one variable*, regarding the other as a constant. So, for example, as in Section 6.6 we could take:

$$\omega = x^2y.\mathbf{d}x\mathbf{d}y = \mathbf{d}(⅓.x^3y.\mathbf{d}y)$$

because

$$\mathbf{d}(⅓.x^3y.\mathbf{d}y) = ⅓.\mathbf{d}(x^3y).\mathbf{d}y = (x^2y.\mathbf{d}x).\mathbf{d}y + ⅓.(x^3\mathbf{d}y).\mathbf{d}y$$

and the second of these terms must be zero because **d**y**d**y = 0. So here ϕ = ⅓.x^3y.**d**y is an acceptable **1-form** with ω = **d**ϕ.

One example of outstanding historical importance, as this was the case that gave rise to the entire subject of the **integral calculus**, is where Ψ is a region of the xy plane bounded below by the x **axis** where y = 0, to the left and right by lines $x = a$ and $x = b$, and above by curve $y = y(x)$.

Then the **area integral** is:

[91] This **mesh** is a **strict 2-chain** by the definition in Section 1.13.
[92] With the usual reservations about the **Converse** of course.

$$\mathbf{dxdy} \mid \Psi = \mathbf{d}(y.\mathbf{dx}) \mid \Psi = y.\mathbf{dx} \mid \mathbf{b}\Psi$$

and since this **line integral** is zero along the x **axis** where $y = 0$, and again for $x = a$ and for $x = b$, since $\mathbf{dx} = 0$ for both of those verticals, it reduces to the simple **integral**:

$$y(x).\mathbf{dx} \mid [a, b].$$

Green's theorem is normally presented for a general **1-form** $f(x, y).\mathbf{dx} + g(x, y).\mathbf{dy}$, when it takes the form:

$$\mathbf{d}(f(x, y).\mathbf{dx} + g(x, y).\mathbf{dy}) \mid \Sigma$$

$$= ((\partial_x f(x, y).\mathbf{dx} + \partial_y f(x, y).\mathbf{dy}).\mathbf{dx} + (\partial_x g(x, y).\mathbf{dx} + \partial_y g(x, y).\mathbf{dy}).\mathbf{dy}) \mid \Sigma$$

$$= \partial_y f(x, y).\mathbf{dydx} + \partial_x g(x, y).\mathbf{dxdy}$$

$$= (\partial_x g(x, y) - \partial_y f(x, y)).\mathbf{dxdy} \mid \Sigma$$

$$= (f(x, y).\mathbf{dx} + g(x, y).\mathbf{dy}) \mid \mathbf{b}\Sigma .$$

The actual theorem comes from the last two lines:

Green's Theorem: $(\partial_x g(x, y) - \partial_y f(x, y)).\mathbf{dxdy} \mid \Sigma = (f(x, y).\mathbf{dx} + g(x, y).\mathbf{dy}) \mid \mathbf{b}\Sigma .$

But it is probably true to say that the separate $\mathbf{dx}$ and $\mathbf{dy}$ forms appear more often in practice.

As an example of the sort of problem that commonly appears in texts, and which might at first seem to have little connection with the **simplex** model, consider an **integral** of the dome-shaped function $(1 - (r/R)^3)$ over the area of the circle centred on the **origin** with radius R.
Here r is the distance from the **origin** of any point, so that $r^2 = x^2 + y^2$. Using the arc length **parameterization** of the circle introduced in Section 4.4, and using conventional notation:

$$x = r.\mathbf{cos}(\theta) = r.\boldsymbol{\theta}_x$$
$$y = r.\mathbf{sin}(\theta) = r.\boldsymbol{\theta}_y$$

gives a transformation of any point (x, y) into **polar coordinates** (r, θ). Now calling the circle of radius R: $\mathbf{C}_R$, the **integral** of our dome-shaped **function** is given by:

$$(1 - (r/R)^3).\mathbf{dxdy} \mid \mathbf{C}_R.$$

Polar coordinates are particularly well suited to this sort of problem. From Section 4.4, $\mathbf{d\theta}_x = -\,\boldsymbol{\theta}_y.\mathbf{d}\theta$ and $\mathbf{d\theta}_y = \boldsymbol{\theta}_x.\mathbf{d}\theta$, and these appear in the conventional notation (this being for θ increasing anti-clockwise):

$$\mathbf{d}(\mathbf{cos}(\theta)) = -\,\mathbf{sin}(\theta).\mathbf{d}\theta$$
$$\mathbf{d}(\mathbf{sin}(\theta)) = \mathbf{cos}(\theta).\mathbf{d}\theta$$

so:

$$\mathbf{d}x = \mathbf{cos}(\theta).\mathbf{d}r - r.\mathbf{sin}(\theta).\mathbf{d}\theta$$
$$\mathbf{d}y = \mathbf{sin}(\theta).\mathbf{d}r + r.\mathbf{cos}(\theta).\mathbf{d}\theta$$

giving:

$$\mathbf{d}x \wedge \mathbf{d}y = (\mathbf{cos}(\theta).\mathbf{d}r - r.\mathbf{sin}(\theta).\mathbf{d}\theta) \wedge (\mathbf{sin}(\theta).\mathbf{d}r + r.\mathbf{cos}(\theta).\mathbf{d}\theta)$$

$$= r.\mathbf{cos}^2(\theta).\mathbf{d}r\mathbf{d}\theta - r.\mathbf{sin}^2(\theta).\mathbf{d}\theta\mathbf{d}r \;=\; r.(\mathbf{cos}^2(\theta) + \mathbf{sin}^2(\theta)).\mathbf{d}r\mathbf{d}\theta = r.\mathbf{d}r\mathbf{d}\theta$$

because $\mathbf{cos}^2(\theta) + \mathbf{sin}^2(\theta) = (\boldsymbol{\theta}_x)^2 + (\boldsymbol{\theta}_y)^2 = 1$.

So our **integral** now takes the form:

$$(1 - (r/R)^3).r.\mathbf{d}r\mathbf{d}\theta \,|\, {}^*\mathbf{C}_R,$$

where ${}^*\mathbf{C}_R$ is the **pullback** of $\mathbf{C}_R$ in **polar coordinates**. This is simply a rectangle defined by $r : [0, R]$ and $\theta : [0, 2\pi]$.

As such, this **integral** is easily evaluated as the product:

$$(1 - (r/R)^3).r.\mathbf{d}r \,|\, [0, R] \times \mathbf{d}\theta \,|\, [0, 2\pi] = 2\pi.(r - r^4/R^3).\mathbf{d}r \,|\, [0, R]$$

$$= 2\pi.\mathbf{d}(\tfrac{1}{2}r^2 - r^5/5R^3) \,|\, [0, R] = 2\pi.(\tfrac{1}{2}R^2 - R^2/5) = (6\pi/10)R^2.$$

But we can also do this directly in the **simplex** form, because this rectangle can be set up as two **2-simplexes** as shown in *Figure 6.8.2*, so putting the **integral** in the form:

$$(1 - (r/R)^3).r.\mathbf{d}r\mathbf{d}\theta \,|\, (\mathbf{p}_0\mathbf{p}_1\mathbf{p}_2 + \mathbf{p}_0\mathbf{p}_2\mathbf{p}_3).$$

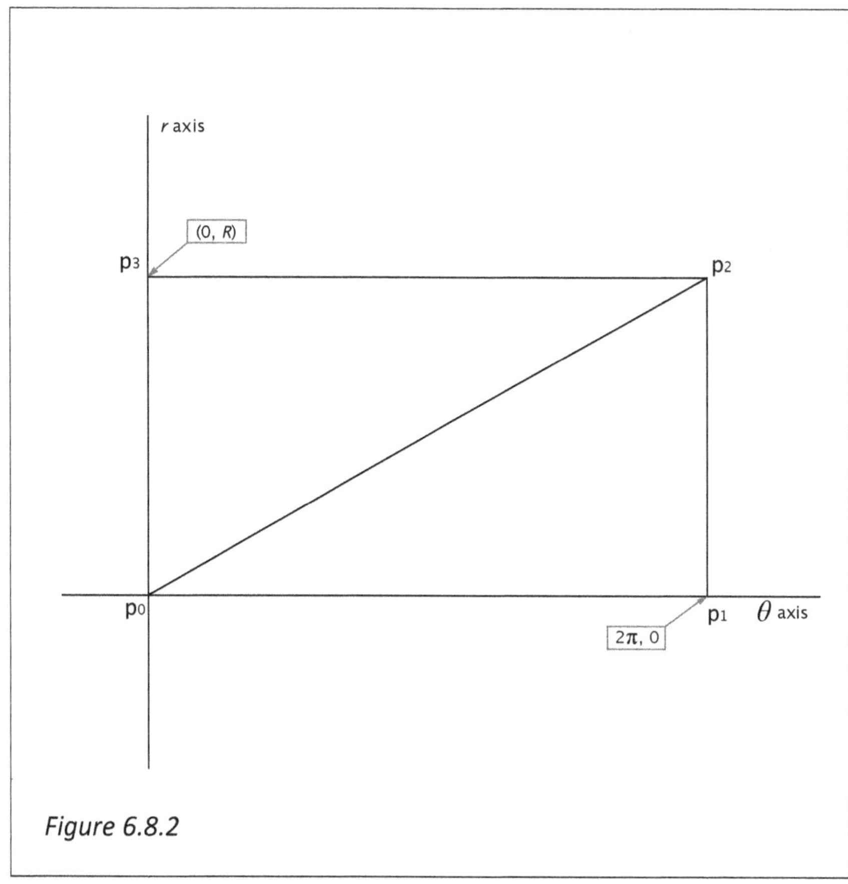

Figure 6.8.2

Again we can find the **antiderivative** in r, since $\mathbf{d}r\mathbf{d}\theta$ is the **pseudoscalar** for the $r\theta$ plane, so getting:

$$\mathbf{d}((\tfrac{1}{2}r^2 - r^5/5R^3).\mathbf{d}\theta) \,|\, (\mathbf{p}_0\mathbf{p}_1\mathbf{p}_2 + \mathbf{p}_0\mathbf{p}_2\mathbf{p}_3)$$
$$= (\tfrac{1}{2}r^2 - r^5/5R^3).\mathbf{d}\theta \,|\, \mathbf{b}(\mathbf{p}_0\mathbf{p}_1\mathbf{p}_2 + \mathbf{p}_0\mathbf{p}_2\mathbf{p}_3)$$
$$= (\tfrac{1}{2}r^2 - r^5/5R^3).\mathbf{d}\theta \,|\, (\mathbf{p}_1\mathbf{p}_2 - \mathbf{p}_0\mathbf{p}_2 + \mathbf{p}_0\mathbf{p}_1 + \mathbf{p}_2\mathbf{p}_3 - \mathbf{p}_0\mathbf{p}_3 + \mathbf{p}_0\mathbf{p}_2)$$
$$= (\tfrac{1}{2}r^2 - r^5/5R^3).\mathbf{d}\theta \,|\, (\mathbf{p}_1\mathbf{p}_2 + \mathbf{p}_0\mathbf{p}_1 + \mathbf{p}_2\mathbf{p}_3 - \mathbf{p}_0\mathbf{p}_3)$$

where the two opposite signed $\mathbf{p}_0\mathbf{p}_2$ contributions cancel. On $\mathbf{p}_1\mathbf{p}_2$ and $\mathbf{p}_0\mathbf{p}_3$, $\mathbf{d}\theta = \mathbf{0}$, so these give no contribution, and on $\mathbf{p}_0\mathbf{p}_1$, $r = 0$, so again this gives zero. So we have just:

$$(\tfrac{1}{2}r^2 - r^5/5R^3).\mathbf{d}\theta \,|\, \mathbf{p}_2\mathbf{p}_3$$

and, **parameterizing** $\mathbf{p}_2\mathbf{p}_3$ with θ, we have $r(\theta) = R$, and since θ runs from 2π to 0 in going from $\mathbf{p}_2$ to $\mathbf{p}_3$, so we get:

$$(\tfrac{1}{2}R^2 - R^5/5R^3).\mathbf{d}\theta \,|\, [2\pi, 0] = -\,2\pi.(\tfrac{1}{2}R^2 - R^5/5) = -(6\pi/10)R^2.$$

The change of sign is basically because the chosen **orientation** of the **simplexes** (anticlockwise) made the leading **edge** $\mathbf{p}_2\mathbf{p}_3$ one with the property that $\mathbf{vec}(\mathbf{p}_2\mathbf{p}_3) = -\kappa\mathbf{e}_\theta$, *negative* in $\mathbf{e}_\theta$.

The point is that the choice of a simple regular space for the **pullback** by an appropriate choice of variables not only trivialized the usual evaluation as a product of two simple **integrals** but also made the **simplex** method fairly simple too. The advantage of the **simplex** method is its analytical reliability: it can be trusted to give the right answer! Above all, it gives the right *signs*.

Also note how here we found the **boundary** of the circle to be given by the $\mathbf{p}_2\mathbf{p}_3$ edge which **maps** into the **boundary** approximated by the *in situ* decomposition of the circle into **2-simplexes** described in the discussion of *Figure 6.8.1*. This is a good example of the equivalence of the two alternative approaches to the handling of **manifolds** – the *in situ* decomposition and the **mapping** approaches mentioned in Section 6.3.

Stokes's Theorem refers to the exceptional case where a **2-form** in *xyz*-space *is* **exact**, when its **integral** over any surface in space will have the peculiar property that *it depends only on the* **boundary** *of the surface*. In effect, in this exceptional case, the **adjoint definition** *will* work, *even for a* **2-form** *in 3-space*. So the **theorem** takes the form:

Stokes's Theorem:

$$[(\partial_x g(x,y,z) - \partial_y f(x,y,z))\mathbf{dxdy} + (\partial_y h(x,y,z) - \partial_z g(x,y,z))\mathbf{dydz} + (\partial_z f(x,y,z) - \partial_x h(x,y,z)]\mathbf{dzdx}\,|\,\Sigma$$

$$= [f(x,y,z)\mathbf{dx} + g(x,y,z)\mathbf{dy} + h(x,y,z)\mathbf{dz}]\,|\,\mathbf{b}\Sigma.$$

Note the **curl** form in the first line.

We can prove this simply by using the defining equations of the surface Σ from Section 6.7:

$$x = x(u,v), \qquad y = y(u,v), \qquad z = z(u,v).$$

From these we obtain:

$$\mathbf{dx} = \partial_u x(u,v).\mathbf{du} + \partial_v x(u,v).\mathbf{dv}$$
$$\mathbf{dy} = \partial_u y(u,v).\mathbf{du} + \partial_v y(u,v).\mathbf{dv}$$
$$\mathbf{dz} = \partial_u z(u,v).\mathbf{du} + \partial_v z(u,v).\mathbf{dv}$$

and

$$\mathbf{dxdy} = (\partial_u x(u,v).\partial_v y(u,v) - \partial_v x(u,v).\partial_u y(u,v))\mathbf{dudv}$$
$$\mathbf{dydz} = (\partial_u y(u,v).\partial_v z(u,v) - \partial_v y(u,v).\partial_u z(u,v))\mathbf{dudv}$$
$$\mathbf{dzdx} = (\partial_u z(u,v).\partial_v x(u,v) - \partial_v z(u,v).\partial_u x(u,v))\mathbf{dudv}.$$

Now we're looking for the form of **Green's theorem** in the *uv* plane, which should be:

$$(\partial_u G(u,v) - \partial_v F(u,v))\mathbf{dudv}\,|\,\Sigma^\dagger = (F(u,v)\mathbf{du} + G(u,v)\mathbf{dv})\,|\,\mathbf{b}\Sigma^\dagger$$

where $\Sigma^{\dagger}$ is the **pullback** of Σ onto the *uv* plane. To see where we should be going, we need to find out the form of the *F* and *G* functions on the right-hand side. So let's convert the provisional **Stokes's theorem**'s RHS (right-hand side) **1-form** above:

$$f(x, y, z)\mathbf{d}x + g(x, y, z)\mathbf{d}y + h(x, y, z)\mathbf{d}z$$

$$= f(x(u,v), y(u,v), z(u, v)).(\partial_u x(u, v).\mathbf{d}u + \partial_v x(u, v).\mathbf{d}v) + g(x(u, v), y(u, v), z(u, v)).(\partial_u y(u, v).\mathbf{d}u + \partial_v y(u, v).\mathbf{d}v) + h(x(u, v), y(u, v), z(u, v)).(\partial_u z(u, v).\mathbf{d}u + \partial_v z(u, v).\mathbf{d}v)$$

which shows that:

$$F(u, v) = f(x(u, v), y(u, v), z(u, v)).\partial_u x(u, v) + g(x(u, v), y(u, v), z(u, v)).\partial_u y(u, v) + h(x(u, v), y(u, v), z(u, v)).\partial_u z(u, v)$$

and

$$G(u, v) = f(x(u, v), y(u, v), z(u, v)).\partial_v x(u, v) + g(x(u, v), y(u, v), z(u, v)).\partial_v y(u, v) + h(x(u, v), y(u, v), z(u, v)).\partial_v z(u, v)$$

so the correct form of $\mathbf{d}(F(u, v)\mathbf{d}u + G(u, v)\mathbf{d}v)$ that we will need for the LHS (left-hand side) will be:

$$(\partial_u G(u, v) - \partial_v F(u, v)).\mathbf{d}u\mathbf{d}v$$

$$=\{[\partial_x f(x, y, z)\partial_u x(u, v) + \partial_y f(x, y, z)\partial_u y(u, v) + \partial_z f(x, y, z)\partial_u z(u, v)].\partial_v x(u, v)$$

$$+ f(x, y, z)\partial_u\partial_v x(u, v)$$

$$+ [\partial_x g(x, y, z)\partial_u x(u, v) + \partial_y g(x, y, z)\partial_u y(u, v) + \partial_z g(x, y, z)\partial_u z(u, v)].\partial_v y(u, v)$$

$$+ g(x, y, z)\partial_u\partial_v y(u, v)$$

$$+ [\partial_x h(x, y, z)\partial_u x(u, v) + \partial_y h(x, y, z)\partial_u y(u, v) + \partial_z h(x, y, z)\partial_u z(u, v)].\partial_v z(u, v)$$

$$+ h(x, y, z)\partial_u\partial_v z(u, v)$$

$$- [\partial_x f(x, y, z)\partial_v x(u, v) + \partial_y f(x, y, z)\partial_v y(u, v) + \partial_z f(x, y, z)\partial_v z(u, v)].\partial_u x(u, v)$$

$$- f(x, y, z)\partial_u\partial_v x(u, v)$$

$$- [\partial_x g(x, y, z)\partial_v x(u, v) + \partial_y g(x, y, z)\partial_v y(u, v) + \partial_z g(x, y, z)\partial_v z(u, v)].\partial_u y(u, v)$$

$$- g(x, y, z)\partial_u\partial_v y(u, v)$$

$$- [\partial_x h(x, y, z)\partial_v x(u, v) + \partial_y h(x, y, z)\partial_v y(u, v) + \partial_z h(x, y, z)\partial_v z(u, v)].\partial_u z(u, v)$$

$$- h(x, y, z)\partial_u\partial_v z(u, v)\ \}.\mathbf{d}u\mathbf{d}v$$

where I've used **Young's theorem** to swap $\partial_u\partial_v$ and $\partial_v\partial_u$ in the terms at the right, which therefore cancel, and this simplifies, with further cancellations, to:

$$(\partial_x g(x, y, z) - \partial_y f(x, y, z))(\partial_u x(u, v).\partial_v y(u, v) - \partial_v x(u, v).\partial_u y(u, v))\mathbf{d}u\mathbf{d}v +$$

$$(\partial_y h(x, y, z) - \partial_z g(x, y, z))(\partial_u y(u, v).\partial_v z(u, v) - \partial_v y(u, v).\partial_u z(u, v))\mathbf{d}u\mathbf{d}v +$$

$$(\partial_z f(x, y, z) - \partial_x h(x, y, z))(\partial_u z(u, v).\partial_v x(u, v) - \partial_v z(u, v).\partial_u x(u, v))\mathbf{d}u\mathbf{d}v.$$

This is exactly the form that we get from plugging in the expressions for **d***x***d***y*, **d***y***d***z* and **d***z***d***x* in terms of **d***u***d***v* given above into the LHS of the provisional **Stokes's theorem** statement. Note that because we have to **pull back** a **surface integral** onto a **parametric** *uv* space to evaluate it at all, the fact just established that the **pulled-back** equation exactly matches that for **Green's theorem** establishes our result.

The **Divergence Theorem** is simply the extension of our **adjoint definition** applied to **3-forms** in 3-space (*xyz*-space) to arbitrary **domains of integration** (or **manifolds**), as these can be arbitrarily well approximated by sums of **tets** just as arbitrary **domains of integration** in the plane could be approximated by suitable **meshes**. So this is the same idea as **Green's Theorem**, but we're now working one **grade** higher. Again the **theorem** takes the form:

$$\mathbf{d}\omega \,|\, \mathbf{V} = \omega \,|\, \mathbf{bV},$$

but here ω is a **2-form** like *f* (*x*, *y*, *z*)**d***y***d***z* + *g*(*x*, *y*, *z*)**d***z***d***x* + *h*(*x*, *y*, *z*)**d***x***d***y* and such a **form** has a particularly simple **derivative** because by our Axiom V in Section 6.6 only *one* term survives from each of **d***f* (*x*, *y*, *z*), **d***g*(*x*, *y*, *z*) and **d***h*(*x*, *y*, *z*), since **d***y***d***y***d***z* = **d***x***d***z***d***x* = **d***y***d***x***d***y* = **0** and so on: any triple product with two **differentials** the same is zero. So

$$\mathbf{d}\{f(x, y, z)\mathbf{d}y\mathbf{d}z + g(x, y, z)\mathbf{d}z\mathbf{d}x + h(x, y, z)\mathbf{d}x\mathbf{d}y\}$$

$$= \mathbf{d}f(x, y, z)\mathbf{d}y\mathbf{d}z + \mathbf{d}g(x, y, z)\mathbf{d}z\mathbf{d}x + \mathbf{d}h(x, y, z)\mathbf{d}x\mathbf{d}y$$

$$= \partial_x f(x, y, z)\mathbf{d}x\mathbf{d}y\mathbf{d}z + \partial_y g(x, y, z)\mathbf{d}y\mathbf{d}z\mathbf{d}x + \partial_z h(x, y, z)\mathbf{d}z\mathbf{d}x\mathbf{d}y$$

$$= (\partial_x f(x, y, z) + \partial_y g(x, y, z) + \partial_z h(x, y, z))\mathbf{d}x\mathbf{d}y\mathbf{d}z$$

since **d***y***d***z***d***x* = − **d***y***d***x***d***z* = +**d***x***d***y***d***z*, and **d***z***d***x***d***y* = − **d***x***d***z***d***y* = +**d***x***d***y***d***z*.

The actual statement of the **Divergence Theorem** is:

The Divergence Theorem: $(\partial_x f(x, y, z) + \partial_y g(x, y, z) + \partial_z h(x, y, z))\mathbf{d}x\mathbf{d}y\mathbf{d}z \,|\, \mathbf{V}$
$= f(x, y, z)\mathbf{d}y\mathbf{d}z + g(x, y, z)\mathbf{d}z\mathbf{d}x + h(x, y, z)\mathbf{d}x\mathbf{d}y \,|\, \mathbf{bV}$.

The name comes from the fact that for a **vector** $\mathbf{v} = f.\mathbf{i} + g.\mathbf{j} + h.\mathbf{k}$, where f, g and h are functions over x, y, z (**v** is more strictly called a **vector field**) the **divergence** of **v** is defined as the **scalar** $(\partial_x f + \partial_y g + \partial_z h)$.

As with the **line integral** in Section 6.2, and with **Green's Theorem** earlier in this section, the extension of the **adjoint definition** to arbitrary **manifolds** uses the fact that the **integrals** over all the internal **faces** of the **tets** making up the arbitrary volume cancel out.

In Appendix 6.A, I will use the **adjoint definition** for a **3-form**, which corresponds to the **divergence theorem**, to calculate the volume of a **tet**.

6.A Appendix A: The Volume of a **Tet**

A **tet** is a **3-simplex** defined by four points in space $\mathbf{p}_0\mathbf{p}_1\mathbf{p}_2\mathbf{p}_3$. Its **boundary** is the signed set of four **faces**:

$$\mathbf{p}_1\mathbf{p}_2\mathbf{p}_3 - \mathbf{p}_0\mathbf{p}_2\mathbf{p}_3 + \mathbf{p}_0\mathbf{p}_1\mathbf{p}_3 - \mathbf{p}_0\mathbf{p}_1\mathbf{p}_2.$$

Its **volume** can be calculated as $\mathbf{d}x\mathbf{d}y\mathbf{d}z \,|\, \mathbf{p}_0\mathbf{p}_1\mathbf{p}_2\mathbf{p}_3$, or by use of the **divergence theorem**, as:

$$x.\mathbf{d}y\mathbf{d}z \,|\, (\mathbf{p}_1\mathbf{p}_2\mathbf{p}_3 - \mathbf{p}_0\mathbf{p}_2\mathbf{p}_3 + \mathbf{p}_0\mathbf{p}_1\mathbf{p}_3 - \mathbf{p}_0\mathbf{p}_1\mathbf{p}_2).$$

The two are essentially equivalent, as we need to construct the bounding triangles anyway to get the limits of integration for the volume. We can **parameterize** these **faces** quite trivially. Suppose we have the **face** $\mathbf{p}_0\mathbf{p}_1\mathbf{p}_2$, with **vertices** $\mathbf{p}_0 = (x_0, y_0, z_0)$, $\mathbf{p}_1 = (x_1, y_1, z_1)$ and $\mathbf{p}_2 = (x_2, y_2, z_2)$. Let's define a **pullback mapping** from the uv plane that maps $\mathbf{p}_0$ from the **origin** (0,0), $\mathbf{p}_1$ from the unit point on the u **axis** (1, 0), and $\mathbf{p}_2$ from the unit point on the v **axis** (0, 1). We'll need **mappings** $x = x(u,v)$, $y = y(u,v)$, $z = z(u,v)$ that do this, but since everything is straight-lined, we can have these **linear**:

$$x = a_x u + b_x v + c_x$$
$$y = a_y u + b_y v + c_y$$
$$z = a_z u + b_z v + c_z.$$

Putting in $(u, v) = (0, 0)$, which we want to map to (x_0, y_0, z_0), so e.g. $x_0 = a_x 0 + b_x 0 + c_x$ or $x_0 = c_x$. Then for $(u, v) = (1, 0)$, we have $x_1 = a_x 1 + b_x 0 + x_0$, giving $a_x = (x_1 - x_0)$ and so on to get:

$$x = (x_1 - x_0)u + (x_2 - x_0)v + x_0$$
$$y = (y_1 - y_0)u + (y_2 - y_0)v + y_0$$

$z = (z_1 - z_0)u + (z_2 - z_0)v + z_0$

as our full set of **mappings** from *uv* to *xyz*, which I will call *T*. Now:

$\mathbf{d}y\mathbf{d}z = ((y_1 - y_0)\mathbf{d}u + (y_2 - y_0)\mathbf{d}v).((z_1 - z_0)\mathbf{d}u + (z_2 - z_0)\mathbf{d}v)$

$= ((y_1 - y_0)(z_2 - z_0) - (y_2 - y_0)(z_1 - z_0))\mathbf{d}u\mathbf{d}v$

and our **integral** for the single **face** $\mathbf{p}_0\mathbf{p}_1\mathbf{p}_2$, which is $(-)\ x.\mathbf{d}y\mathbf{d}z \mid \mathbf{p}_0\mathbf{p}_1\mathbf{p}_2$, expands to:

$((x_1 - x_0)u + (x_2 - x_0)v + x_0)((y_1 - y_0)(z_2 - z_0) - (y_2 - y_0)(z_1 - z_0))\mathbf{d}u\mathbf{d}v \mid T^{-1}(\mathbf{p}_0\mathbf{p}_1\mathbf{p}_2).$

(I'll leave out the minus sign for this **face** for now.) This evaluates to:

$$(x_1 - x_0)((y_1 - y_0)(z_2 - z_0) - (y_2 - y_0)(z_1 - z_0))\ (\ u\mathbf{d}u\mathbf{d}v \mid T^{-1}(\mathbf{p}_0\mathbf{p}_1\mathbf{p}_2)) +$$
$$(x_2 - x_0)((y_1 - y_0)(z_2 - z_0) - (y_2 - y_0)(z_1 - z_0))\ (\ v\mathbf{d}u\mathbf{d}v \mid T^{-1}(\mathbf{p}_0\mathbf{p}_1\mathbf{p}_2)) +$$
$$x_0((y_1 - y_0)(z_2 - z_0) - (y_2 - y_0)(z_1 - z_0))\ (\ \mathbf{d}u\mathbf{d}v \mid T^{-1}(\mathbf{p}_0\mathbf{p}_1\mathbf{p}_2)).$$

$T^{-1}(\mathbf{p}_0\mathbf{p}_1\mathbf{p}_2)$ is the simple (0, 0)−(1, 0)−(0, 1) triangle − sometimes called the **standard 2-simplex** − in the *uv* plane, so all three **integrals** are quite easy to evaluate. I'll call $T^{-1}(\mathbf{p}_0\mathbf{p}_1\mathbf{p}_2)$ in the *uv*-plane $\mathbf{u}_0\mathbf{u}_1\mathbf{u}_2$.

Our **integrals** are:

(1.) $u.\mathbf{d}u\mathbf{d}v \mid \mathbf{u}_0\mathbf{u}_1\mathbf{u}_2 = ½.\mathbf{d}(u^2\mathbf{d}v) \mid \mathbf{u}_0\mathbf{u}_1\mathbf{u}_2 = ½.u^2\mathbf{d}v \mid \mathbf{b}(\mathbf{u}_0\mathbf{u}_1\mathbf{u}_2)$

$= ½.u^2\mathbf{d}v \mid (\mathbf{u}_0\mathbf{u}_1 + \mathbf{u}_1\mathbf{u}_2 + \mathbf{u}_2\mathbf{u}_0).$

Here, **parameterize** each **edge** over t : [0, 1]. So $\mathbf{u}_0\mathbf{u}_1$, from (0, 0) to (1, 0), can be given by:

$u = t, \quad v = 0,$

so that $\mathbf{d}v = \mathbf{0}$, and the **integral** here is zero. For $\mathbf{u}_1\mathbf{u}_2$, from (1, 0) to (0, 1), use:

$u = 1 - t, \qquad v = t,$

so we have:

$½.(1 - t)^2\mathbf{d}t \mid [0, 1] = -½.⅓.\mathbf{d}(1 - t)^3 \mid [0, 1] = -(1/6)[(1 - 1)^3 - (1 - 0)^3]$

$= -(1/6)(0 - 1) = 1/6.$

For $\mathbf{u}_2\mathbf{u}_0$, from (0, 1) to (0, 0), use:

$u = 0, \quad v = 1 - t,$

so we get $-½.0.\mathbf{d}t \mid [0, 1] = 0$ again, so this **integral** over the whole **simplex** is 1/6.

(2.) For $v\mathbf{d}u\mathbf{d}v$, we need to remember Axiom IV, and substitute:

$$-\mathbf{d}(½.v^2.\mathbf{d}u) = -2×½.v\mathbf{d}v\mathbf{d}u = -v\mathbf{d}v\mathbf{d}u = +v\mathbf{d}u\mathbf{d}v.$$

So our **integral** is:

$$-½.\mathbf{d}(v^2.\mathbf{d}u) \mid \mathbf{u}_0\mathbf{u}_1\mathbf{u}_2 = -½.(v^2.\mathbf{d}u) \mid (\mathbf{u}_1\mathbf{u}_2 - \mathbf{u}_0\mathbf{u}_2 + \mathbf{u}_0\mathbf{u}_1).$$

We use the same **parameterizations** as before to give 0 for the $\mathbf{u}_0\mathbf{u}_1$ part as $v = 0$ there, and 0 for the $\mathbf{u}_2\mathbf{u}_0$ part as $u = 0$, so $\mathbf{d}u = 0$ there, and for $\mathbf{u}_1\mathbf{u}_2$, where $u = 1 - t$, and $v = t$, we get:

$$-½.(t)^2.\mathbf{d}(1 - t) \mid [0, 1] = +½.(t)^2.\mathbf{d}t \mid [0, 1] = (1/6).\mathbf{d}(t^3) \mid [0, 1]$$

$$= (1/6)[1^3 - 0^3] = 1/6 \text{ again.}$$

(3.) $\mathbf{d}u\mathbf{d}v \mid \mathbf{u}_0\mathbf{u}_1\mathbf{u}_2 = \mathbf{d}(u\mathbf{d}v) \mid \mathbf{u}_0\mathbf{u}_1\mathbf{u}_2 = u\mathbf{d}v \mid (\mathbf{u}_1\mathbf{u}_2 - \mathbf{u}_0\mathbf{u}_2 + \mathbf{u}_0\mathbf{u}_1)$

and using the same **parameterizationns** yet again, only the $\mathbf{u}_1\mathbf{u}_2$ term contributes, with:

$$(1 - t).\mathbf{d}t \mid [0, 1] = -½.\mathbf{d}(1 - t)^2 \mid [0, 1] = -½[(0 - 0)^2 - (1 - 0)^2] = +½.$$

So altogether we have:

$$((1/6)(x_1 - x_0) + (1/6)(x_2 - x_0) + ½x_0)((y_1 - y_0)(z_2 - z_0) - (y_2 - y_0)(z_1 - z_0)).$$

But $(1/6)(x_1 - x_0) + (1/6)(x_2 - x_0) + ½x_0 = (1/6)(x_1 + x_2) + ½x_0 - ⅓x_0 = (1/6)(x_1 + x_2 + x_0)$, so we have:

$$(1/6)(x_1 + x_2 + x_0)((y_1 - y_0)(z_2 - z_0) - (y_2 - y_0)(z_1 - z_0))$$

for the contribution of the $\mathbf{p}_0\mathbf{p}_1\mathbf{p}_2$ **face**. Analogously, we can evaluate all the other three **faces**, just altering the indices appropriately, and we get for the entire **integral**, *now allowing for the signs of the* **faces**:

$$x.\mathbf{d}y\mathbf{d}z \mid (\mathbf{p}_1\mathbf{p}_2\mathbf{p}_3 - \mathbf{p}_0\mathbf{p}_2\mathbf{p}_3 + \mathbf{p}_0\mathbf{p}_1\mathbf{p}_3 - \mathbf{p}_0\mathbf{p}_1\mathbf{p}_2) =$$

$$\begin{aligned}(1/6)\{ &(x_1 + x_2 + x_3)((y_2 - y_1)(z_3 - z_1) - (y_3 - y_1)(z_2 - z_1)) \\ &- (x_0 + x_2 + x_3)((y_2 - y_0)(z_3 - z_0) - (y_3 - y_0)(z_2 - z_0)) \\ &+ (x_0 + x_1 + x_3)((y_1 - y_0)(z_3 - z_0) - (y_3 - y_0)(z_1 - z_0)) \\ &- (x_0 + x_1 + x_2)((y_1 - y_0)(z_2 - z_0) - (y_2 - y_0)(z_1 - z_0)).\end{aligned}$$

Now to see what will cancel, all I can do is multiply this whole thing out into $x_iy_jz_k$ products, which take the form, writing *only* the *ijk* indices:

$$\begin{aligned}(1/6)\{&123 - 121 - 113 - 132 + 131 + 112 + 223 - 221 - 213 \\ &- 232 + 231 + 212 + 323 - 321 - 313 - 332 + 331 + 312\end{aligned}$$

− (023 − 020 − 003 − 032 + 030 + 002 + 223 − 220 − 203

− 232 + 230 + 202 + 323 − 320 − 303 − 332 + 330 + 302)

+ 013 − 010 − 003 − 031 + 030 + 001 + 113 − 110 − 103

− 131 + 130 + 101 + 313 − 310 − 303 − 331 + 330 + 301

− (012 − 010 − 002 − 023 + 020 + 001 + 112 − 110 − 102

− 121 + 120 + 101 + 212 − 210 − 202 − 221 + 220 + 201)}

where I've put the negative **faces** in bold. The cancellations are perhaps best done by copying the page and crossing through on the copy! The result, which *is* the volume of a **tet**, is:

$$(1/6)\{x_1y_2z_3 - x_1y_3z_2 - x_2y_1z_3 + x_2y_3z_1 - x_3y_2z_1 + x_3y_1z_2 - x_0y_2z_3 + x_0y_3z_2$$
$$+ x_2y_0z_3 - x_2y_3z_0 + x_3y_2z_0 + x_0y_1z_3 - x_0y_3z_1 - x_1y_0z_3 + x_1y_3z_0 - x_3y_1z_0$$
$$+ x_3y_1z_0 + x_3y_0z_1 - x_0y_1z_2 + x_0y_2z_1 + x_1y_0z_2 - x_1y_2z_0 + x_2y_1z_0 - x_2y_0z_1\}.$$

We can check that this agrees with the formula for (1/6) of the volume of a parallelepiped:

$$(1/6)((x_1 - x_0)\mathbf{e}_x + (y_1 - y_0)\mathbf{e}_y + (z_1 - z_0)\mathbf{e}_z) \wedge ((x_2 - x_0)\mathbf{e}_x + (y_2 - y_0)\mathbf{e}_y + (z_2 - z_0)\mathbf{e}_z)$$
$$\wedge ((x_3 - x_0)\mathbf{e}_x + (y_3 - y_0)\mathbf{e}_y + (z_3 - z_0)\mathbf{e}_z)$$

$$= (1/6)[(x_1 - x_0)(y_2 - y_0)(z_3 - z_0)\mathbf{e}_x\mathbf{e}_y\mathbf{e}_z + (x_1 - x_0)(y_3 - y_0)(z_2 - z_0)\mathbf{e}_x\mathbf{e}_z\mathbf{e}_y$$
$$+ (x_2 - x_0)(y_1 - y_0)(z_3 - z_0)\mathbf{e}_y\mathbf{e}_x\mathbf{e}_z + (x_3 - x_0)(y_1 - y_0)(z_2 - z_0)\mathbf{e}_y\mathbf{e}_z\mathbf{e}_x$$
$$+ (x_3 - x_0)(y_2 - y_0)(z_1 - z_0)\mathbf{e}_z\mathbf{e}_y\mathbf{e}_x + (x_2 - x_0)(y_3 - y_0)(z_1 - z_0)\mathbf{e}_z\mathbf{e}_x\mathbf{e}_y]$$

$$= (1/6)[(x_1 - x_0)(y_2 - y_0)(z_3 - z_0) - (x_1 - x_0)(y_3 - y_0)(z_2 - z_0)$$
$$- (x_2 - x_0)(y_1 - y_0)(z_3 - z_0) + (x_3 - x_0)(y_1 - y_0)(z_2 - z_0)$$
$$- (x_3 - x_0)(y_2 - y_0)(z_1 - z_0) + (x_2 - x_0)(y_3 - y_0)(z_1 - z_0)]\mathbf{e}_x\mathbf{e}_y\mathbf{e}_z.$$

This multiplies out to give $x_iy_jz_k$ products, writing *only* the *ijk* indices:

123 − 120 − 103 + 100 − 023 + 020 + 003 − 000

− 213 + 210 + 203 − 200 + 013 − 010 − 003 + 000

− 132 + 130 + 102 − 100 + 032 − 030 − 002 + 000

+ 312 − 310 − 302 + 300 − 012 + 010 + 002 − 000

− 321 + 320 + 301 − 300 + 021 − 020 − 001 + 000

+ 231 − 230 − 201 + 200 − 031 + 030 + 001 − 000

also omitting the (1/6) factor. Cancellations again remove the terms with repeating factors like 000 or 100 or 030 (all of which repeat the 0's) and reduce these 48 terms to precisely the 24 we had before.

6.B Appendix B: The Conventional Evaluation of the Volume of a Pyramid

A **tet** is basically a triangular pyramid, differing from the Egyptian type in having a *triangular* rather than a square base. We can use this concept to give a general justification that the volume of *any* **tet** is 1/6 of the volume of the parallelepiped with the same three defining sides at the same angles of splay from a single **vertex**. But in so doing it will be necessary to use a somewhat "infinitesimal" line of argument, and this is obviously something I'd prefer to have avoided! So you can look on this Appendix as an introduction to the traditional kind of argument that dominated the early development of the **calculus**.

Figure 6.B.1 shows an approximation to the shape of a pyramid with an *arbitrary* shape of base, which I'll refer to as Д ("de"). As you can see, a pile of platters of shape Д but ever-decreasing size looks roughly like a Д-based pyramid. If the height of each platter is halved, but an extra platter is inserted of dimensions just halfway between those of the platter below and the platter above, but of this same half height, as shown in *Figure 6.B.1b*, the approximation to a smooth-sided Д-shaped pyramid is improved.

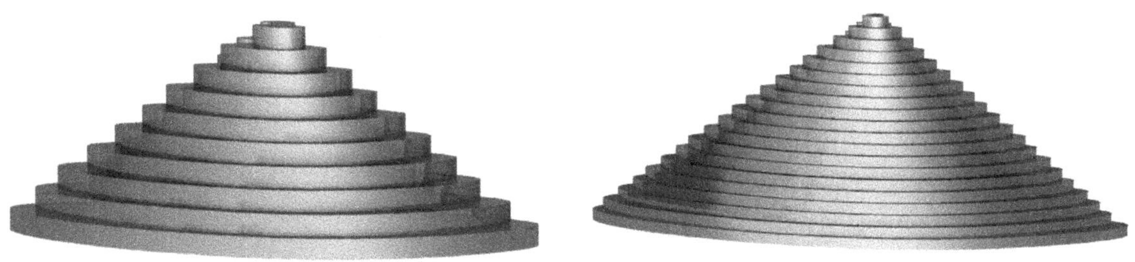

Figure 6.B.1

If we now push the pile of platters across so as still to keep each platter approximately over the one below, but giving the pile as a whole a definite "skewness", the *volume* must still be the same, as it's the volume of all the platters summed up. So a pile of platters of the same total *height* but slid out of true has the same volume as the pile where they all lie neatly on top of each other. This is shown in *Figure 6.B.2*.

Figure 6.B.2

This is the three-dimensional generalization of the argument in Section 6.4 that showed that:

Parallelograms on the same base and between the same parallels are equal in area.

It is a heuristic argument, as here presented, but, granted that the platter model reasonably approximates the pyramid, it *is* based on a decomposition, and can be stated as:

Pyramids of the same base (or cross-sectional) shape and of the same vertical height are equal in volume.

Now comes the "infinitesimal" part of the argument, which is used to establish how the platter model can model the pyramid *exactly*.

Call the base area – however it's evaluated – of the bottom platter A. Let there be n platters in the pile. If we had $n+1$ altogether, the extra one would have area zero to conform with the pyramid structure. So let's number the *top* platter #1 and the bottom n and then the k'th platter must have area $A(k) = A.(k/n)^2$, giving $A(0) = 0$ for the one above the topmost, and $A(n) = A$. The height of each platter is h/n, so the total volume of the pile is:

$$A.\sum_{k=1}^{n}\frac{h}{n}\left(\frac{k}{n}\right)^2.$$

This figure is correct both for a "true" pile and for a "skew" one as we know from the foregoing argument. Only the base area and the total vertical height determine the volume.

To approximate a smooth-sided pyramid we simply need more and more platters, so our actual figure for the volume of a pyramid will be:

$$\lim_{n\to\infty} A.\sum_{k=1}^{n}\frac{h}{n}\left(\frac{k}{n}\right)^2.$$

How do we evaluate such a result?

Basically we use simple algebra. It can be shown by **induction** that:

$$1^2 + 2^2 + 3^2 + \ldots + (n-1)^2 = (1/6)(n-1)n(2n-1).$$

So putting $m = n - 1$, $n = m + 1$, so $2n - 1 = 2(m + 1) - 1 = 2m + 1$, giving:

$$1^2 + 2^2 + 3^2 + \ldots + m^2 = (1/6)m(m+1)(2m+1),$$

so

$$\sum_{k=1}^{n}\frac{k^2}{n^3} = (1/6n^3)n(n+1)(2n+1) = (1/6)(n/n)((n+1)/n)((2n+1)/n)$$

$$= (1/6)(1)(1 + 1/n)(2 + 1/n),$$

and clearly, since $1/n \to 0$ as $n \to \infty$,

$\mathbf{lim}_{n\to\infty}\sum_{k=1}^{n}\frac{k^2}{n^3}$ = (1/6).1.1.2 = ⅓.

So our final result for the volume of the pile of platters is ⅓*Ah*.

Now for Д ≡ Δ, a triangle, we know its area = ½ × its base length (say *l*) × *its* "vertical height" (i.e. its diameter or maximum breadth at right angles to the base side (say *d*)), so giving the volume of a **tet** as:

½⅓*ldh* = (1/6).*ldh*,

giving the required factor of (1/6) but provided we use *d* and *h* as *diameters perpendicular* to the base side of length *l*.

6.C Appendix C: The Numerical Interpretation of Differentials

The numerical concept is easily recovered using the **dual space** definition of a **differential**:

d*f* (P, **v**) := (**grad** *f*)(P) · **v**.

We define the **numerical form** of a **differential** to be precisely the value of this expression for an arbitrary **vector** **v**. In other words, at any given point P = (*x*, *y*, *z*), any **3-vector v** defines unique numerical values for each **coordinate differential** as **d***x*(P, **v**) = **grad**(*x*) · **v** = v_x, **d***y*(P, **v**) = **grad**(*y*) · **v** = v_y, and **d***z*(P, **v**) = **grad**(*z*) · **v** = v_z, and for any *f* we have:

d*f* (P, **v**) = (**grad** *f*)(P) · **v** = $\partial_x f$ (P).**d***x*(P, **v**) + $\partial_y f$ (P).**d***y*(P, **v**) + $\partial_z f$ (P).**d***z*(P, **v**).

If we now regard it as legitimate to write **d***f* (P, **v**) as just **d***f*, and **d***x*(P, **v**) as just **d***x* and so on, *on the understanding that these are now arbitrary numerical values*, we can write this just as we did in Chapter 2 as:

d*f* = $\partial_x f$ (P).**d***x* + $\partial_y f$ (P).**d***y* + $\partial_z f$ (P).**d***z*

so recovering the original formulation.[93]

Be careful not to confuse the two functional roles of the **differential**. For the **coordinate differentials**, there's no problem because there the two functions coincide. If **v** is regarded now as **vec**($\mathbf{p}_0\mathbf{p}_1$), we have:

d*x*(P, **v**) = **grad**(*x*) · **v** = **grad**(*x*) · **vec**($\mathbf{p}_0\mathbf{p}_1$) = v_x = **d***x* | $\mathbf{p}_0\mathbf{p}_1$ = *x*($\mathbf{p}_1$) − *x*($\mathbf{p}_0$) = $x_1 - x_0 = v_x$

d*y*(P, **v**) = **grad**(*y*) · **v** = **grad**(*y*) · **vec**($\mathbf{p}_0\mathbf{p}_1$) = v_y = **d***y* | $\mathbf{p}_0\mathbf{p}_1$ = *y*($\mathbf{p}_1$) − *y*($\mathbf{p}_0$) = $y_1 - y_0 = v_y$

d*z*(P, **v**) = **grad**(*z*) · **v** = **grad**(*z*) · **vec**($\mathbf{p}_0\mathbf{p}_1$) = v_z = **d***z* | $\mathbf{p}_0\mathbf{p}_1$ = *z*($\mathbf{p}_1$) − *z*($\mathbf{p}_0$) = $z_1 - z_0 = v_z$

[93] In doing this we're using the traditional, and somewhat disparaged, habit of using the same name for a function and its value: **d***f* = **d***f* (P, **v**) is analogous to *x* = *x*(*u*, *v*).

and again we can still write these as **d**x, **d**y, **d**z. So under this interpretation, an arbitrary set of **coordinate differentials** (**d**x, **d**y, **d**z) actually corresponds to an arbitrary **1-simplex** displacement from the point $\mathbf{p}_0$ to the point $\mathbf{p}_1$. Regarding $\mathbf{p}_0 = \mathrm{P} = (x_0, y_0, z_0)$ as the **origin** of a new **3-vector space** specific to P, we recover the original numerical concept.

But for any **scalar function** *other than* the **coordinates**, this identity of the two functional forms reduces to an approximation:

$$\mathbf{d}f(\mathrm{P}, \mathbf{v}) = \mathbf{grad}(f) \cdot \mathbf{v} = \mathbf{grad}(f) \cdot \mathbf{vec}(\mathbf{p}_0\mathbf{p}_1)$$

$$= \partial_x f(\mathrm{P}).v_x + \partial_y f(\mathrm{P}).v_y + \partial_z f(\mathrm{P}).v_z$$

$$= \partial_x f(\mathrm{P}).(x_1 - x_0) + \partial_y f(\mathrm{P}).(y_1 - y_0) + \partial_z f(\mathrm{P}).(z_1 - z_0)$$

but

$$\mathbf{d}f\big|\,\mathbf{p}_0\mathbf{p}_1 = f(\mathbf{p}_1) - f(\mathbf{p}_0) \approx \partial_x f(\mathrm{P}).(x_1 - x_0) + \partial_y f(\mathrm{P}).(y_1 - y_0) + \partial_z f(\mathrm{P}).(z_1 - z_0).$$

So $\mathbf{d}f(\mathrm{P}, \mathbf{v}) \approx \mathbf{d}f\big|\,\mathbf{p}_0\mathbf{p}_1$ only: they are only *approximately* equal, and then only for $|\mathbf{p}_0\mathbf{p}_1|$ small. Because of this inequality, to define a **numerical form** of any *other* **differential** we need to set it as:

$$\mathbf{d}f_{\mathrm{num}}(\mathrm{P}) := \partial_x f(\mathrm{P}).\mathbf{d}x_{\mathrm{num}} + \partial_y f(\mathrm{P}).\mathbf{d}y_{\mathrm{num}} + \partial_z f(\mathrm{P}).\mathbf{d}z_{\mathrm{num}}$$

where $\mathbf{d}x_{\mathrm{num}}$, $\mathbf{d}y_{\mathrm{num}}$, $\mathbf{d}z_{\mathrm{num}}$ are **numerical forms** – i.e. arbitrary numbers – or, put in the more compact notation:

$$\mathbf{d}f_{\mathrm{num}}(\mathrm{P}, \mathbf{vec}(\mathbf{p}_0\mathbf{p}_1)) = (\mathbf{grad}\, f)(\mathrm{P}) \cdot \mathbf{vec}(\mathbf{p}_0\mathbf{p}_1).$$

This does not contradict the fact that the **vectorial form** of **differentials** as operating through the **integral operator** still obeys

$$\mathbf{d}f(x, y, z) = \partial_x f(x, y, z).\mathbf{d}x + \partial_y f(x, y, z).\mathbf{d}y + \partial_z f(x, y, z).\mathbf{d}z$$

which clearly *is* still an identity. So for example, in an **integration**, the left-hand and right-hand sides of this equation are interchangeable.

6.D Appendix D: Heuristic Justification of the Alternating Product

To see most simply just why we use the **alternating product** between **differentials**, remember how in Section 2.8 I argued that the fact that the **boundary** of a **boundary** is zero or that **bb** = 0 needs to be echoed in the properties of the **adjoint operator d**, so we also want **dd** = 0.

Let's see how this rule affects a **1-form** like

$$\phi_x.\mathbf{d}x + \phi_y.\mathbf{d}y + \phi_z.\mathbf{d}z.$$

Applying **d**:

$$\mathbf{d}(\phi_x.\mathbf{d}x + \phi_y.\mathbf{d}y + \phi_z.\mathbf{d}z) = \mathbf{d}\phi_x.\mathbf{d}x + \mathbf{d}\phi_y.\mathbf{d}y + \mathbf{d}\phi_z.\mathbf{d}z$$

as any terms like $\mathbf{d}\mathbf{d}x$ drop out *ex hypothesi*. Expanding the $\mathbf{d}\phi$ bits gives:

$$(\partial_x\phi_x.\mathbf{d}x + \partial_y\phi_x.\mathbf{d}y + \partial_z\phi_x.\mathbf{d}z).\mathbf{d}x + (\partial_x\phi_y.\mathbf{d}x + \partial_y\phi_y.\mathbf{d}y + \partial_z\phi_y.\mathbf{d}z).\mathbf{d}y$$

$$+(\partial_x\phi_z.\mathbf{d}x + \partial_y\phi_z.\mathbf{d}y + \partial_z\phi_z.\mathbf{d}z).\mathbf{d}z. \qquad (6.C.1)$$

Now if our original $\phi_x.\mathbf{d}x + \phi_y.\mathbf{d}y + \phi_z.\mathbf{d}z$ is itself **exact**, so that there is some ϕ such that

$$\mathbf{d}\phi = \phi_x.\mathbf{d}x + \phi_y.\mathbf{d}y + \phi_z.\mathbf{d}z,$$

we want 6.C.1 above to be zero.

6.C.1 consists entirely of terms of two forms:

- terms like $\partial_x\phi_x.\mathbf{d}x.\mathbf{d}x$ with the *same* **differential** repeated
- and terms like $\partial_z\phi_x.\mathbf{d}z.\mathbf{d}x$ with two *distinct* **differentials**.

In the product $\mathbf{d}z.\mathbf{d}x$ we have just two terms

$$\partial_z\phi_x.\mathbf{d}z.\mathbf{d}x \text{ and } \partial_x\phi_z.\mathbf{d}x.\mathbf{d}z.$$

If the original **form** is **exact** these are:

$$\partial_z\partial_x\phi.\mathbf{d}z.\mathbf{d}x \text{ and } \partial_x\partial_z\phi.\mathbf{d}x.\mathbf{d}z.$$

By **Young's theorem**, $\partial_z\partial_x\phi = \partial_x\partial_z\phi$, so these two terms would cancel if we required that:

$$\mathbf{d}x.\mathbf{d}z = -\mathbf{d}z.\mathbf{d}x$$

for this would give:

$$\partial_z\phi_x.\mathbf{d}z.\mathbf{d}x + \partial_x\phi_z.\mathbf{d}x.\mathbf{d}z = (\partial_z\partial_x\phi - \partial_x\partial_z\phi).\mathbf{d}z.\mathbf{d}x = 0.$$

But this would also give $\mathbf{d}x.\mathbf{d}x = 0$ by the argument at the end of Section 5.5, and so we would get overall the required

$$\mathbf{d}\mathbf{d}\phi = 0.$$

This argument is generally regarded as sufficient justification for the rule that $\mathbf{d}x.\mathbf{d}z = -\mathbf{d}z.\mathbf{d}x$ or that $\mathbf{d}x.\mathbf{d}z \equiv \mathbf{d}x \wedge \mathbf{d}z$, and was probably Cartan's original justification (it's the argument given, for example, in https://en.wikipedia.org/wiki/Differential_form).

I still feel that there's some place for an argument based on the volume of a **parallelepiped** defined by four **points** $\mathbf{p}_0$, $\mathbf{p}_1$, $\mathbf{p}_2$ and $\mathbf{p}_3$ being given by $\mathbf{p}_0\mathbf{p}_1 \wedge \mathbf{p}_0\mathbf{p}_2 \wedge \mathbf{p}_0\mathbf{p}_3$, and although this does seem to end up a bit laboured, I sketch it below as a relic from the first edition of this work, doing it in the simpler form for calculating the *area* of a surface.

First, we go back to the idea of **meshes** introduced in Section 1.15. Any three points in space can describe a plane triangle, but any four points *cannot necessarily* describe a plane quadrilateral, let alone a parallelogram. So *any* surface can be approximated by a **mesh** of sufficiently small triangles or **2-simplexes** *all the points of which actually lie in the surface*, assuming certain "smoothness" assumptions are met, which in practice always are. The exceptions are some peculiarly pathological surfaces invented by mathematicians precisely to find where these criteria can break down. An example of a surface represented as a **mesh** is shown in *Figure 6.C.1*. This example actually makes extensive use of "**quads**" – little *rectangular* facets – as well as triangular **facets**, because the extrusion logic of the software arranges that these can be well-defined, but in general, each **quad** would be split into two plane triangles.

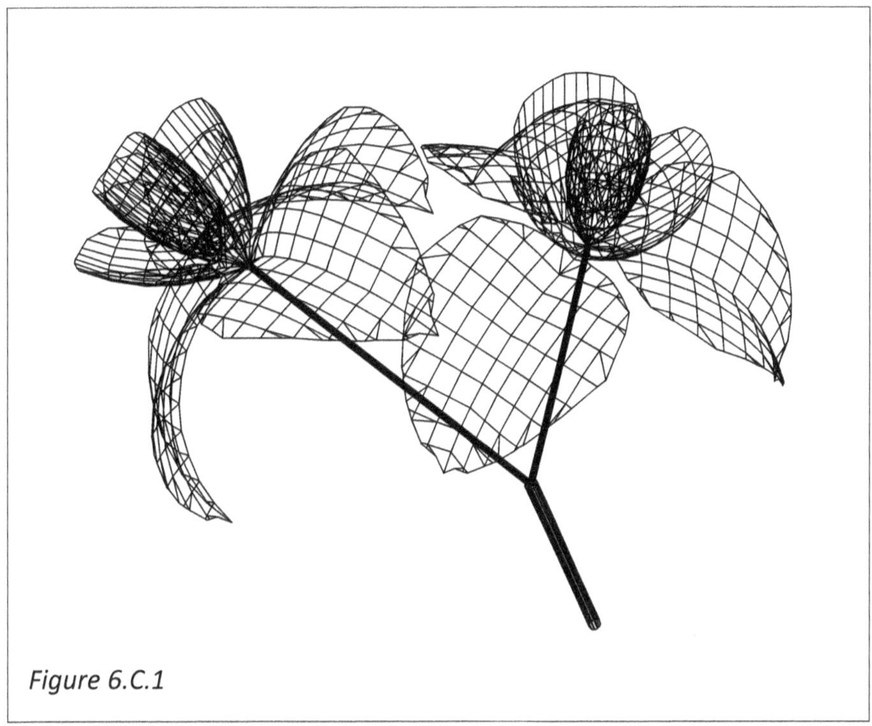

Figure 6.C.1

Look at any computer-generated images of objects in architectural or design work, or in cinema. The quite extraordinary realism of these scenes is achieved by representing all the surfaces involved as **meshes**. In computer graphics, to save memory and processing time, the triangles are larger where the surface is more planar, and become smaller and smaller in areas of high curvature. This technique, called **adaptive subdivision**, is very effective.

For historical reasons, mathematical discourse has preferred to represent surfaces using myriads of parallelograms all of the same "infinitesimal" size, an approach which is less practical. Only rather special surfaces (primarily "linear extrusions" for those familiar with computer graphics systems) satisfactorily admit of such a subdivision. Möbius and Kronecker understood this, which is why they developed the theory of **simplexes**, but the textbooks still stick to their parallelograms.

Just as a surface can be represented to arbitrary accuracy by a **mesh** of triangular **facets**, so its **area** can be calculated to arbitrary accuracy by adding the areas of the **mesh facets.**

To do this, imagine that one of the little triangular **facets** in a **mesh** representing a surface is itself defined as follows.

Consider a triangle – a **2-simplex** – in the *uv* plane, with three points $\mathbf{u}_0 = (u_0, v_0)$, $\mathbf{u}_1 = (u_1, v_1)$ and $\mathbf{u}_2 = (u_2, v_2)$. This translates into three points in space which must all lie in the surface, which I will again call Σ for Σurface. They can be labeled $\mathbf{x}_0 = (x_0, y_0, z_0)$, $\mathbf{x}_1 = (x_1, y_1, z_1)$ and $\mathbf{x}_2 = (x_2, y_2, z_2)$ with $x_0 = x(u_0, v_0)$, $y_0 = y(u_0, v_0)$ and $z_0 = z(u_0, v_0)$, and similarly for the other two points.

Now, from Section 6.4, the **area** of the $\mathbf{u}_0\mathbf{u}_1\mathbf{u}_2$ **simplex** in the *uv* plane is given by:

$(\mathbf{vec}(\mathbf{u}_0\mathbf{u}_1)) \wedge (\mathbf{vec}(\mathbf{u}_0\mathbf{u}_2)) = ½((u_1 - u_0)\mathbf{e}_u + (v_1 - v_0)\mathbf{e}_v) \wedge ((u_2 - u_0)\mathbf{e}_u + (v_2 - v_0)\mathbf{e}_v)$

$= ½((u_1 - u_0)(v_2 - v_0) - (u_2 - u_0)(v_1 - v_0))\ \mathbf{e}_u \wedge \mathbf{e}_v.$

But again from Section 6.4, the **area** of the $\mathbf{x}_0\mathbf{x}_1\mathbf{x}_2$ **simplex** in Σ in the *xyz*-space is given by:

$(\mathbf{vec}(\mathbf{x}_0\mathbf{x}_1)) \wedge (\mathbf{vec}(\mathbf{x}_0\mathbf{x}_2))$

$= ½((x_1 - x_0)\mathbf{e}_x + (y_1 - y_0)\mathbf{e}_y + (z_1 - z_0)\mathbf{e}_z) \wedge ((x_2 - x_0)\mathbf{e}_x + (y_2 - y_0)\mathbf{e}_y + (z_2 - z_0)\mathbf{e}_z)$

$= ½\{((x_1 - x_0)(y_2 - y_0) - (x_2 - x_0)(y_1 - y_0))\ \mathbf{e}_x \wedge \mathbf{e}_y +$
$((y_1 - y_0)(z_2 - z_0) - (y_2 - y_0)(z_1 - z_0))\ \mathbf{e}_y \wedge \mathbf{e}_z +$
$((z_1 - z_0)(x_2 - x_0) - (z_2 - z_0)(x_1 - x_0))\ \mathbf{e}_z \wedge \mathbf{e}_x\}.$

We want to ensure that the **pullback** for a **2-form** in 3-space, which is the basic case treated in Section 6.7, the standard **surface integral** will correctly convert the $(\mathbf{vec}(\mathbf{x}_0\mathbf{x}_1)) \wedge (\mathbf{vec}(\mathbf{x}_0\mathbf{x}_2))$ expression above to the $(\mathbf{vec}(\mathbf{u}_0\mathbf{u}_1)) \wedge (\mathbf{vec}(\mathbf{u}_0\mathbf{u}_2))$ one.

If the triangular **facets** are sufficiently small (which they must be to provide a correct representation of the surface), then we can approximate the differences using the values of the **partial derivatives** at $\mathbf{u}_0$:

$(x_1 - x_0) \approx \partial_u x(u_0, v_0).(u_1 - u_0) + \partial_v x(u_0, v_0).(v_1 - v_0)$
$(y_1 - y_0) \approx \partial_u y(u_0, v_0).(u_1 - u_0) + \partial_v y(u_0, v_0).(v_1 - v_0)$
$(z_1 - z_0) \approx \partial_u z(u_0, v_0).(u_1 - u_0) + \partial_v z(u_0, v_0).(v_1 - v_0),$

and similar terms for the $\mathbf{x}_2 - \mathbf{x}_0$ factors.

Multiplying these factors for the $\mathbf{e}_x \wedge \mathbf{e}_y$ term above, we get:

$((x_1 - x_0)(y_2 - y_0) - (x_2 - x_0)(y_1 - y_0))$

$= \{\partial_u x(u_0, v_0).(u_1 - u_0) + \partial_v x(u_0, v_0).(v_1 - v_0)\}\{\partial_u y(u_0, v_0).(u_2 - u_0) + \partial_v y(u_0, v_0).(v_2 - v_0)\}$

$-\{\partial_u x(u_0, v_0).(u_2 - u_0) + \partial_v x(u_0, v_0).(v_2 - v_0)\}\{\partial_u y(u_0, v_0).(u_1 - u_0) + \partial_v y(u_0, v_0).(v_1 - v_0)\}.$

These multiply out to give:

$\partial_u x(u_0, v_0).(u_1 - u_0).\partial_u y(u_0, v_0).(u_2 - u_0) + \partial_u x(u_0, v_0).(u_1 - u_0).\partial_v y(u_0, v_0).(v_2 - v_0)$

$+ \partial_v x(u_0, v_0).(v_1 - v_0).\partial_u y(u_0, v_0).(u_2 - u_0) + \partial_v x(u_0, v_0).(v_1 - v_0).\partial_v y(u_0, v_0).(v_2 - v_0)$

$-\partial_u x(u_0, v_0).(u_2 - u_0).\partial_u y(u_0, v_0).(u_1 - u_0) - \partial_u x(u_0, v_0).(u_2 - u_0).\partial_v y(u_0, v_0).(v_1 - v_0)$

$-\partial_v x(u_0, v_0).(v_2 - v_0).\partial_u y(u_0, v_0).(u_1 - u_0) - \partial_v x(u_0, v_0).(v_2 - v_0).\partial_v y(u_0, v_0).(v_1 - v_0)$

=

$\partial_u x(u_0, v_0).(u_1 - u_0).\partial_u y(u_0, v_0).(u_2 - u_0) - \partial_u x(u_0, v_0).(u_2 - u_0).\partial_u y(u_0, v_0).(u_1 - u_0)$

$+ \partial_u x(u_0, v_0).(u_1 - u_0).\partial_v y(u_0, v_0).(v_2 - v_0) - \partial_v x(u_0, v_0).(v_2 - v_0).\partial_u y(u_0, v_0).(u_1 - u_0)$

$+ \partial_v x(u_0, v_0).(v_1 - v_0).\partial_v y(u_0, v_0).(v_2 - v_0) - \partial_v x(u_0, v_0).(v_2 - v_0).\partial_v y(u_0, v_0).(v_1 - v_0)$

$+ \partial_v x(u_0, v_0).(v_1 - v_0).\partial_u y(u_0, v_0).(u_2 - u_0) - \partial_u x(u_0, v_0).(u_2 - u_0).\partial_v y(u_0, v_0).(v_1 - v_0).$

The first and third lines here go to zero, leaving only the cross-products:

$\{\partial_u x(u_0, v_0).\partial_v y(u_0, v_0) - \partial_v x(u_0, v_0).\partial_u y(u_0, v_0)\}(u_1 - u_0).(v_2 - v_0)\}$

$+ \{\partial_v x(u_0, v_0).\partial_u y(u_0, v_0) - \partial_u x(u_0, v_0).\partial_v y(u_0, v_0)\}.(u_2 - u_0).(v_1 - v_0).$

$= \{\partial_u x(u_0, v_0).\partial_v y(u_0, v_0) - \partial_v x(u_0, v_0).\partial_u y(u_0, v_0)\}[(u_1 - u_0).(v_2 - v_0) - (u_2 - u_0).(v_1 - v_0)].$

These are correctly in the $(u_1 - u_0).(v_2 - v_0) - (u_2 - u_0).(v_1 - v_0)$ form with an appropriate scaling factor for the change of **coordinates** in taking the **pullback**.

We can obtain this result directly from the **differentials** by evaluating this expression:

$\mathbf{dxdy}.(\mathbf{e}_x \wedge \mathbf{e}_y) + \mathbf{dydz}.(\mathbf{e}_y \wedge \mathbf{e}_z) + \mathbf{dzdx}.(\mathbf{e}_z \wedge \mathbf{e}_x)$

and substituting in

$\mathbf{dxdy} = \mathbf{dx} \wedge \mathbf{dy} = (\partial_u x(u, v)\mathbf{du} + \partial_v x(u, v)\mathbf{dv}) \wedge (\partial_u y(u, v)\mathbf{du} + \partial_v y(u, v)\mathbf{dv})$

$= (\partial_u x(u, v)\partial_v y(u, v) - \partial_v x(u, v)\partial_u y(u, v))\mathbf{dudv}$

and

$\mathbf{dydz} = (\partial_u y(u, v)\partial_v z(u, v) - \partial_v y(u, v)\partial_u z(u, v))\mathbf{dudv}$

$\mathbf{dzdx} = (\partial_u z(u, v)\partial_v x(u, v) - \partial_v z(u, v)\partial_u x(u, v))\mathbf{dudv}$

or, all told:

$\mathbf{dxdy}.(\mathbf{e}_x \wedge \mathbf{e}_y) + \mathbf{dydz}.(\mathbf{e}_y \wedge \mathbf{e}_z) + \mathbf{dzdx}.(\mathbf{e}_z \wedge \mathbf{e}_x)$

$= \mathbf{dudv}.[(\partial_u x(u, v)\partial_v y(u, v) - \partial_v x(u, v)\partial_u y(u, v)).(\mathbf{e}_x \wedge \mathbf{e}_y) +$

$(\partial_u y(u, v)\partial_v z(u, v) - \partial_v y(u, v)\partial_u z(u, v)).(\mathbf{e}_y \wedge \mathbf{e}_z) +$

$(\partial_u z(u, v)\partial_v x(u, v) - \partial_v z(u, v)\partial_u x(u, v)).(\mathbf{e}_z \wedge \mathbf{e}_x)].$

This is the key result: only by treating the product of **differentials** *as an* **alternating product** *do we get the correct formula for the* **pullback** *from a triangular* **facet** *in an arbitrary surface to one in the* **pullback** *space.*

If we want simply the *magnitude* of the **area** — the normal concept of "area" — we take the **magnitude** of this **bivector** using the formula for any **vector**:

$|\mathbf{v}| = \sqrt{(v_x^2 + v_y^2 + v_z^2)}$

$= \mathbf{d}u\mathbf{d}v.\sqrt{}[(\partial_u x(u, v)\partial_v y(u, v) - \partial_v x(u, v)\partial_u y(u, v))^2 +$

$(\partial_u y(u, v)\partial_v z(u, v) - \partial_v y(u, v)\partial_u z(u, v))^2 +$

$(\partial_u z(u, v)\partial_v x(u, v) - \partial_v z(u, v)\partial_u x(u, v))^2]$

and this is the formula you will find in the textbooks for the **differential form** (a **2-form**) giving the area of a surface.[94]

This result, in its original **bivector** form, also gives the proper derivation of the transformation of **coordinates** from (x, y) to (u, v) mentioned in an earlier section. All we need to obtain that is to assume that $z(u, v)$ is constant, so the terms containing $\partial_u z$ and $\partial_v z$ drop out and we get the conversion formula:

$\mathbf{d}x\mathbf{d}y = \mathbf{d}u\mathbf{d}v.(\partial_u x(u, v)\partial_v y(u, v) - \partial_v x(u, v)\partial_u y(u, v)).$

6.X Exercises

1. There are 8!/(4!×4!) = (8 × 7 × 6 × 5)/(4 × 3 × 2 × 1) = (2 × 7 × 5) = 70 possible "**simplexes**" defined by four points from the set of vertices of the parallelepiped ABCDEFGH shown in *Figure 6.4.2*. Write a computer program to generate all of these, and then eliminate all those with two or more sides equal as in AB = CD or EH = BC or CH = EB, because any such sets will be coplanar and so cannot define simplexes. Use the program to show that there's no way *any* set of six **tet simplexes** with all six sides equal, and so congruent, can be obtained. Are there in fact *any* **simplexes** that are congruent?

2. Evaluate the **line integral** $(x^2y^2\mathbf{d}x + 2yz\mathbf{d}y - xyz^2\mathbf{d}z)\,|\,C$, where C is the arc of the cubic $y = x^3 - x$, in the ramp $z = 3y$.

3. Evaluate $(x^2 + y^2).\mathbf{d}x\mathbf{d}y\,|\,\Omega$, where Ω is the circular disc about the **origin**, radius R. Do the calculation using **polar coordinates**. Take points $\mathbf{p}_0 = (0, 0)$, $\mathbf{p}_1 = (1, 1)$, $\mathbf{p}_2 = (2, 0)$, $\mathbf{p}_3 = (1, -1)$. Integrate the same **integrand** (i.e.$(x^2 + y^2).\mathbf{d}x\mathbf{d}y$) over the square enclosed by these points (there's a very nifty transformation of variables from (x,y) to (u,v) that trivializes this). (E. Kreyszig, *Advanced Engineering Mathematics*, Wiley 1967.)

4. **Spherical coordinates** in 3-space are normally defined as:

   a. $x = r.\mathbf{cos}\varphi.\mathbf{sin}\theta$

   b. $y = r.\mathbf{sin}\varphi.\mathbf{sin}\theta$

[94] Usually in a somewhat unrecognisable $\sqrt{}(EG - F^2)$ form, due actually to Gauss. e.g. E.Kreyszig, *Advanced Engineering Mathematics*, 2nd ed. Wiley 1968, section 6.6.

c. $z = r.\cos\theta$

in terms of (r, θ, φ) where r is the radial distance from the **origin**, φ is the **longitude**, and θ is the **colatitude** – the angle measured down from the North Pole. Note that φ runs from 0 to 2π, and θ from 0 to π, unlike normal **latitude** which runs from $\pi/2$ (+90°) to $-\pi/2$ (−90°). Expand this **coordinate transform** in terms of the **differentials** {**d**x, **d**y, **d**z} expressed as functions of { **d**r, **d**θ, **d**φ} and also, using an arbitrary auxiliary function $g(\mathbf{x})$, evaluate the **tangent vectors** $\{\partial_r, \partial_\theta, \partial_\varphi\}$ in terms of $\{\partial_x, \partial_y, \partial_z\}$. The two **transforms**, viewed as **matrices**, should be **transposes** of each other, as they represent **transforms** in opposite directions. Are the **vectors** $\{\partial_r, \partial_\theta, \partial_\varphi\}$ **unit vectors**?

5. Do the same for **cylindrical coordinates** defined as:

    a. $x = \rho.\cos(\varphi)$

    b. $y = \rho.\sin(\varphi)$

    c. $z = z$.

6. In all these **coordinates**, the **gradient**, **grad**(φ) or $\nabla\varphi$ for an arbitrary function φ can be expressed as $(1/h_1)\,\partial_\xi\varphi.\mathbf{e}_1 + (1/h_2)\,\partial_\psi\varphi.\mathbf{e}_2 + (1/h_3)\,\partial_\zeta\varphi.\mathbf{e}_3$ where $\{\mathbf{e}_1, \mathbf{e}_2, \mathbf{e}_3\}$ are **unit vectors tangent** to the respective **coordinate curves**, and I write the **coordinates** themselves generically as $\{\xi, \psi, \zeta\}$. $\{h_1, h_2, h_3\}$ are called **scale factors**. So for **spherical coordinates**, $h_1 = 1$, $h_2 = r$ and $h_3 = r.\sin(\theta)$. Show that the general form of the **Laplacian operator**, which in **Cartesian coordinates** is $\frac{\partial^2}{\partial x^2} + \frac{\partial^2}{\partial y^2} + \frac{\partial^2}{\partial z^2}$, is: $(1/(h_1h_2h_3))\left\{\frac{\partial}{\partial\xi}\left(\frac{h_2\,h_3}{h_1}\frac{\partial}{\partial\xi}\right) + \frac{\partial}{\partial\psi}\left(\frac{h_3\,h_1}{h_2}\frac{\partial}{\partial\psi}\right) + \frac{\partial}{\partial\zeta}\left(\frac{h_1\,h_2}{h_3}\frac{\partial}{\partial\zeta}\right)\right\}$. (Murray Spiegel, *Advanced Calculus*, Schaum's Outlines 1963.)

7. Going back to normal **Cartesian coordinates**, the **curl** of a **vector** defined at $\mathbf{x} = (x, y, z)$ as $\mathbf{v} = v_x(\mathbf{x}).\mathbf{i} + v_y(\mathbf{x}).\mathbf{j} + v_z(\mathbf{x}).\mathbf{k}$ is the **vector**: $\mathbf{curl}(\mathbf{v}) = (\partial_y v_z - \partial_z v_y)\mathbf{i} + (\partial_z v_x - \partial_x v_z)\mathbf{j} + (\partial_x v_y - \partial_y v_x)\mathbf{k}$, which clearly echoes the form of the **grade-2 form** appearing in **Stokes's Theorem**. **curl**(**v**) is sometimes written as $\nabla \times \mathbf{v}$. Obviously if **v** is itself a **grad**(φ) form, we have **curl**(**grad**(φ)) = 0, reflecting the **dd**φ nature of this expression. Similarly, defining div(**v**) as the **scalar** $\partial_x v_x + \partial_y v_y + \partial_z v_z$ we have div(**curl**(**v**)) = 0, again reflecting the **dd**(**v**) nature one **grade** up. div(**v**) is sometimes written as $\nabla \cdot \mathbf{v}$. But show that:

    a. **curl**(**curl**(**v**)) = **grad**(div(**v**)) − ∇^2(**v**), where ∇^2 is the **Laplacian** defined in the last question, and

    b. **div**(φ**v**) = φ.div(**v**) + **v** · **grad**φ

    c. **curl**(φ**v**) = φ.**curl**(**v**) + **grad**φ × **v**.

    Note the **Leibnitz** form of these last two results.

8. Express div(**v**) and **curl**(**v**) in terms of **spherical coordinates** using ∂_r, ∂_θ, ∂_φ, and in terms of **cylindrical coordinates** using ∂_ρ, ∂_φ, ∂_z.

9. Given a function φ over the *xy*-plane. Define $f := -\partial_x\varphi$ and $g := \partial_y\varphi$, then $\partial_y g - \partial_x f = \partial_x^2\varphi + \partial_y^2\varphi = \nabla^2(\varphi)$ is the **Laplacian** of φ in the plane. Use **Green's theorem** to express the **integral** of this over a region *R* in the plane as a **line integral** ove **b***R*. (E. Kreyszig, *Advanced Engineering Mathematics*, Wiley1967.)

10. Express **Stokes's theorem** in **cylindrical** and in **spherical coordinates.**

11. A surface in 3-space, **Σ**, can be defined by a **parameterization** using two parameters *u* and *v* as: $\mathbf{x}(u, v) = x(u, v).\mathbf{i} + y(u, v).\mathbf{j} + z(u, v).\mathbf{k}$, somewhat improperly using {**i**, **j**, **k**} as a **basis** for points of the space. Define $\mathbf{dx} = \partial_u\mathbf{x}.\mathbf{d}u + \partial_v\mathbf{x}.\mathbf{d}v$, where $\partial_u\mathbf{x}$ and $\partial_v\mathbf{x}$ are **3-vectors** in space. These are commonly written as $\mathbf{x}_u$ and $\mathbf{x}_v$. Using the numerical concept of **differentials** which allows $(\mathrm{d}u)^2 \neq 0$, where I've chosen to write d*u* to emphasize this interpretation rather that **d***u*, the expression $(\mathbf{x}_u \cdot \mathbf{x}_u)(\mathrm{d}u)^2 + 2(\mathbf{x}_u \cdot \mathbf{x}_v)\mathrm{d}u\mathrm{d}v + (\mathbf{x}_v \cdot \mathbf{x}_v)(\mathrm{d}v)^2$ is called the **first fundamental form** on the surface. Following Gauss, who invented all this stuff, we'll write $E := (\mathbf{x}_u \cdot \mathbf{x}_u)$, $F := (\mathbf{x}_u \cdot \mathbf{x}_v)$ and $G := (\mathbf{x}_v \cdot \mathbf{x}_v)$. So the "**fff**" is usually written $E.(\mathrm{d}u)^2 + 2F.\mathrm{d}u\mathrm{d}v + G.(\mathrm{d}v)^2$. (We won't actually need this form itself, only the *E*, *F* and *G* definitions.) If a curve is defined in the *uv*-plane as $\Gamma:[u(t), v(t)]$, show that the **arc length** of the curve in **Σ** over $t : [a, b]$ is given by $\sqrt{}(E.(\partial_t u)^2 + 2F.\partial_t u\partial_t v + G.(\partial_t v)^2) \,|\, [a, b]$.

12. Show that the **area** in **Σ** of the mapping of the region *R* in the *uv*-plane is given by the **integral**: $\sqrt{}(EG - F^2)$.**d***u***d***v* | *R*. If **Σ** is represented $z = g(u, v)$, so that $\mathbf{x} = x.\mathbf{i} + y.\mathbf{j} + g(x, y).\mathbf{k}$, then *u* becomes *x* and *v* becomes *y*, and this **integral** takes the form $\sqrt{}(1 + (\partial_x g)^2 + (\partial_y g)^2)$.**d***x***d***y* | *R*. Show this. Note that $(EG - F^2) = |\mathbf{x}_u \times \mathbf{x}_v|^2$. (E. Kreyszig, *Advanced Engineering Mathematics*, Wiley 1967.)

Chapter 7

FUNCTIONAL VECTOR SPACES

7.1 Functions as Vectors

Once or twice before, for example in Section 5.1, I suggested that the geometrical entities typified as **standard vectors** were not the only mathematical objects that obeyed the axioms of a **vector space**, and that **functions** defined over particular **domains** could also be treated as **vectors**. This Chapter now develops this idea at some length.

The theory of **functions** as **vector spaces**, which for now I will call the theory of **function spaces** or **spaces of functions**, is rather fragmented by the imposition of different defining conditions that specify the actual **functions** admitted as members of each **space**, but one particular condition dominates. This is the specification of the **domain** over which the **functions** are defined.[95] Historically, among the various possible **domains**, one particular one was of outstanding importance, as this was the **domain** on which the idea of **spaces of functions** was first realized, by Joseph Fourier (1768-1830): the **domain** in question is the **unit circle** that was the foundation of Chapter 4.

When our **vectors** are **functions**, we define the two basic **operations** of **vector algebra** as follows:

Vector Addition: $(\phi + \psi)(x_1, x_2, \ldots x_n) := \phi(x_1, x_2, \ldots x_n) + \psi(x_1, x_2, \ldots x_n)$

Scalar Multiplication: $(\lambda\phi)(x_1, x_2, \ldots x_n) := \lambda.\phi(x_1, x_2, \ldots x_n)$

where $(x_1, x_2, \ldots x_n)$ indicates the **arguments** of the **function**, and λ is a **real number** or **scalar**.[96] ":=" is our usual symbol for "is defined as". With these two definitions, the sum and **scalar multiple** of **functions** are well-defined and obey the axioms of Section 5.3. So, for example, we now have A**xiom** V:

$$c(\phi + \psi) = c\phi + c\psi.$$

For **function spaces**, the **scalar product** is usually defined by an **integral**, and this is always over the **domain** of the **arguments**, so if, for example, the **domain** is defined as the **interval** [0, 1] and therefore our **functions** have only a single **argument**, the **scalar product** of ϕ and ψ would be defined as:

[95] The **domain** of a **function**, again, is the set of values over which its **arguments** are allowed.
[96] It is sometimes a **complex number**.

$\phi \cdot \psi := \phi(x)\psi(x)\mathbf{d}x \,|\, [0, 1]$.

Note how this involves the **integral** of a **1-form** that is *itself* formed using the ordinary product of the two **functions**. This **scalar product** will satisfy all the axioms given in Section 5.4:

Axiom I:	For all ϕ and ψ in the **vector space**,	$\phi \cdot \psi$ is a **real number.**
Axiom II:	**Bilinearity**	$(\alpha\phi + \beta\psi) \cdot \omega = \alpha(\phi \cdot \omega) + \beta(\psi \cdot \omega)$ $\phi \cdot (\alpha\psi + \beta\omega) = \alpha(\phi \cdot \psi) + \beta(\phi \cdot \omega)$
AxiomIII:	**Symmetry (Commutativity)**	$\phi \cdot \psi = \psi \cdot \phi$
Axiom IV:	**Positive Definiteness**	$\phi \cdot \phi \geq 0$.

Because of the historical pre-eminence of the **unit circle** in the theory of **spaces of functions**, and the fact, as shown in Section 4.5 on, that the geometry of the plane, and of the **unit circle** in particular, is especially easily handled by **complex numbers**, we will give much attention to the case where the **functions** ϕ are defined in the **complex plane** and so take the form $\phi(x + jy)$ or $\phi(x, y)$.

Central to the discussion of **function spaces** will be their **dual spaces**, which from Section 5.4 are defined as the spaces of all **linear scalar functions** on the basic underlying space. As in Chapter 5, it will turn out that there is a strong parallel between the **scalar product** on a **function space**, and the action of elements of the **dual space**. For suppose ϕ is a **function** in the basic **function space**, and let it act upon another **function** ψ through the **scalar product** to give a **scalar** value. Then because of the linearity of the **scalar product**, this action actually defines an element of the **dual space** ϕ_D:

$\phi_D(\psi) := \phi \cdot \psi = \phi(x)\psi(x)\mathbf{d}x \,|\, [0, 1]$.

But now there's a big difference. In **function spaces**, there are *more* elements in the **dual space** that *cannot* be defined by the action of the **scalar product** with another member of the basic **function space**. **Functions** have a richer structure than the **standard vectors** of Chapter 5, and this is reflected in the breakdown of this analogy.

The "extra" elements of the **dual space** we will call **singular functionals** below. Traditionally, their existence is seen as due to the idea that there are elements of the space of functions that are analogues of **irrational numbers** like √2. These are numbers that legitimately solve simple polynomial equations (√2 solves $x^2 - 2 = 0$), but which can be proved not to exist as **rational numbers**, which are numbers that can be expressed as the **ratio** (hence "**rational**") of two **integers** like 12954/1120 or 22/7 (a coarse approximation to π, which is one of these numbers).

The **singular functionals** are definable members of the **dual space** that cannot be represented by actual *functions* acting in the **scalar product** above.

In both cases, the traditional solution is to suggest that we can develop sequences of either **rational numbers** or of well-defined functions that can *approximate* the "true" √2, or the "true" **singular functional** ever closer and closer as we extend the sequence *without ever actually getting there*.

In such cases we speak of the "true" value as an imaginary unrealizable entity called the **limit** of the sequence, and say that by postulating the "existence" of all **limits** of all possible sequences, we have "**completed**" the space, be it of numbers or of functions.

The "**completion**" means that all possible equations, including those that don't have actual solutions, now *do* when we include the extra imaginary values.

Now it's well-known that we don't actually have to do this for things like $\sqrt{2}$ as there's an algebraic construction directly analogous to that used to create **complex numbers** from **reals** that allows us to add in all these solutions for **polynomials.**[97]

In conformity with my general paradigm in this book, I'll show that such an approach isn't necessary here either, and we can trivially and legitimately extend the concept of the **differential operator** to act on functions with "breaks" or **discontinuities** that will give our **singular functionals** just as easily.

Discontinuities are one form of local anomaly of a function that are generally described as **singularities.** A **discontinuity** in a function $\phi(x)$ at a point x_0 is a "jump" in the value of ϕ so that $\phi(x_0 - \varepsilon)$ has one value, however close ε is to zero, but $\phi(x_0 + \varepsilon)$ has a completely different value. Another kind of **singularity** is where $\phi(x)$ runs off to infinity at a particular point, as for example $\phi(x) = 1/x$ does at zero.

Singularities or **singular points** will be a recurrent theme of this Chapter.

The approach to their handling that I will develop here doesn't dispense with **limits** altogether, but keeps them well within a simple context so that hopefully the reader, far from being deterred by them, will hardly notice their appearance!

The simplest example of a **singular functional** is where the **dual space** element abstracts the value at a specific point of the **function** on which it operates. So if the **function** ϕ always evaluates to a **real** or **complex number**, then:

$$\delta_a(\phi) := \phi(a)$$

defines δ_a as a valid member of the **dual space** of the set of **functions** to which ϕ belongs. $\delta_a(\phi)$ is a **number**, and $\delta_a()$ is **linear** over its **arguments**:

$$\delta_a(p.\phi + q.\psi) = p.\phi(a) + q.\psi(a) = p.\delta_a(\phi) + q.\delta_a(\psi),$$

where p and q are **numbers**. So δ_a satisfies the requirements that place it in the **dual space**.

But there is no **function** $\delta(x)$ which will give the result:

[97] Solomon Feferman, *The Number Systems*, Addison-Wesley 1964, p.222.

$\delta(x).\phi(x).\mathbf{d}x \mid \mathbf{domain}(\phi) = \delta_a(\phi)$.[98]

In effect, although the **integral scalar product** may give one class of **dual space** elements, and a very large one, we shouldn't be surprised that it cannot exhaust the possibilities when our **basic space** elements are things as subtle as **functions**.[99]

Still, the existence of these extra elements of the **dual space** has motivated a distinct name for the **dual space** elements for **function spaces**: they are called **distributions**. The term **functionals**, a general name for **functions** that take other **functions** as **arguments**, is also used, and **distributions** are commonly called **generalized functions** in the UK, the US and in works translated from the former Soviet Union.

Distributions that can be rigorously realized as operating through the **integral scalar product** are called **regular distributions** or **regular functionals**. **Distributions** like δ_a above, which is called the **Dirac delta distribution** or less rigorously the **Dirac delta function** after the physicist who popularized it, are known as **singular distributions** or **singular functionals**, the term I've used so far. The term **generalized functions** itself is sometimes used as if it just meant the **singular** ones, but strictly should cover *any* elements of the **dual space** of the **functions** currently under discussion.

The main line of development of this theory was traditionally built around a determination to railroad the **singular distributions** into being treated as if they *were* definable in terms of the **integral scalar product**. This has led to a theory of quite exceptional difficulty, loaded with unnecessary philosophical baggage. Worse, the determination to hide away and regularize the "anomalies" has obscured a key feature of these **scalar products**: that when we develop the higher **grade** analogue of **integration by parts**, it *automatically* gives many of our "**singular distributions**" as **integrals** over a lower **grade**. I'll discuss this at some length in Section 7.5.

The heyday of the theory of **distributions** was in the nineteen-fifties and sixties, when analogue signal analysis was an area of intensive research. Since then, interest has waned somewhat, which is perhaps a pity because it is important to much of the mathematics used in physics. The most highly developed formalism of the subject was that pioneered in France by Laurent Schwartz but it was one that never really caught on in Britain, where a more traditional approach was developed by M.J.Lighthill and others. Lighthill's book on the subject is a very concise text. A superb, and more intuitive, presentation of the *application* of the theory is Ronald Bracewell's *The Fourier Transform and its Applications*, but this is actually a much more demanding text than it looks. This Chapter draws extensively, albeit not without reservations, on Schwartz's more popular book which was translated into English under the broad title *Mathematics for the Physical Sciences*. His definitive work was a huge two-volume tome which is only available in French. The best treatment I know of covering the underlying mathematics pertinent to this area is still that by H.K.Nickerson, D.C.Spencer and N.E.Steenrod at Princeton (1959), called simply *Advanced Calculus*. This splendid work has finally been reprinted by Dover in 2011.

Before developing the theory of **distributions**, I will give an elementary treatment of the theory of **complex functions**. This theory leads to an excellent introduction to the handling of **integration** of a

---

[98] I have written the space over which the **scalar product integral** is defined here as just **domain**(ϕ) to cover all cases. It may be [0, 1] or [$-\infty$, ∞], or a variety of other spaces.

[99] Henceforth I'll refer to the underlying **function space** as the **basic space** to distinguish it from its **dual space**.

differential form with a **singularity**. The case will appear quite naturally, and indeed you may not even notice it, but it will serve as an exemplar of what is to follow.

7.2 Elements of Complex Analysis

Complex analysis is the theory of the **calculus** applied to **functions** that **map** from the plane *into* (or onto) the plane. We have already encountered several ways of treating points in the plane. Examples are:

- Defining a point **p** in the plane by its Cartesian **coordinates** x and y, usually expressed as an **ordered pair** (x, y).
- Expressing (x, y) as a **2-vector** $x\mathbf{e}_1 + y\mathbf{e}_2$.
- Pre-multiplying the **2-vector** by $\mathbf{e}_1$ to give $x\mathbf{e}_1\mathbf{e}_1 + y\mathbf{e}_1\mathbf{e}_2 = x + jy$, where we define j to be the **2-pseudoscalar**: $j := \mathbf{e}_1\mathbf{e}_2$, so that $j^2 = -1$.
- Expressing $x + jy$ as a **magnitude** r and an **angle** $\boldsymbol{\theta}$, which can itself be written either as $\theta_x + j\theta_y$, or as $\mathbf{e}^{j\theta} \equiv \mathbf{e}\uparrow j\theta$, which gives $x + jy$ as $r\mathbf{e}^{j\theta}$.

Both the $x + jy$ and $r\mathbf{e}^{j\theta}$ forms are called **complex numbers**, and because of its built-in ability to handle rotations in the plane simply by multiplying by $\theta_x + j\theta_y$ to rotate a point by an *anticlockwise* **angle** $\boldsymbol{\theta}$, or by the **complex conjugate** of $\boldsymbol{\theta}$, which is $\theta_x - j\theta_y$, to turn a point through a *clockwise* **angle** $-\boldsymbol{\theta}$, and for its general power and elegance, the **complex number** formulation is the preferred way to handle points in the plane.

Since x and y are commonly used for the **coordinates** of a point in the plane, the letter z is preferred for the points (or **complex numbers**) themselves, and it is used in various guises, as plain z, as z_1, z_2, and so on, and as Greek ζ. So we will often write $z = x + jy$, and therefore **functions** defined with a point in the plane as **argument** appear as, for example, $\phi(z)$ or $\phi(x + jy)$ rather than $\phi(x, y)$. Because I cannot enter overscores – the usual notation for **complex conjugates** – on this word processor, I will use the alternative, and slightly more rare, asterisk notation, so:

$$z = x + jy \longrightarrow z^* := x - jy.$$

For **functions** that **map** from the plane *into* (or onto) the plane, the *resulting value* of ϕ is *also* a **complex number**, so that ϕ takes the form:

$$\phi(x, y) = \phi(x + jy) = u(x + jy) + jv(x + jy)$$

so that ϕ *itself* has both an x and a y component, with

$$\phi_x(x, y) = u(x + jy)$$
$$\phi_y(x, y) = v(x + jy).$$

The u and v here are simply the parts of the *double* **function** ϕ, which takes a point in the *plane* (x, y) and **maps** it into another point in the plane $(u(x, y), v(x, y))$. Put another way, now treating the plane as the space of **2-vectors**, this is:

$$\phi(x\mathbf{e}_x + y\mathbf{e}_y) = u(x, y)\mathbf{e}_x + v(x, y)\mathbf{e}_y.$$

Bringing in the z notation, the **function** ϕ might be written as:

$$\zeta = \phi(z) = u(z) + jv(z) \equiv u(z)\mathbf{e}_x + v(z)\mathbf{e}_y = u(x, y)\mathbf{e}_x + v(x, y)\mathbf{e}_y.$$

I may refer to these as **complex-to-complex functions**, or to emphasize that the **complex number** aspect is simply an *interpretation*, as **plane-to-plane functions**.

The key theorem that gives the foundation of **complex analysis** is **Green's theorem** from Section 6.8, which I will repeat here:

$$(\partial_x g(x, y) - \partial_y f(x, y)).\mathbf{dxdy}\,|\,\Sigma = (f(x, y).\mathbf{dx} + g(x, y).\mathbf{dy})\,|\,\mathbf{b}\Sigma$$

where Σ is a bounded area in the plane, with **boundary b**Σ.

In the **complex** formulation, this theorem is given a rather elegant form by the introduction of two **derivative operators** rather than using the **differential operator** directly. These two **operators** are defined as:

$$\partial_z := \tfrac{1}{2}(\partial_x - j\partial_y) \quad \text{and} \quad \partial_{z^*} := \tfrac{1}{2}(\partial_x + j\partial_y).$$

These **operators** are not actually numbers, but are treated as obeying the regular algebra of **complex numbers** and distributing across their arguments accordingly. They have a *pseudo*-**complex conjugate** form, and *note which way round they are*: the one with the **complex conjugate** symbol "*" is now the one with the *plus sign*.

Their constituent **derivatives** ∂_x and ∂_y just mean what they always mean – they define the **partial derivatives** or **PD**'s with respect to x and y. These **operators** really only affect $\partial_z z$, $\partial_z z^*$, $\partial_{z^*} z$, and $\partial_{z^*} z^*$ because, like all **derivatives**, they simply cascade down through functions of z and z^* by the **chain rule** of Section 2.2, remembering that $z = z(x, y) = x + jy$, and $z^* = z^*(x, y) = x - jy$. So, for example:

$$\partial_z(z^3) = 3z^2.\partial_z z.$$

They are written the "wrong" way round with good reason, as we see if we expand the four basic forms:

$$\partial_z z = \tfrac{1}{2}(\partial_x - j\partial_y)(x + jy) = \tfrac{1}{2}(\partial_x x + j\partial_x y - j\partial_y x - j^2\partial_y y) = \tfrac{1}{2}(\partial_x x + \partial_y y) = \tfrac{1}{2}(1 + 1) = 1$$
$$\partial_z z^* = \tfrac{1}{2}(\partial_x - j\partial_y)(x - jy) = \tfrac{1}{2}(\partial_x x - j\partial_x y - j\partial_y x + j^2\partial_y y) = \tfrac{1}{2}(\partial_x x - \partial_y y) = \tfrac{1}{2}(1 - 1) = 0$$
$$\partial_{z^*} z = \tfrac{1}{2}(\partial_x + j\partial_y)(x + jy) = \tfrac{1}{2}(\partial_x x + j\partial_x y + j\partial_y x + j^2\partial_y y) = \tfrac{1}{2}(\partial_x x - \partial_y y) = \tfrac{1}{2}(1 - 1) = 0$$
$$\partial_{z^*} z^* = \tfrac{1}{2}(\partial_x + j\partial_y)(x - jy) = \tfrac{1}{2}(\partial_x x - j\partial_x y + j\partial_y x - j^2\partial_y y) = \tfrac{1}{2}(\partial_x x + \partial_y y) = \tfrac{1}{2}(1 + 1) = 1$$

where we remember that $\partial_x y = \partial_y x = 0$, and $\partial_x x = \partial_y y = 1$. So likewise, these **complex derivative operators** give $\partial_z z^* = \partial_{z*} z = 0$ and $\partial_z z = \partial_{z*} z^* = 1$.

Just to show that these **operators** are consistent, I'll give another example:

$$\begin{aligned}\partial_z(z^2) &= ½(\partial_x - j\partial_y)(x + jy)^2 = ½(\partial_x - j\partial_y)(x^2 + 2jxy + j^2y^2)\\ &= ½(2x + 2jy - 2j^2x - 2j^3y) = x + jy + x + jy\\ &= 2(x + jy) = 2z.\end{aligned}$$

Applying ∂_{z*} to a **complex-to-complex function** $\phi(z) = u(z) + jv(z)$, we find:

$$\begin{aligned}\partial_{z*}\phi(z) &= ½(\partial_x + j\partial_y)(u(z) + jv(z)) = ½[\partial_x u(z) + j^2.\partial_y v(z) + j\partial_y u(z) + j\partial_x v(z)]\\ &= ½[\partial_x u(z) - \partial_y v(z) + j.[\partial_y u(z) + \partial_x v(z)]].\end{aligned}$$

So: $\partial_{z*}\phi(z) = 0$ if and only if (1) $\partial_x u(z) = \partial_y v(z)$ and (2) $\partial_y u(z) = -\partial_x v(z)$.

These two subsidiary equations in terms of ∂_x and ∂_y on $u(z)$ and $v(z)$ are called the **Cauchy-Riemann equations**. What makes them important is that they give a critical condition for the application of **Green's theorem**. For if we take a **1-form** in the plane of the form:

$$\phi(z).\mathbf{d}z = (u(z) + jv(z))(\mathbf{d}x + j\mathbf{d}y)$$

where we have simply applied the **differential operator** in the usual way to $z = x + jy$ to get:

$$\mathbf{d}z = \mathbf{d}x + j\mathbf{d}y,$$

the expression for $\phi(z).\mathbf{d}z$ expands to give:

$$\begin{aligned}\phi(z).\mathbf{d}z &= (u(z)\mathbf{d}x + j^2v(z)\mathbf{d}y) + j(u(z)\mathbf{d}y + v(z)\mathbf{d}x)\\ &= (u(z)\mathbf{d}x - v(z)\mathbf{d}y) + j.[u(z)\mathbf{d}y + v(z)\mathbf{d}x].\end{aligned}$$

Now, replacing z with x, y, and applying the **1-form** to a **boundary** of a closed region Σ, the *second* term becomes:

$$j.(v(x, y).\mathbf{d}x + u(x, y).\mathbf{d}y) \,|\, \mathbf{b}\Sigma = j.(\partial_x u(x, y) - \partial_y v(x, y)).\mathbf{d}x\mathbf{d}y \,|\, \Sigma.$$

Applying **Green's theorem**, if $\partial_{z*}\phi(z) = 0$, then $\partial_x u(z) = \partial_y v(z)$ by the first **Cauchy-Riemann equation**, and so the **integral** is zero.

Likewise for the *first* term, we obtain:

$$(u(x, y).\mathbf{d}x - v(x, y).\mathbf{d}y) \,|\, \mathbf{b}\Sigma = (-\partial_x v(x, y) - \partial_y u(x, y)).\mathbf{d}x\mathbf{d}y \,|\, \Sigma,$$

and again, if $\partial_{z*}\phi(z) = 0$, then $\partial_y u(z) = -\partial_x v(z)$ by the second **Cauchy-Riemann equation**, and so the **integral** is zero.

This proves **Cauchy's integral theorem**: that if $\partial_{z*}\phi(z) = 0$, then $\phi(z).\mathbf{d}z \,|\, \mathbf{b}\Sigma = 0$.

Complex-to-complex functions which obey $½(\partial_x + j\partial_y)\phi(z) = \partial_{z*}\phi(z) = 0$ or the **Cauchy-Riemann equations**, are said to be **analytic** or **holomorphic**[100] because they *do* have a valid $\partial_z\phi(z)$ **derivative.** *These terms are very important.*

We will also need a second result, known as **Cauchy's integral formula**. For this, we need first to evaluate the **line integral**:

$$\phi(z).\mathbf{d}z \,|\, \mathbf{C} = (z - z_0)^m \mathbf{d}z \,|\, \mathbf{C}$$

where **C** is a small circle of radius ρ about the point z_0. To evaluate this, we **parameterize C** by t on the **interval** [0, 2π], giving:

$$z(t) = z_0 + \rho\mathbf{e}^{jt}.$$

This is of course using the convenient **parameterization** of a circle introduced in Sections 4.4 and 4.6, this time around a point z_0 away from the **origin**. Now:

$$(z - z_0)^m = \rho^m\mathbf{e}^{jmt} \qquad \text{and} \qquad \mathbf{d}z = j\rho\mathbf{e}^{jt}.\mathbf{d}t,$$

so the **line integral** around the circle, [0, 2π] being the **pullback** of **C**, becomes:

$$(z(t) - z_0)^m.\mathbf{d}z \,|\, \mathbf{C} = (\rho^m\mathbf{e}^{jmt} \times j\rho\mathbf{e}^{jt}).\mathbf{d}t \,|\, [0, 2\pi] = j\rho^{(m+1)} \times \mathbf{e}^{j(m+1)t}\mathbf{d}t \,|\, [0, 2\pi]\ .$$

When $m = -1$, this gives:

$$j\rho^0 \times \mathbf{e}^0\mathbf{d}t \,|\, [0, 2\pi] = j.\mathbf{d}t \,|\, [0, 2\pi] = j.[2\pi - 0] = 2\pi j$$

because $x^0 = 1$ always, and for $m \neq -1$,

$$\mathbf{e}^{j(m+1)t}\mathbf{d}t \,|\, [0, 2\pi] = \mathbf{e}^{j(m+1)t}/(j(m+1)) \,|\, \mathbf{b}[0, 2\pi] = 0,$$

because $\mathbf{e}^0 = \mathbf{e}^{n2\pi} = 0$.

Now we can proceed straight to **Cauchy's integral formula**, which states that

[100] As e.g. Nickerson, Spencer and Steenrod p.510.

> if $\phi(z)$ is **analytic** in a region including a closed path **C**, (i.e. $\partial_{z*}\phi(z) = 0$ in this region), then for any point z_0 enclosed by the path **C**:
>
> $[\phi(z)/(z - z_0)].\mathbf{dz}\,|\,\mathbf{C} = 2\pi j.\phi(z_0)$.

First we evaluate:

$$\partial_{z*}[\phi(z)/(z - z_0)] = \partial_{z*}[\phi(z)(z - z_0)^{-1}] = (\partial_{z*}\phi(z))(z - z_0)^{-1} - \phi(z)(z - z_0)^{-2}\partial_{z*}(z - z_0)$$

$$= 0.(z - z_0)^{-1} - \phi(z)(z - z_0)^{-2}\partial_{z*}z = 0 - \phi(z)(z - z_0)^{-2}.0 = 0$$

so, defining $\psi(z) := \phi(z)/(z - z_0)$, we can say that $\psi(z)$ is also **analytic** *wherever* $z \neq z_0$, so **Cauchy's integral theorem** holds for it.

Now we use the device indicated in *Figure 7.2.1*, putting an arbitrarily small ring $\mathbf{C}_0$ around z_0 and in the annular space – shown shaded in the *Figure* – between **C** and $\mathbf{C}_0$,
$\psi(z) = \phi(z_0)/(z - z_0)$ is **analytic** there and so, by **Cauchy's integral theorem**, the combined **integral**:

$$[\phi(z)/(z - z_0)].\mathbf{dz}\,|\,(\mathbf{C} + \mathbf{C}_0)$$

over the *two* curves is zero. Note that here we go round **C** *anticlockwise* and $\mathbf{C}_0$ *clockwise* to give a consistently orientated **boundary** to the annulus.

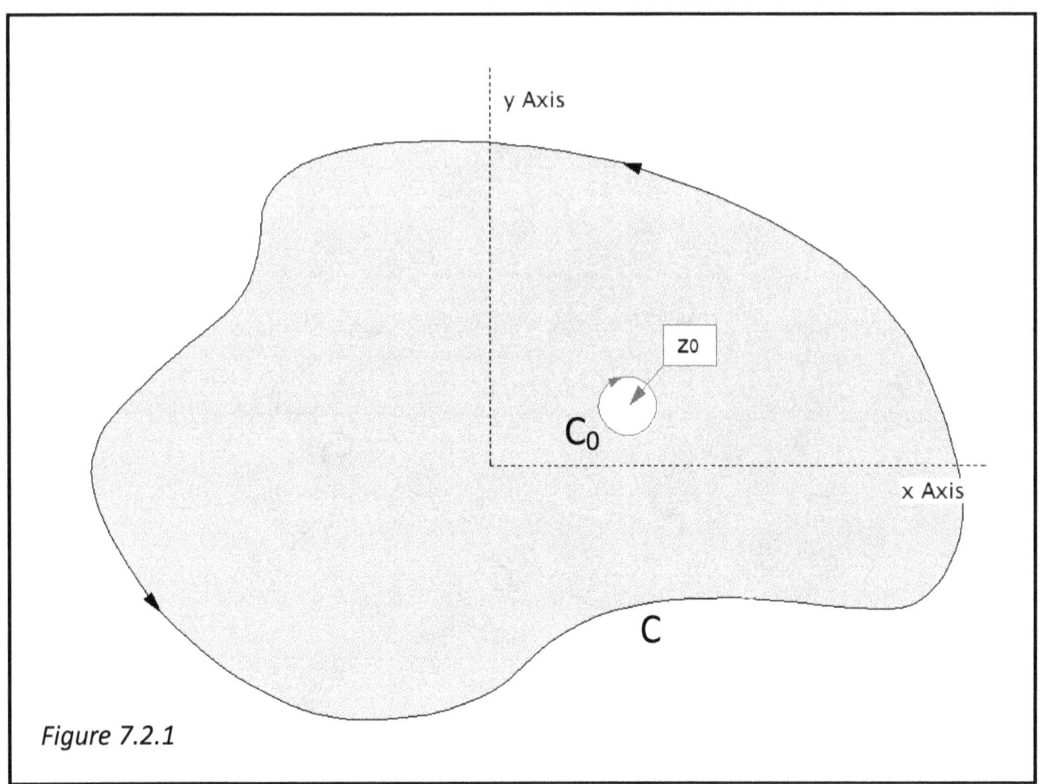

Figure 7.2.1

So it must be that:

$$[\phi(z)/(z - z_0)].\mathbf{d}z \,|\, \mathbf{C} = -[\phi(z)/(z - z_0)].\mathbf{d}z \,|\, \mathbf{C}_0.$$

If $\mathbf{C}_0$ is small enough – and we can make it as small as we choose – then $\phi(z) \approx \phi(z_0)$ throughout the area of $\mathbf{C}_0$, and so:

$$-[\phi(z)/(z - z_0)].\mathbf{d}z \,|\, \mathbf{C}_{0(clockwise)} = [\phi(z)/(z - z_0)].\mathbf{d}z \,|\, \mathbf{C}_{0(anticlockwise)}$$

$$\approx [\phi(z_0)/(z - z_0)].\mathbf{d}z \,|\, \mathbf{C}_{0(anticlockwise)} = \phi(z_0).[(z - z_0)^{-1}.\mathbf{d}z \,|\, \mathbf{C}_0] = 2\pi j.\phi(z_0).$$

This **Cauchy integral formula** opens up tremendous possibilities, because these **line integrals** in the plane come up again and again, and they usually involve **functions** that are **analytic** *everywhere except at a few singular points*. If such is the case, we can "hive off" the **singularities** as shown in *Figure 7.2.2* and evaluate the **line integrals** around them by **Cauchy's integral formula**. Because the **function** is **analytic**, the **line integral** *overall* must be zero, so that over the enveloping curve – **C** in the *Figure* – it must be *equal and opposite* to the sum of the **line integrals** around the **singularities**. This trick is called **integration by the method of residues**.

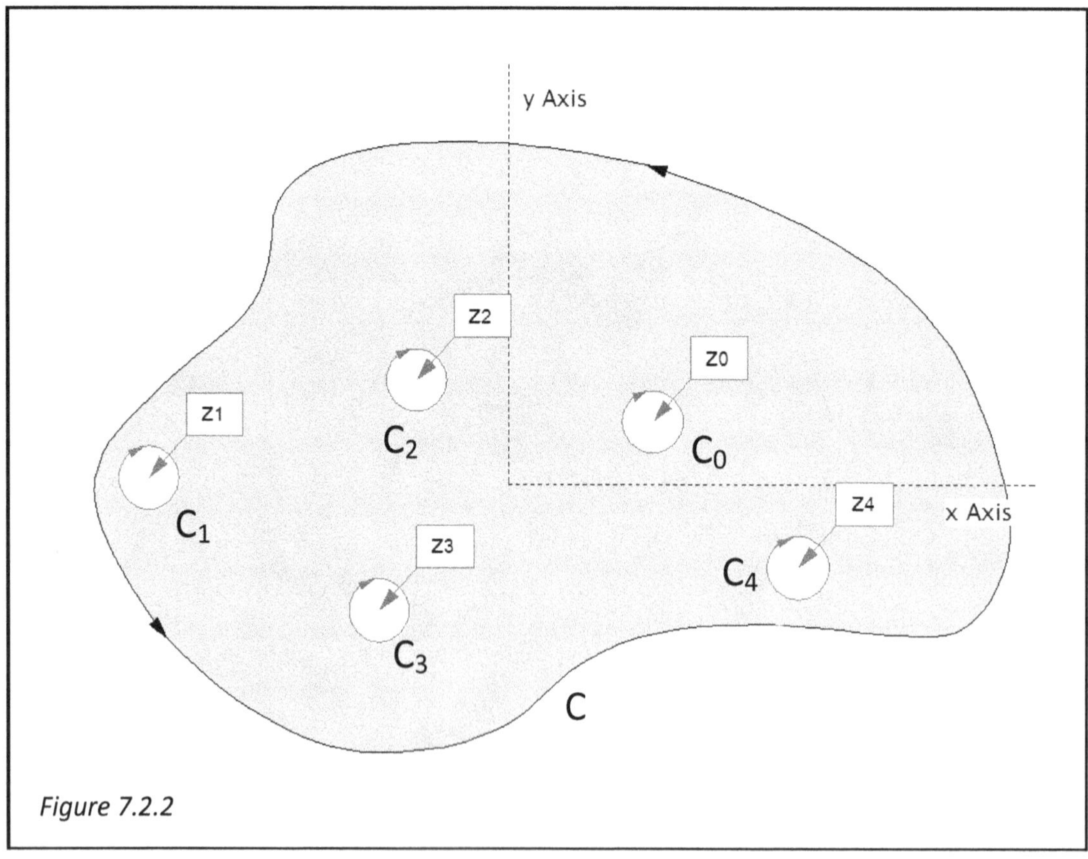

Figure 7.2.2

As an example, suppose we have $\phi(z) = (4 - 3z)/(z^2 - z)$. This can be expanded by **partial fractions** as introduced in Section 1.9:

$$(4 - 3z)/z(z - 1) = A/z + B/(z - 1) = (A(z - 1) + Bz)/z(z - 1).$$

Equating like powers of z, $A + B = -3$, $-A = 4$, so $A = -4$, $B = 1$, and

$$(4 - 3z)/(z^2 - z) = -4/z + 1/(z - 1).$$

Partial fractions, by the way, are much used in evaluating "**integrals**" (in the sense of finding **antiderivatives**).

Now $\phi(z)$ appears as the sum of two **functions** of the type in **Cauchy's integral formula** with **singularities** – points where we get a division by zero – at $z = 0$ and at $z = 1$, and so a **line integral** in $\phi(z)$ can be expanded as:

$$\phi(z)\mathbf{d}z\,|\,\mathbf{C} = [-4/(z - 0)]\mathbf{d}z\,|\,\mathbf{C} + [1/(z - 1)]\mathbf{d}z\,|\,\mathbf{C}$$

$$= [v_0(z_0)/(z - 0)]\mathbf{d}z\,|\,\mathbf{C} + [v_1(z_1)/(z - 1)]\mathbf{d}z\,|\,\mathbf{C}.$$

Here I write 4 as $v_0(z_0)$ with $z_0 = 0$, and similarly 1 as $v_1(z_1)$ with $z_1 = 1$ to emphasize the analogy with the **Cauchy formula**. So if **C** encloses only the **origin** ($z = 0$), the **integral** will be $2\pi j.v_0(z_0) = 2\pi j.(-4) = -8\pi j$; if it encloses only the point $z = 1$, the **integral** will be $2\pi j.v_1(z_1) = 2\pi j.(1) = 2\pi j$; if it encloses *both* it will be $\pi j(2 - 8) = -6\pi j$.

In practice, **integration by residues** has a few more subtleties than this,[101] but this gives the general idea.

Before leaving this Section, I'll mention an interesting historical note, which highlights the influence of fashion even in advanced mathematical work.

Because **complex numbers** are *numbers* and constitute an extension of the normal number system of **real numbers**, it occurred to the hugely influential German mathematician David Hilbert that one could set up **vector spaces** with **complex numbers** acting as the **scalars** of the space. So, in a way, we'd have **n-vector spaces** with **2-vector scalars**! Such spaces are called **Hilbert spaces**, and they are widely used. In particular, because Hilbert made this suggestion at just about the time that **quantum mechanics** was starting to be developed, they came to be used as a foundation for the mathematics of **quantum** physics. This led to the notion that there was something *unavoidably* "complex numberish" about **quantum** theory. This notion has recently been challenged by the Geometric Algebra school, as, for example, GA shows that one of the central quantum algebras, the Pauli algebra, described in Section 5.9, appears naturally in 3-space when the **geometric product** is used. As it happens, this was apparently known to Pauli himself, who knew of Clifford's work.

[101] See also the Appendix on Methods of Integration Section A.6

7.3 Singularities

Up until now, I've avoided the problems where **integration** involves a **differential form** that has **singular points** or **singularities**. But the existence of **singular points** can be used to suggest a more rigorous foundation for incorporating **singular distributions** within the conventional framework and

I'll choose a definition of them that opens up this possibility.

> A **singular point** of a **differential form** ω is a point **p** such that **integration** of the **form** over any volume V including **p**, however small, is non-zero.

You won't see this definition in the textbooks, but it serves our purpose admirably. The last Section already introduced one example, where, if $\partial_z{*}\phi(z) = 0$, the **integral** of any **form** that can be written **d**([$\phi(z)/(z - z_0)$].**d**z) over any area in the plane that includes the point z_0 gives the value $2\pi j\phi(z_0)$. In this case **p** = z_0, and the **2-form** can readily be **integrated** as a **1-form** over the **boundary** of the region: [$\phi(z)/(z - z_0)$].**d**z | **b**Σ = [$\phi(z)/(z - z_0)$].**d**z | **C**.[102]

This example illustrates perhaps the most common origin of **singular points** in a **form** as points where the values of the **coefficient functions** of the **form**, and so the **magnitude** of the **form**, becomes $\pm\infty$. Here we see the common case where a **coefficient function** $\phi(z)/(z - z_0)$ involves a **division by zero**.

This is not the only cause of a **form** having a **singular point**. By contrast, consider the **function** $H(x)$ defined as:

$$H(x) = 1 \text{ if } x > 0,\ H(x) = 0 \text{ if } x < 0.$$

This **function** is called the **Heaviside step function** after Oliver Heaviside – mentioned in Section 5.2 – who, presumably, introduced it. The **function** is commonly regarded as being undefined at $x = 0$, but I suspect that some mileage might be gained by regarding it as **mapping** onto the entire **interval** [0, 1] there: $H(0) = [0, 1]$. I've not seen this suggested. The "break" in the value of the function at zero is an example of a **discontinuity**.

> The **Heaviside function** defines a perfectly valid **differential form** by:
>
> **d**$H(x)$ | $[a, b]$ = $H(x)$ | **b**$[a, b]$ = $H(b) - H(a)$.

If $a < 0$ and $b > 0$, then $H(b) - H(a) = 1$ *always*, however small $|a|$ and $|b|$ may be. So if $\varepsilon > 0$, then:

d$H(x)$ | $[-\varepsilon, \varepsilon]$ = $H(x)$ | **b**$[-\varepsilon, \varepsilon]$ = $H(\varepsilon) - H(-\varepsilon) = 1$

[102] In case you might suspect that this result depends on the choice of a circular path around z_0, Appendix 7.A shows how the **integration** proceeds for a more "fundamental" **simplex** to give the same result.

however small ε may be, precisely fitting my definition above.

The decisive trigger that led to the subject of the present Chapter, the theory of **distributions**, was the realization that *there is no way of representing this* **form d***H*(*x*) as a **linear combination** in **d***x*.

But in the 1920s, a physicist of the first rank, Paul Adrien Maurice Dirac, who was actually brought up in England despite the French name, started writing **d***H*(*x*) as a "normal" **1-form** $\delta(x)$.**d***x*, and this notation caught on in the physics community.

Note that over any **interval** not spanning zero, the **integral** of **d***H* is zero: so for $[a, b]$ where $0 < a < b$, $\mathbf{d}H(x)\,|\,[a, b] = H(x)\,|\,\mathbf{b}[a, b] = H(b) - H(a) = 1 - 1 = 0$, and likewise if $a < b < 0$.

$H(x)$ also has the useful variants $H(x - a)$, which equals 1 if $x - a > 0$, and equals 0 if $x - a < 0$, and so:

$$H(x-a) = 1 \text{ if } x > a, \qquad H(x-a) = 0 \text{ if } x < a.$$

Also important is the *reflected* form $H(a - x)$, equal to 1 if $a - x > 0 \equiv x < a$, and equal to 0 if $a - x < 0$, so:

$$H(a-x) = 1 \text{ if } x < a, \qquad H(a-x) = 0 \text{ if } x > a.$$

I said earlier in this Section that the "extra" elements of the **dual space**, whose operation on **functions** of the underlying **function space** cannot be represented themselves as **functions** working through the **integral scalar product**, can nevertheless be fitted in to the conventional framework by use of the existence of **singularities**. We've now got enough machinery to get a preview of the basic device used to do this in **distribution theory**.

We can *define* Dirac's $\delta(x)$ by the equation $\delta(x)\mathbf{d}x := \mathbf{d}H(x)$. Then in an **integral scalar product** on the **real line** *R* between a **function** $g(x)$, chosen to be zero at $x = \pm\infty$, with $\delta(x)$:

$$g(x)\delta(x)\mathbf{d}x\,|\,[-\infty, \infty] := g(x)\mathbf{d}H(x)\,|\,[-\infty, \infty].$$ [103]

Now by our axiom II of Section 2.1, we can evaluate this by **integration by parts**:

$$\mathbf{d}(g(x)H(x)) = \mathbf{d}g(x).H(x) + g(x).\mathbf{d}H(x)$$

so $$g(x)\mathbf{d}H(x)\,|\,[-\infty, \infty] = \mathbf{d}(g(x)H(x))\,|\,[-\infty, \infty] - \mathbf{d}g(x).H(x)\,|\,[-\infty, \infty].$$

Now

$$\mathbf{d}(g(x)H(x))\,|\,[-\infty, \infty] = (g(x)H(x))\,|\,\mathbf{b}[-\infty, \infty] = g(\infty)H(\infty) - g(-\infty)H(-\infty) = 0$$

by the chosen properties of *g*.[104]

[103] Traditionally, the **domain of integration** represented as $[-\infty, \infty]$ is taken to mean the **real line *R***. But any **integral** $\mathbf{d}\omega\,|\,\boldsymbol{R}$ over ***R*** is *always undefined* because ***R*** has no **boundary**. In the present context take it as a shorthand to mean integration over an **interval** sufficiently large that $g(x) = 0$ at its end points.

The remaining term, $-\,\mathbf{d}g(x).H(x)\,|\,[-\infty, \infty]$, can be split into two by the definition of H:

$$-\,\mathbf{d}g(x).0\,|\,[-\infty, 0] - \mathbf{d}g(x).1\,|\,[0, \infty].$$

The first of these gives zero because $0.\mathbf{d}g(x)$ is identically zero, and the second gives:

$$-\,\mathbf{d}g(x)\,|\,[0, \infty] = -\,g(x)\,|\,\mathbf{b}[0, \infty] = -\,g(\infty) + g(0) = g(0).$$

So $\quad g(x)\delta(x)\mathbf{d}x\,|\,[-\infty, \infty] = g(0),$

which identifies this artificial "**scalar product**" with the perfectly well-defined **functional** δ_0 defined as:

$$\delta_0(g) := g(0).$$

In much the same way, we can define a **1-form** $\delta(x-a)\mathbf{d}x := \mathbf{d}H(x-a)$, and on the same g **function**, this gives:

$$g(x)\delta(x-a)\mathbf{d}x\,|\,[-\infty, \infty] := g(x)\mathbf{d}H(x-a)\,|\,[-\infty, \infty]$$
$$= \mathbf{d}(g(x)H(x-a))\,|\,[-\infty, \infty] - \mathbf{d}g(x).H(x-a)\,|\,[-\infty, \infty]$$
$$= 0 - \mathbf{d}g(x).0\,|\,[-\infty, a] - \mathbf{d}g(x).1\,|\,[a, \infty]$$
$$= -g(x)\,|\,\mathbf{b}[a, \infty] = g(a).$$

So $g(x)\delta(x-a)\mathbf{d}x\,|\,[-\infty, \infty] \equiv \delta_a(g) := g(a)$, again a perfectly well-defined **functional**.

It is not too much of an exaggeration to say that the theory of **distributions** or **generalized functions** develops almost entirely from this basic device.

For $H(a-x)$ we should strictly say that $\mathbf{d}H(a-x) = \delta(a-x)\mathbf{d}(a-x) = -\delta(a-x)\mathbf{d}x$. This would give:

$$g(x)\delta(a-x)\mathbf{d}x\,|\,[-\infty, \infty] := -g(x)\mathbf{d}H(a-x)\,|\,[-\infty, \infty]$$
$$= 0 + \mathbf{d}g(x).1\,|\,[-\infty, a] + \mathbf{d}g(x).0\,|\,[a, \infty] = g(x)\,|\,\mathbf{b}[-\infty, a] = -g(a).$$

But BEWARE. The convention used in this context is that forms such as $\delta(\phi(x))$ are *defined* as $g(x)\delta(\varphi(x))\mathbf{d}x\,|\,[-\infty, \infty] := (\phi(x_0)/|\partial\phi(x_0)|)$ where $\phi(x_0) = 0$ and we use the **absolute value** $|\partial\phi|$ precisely to avoid this sign change.

More consideration is given to this question in Appendix D of this Chapter.

[104] Lighthill refers to **functions** with such suitably chosen terminal conditions as "**good functions**". The specific conditions that make a **function** "**good**" depend on the context. This is the simplest case.

In both this treatment of the **Heaviside function** and in the **Cauchy integral formula**, much the same basic ploy has been used, that we break up the **domain of integration** so as to isolate the **singular point**.

This device exemplifies the way to handle them all. What we did in the **Cauchy formula** was to take as a specific case of **C** a little circle of radius ρ about the point z_0, which we called $\mathbf{C}_0$, and "hive off" the offending **singularity**. The **integral** about z_0 along $\mathbf{C}_0$ was then quite easily evaluated and turned out to be zero except in the one case where $m = -1$, when its value came to $2\pi j$.

In the **real line**, the analogous structure to a little *circle* about a point z_0 is a symmetrical **interval** about the **singular** point x_0 where instead of the *radius* ρ, we write the half-interval distance ε. So we "hive off" the **interval** $[x_0 - \varepsilon, x_0 + \varepsilon]$.

An example is:

$$[1/(x-1)^3].\mathbf{d}x \,|\, [0, 4]$$

where $1/(x-1)^3$ involves a **division by zero** at $x = 1$. We can separate it into two **integrals**:

$$[1/(x-1)^3].\mathbf{d}x \,|\, [0, 1-\varepsilon_1] + [1/(x-1)^3].\mathbf{d}x \,|\, [1+\varepsilon_2, 4].$$

Since $(x-1)^{-3}\mathbf{d}x = -½\mathbf{d}(x-1)^{-2}$, we get:

$$\begin{aligned}
&-½\mathbf{d}(x-1)^{-2} \,|\, [0, 1-\varepsilon_1] - ½\mathbf{d}(x-1)^{-2} \,|\, [1+\varepsilon_2, 4] \\
&= -½(x-1)^{-2} \,|\, \mathbf{b}[0, 1-\varepsilon_1] - ½(x-1)^{-2} \,|\, \mathbf{b}[1+\varepsilon_2, 4] \\
&= -½\{(1-\varepsilon_1-1)^{-2} - (0-1)^{-2}\} - ½\{(4-1)^{-2} - (1+\varepsilon_2-1)^{-2}\} \\
&= -½\{1/(\varepsilon_1)^2 - 1\} - ½\{1/9 - 1/(\varepsilon_2)^2\}.
\end{aligned}$$

Now this indicates a characteristic pattern. Taken as two separate **integrals**, these would both be infinite as $\varepsilon_1 \to 0$ and $\varepsilon_2 \to 0$:

$$= -½\{\infty - 1\} - ½\{1/9 - \infty\}.$$

But if we equate the two ε's as suggested in specifying the **interval** $[x_0 - \varepsilon, x_0 + \varepsilon]$ around the **singularity**, which would here be $[1-\varepsilon, 1+\varepsilon]$, we get overall:

$$-½\{1/(\varepsilon)^2 - 1 + 1/9 - 1/(\varepsilon)^2\} = -½\{1/(\varepsilon)^2 - 1/(\varepsilon)^2 - 1 + 1/9\} = ½ - 1/18 = 8/18.$$

This use of a *symmetrical* **interval** about the **singular** point was suggested by the same Augustin Cauchy (1789 – 1857) who came up with the **Cauchy integral theorem** and **Cauchy integral formula**, and this version, in the **real domain**, is called the **Cauchy principal value**. The principle is much the same.

More commonly, the **magnitude** of a **differential form** will run off to $\pm\infty$ as the **argument** itself goes off to infinity. The case above, where there is a **singular point** at which the **magnitude** of the **differential form** – here the **form** $(x - 1)^{-3}$.**d**x – becomes infinite at a *finite* value of an **argument**, is called an **improper integral of the second kind**.

When the **magnitude** of the **form** goes off to infinity as the **argument** – or *an* **argument** – goes off to infinity, the **integral** is called an **improper integral of the first kind**. This case could never have occurred with our treatment of **integration** in Chapter 6, because there we always **integrated** over **simplexes** or **manifolds** which were necessarily limited to a *finite* extent. In the **theory of distributions**, it is customary to imagine the possibility of **integrals** over an *infinite* **domain**.

I will, however, mention in passing that much the same technique as above can work here. One can always replace the "infinite" **boundary** point with a finite one and see how the **integral** would behave as this grows arbitrarily.

For example, say we have $(1/x^2)$.**d**$x \mid [1, \infty]$. Replace this by:

$$x^{-2}.\mathbf{d}x \mid [1, 1/\varepsilon] = -\mathbf{d}(x^{-1}) \mid [1, 1/\varepsilon] = -(x^{-1}) \mid \mathbf{b}[1, 1/\varepsilon] = -(1/\varepsilon)^{-1} + 1/1 = -\varepsilon + 1.$$

Now as $\varepsilon \to 0$, so that $1/\varepsilon \to \infty$, this result tends to the value 1, so the **integral** has a valid **limit** and is well-defined.

An **integral** that *doesn't* have a valid **limit** value is called **divergent**. In dealing with these kinds of **singularity**, another useful term is that of a **pole**: a **pole** of a **scalar function** is a point where the **function**'s value becomes $\pm\infty$, usually as a consequence of a division by zero – i.e. of a **denominator** in some term becoming zero. By extension we can talk of a **pole** of a **differential form**.

I will refer to the trick of "hiving off" a **singular point** by forming a little **interval** of "radius" ε, or **circle** of radius ε (or in higher-dimensional spaces, an n-dimensional *sphere* of radius ε), as "forming an ε-**sphere** around the **singular point**".

Where the **singularity** is at infinity, the ε-**sphere** around infinity manifests itself as a **sphere** of radius $1/\varepsilon$ *around everything else*! But the idea still works.

Infinite values are not the only kind of **singularity**. The **singularity** in the **Heaviside step function** is a **discontinuity**. On a line or **interval**, this is an easy concept:

> a point x_0 in the **argument** x of a **function** ϕ is a **point of discontinuity** of ϕ if
>
> $$\mathbf{lim}_{\varepsilon\to 0} \mid \phi(x_0 + \varepsilon) - \phi(x_0 - \varepsilon) \mid \neq 0.$$

In other words, two points arbitrarily close on either side of x_0 still have a finite difference in value, no matter how close to x_0 they are. This is what gave the finite **integral** over an arbitrarily small **interval** in the case of the **Heaviside step function**, where the "step" *is* the **discontinuity**.

Very commonly in this theory, we will be interested in **functions** which, although they do not have **discontinuities** themselves, have them in their **derivatives**. In this context, the **derivatives** in question are the traditional **derivatives** that I've sometimes called the **coefficient functions** of the **1-form** derived from the **function** itself. These are the **OD** or **ordinary derivative** for **functions** of one **argument**, or the **PD**'s or **partial derivatives** otherwise. Look at the triangular **function** shown in *Figure 7.3.1*. This is defined as

$$y = 1 - |x| \text{ for } |x| < 1, \text{ or } 1 - x \text{ if } x > 0 \text{ and } x < 1, \text{ or } 1 - (-x) \text{ if } x < 0 \text{ but } x > -1.$$

Its **OD** $\partial_x y$ is 1 for the $x < 0$ part, and -1 for the $x > 0$, so although at $x = 0$, the **function** *itself* is **continuous** with value 1, its **OD** has a **discontinuity** of magnitude 2 there.

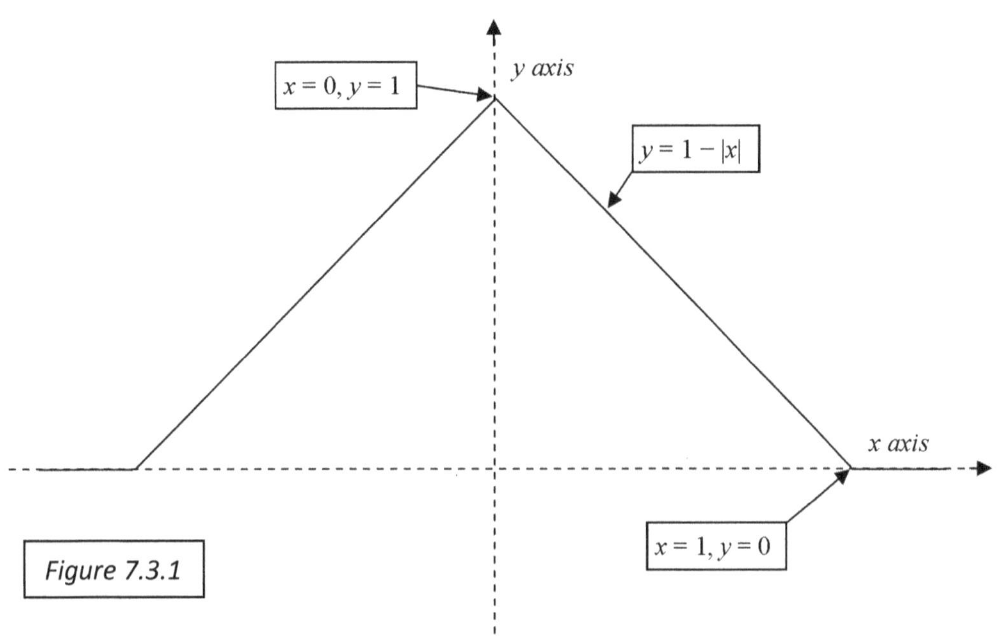

Figure 7.3.1

In higher-dimensional spaces, the concept is not quite so intuitive. Consider the two-**argument** analogue of the **function** we've just looked at, shown in *Figure 7.3.2*. Defining $r = \sqrt{(x^2 + y^2)}$, this **function** is $\phi(x, y) = \phi(r) = 1 - r$, for $0 < r < 1$, and again we'll make it 0 for $r > 1$.
Now $\partial_x\phi = -\frac{1}{2}(x^2 + y^2)^{-\frac{1}{2}}.2x$ which equals $-x/r$.

Now again along the line $y = 0$ – the x **axis** – we have the same pattern as before, because there $r = |x|$, so $\partial_x\phi$ has a step of magnitude 2 at $x = 0$ as $-x/|x|$ goes from 1 to -1.

If this function seems a rather artificial and contrived example, functions with precisely this sort of **discontinuity** *do* in fact arise as **Green's functions** in Section 7.7 later in this Chapter.

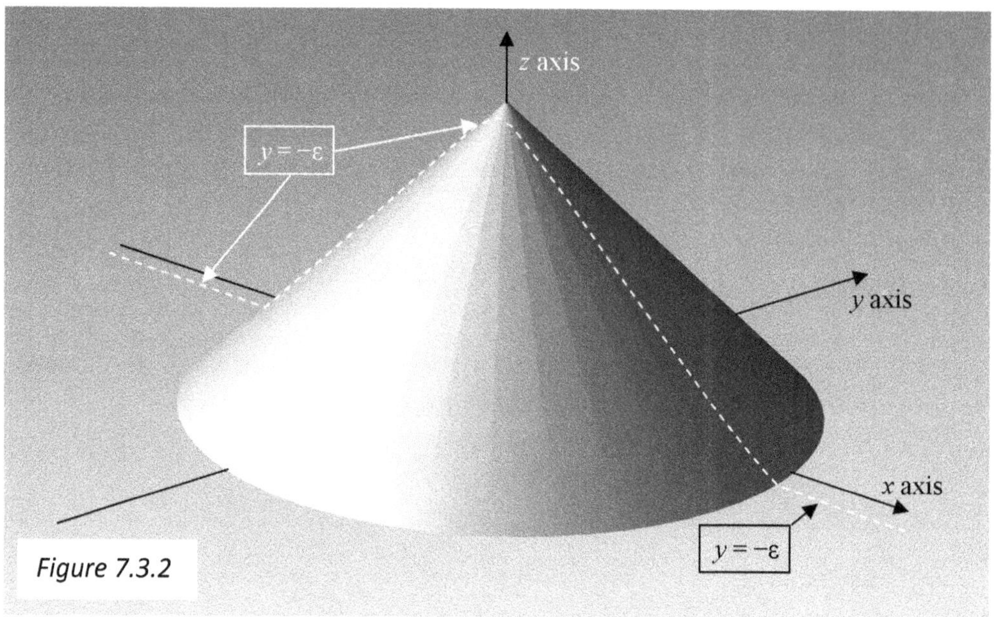

Figure 7.3.2

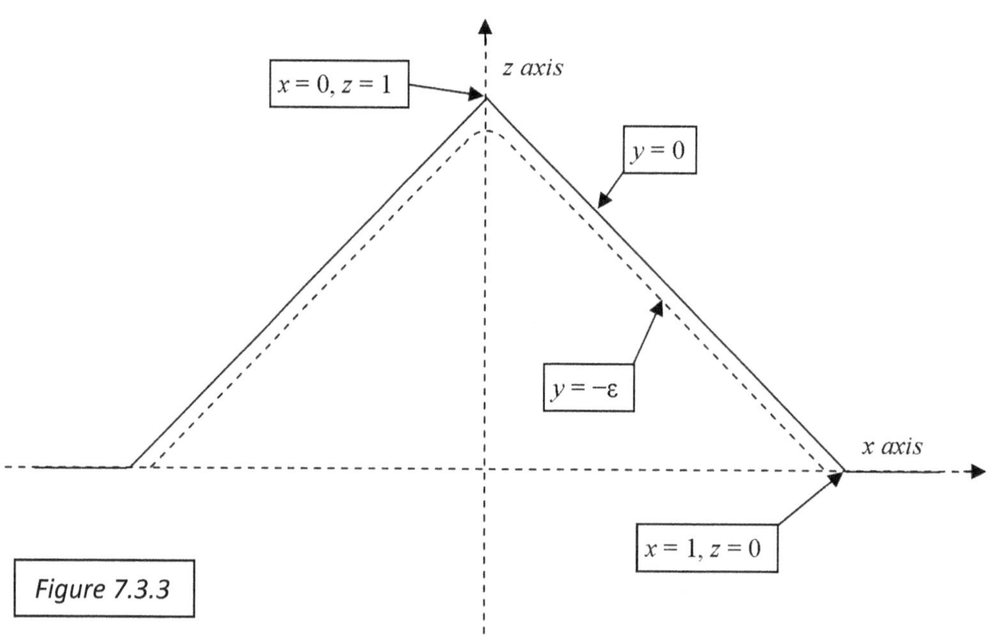

Figure 7.3.3

But along the line $y = -\varepsilon$, we just bypass the **origin** and at $x = 0$, $r = \sqrt{(y^2)} = \varepsilon$, and so $\partial_x\phi = -x/r = 0/\varepsilon = 0$. Along this line, $\partial_x\phi$ varies *smoothly* with the sharp but smooth 90° turn at the top shown as the dotted curve in *Figures 7.3.2* and *7.3.3*. There is now no **discontinuity**. But then along a 45° line, $y = x$, so $r = \sqrt{(2x^2)} = \sqrt{2}.|x|$, and so $\partial_x\phi = -x/r = -x/\sqrt{2}|x|$ and again we do have a **discontinuity**.

Figure 7.3.4 shows another problem area – a **branch line** where the value of ϕ splits into two *sheets* along a line, the *y* axis shown in the illustration running towards the viewer, with a **discontinuity** in ϕ along any line in the plane crossing the **branch line**. (The *Figure* shows the same surface illustrated in different ways.)

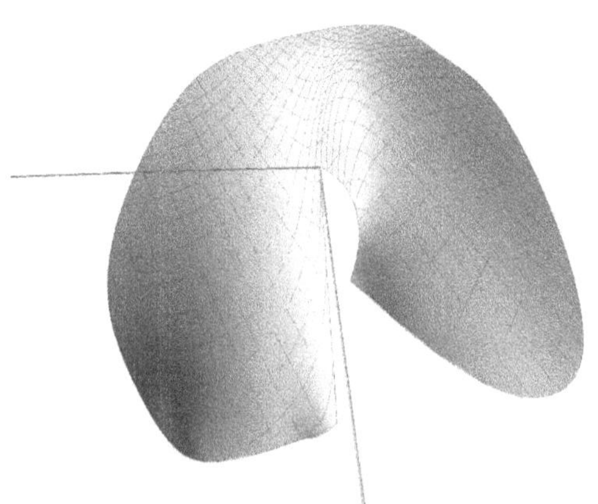

Figure 7.3.4

All of these **singularities** will give unexpected results when we **integrate** the **functions** involved unless we take care to allow for them, because they can all give non-zero **integrals** for arbitrarily small **domains of integration**.

The normal approach is not to define **discontinuity** as I have done, but **continuity** using the following rather counter-intuitive definition:

> A **function** $\phi(x)$ is said to be **continuous** at a point x_0 if, for any **real number** $\varepsilon > 0$, however small, there is a **real number** $\delta > 0$ such that $\Delta(x, x_0) < \delta$ implies that $\Delta(\phi(x), \phi(x_0)) < \varepsilon$.[105]

What is Δ? Δ is just a **function** that takes two **arguments** and returns a **real number** as the **distance** between them. Mathematically, this is called a **metric**. In the present case, we would just take $\Delta(x, x_0) := |x - x_0|$. In a **vector space**, we take the **norm** or **magnitude** of $\mathbf{v} - \mathbf{v}_0$, also written $|\mathbf{v} - \mathbf{v}_0|$.

The idea is that all values of $\phi(x)$ for values of x sufficiently close to x_0 are themselves close.[106]

[105] G.F.Simmons, *Intoduction to Topology and Modern Analysis*, McGraw-Hill 1963, p.76. It's so long since I've used this infernal notion, I had to look it up.

[106] Most texts on the **calculus** would have introduced this right at the start of my Chapter 2, in fact would probably have devoted about three Chapters to this and related matters before starting on the **calculus**. Because the concept *is* difficult, I've been at pains to avoid doing so, but in the present theory these anomalies start to become important. Indeed, one of the great strengths of the theory of **distributions** is that it gives general methods for dealing with these problems.

7.4 A Scalar Product for Differential Forms

As it is normally presented, [107] the **theory of distributions** is not combined with the theory of **differential forms** that is the backbone of most of our work. To bring the two together, the first requirement is to extend the **integral scalar product** introduced in Section 7.1 to higher **grades**. We do this by use of the **Hodge duality** mentioned informally in Section 5.9. We first formalize this concept by the introduction of the **Hodge $*$ operator**, spoken as the "Hodge star operator".

First note that the **dimension** of the **p-grade** space – the space of **p-forms** – in n-space is $({}^{n}{}_{p})$ which is defined as $n!/[p!(n - p)!]$, and is the number of combinations of n things taken p at a time.[108] This always equals $({}^{n}{}_{n-p})$, so the space of **p-forms** and the space of **(n−p)-forms** have the *same* **dimension.**

So in 3-space, say, there are 3×2/(2×1).1 = 3 **linearly independent 2-forms** and the same number of **linearly independent 1-forms.** So we can set up a 1:1 correspondence between them by defining:[109]

$$*(\mathbf{d}x_{i1}\mathbf{d}x_{i2}\ldots\mathbf{d}x_{ip}) := \mathbf{d}x_{j1}\mathbf{d}x_{j1}\ldots\mathbf{d}x_{j(n-p)}$$

where $(i_1, i_2, \ldots i_p, j_1, j_2, \ldots j_{(n-p)})$ is an **even** or **cyclic permutation** of $(1, 2, \ldots, n)$. Thus in 3-space we have from **1-forms** to **2-forms**:

$$*\mathbf{d}x = \mathbf{d}y\mathbf{d}z \qquad *\mathbf{d}y = \mathbf{d}z\mathbf{d}x \qquad *\mathbf{d}z = \mathbf{d}x\mathbf{d}y$$

and in 2-space, we have from **1-forms** to **1-forms**:

$$*\mathbf{d}x = \mathbf{d}y \qquad *\mathbf{d}y = -\mathbf{d}x.$$

The **operator** reverses symmetrically:

$$*\mathbf{d}y\mathbf{d}z = \mathbf{d}x \qquad *\mathbf{d}z\mathbf{d}x = \mathbf{d}y \qquad *\mathbf{d}x\mathbf{d}y = \mathbf{d}z.$$

Now the **exterior** or **alternating product** $\wedge$ of any **p-form** with an **(n−p)-form** is an **n-form** expressible in terms of the unique **pseudoscalar** $\mathbf{d}x_1\mathbf{d}x_2\ldots\mathbf{d}x_n$, which is also sometimes called the **volume element** of the n-space. So where **D** represents either a specific finite n-dimensional subspace of the n-space, or the entire (infinite) n-space itself,

[107] These Sections draw largely on the excellent text by H.K.Nickerson, D.C.Spencer and N.E.Steenrod, *Advanced Calculus*, Van Nostrand, Princeton 1959. Until the year 2011, this superb book was virtually unobtainable. But it has now been reissued by Dover.

[108] $n!$ here is the **factorial function** defined: $n! := n \times (n-1) \times (n-2) \times \ldots \times 2 \times 1$.

[109] This is the formal definition of **Hodge duality** which differs in sign from the **geometric algebra** analogue given in Section 5.9, as was mentioned there.

we can define a (**functional**) **scalar product** of any two **p-forms** ω, θ by:

$$\langle\omega, \theta\rangle := (\omega \wedge *\theta) \,|\, \mathbf{D},$$

giving a **real number** as a result.

Here $\omega \wedge *\theta$ is an **n-form** or a **form** on the **pseudoscalar**. Thus the **scalar product** of two **1-forms** in 2-space might use the **2-form**:

$$(E_{1x}\mathbf{d}x + E_{1y}\mathbf{d}y) \wedge *(E_{2x}\mathbf{d}x + E_{2y}\mathbf{d}y) = (E_{1x}\mathbf{d}x + E_{1y}\mathbf{d}y) \wedge (E_{2x}\mathbf{d}y - E_{2y}\mathbf{d}x)$$

$$= (E_{1x}E_{2x} + E_{1y}E_{2y})\mathbf{d}x\mathbf{d}y$$

which is familiar in the theory of waveguides in the **integral**:

$$\boldsymbol{E}_1 \cdot \boldsymbol{E}_2.\mathbf{d}\Sigma \,|\, \Sigma$$

across the transverse section Σ of a waveguide.

The angular bracket notation for the **integral scalar product** is the normal one used here, *not* the dot notation. It is also customary to extend this notation to cases where one argument is actually a **singular distribution** and so where no actual **integral** is involved. So we commonly write

$$\langle\delta(x - a), \phi(x)\rangle = \phi(a)$$

and this is probably a better habit than "pretending" that there is some **function** $\delta(x - a)$ that will actually realize this as $\delta(x - a)\phi(x).\mathbf{d}x \,|\, [-\infty, \infty]$. That said, remember that we *define* $\langle\delta(x - a), \phi(x)\rangle$ by using $\delta(x - a)\mathbf{d}x := \mathbf{d}H(x - a)$ giving $\langle\delta(x - a), \phi(x)\rangle = \mathbf{d}H(x-a).\phi(x) \,|\, \mathbf{D}$, and this does conform to our definition.

In Sections 8.2 and 8.3 I will also use the bracket notation to extend the product to the case of two *actual*, and so **regular**, **functions** which have a **divergent scalar product** which can nevertheless be construed as a **delta function**.

So the angle-bracket notation will often be used as a "catch-all" to cover the cases where some **singular functional** is involved.

On the other hand, there will be times when I want to make it clear that I *am* using the **integral scalar product** *explicitly* even though the angular bracket notation seems natural. To do this I'll write the **dot** of the **dot product** instead of the comma:

$$\langle\phi \cdot \psi\rangle = \phi \wedge *\psi \,|\, \mathbf{D}$$

and this expressly disallows the possibility that ϕ or ψ is a **singular distribution**.

So I will *not* write $\langle \delta_a \cdot \phi \rangle$.

Armed with a general concept of an **integral scalar product** for **differential forms**, we'll try to translate the theory of Chapter 5 from a theory of *discrete* **functions** with **integer arguments** into a theory of *continuous* **functions** with **real** (or **complex**) **number arguments**. Because the natural higher **grade** objects here are **differential forms**, they will be the main elements that we now use.

Our theory will differ from the discrete version primarily because **singular points** in the **forms** necessitate that we allow elements of the **dual spaces** that cannot be represented as ordinary **functions**.

It has to be said that the textbooks generally make a frightful pig's ear of this theory, and I probably won't do much better! In the hope of making less of a hash of things, however, I will from time to time refer back to the discrete theory as a guide, and to illustrate what we're trying to do.

So to reiterate: to distinguish it from its **dual space**, I will henceforth refer to the underlying **vector space** of **functions** (or **differential forms**) over a defined **domain** as the **basic space**, as mentioned in a footnote in Section 7.1. So if g is an element of the **basic space** and ϕ an element of the **dual space**, $\phi(g)$ is a **scalar** result, either a **real** or a **complex number**, and $\phi(g)$ may also be written $\langle \phi, g \rangle$, *whether or not it can be realized as an* **integral scalar product**.

Elements of the **functional dual space** that *can* be represented by members of the **basic space** operating through the **integral scalar product** are **regular functionals**, and those that *can't* are **singular functionals**.

As mentioned earlier, a common approach to the definition of **singular functionals** like the **Dirac delta functional** is to regard them as a **limiting** case of a series of **regular functionals**. This is the preferred approach in most British textbooks, but one I will try to avoid using. We could use it, for example, to define **singular p-form functionals**, by requiring these to be the limit of a sequence of **regular p-form functionals**:

$$T(\theta) = \mathbf{lim}_{(i \to \infty)} \langle \omega_i, \theta \rangle$$

for some sequence $\{\omega_i\}$. (The classic example of such a sequence is the approximation of the **Dirac delta functional** $\delta(x - x_0)$ by a sequence of rectangular **functions** centred about x_0, of ever decreasing width (say $1/h$) but increasing height (say h) so that their area is constant = 1.)

In choosing to base the theory on **differential forms** rather than on a simple relationship between **functions** in the **basic space** and **functionals** in the **dual space**, I may be seeming to add unnecessarily to the difficulties of the subject area, because we now have a *continuous* vectorial aspect to the **functions** combined with the established *discrete* vectorial aspect of the **graded algebra** of **differential forms**. But only by using **forms** can we see the correct application of the **Green's formulae**, as we shall see in the following Sections.

So what are the **Green's formulae**?

In Section 7.3, I showed how **integration by parts** (using axiom II of Chapter 2) can be used to give a meaning to the **singular functionals** by casting them as the "odd" term of the switch:

$$u\mathbf{d}v = -\, v\mathbf{d}u + \mathbf{d}(uv).$$

Here the last term is the "odd" term, because this can easily be switched onto the **boundary** of the **function domain** in the **integral scalar product**. This simplest case we can now write as:

$$\langle u, \mathbf{d}v\rangle = u\mathbf{d}v \,|\, \Sigma = -\,\langle \mathbf{d}u, v\rangle + \mathbf{d}(uv) \,|\, \Sigma = -\, v\mathbf{d}u \,|\, \Sigma + uv \,|\, \mathbf{b}\Sigma.$$

Green's formulae extend this idea.

7.5 Green's Formulae and The Functional Derivative

Given a **p-form** ϕ, on an **n-manifold M**, the **co-differential operator**[110] $\boldsymbol{\delta}$ acting on ϕ gives the **(p−1)-form** defined:

$$\boldsymbol{\delta}\phi := (-1)^{np+n+1} {*}\mathbf{d}{*}\phi,$$

where $*$ is the **Hodge $*$ operator**.

The importance of the **co-differential operator** is that it gives the general **Green's formulae** which are related to **integration by parts**[111] much as the general **Stokes's theorem** is to the **fundamental theorem of calculus** – they extend it to higher **grades**. It also enables us to introduce a wider variety of **differential operators** because the $*$ **operator** can bring the **grade** of an already **differentiated form** back down, and so enable us to apply the basic **differential operator** again.

In the statements of the **Green's formulae** which follow, the angular bracket notation always refers to *actual* **integral scalar products**. **Singular distributions** don't come into it. So here I'll use the dot rather than the comma between the arguments. We're looking here at the real theoretical underpinnings that are generally obscured in **distribution** theory.

Let ϕ be a **(p−1)-form** which is **non-singular** – i.e. free of **singularities** – over Σ, and ψ a **p-form** likewise **non-singular** over Σ. Then the basic **Green's formula** is:

I.	$\langle \mathbf{d}\phi \cdot \psi\rangle - \langle \phi \cdot \boldsymbol{\delta}\psi\rangle = (\phi \wedge {*}\psi) \,|\, \mathbf{b}\Sigma$

[110] In choosing this term I follow Theodore Frankel in *The Geometry of Physics*, CUP 2004, p. 364, although I use the traditional $\boldsymbol{\delta}$ rather than his $\mathbf{d}^*$, partly to avoid any suggestion of **complex conjugation**.

[111] **Integration by parts**, again, is the use of axiom II in Section 2.1 "inside out" to switch a form $v.\mathbf{d}u = \mathbf{d}(uv) - u.\mathbf{d}v$.

where the **scalar products** are over Σ, and **b**Σ is the **boundary** of Σ. Note that the first **scalar product** is a product of **p-forms**, the second of **(p−1)-forms**, but both are evaluated over the **n-manifold** Σ, whereas the **integral** on the right is over the **(n−1)-manifold bΣ**.

To grasp what's involved here, suppose we're in 3-space, and $\phi = \phi_x\mathbf{d}x + \phi_y\mathbf{d}y + \phi_z\mathbf{d}z$, and $\psi = \psi_x\mathbf{d}y\mathbf{d}z + \psi_y\mathbf{d}z\mathbf{d}x + \psi_z\mathbf{d}x\mathbf{d}y$. Then:

$$(\phi \wedge *\psi) = (\phi_x\mathbf{d}x + \phi_y\mathbf{d}y + \phi_z\mathbf{d}z) \wedge *(\psi_x\mathbf{d}y\mathbf{d}z + \psi_y\mathbf{d}z\mathbf{d}x + \psi_z\mathbf{d}x\mathbf{d}y)$$

$$= (\phi_x\mathbf{d}x + \phi_y\mathbf{d}y + \phi_z\mathbf{d}z) \wedge (\psi_x\mathbf{d}x + \psi_y\mathbf{d}y + \psi_z\mathbf{d}z)$$

$$= (\phi_x\psi_y - \phi_y\psi_x)\mathbf{d}x\mathbf{d}y + (\phi_z\psi_x - \phi_x\psi_z)\mathbf{d}z\mathbf{d}x + (\phi_y\psi_z - \phi_z\psi_y)\mathbf{d}y\mathbf{d}z.$$

Now evaluate **d** on this by our established method:

$$\mathbf{d}(\phi \wedge *\psi) = \partial_z(\phi_x\psi_y - \phi_y\psi_x)\mathbf{d}z\mathbf{d}x\mathbf{d}y + \partial_y(\phi_z\psi_x - \phi_x\psi_z)\mathbf{d}y\mathbf{d}z\mathbf{d}x$$

$$+ \partial_x(\phi_y\psi_z - \phi_z\psi_y)\mathbf{d}x\mathbf{d}y\mathbf{d}z$$

$$= [\partial_z(\phi_x\psi_y - \phi_y\psi_x) + \partial_y(\phi_z\psi_x - \phi_x\psi_z) + \partial_x(\phi_y\psi_z - \phi_z\psi_y)]\mathbf{d}x\mathbf{d}y\mathbf{d}z. \quad (7.5.1)$$

But we can now use the general formula "axiom IVb" introduced in Section 6.6, for any two **forms** φ and ψ, where φ is a **p-form**:

$$\mathbf{d}(\phi \wedge \psi) = \mathbf{d}\phi \wedge \psi + (-1)^p(\phi \wedge \mathbf{d}\psi)$$

which is the extension of our original axiom II of Section 2.1 to arbitrary **grade forms**. This gives a second method for evaluating a result like $\mathbf{d}(\phi \wedge *\psi)$ above.

To see this for our present example, first note that φ is a **1-form**, so the new formula gives:

$$\mathbf{d}(\phi \wedge *\psi) = \mathbf{d}\phi \wedge *\psi - (\phi \wedge \mathbf{d}(*\psi))$$

$$= [(\partial_x\phi_y - \partial_y\phi_x)\mathbf{d}x\mathbf{d}y + (\partial_z\phi_x - \partial_x\phi_z)\mathbf{d}z\mathbf{d}x + (\partial_y\phi_z - \partial_z\phi_y)\mathbf{d}y\mathbf{d}z] \wedge (\psi_x\mathbf{d}x + \psi_y\mathbf{d}y + \psi_z\mathbf{d}z)$$

$$- (\phi_x\mathbf{d}x + \phi_y\mathbf{d}y + \phi_z\mathbf{d}z) \wedge [(\partial_x\psi_y - \partial_y\psi_x)\mathbf{d}x\mathbf{d}y + (\partial_z\psi_x - \partial_x\psi_z)\mathbf{d}z\mathbf{d}x + (\partial_y\psi_z - \partial_z\psi_y)\mathbf{d}y\mathbf{d}z]$$

$$= [(\partial_x\phi_y.\psi_z - \partial_y\phi_x.\psi_z) + (\partial_z\phi_x.\psi_y - \partial_x\phi_z.\psi_y) + (\partial_y\phi_z.\psi_x - \partial_z\phi_y.\psi_x)]\mathbf{d}x\mathbf{d}y\mathbf{d}z$$

$$- [(\phi_z.\partial_x\psi_y - \phi_z.\partial_y\psi_x) + (\phi_y.\partial_z\psi_x - \phi_y.\partial_x\psi_z) + (\phi_x.\partial_y\psi_z - \phi_x.\partial_z\psi_y)]\mathbf{d}x\mathbf{d}y\mathbf{d}z.$$

Careful comparison with (7.5.1) above shows these to be the same!

In the statement of the basic **Green's formula**, φ is a **(p−1)-form** and ψ a **p-form**, so we have:

$$\mathbf{d}(\phi \wedge *\psi) = \mathbf{d}\phi \wedge \psi + (-1)^{p-1}(\phi \wedge \mathbf{d}(*\psi)) = \mathbf{d}\phi \wedge \psi - (-1)^{p}(\phi \wedge \mathbf{d}(*\psi))$$

but $*\boldsymbol{\delta}\psi = (-1)^{np+n+1}**\mathbf{d}*\psi = (-1)^{np+n+1}(-1)^{n(n-p+1)+n-p+1}\mathbf{d}*\psi = (-1)^{p}\mathbf{d}*\psi$, where we've used the formula $**\phi = (-1)^{np+p}\phi$, knowing that with ψ a **p-form**, $*\psi$ is an **(n−p)-form**, so $\mathbf{d}(*\psi)$ is an **(n−p+1)-form**. The details of the sign calculation are given by:

$$\begin{aligned}&np + n + 1 + n(n - p + 1) + n - p + 1\\ &= np + n + 1 + n^2 - np + n + n - p + 1 = n^2 + 3n - p + 2\\ &= n(n + 3) + 2 - p.\end{aligned}$$

$n(n + 3)$ is the product of an even and an odd number and so must be even, 2 is of course even, so the final sign of (-1) is given by $-p$.

Now we just form the **integrals**:

$$\mathbf{d}(\phi \wedge *\psi)\,|\,\Sigma = (\phi \wedge *\psi)\,|\,\mathbf{b}\Sigma = (\mathbf{d}\phi \wedge *\psi)\,|\,\Sigma - (\phi \wedge *\boldsymbol{\delta}\psi)\,|\,\Sigma$$

$$= \langle \mathbf{d}\phi, \psi\rangle - \langle \phi, \boldsymbol{\delta}\psi\rangle.$$

Various other **Green's formulae** derive from this, perhaps the most important being the **Laplacian** one for the **Laplace-Beltrami operator** $\boldsymbol{\Delta} := \mathbf{d}\boldsymbol{\delta} + \boldsymbol{\delta}\mathbf{d}$ where <u>now ϕ and ψ are **forms** of the same **grade**</u>:[112]

V. $\quad \langle \boldsymbol{\Delta}\phi \cdot \psi\rangle - \langle \phi \cdot \boldsymbol{\Delta}\psi\rangle = (\boldsymbol{\delta}\phi \wedge *\psi - \boldsymbol{\delta}\psi \wedge *\phi + \phi \wedge *\mathbf{d}\psi - \psi \wedge *\mathbf{d}\phi)\,|\,\mathbf{b}\Sigma.$

I label this as (V) because there are three other **Green's formulae** that lead to the derivation of this one, which will appear at the end of this Section.

To put this **Δ operator** into a more familiar form, consider $\Delta\phi(x, y, z)$ where $\phi(x, y, z)$ is simply a **scalar function** or **0-form** on 3-space. Now $*\phi(x, y, z) = \phi(x, y, z)\mathbf{d}x\mathbf{d}y\mathbf{d}z$, so
$\mathbf{d}*\phi = 0$, and the $\mathbf{d}\boldsymbol{\delta}$ term goes no further.

$\mathbf{d}\phi(x, y, z) = \partial_x\phi(x, y, z)\mathbf{d}x + \partial_y\phi(x, y, z)\mathbf{d}y + \partial_z\phi(x, y, z)\mathbf{d}z$, however, and this being a **1-form**, $p = 1$, so $np + n + 1 = 7$, and we get:

$$*\mathbf{d}\phi = \partial_x\phi\mathbf{d}y\mathbf{d}z + \partial_y\phi\mathbf{d}z\mathbf{d}x + \partial_z\phi\mathbf{d}x\mathbf{d}y$$

$$\mathbf{d}*\mathbf{d}\phi = (\partial_x^2\phi + \partial_y^2\phi + \partial_z^2\phi)\mathbf{d}x\mathbf{d}y\mathbf{d}z$$

$$-*\mathbf{d}*\mathbf{d}\phi = -(\partial_x^2\phi + \partial_y^2\phi + \partial_z^2\phi).$$

[112] Easily missed one line from the bottom of p. 434 in Nickerson, Spencer and Steenrod!

This is the usual form of the **Laplacian**, albeit with the wrong sign. The difference in sign is intentional.[113]

Integration by parts is subsumed under this formula if we work in R ($n = 1$) with **0-forms** f and g. In R, the **pseudoscalar** is just $\mathbf{d}x$, and p can only be 0 or 1, so the **Hodge duality** is simply between the **scalar 0-forms** and the **pseudoscalar 1-form** and appears as $*\mathbf{d}x = 1$ and $*1 = \mathbf{d}x$. So, putting f for the **0-form** ϕ and $*g = g\mathbf{d}x$ for the **1-form** ψ, so that $*\psi = **g = g$, and $p = 1$ and so $np + n + 1 = 1 + 1 + 1 = 3$, we obtain, for **integration** over an **interval I**:

$$\langle \mathbf{d}\phi \cdot \psi\rangle = \langle \mathbf{d}f \cdot g\mathbf{d}x\rangle = [\partial_x f.\mathbf{d}x \wedge *(g\mathbf{d}x)]\,\big|\,\mathbf{I} = [g.\partial_x f.\mathbf{d}x]\,\big|\,\mathbf{I}$$

$$= \langle f \cdot \boldsymbol{\delta}\psi\rangle + (f \wedge **g)\,\big|\,\mathbf{bI} = \langle f \cdot \boldsymbol{\delta}(*g)\rangle + (f.g)\,\big|\,\mathbf{bI}$$

$$= (-1)^3\langle f \cdot *\mathbf{d}g\rangle + fg\,\big|\,\mathbf{bI} = -\langle f \cdot \partial_x g(*\mathbf{d}x)\rangle + (f.g)\,\big|\,\mathbf{bI} = -[f.\partial_x g.\mathbf{d}x]\,\big|\,\mathbf{I} + (f.g)\,\big|\,\mathbf{bI}.$$

So, all told:

$$[g.\partial_x f.\mathbf{d}x]\,\big|\,\mathbf{I} = -[f.\partial_x g.\mathbf{d}x]\,\big|\,\mathbf{I} + (f.g)\,\big|\,\mathbf{bI} = -[f.\partial_x g.\mathbf{d}x]\,\big|\,\mathbf{I} + \mathbf{d}(f.g)\,\big|\,\mathbf{I}$$

which we have obtained before from our original axiom II as:

$$g.\partial_x f.\mathbf{d}x = -f.\partial_x g.\mathbf{d}x + \mathbf{d}(f.g).$$

If $\mathbf{I} = [a, b]$, but, say, f is **discontinuous** at x_0, with $a < x_0 < b$, then we form an **ε-sphere** around x_0 and this means we take the **principal value** form:

$$\mathbf{lim}_{(\varepsilon\to 0)}\,\{\omega\,\big|\,[a, x_0 - \varepsilon] + \omega\,\big|\,[x_0 + \varepsilon, b]\}$$

for any **1-form** ω being **integrated** over **I**. In practice, this means that $\mathbf{I}_\varepsilon$ has *four* **boundary** points.

The crucial consequence of the **Green's formula** for our present purposes is this: <u>**derivative operators** *in general do not have a simple* **adjoint** *over the* **scalar product**</u>. You can see it above:

1. $\langle \mathbf{d}\phi \cdot \psi\rangle \neq \langle \phi \cdot \boldsymbol{\delta}\psi\rangle$
2. $\langle \boldsymbol{\Delta}\phi \cdot \psi\rangle \neq \langle \phi \cdot \boldsymbol{\Delta}\psi\rangle$.

In both cases, an extra term over the **boundary** of the region of **integration** is thrown up.

A key trick of the theory of **distributions** is to define a synthetic **adjoint** using the **Green's formulae**. We define the **adjoint exterior derivative** (my terminology[114]) of a **p-form** ϕ as the **(p+1)-form functional** $\mathbf{d}^\dagger\phi$ defined:

$$\mathbf{d}^\dagger\phi(\psi) := \langle \phi \cdot \boldsymbol{\delta}\psi\rangle$$

---

[113] Nickerson, Spencer and Steenrod again, p.436

[114] Remember the **exterior derivative** is the formal name for the application of the **differential operator** to a **p-form** to generate a **(p+1)-form**.

$$= \langle \mathbf{d}\phi \cdot \psi\rangle - (\phi \wedge *\psi)\,|\,\mathbf{b}\Sigma,$$

for any **(p+1)-form** ψ.

This is evidently a **singular functional**, since there is no **(p+1)-form** η obeying $\mathbf{d}^{\dagger}\phi(\psi) = \langle \eta \cdot \psi\rangle$ for any ψ. Similarly we can define the **adjoint co-derivative** as the **(p−1)-form functional** $\boldsymbol{\delta}^{\dagger}\phi$ defined:

$$\boldsymbol{\delta}^{\dagger}\phi(\Sigma) := \langle \mathbf{d}\Sigma \cdot \phi\rangle$$

$$= \langle \Sigma \cdot \boldsymbol{\delta}\phi\rangle + (\Sigma \wedge *\phi)\,|\,\mathbf{b}\Sigma$$

for any **(p−1)-form** Σ.

Now for **0-forms** f, g on R (n = 1), this (first) definition takes the form:

$$\mathbf{d}^{\dagger}f\,(g\mathbf{d}x) = \langle f \cdot \boldsymbol{\delta}(g\mathbf{d}x)\rangle = -\,\langle f \cdot \partial_x g\rangle = \langle \mathbf{d}f \cdot g\mathbf{d}x\rangle - (f \wedge *g)\,|\,\mathbf{bI}$$

$$= g.\partial_x f.\mathbf{d}x\,|\,\mathbf{I} - fg\,|\,\mathbf{bI}$$

where **bI** may "ring off" any **singularities** of f or g. This is the operation which Schwartz calls **differentiation in the sense of distributions**, and he presents it a being the "natural extension" of **differentiation** to **distributions**, subsequently using it implicitly whenever **distributions** are involved. As this approach can cause confusion, especially when a form like $\langle T, f\rangle$ may involve a **distribution** *or* a **function** (hence my preference for $T(f)$ when T *is* a **distribution**), I find it more acceptable to define it, as here, as a *distinct* operation.

Suppose f is the **Heaviside step function** $H_{x0}(x)$ or $H(x - x_0)$, which is 1 for $x > x_0$, and otherwise zero. This has a **discontinuity** at x_0. Now say **I** runs from $-T$ to $+T$ where T is large enough that we know $g(-T) = g(T) = 0$, and we break **I** at x_0 by hiving off the **singular point**. Call this modified interval $\mathbf{I}_\varepsilon$ = $[-T, x_0 - \varepsilon] + [x_0 + \varepsilon, T]$. Then:

$$\mathbf{d}^{\dagger}H_{x0}(g\mathbf{d}x) = \mathbf{lim}_{(\varepsilon\to 0)}\{g.\partial_x H_{x0}.\mathbf{d}x\,|\,\mathbf{I}_\varepsilon - H_{x0}(x).g(x)\,|\,\mathbf{bI}_\varepsilon\}$$

from the foregoing, which, expanding $\mathbf{I}_\varepsilon$ as $[-T, x_0 - \varepsilon] + [x_0 + \varepsilon, T]$, gives:

$$\mathbf{lim}_{(\varepsilon\to 0)}\{g.\partial_x H_{x0}.\mathbf{d}x\,|\,[-T, x_0 - \varepsilon] + g.\partial_x H_{x0}.\mathbf{d}x\,|\,[x_0 + \varepsilon, T]$$

$$- H_{x0}(x).g(x)\,|\,(x_0 - \varepsilon) + H_{x0}(x).g(x)\,|\,(-T)$$

$$- H_{x0}(x).g(x)\,|\,(T) + H_{x0}(x).g(x)\,|\,(x_0 + \varepsilon)\}.$$

Now in the two **intervals** $[-T, x_0 - \varepsilon]$ and $[x_0 + \varepsilon, T]$, $H_{x0}(x)$ is constant, so $\partial_x H_{x0}(x) = 0$, so both the terms on the parts of $\mathbf{I}_\varepsilon$ are zero, and $H_{x0}(x) = 0$ at $-T$ and at $x_0 - \varepsilon$, and $g(x) = 0$ at T, so of the **boundary** terms, all vanish except for $H_{x0}(x).g(x) \,|\, (x_0 + \varepsilon)$.

So $\quad \mathbf{d}^\dagger H_{x0}(g\mathbf{d}x) = \mathbf{lim}_{(\varepsilon\to 0)} H_{x0}(x).g(x) \,|\, (x_0 + \varepsilon) = \mathbf{lim}_{(\varepsilon\to 0)} 1.g(x_0 + \varepsilon) = g(x_0)$.

If we now put $\mathbf{d}^\dagger H_{x0}$ formally as $\partial^\dagger_x H_{x0}.\mathbf{d}x$, we can write this as:

$$\partial^\dagger_x H_{x0}(g) = g(x_0) = \delta_{x0}(g)$$

or

$$\partial^\dagger_x H(x - x_0) = \delta(x - x_0)$$

which gives Schwartz's famous result, obtained in a simpler context in Section 7.3, that the **derivative** (**OD**) of the **Heaviside function** is the **Dirac distribution** (or **functional**). So this more formal (and general) approach still gives our definitive case correctly.

We can extend this notion to defining $\mathbf{d}^\dagger$ etc. on **functionals** themselves. Suppose we have some **distribution** or **p-form functional** T, and an arbitrary **(p+1)-form** ψ, then we define

$$\mathbf{d}^\dagger T(\psi) := \langle T, \boldsymbol{\delta}\psi\rangle = T(\boldsymbol{\delta}\psi).$$

So if T is the **singular 0-form** $\delta(x - a)$, and ϕ is a simple **1-form** $\phi(x)\mathbf{d}x$, we have
$\boldsymbol{\delta}\phi := (-1)^{np+n+1} *\mathbf{d}*\phi$ with $n = 1$ and $p = 1$, so here:

$$\boldsymbol{\delta}\phi = (-1)^3 *\mathbf{d}\phi(x) = - *\partial_x\phi(x)\mathbf{d}x = -\partial_x\phi(x).$$

Now, *bringing back the use of the ordinary angular bracket notation for* **singular distributions**, we have that:

$$\langle \mathbf{d}^\dagger\delta(x - x_0), \phi(x)\mathbf{d}x\rangle = \langle\delta(x - x_0), -\partial_x\phi(x)\rangle = -\partial_x\phi(x_0),$$

a result you will more commonly see written as: $\langle\partial_x\delta(x - a), f\,\rangle := -\,\partial_x f\,(a)$. Note that here I am writing $\delta(x - x_0)$ as a **functional** acting on $-\partial_x\phi(x)$. We could put it as:

$$\partial_x\delta(x - a)\{\phi(x)\} = \delta(x - a)\{-\partial_x\phi(x)\} = -\partial_x\phi(a),$$

but Schwartz writes it as (e.g. his p.80):

$$\langle\partial_x\delta(x - a), \phi(x)\rangle = \langle\delta(x - a), -\partial_x\phi(x)\rangle = -\partial_x\phi(a).$$

In other words, the **derivative** (**OD**) of the **delta functional** acting on f is defined as the *negative* of the **delta functional** (at the same a) acting on the **derivative** $\partial_x f$.

This is the **functional derivative**.

Note too that if we have a **function** f such that $f(x) = H(x).f(x)$ and f is well-behaved for $x > 0$, we may define the **Laplace transform** of f as:

$$\mathfrak{L}(f) := \langle f, \mathbf{e}^{-sx}\rangle_{\mathbf{I}} \quad \text{for } s > 0,$$

where the domain of the **scalar product, I**, is from 0 to $+\infty$: i.e. $\mathbf{I} = [0, \infty]$. Hence for a **distribution**, we can define it as $T(\mathbf{e}^{-sx})$ by analogy, and for a **1-form functional** as $T(\mathbf{e}^{-sx}\mathbf{d}x)$.

So now we have:

$$\mathfrak{L}(\mathbf{d}^{\dagger}f) = \mathbf{d}^{\dagger}f\,(\mathbf{e}^{-sx}\mathbf{d}x)$$

$$= \langle \mathbf{d}f, \mathbf{e}^{-sx}\mathbf{d}x\rangle_{\mathbf{I}} - [\,f \wedge *(\mathbf{e}^{-sx}\mathbf{d}x)]\,\big|\,\mathbf{bI}$$

$$= \partial_x f.\mathbf{e}^{-sx}\mathbf{d}x\,\big|\,\mathbf{I} - (f.\mathbf{e}^{-sx})(\infty) + (f.\mathbf{e}^{-sx})(0).$$

$$= \partial_x f.\mathbf{e}^{-sx}\mathbf{d}x\,\big|\,\mathbf{I} - 0 + f\,(0).$$

But **integrating** the first term **by parts,**

$$\partial_x f.\mathbf{e}^{-sx}\mathbf{d}x\,\big|\,\mathbf{I} = \partial_x f.\mathbf{e}^{-sx}\mathbf{d}x\,\big|\,[0, \infty] = \mathbf{d}(f.\mathbf{e}^{-sx})\,\big|\,[0, \infty] - (f.\partial_x\mathbf{e}^{-sx})\,\big|\,[0, \infty]$$

$$= (f.\mathbf{e}^{-sx})(\infty) - (f.\mathbf{e}^{-sx})(0) + s.(f.\partial_x\mathbf{e}^{-sx})\,\big|\,[0, \infty]$$

$$= 0 - f\,(0) + s.\mathfrak{L}(f)$$

so all told:

$$\mathfrak{L}(\mathbf{d}^{\dagger}f) = s.\mathfrak{L}(f) - f\,(0) + f\,(0) = s.\mathfrak{L}(f\,).$$

So with **differentiation in the sense of distributions** or **adjoint differentiation** the $f(0)$ terms produced by the usual **Laplace transform** *cancel out*.

All I can say is, for all the mathematical elegance, *don't do it* this way! Those $f\,(0)$ terms are important.

Three other **Green's formulae** are sometimes used, which I simply state here. They can be easily derived from the basic one. They all apply where *both* ϕ and ψ are **p-forms**.

II.	$\langle \mathbf{d}\boldsymbol{\delta}\phi \cdot \psi\rangle - \langle \boldsymbol{\delta}\phi \cdot \mathbf{d}\psi\rangle = \boldsymbol{\delta}\phi \wedge *\psi\,\big|\,\mathbf{b}\Sigma$
III.	$\langle \mathbf{d}\phi \cdot \mathbf{d}\psi\rangle - \langle \phi \cdot \boldsymbol{\delta}\mathbf{d}\psi\rangle = \phi \wedge *\mathbf{d}\psi\,\big|\,\mathbf{b}\Sigma$
IV.	$\langle \Delta\phi \cdot \psi\rangle - \langle \mathbf{d}\phi \cdot \mathbf{d}\psi\rangle - \langle \boldsymbol{\delta}\phi \cdot \boldsymbol{\delta}\psi\rangle = (\boldsymbol{\delta}\phi \wedge *\psi - \psi \wedge \mathbf{d}\phi)\,\big|\,\mathbf{b}\Sigma.$

7.6 Tensor Functionals

I introduced **tensors** in Section 5.5 in terms of the **tensor product**. This is the usual modern style. The traditional style defined them in terms of the way their components change under a **transformation of coordinates**. Neither approach is very intuitive, so I'm going to break with convention and simply define a **tensor** in this way:

I. a **tensor** is a **linear functional** that **maps vector objects** of a particular **grade** into **vector objects** of the same or of another **grade**.

So a member of the **dual space** of a **basic vector space maps** the **basic vectors** (as **arguments**) **linearly** into **scalars** – generally **real numbers**. The concept of a **tensor** simply takes this idea up into **functions** whose **range** is itself a **vector space** of *arbitrary* **grade**. So in 3-space, we might have a **tensor** of the form:

$$\mathbf{T} = a_{23}{}^{1}\mathbf{e}^{23}\mathbf{e}_1 + a_{23}{}^{2}\mathbf{e}^{23}\mathbf{e}_2 + a_{23}{}^{3}\mathbf{e}^{23}\mathbf{e}_3 + a_{31}{}^{1}\mathbf{e}^{31}\mathbf{e}_1 + a_{31}{}^{2}\mathbf{e}^{31}\mathbf{e}_2 + a_{31}{}^{3}\mathbf{e}^{31}\mathbf{e}_3$$

$$+ a_{12}{}^{1}\mathbf{e}^{12}\mathbf{e}_1 + a_{12}{}^{2}\mathbf{e}^{12}\mathbf{e}_2 + a_{12}{}^{3}\mathbf{e}^{12}\mathbf{e}_3,$$

where $\mathbf{e}^{12}$ is the shorthand for $\mathbf{e}^1 \wedge \mathbf{e}^2$, the **bivector** formed from the *x* and *y* elements of the **reciprocal basis**, and the product of the **basis vectors** as in $\mathbf{e}^{23}\mathbf{e}_3$ is a **tensor product** as defined in Section 5.5:

$$\mathbf{e}^{23}\mathbf{e}_3 = \mathbf{e}^2 \wedge \mathbf{e}^3 \otimes \mathbf{e}_3 = \tfrac{1}{2}(\mathbf{e}^2 \otimes \mathbf{e}^3 - \mathbf{e}^3 \otimes \mathbf{e}^3) \otimes \mathbf{e}_3.$$

This **tensor** would operate on a **basic bivector** and a **dual vector** together to give a **scalar**. So on $\mathbf{B} = b^{23}\mathbf{e}_{23} + b^{31}\mathbf{e}_{31} + b^{12}\mathbf{e}_{12}$ and $\mathbf{a} = a_1\mathbf{e}^1 + a_2\mathbf{e}^2 + a_3\mathbf{e}^3$, because $\mathbf{e}^{km}\mathbf{e}_{ij} = \delta_{ki}\delta_{mj}$, only products from terms of the form $\mathbf{e}^{ab}\mathbf{e}_{ab}\mathbf{e}_d\mathbf{e}^d$ will be non-zero and we'd get:

$$a_{23}{}^{1}b^{23}a_1 + a_{23}{}^{2}b^{23}a_2 + a_{23}{}^{3}b^{23}a_3 + a_{31}{}^{1}b^{31}a_1 + a_{31}{}^{2}b^{1}a_2 + a_{31}{}^{3}b^{31}a_3$$

$$+ a_{12}{}^{1}b^{12}a_1 + a_{12}{}^{2}b^{12}a_2 + a_{12}{}^{3}b^{12}a_3$$

as the **scalar** result. But much more interesting is if we operate *only* on **B** or *only* on **a**. These *partial* evaluations are called **contractions**, and it is the operation of **contraction** that gives the effect of my original definition. Thus, operating on **B** the $\mathbf{e}^{23}$ terms would give zero on $b^{31}\mathbf{e}_{31} + b^{12}\mathbf{e}_{12}$, but abstract the $b^{23}\mathbf{e}_{23}$ alone by $\mathbf{e}^{23}(b^{23}\mathbf{e}_{23}) = b^{23}$ to give:

$$\mathbf{T}(\mathbf{B}) = a_{23}{}^{1}b^{23}\mathbf{e}_1 + a_{23}{}^{2}b^{23}\mathbf{e}_2 + a_{23}{}^{3}b^{23}\mathbf{e}_3 + a_{31}{}^{1}b^{31}\mathbf{e}_1 + a_{31}{}^{2}b^{31}\mathbf{e}_2 + a_{31}{}^{3}b^{31}\mathbf{e}_3$$

$$+ a_{12}{}^{1}b^{12}\mathbf{e}_1 + a_{12}{}^{2}b^{12}\mathbf{e}_2 + a_{12}{}^{3}b^{12}\mathbf{e}_3$$

$$= (a_{23}{}^{1}b^{23} + a_{31}{}^{1}b^{31} + a_{12}{}^{1}b^{12})\mathbf{e}_1 + (a_{23}{}^{2}b^{23} + a_{31}{}^{2}b^{31} + a_{12}{}^{2}b^{12})\mathbf{e}_2$$

$$+ (a_{23}{}^{3}b^{23} + a_{31}{}^{3}b^{31} + a_{12}{}^{3}b^{12})\mathbf{e}_3.$$

This is an ordinary **vector**. So this **tensor, contracted** on a **bivector** of the **basic** space has returned a **vector** of the **basic space**. The same **tensor, contracting** with a **dual space vector**, would give a **dual space bivector**: it's whatever bit of the **tensor** itself that is not **contracted** out that determines the type of the result.

The analogous operation in terms of **differential forms** is to use the **integral scalar product** to **contract** out certain variables to leave a new **form** in the remaining variables. The **tensor forms** that do this are strictly called **double forms** because, like the discrete **tensors** above, they sit in two spaces at once, but this suggestion of daunting complexity is easily avoided. You've seen one such **tensor form** in action already – the **Laplace transform**, which operates on a **0-form** in one space to convert it to a **0-form** in another.

Thus if $s > a$,

$$\mathfrak{L}(\uparrow at) = \mathfrak{L}(\mathbf{e}^{at}) = \mathbf{e}^{-st}\mathbf{e}^{at}.\mathbf{d}t \,\big|\, [0, \infty] = (1/(a-s)).\mathbf{d}(\mathbf{e}^{(a-s)t}) \,\big|\, [0, \infty] = 1/(s-a)$$

or:

$$\mathbf{e}^{at} \longrightarrow 1/(s-a).$$

These **tensor transforms** which **transform** one **function** into another using an **integral** of the form of our **integral scalar product** are generally called, reasonably enough, **integral transforms**. They generally appear simply as **function** to **function transforms**, or in our terminology, **0-form** to **0-form transforms**. They are especially important in three contexts: **Green's Functions,**[115] **Fourier-Laplace transforms** and **convolution integrals.**[116]

The latter two are the subject of Chapter 8, but I will briefly cover **Green's Functions** in the next Section.

Note that the tensorial nature of **double forms** means that they are not limited to the **grade** of the **pseudoscalar**, but can involve a product of more **differentials** than the **dimension** of the space. So in 3-space, a sixteen-term **double form** involving terms like:

$$\phi(x, y, z, \xi, \psi, \zeta).\mathbf{d}x\mathbf{d}y \otimes \mathbf{d}\xi\mathbf{d}\zeta$$

is a valid possibility. Note that here we "rename" the second product's variables to ξ, ψ, ζ as these will be **contracted** out when we form the **scalar product** with a **bivector** in $\mathbf{d}\xi$, $\mathbf{d}\psi$ and $\mathbf{d}\zeta$ to leave a **bivector** in $\mathbf{d}x$, $\mathbf{d}y$ and $\mathbf{d}z$. It sounds awful, but it's quite comprehensible in practice!

---

[115] George Green (1793–1841) has rather a lot to answer for! He also died relatively young but is a most interesting character as he appears to have been almost completely self-taught. Until his mid-thirties he operated the family windmill (which still exists and is now a tourist attraction of Nottingham).

[116] There is also a discrete form of **convolution** but I do not consider this here. It will be touched on in an exercise in the next Chapter.

7.7 Green's Functions

Green's functions are a device for *inverting* a generalized **linear differential operator** of a certain kind, which we may write as **L**, and so solving equations of the form:

$$\mathbf{L}\phi(\boldsymbol{x}) = f(\boldsymbol{x}).$$

Here $\boldsymbol{x} := (x_1, x_2, \ldots x_n)$ indicates a point in an n-dimensional space, so ϕ might be defined on x, y, z, for example. **Linearity** of **L** is defined as usual:

$$\mathbf{L}(\alpha u(\boldsymbol{x}) + \beta v(\boldsymbol{x})) = \alpha\mathbf{L}u(\boldsymbol{x})) + \beta\mathbf{L}v(\boldsymbol{x})).$$

The "of a certain kind" is more subtle. **Green's functions** exist only for **elliptic differential operators.** If **L** is a **partial derivative operator** acting on a **function** and of the form:

$$\mathbf{L} \equiv a\partial_x^{\,2} + 2b\partial_x\partial_y + c\partial_y^{\,2}$$

then **L** is **elliptic** if $b^2 < ac$, a condition clearly related to the **discriminant** of a **quadratic equation** defined in Section 1.6.[117] The definitive **elliptical operator** is the **Laplacian**, which has $b = 0$, and appears in two-dimensional and three-dimensional form as:

$$\partial_x^{\,2} + \partial_y^{\,2} \qquad\qquad \partial_x^{\,2} + \partial_y^{\,2} + \partial_z^{\,2}$$

and as we saw above, this equals $-\boldsymbol{\Delta} \equiv -(\mathbf{d\delta} + \mathbf{\delta d})$. Henceforth in this Section, I will focus on the case where **L** is the **Laplacian operator** $\boldsymbol{\Delta}$. You may ask whether $\partial_x^{\,2}$ on $\boldsymbol{R}$ is a **Laplacian**. Here $n = 1$, and for a simple **function** $\phi(x)$, p is 0 – it's a **0-form**. So, as $*\phi = \phi.\mathbf{d}x$:

$$-\mathbf{d\delta}\phi(x) = -\mathbf{d}(-1)^{1\times 0+1+1}*\mathbf{d}(\phi\mathbf{d}x) = -\,\mathbf{d}(\partial_x\phi.\mathbf{d}x\mathbf{d}x) = 0.$$

$$-\mathbf{\delta d}\phi(x) = -\,\mathbf{\delta}(\partial_x\phi.\mathbf{d}x) = -(-1)^{1\times1+1+1}*\mathbf{d}*(\partial_x\phi.\mathbf{d}x) = *\mathbf{d}(\partial_x\phi) = *[\partial_x^{\,2}\phi(x).\mathbf{d}x]$$

$$= \partial_x^{\,2}\phi(x).$$

So, yes it is, and this gives us a simple case to refer to in time of trouble, an option scrupulously avoided by all the standard texts, which aim for the most complicated possible examples from the start.

A **Green's function** for **L** is a **tensor form** $G(\xi, \boldsymbol{x})$ with the property that:

$$\langle\mathbf{L}\phi(\xi), G(\xi, \boldsymbol{x})\rangle = \phi(\boldsymbol{x}).$$

To see that this actually solves the original equation, just replace $\mathbf{L}\phi(\xi)$ with $f(\xi)$ as at the start of this Section, and so obtain $\phi(\boldsymbol{x}) = \langle f(\xi), G(\xi, \boldsymbol{x})\rangle$, so G does indeed "invert" **L**.

---

[117] This form results from the $2b$ coefficient in front of $\partial_x\partial_y$. This follows Mathews and Walker. Other books write this simply as b, giving the more familiar $b^2 - 4ac$.

The **Green's function** property that:

$$\langle \mathbf{L}\phi(\boldsymbol{\xi}), G(\boldsymbol{\xi}, \boldsymbol{x})\rangle = \phi(\boldsymbol{x})$$

has strong suggestions of a **delta functional** form, but this needs to be treated with great care, because what we're asking of the **Green's function** is that the **adjoint** of **L** acts on the **Green's function** to abstract $\phi(\mathbf{x})$, or that, writing the **adjoint** as **L***:

$$\langle \phi(\boldsymbol{\xi}), \mathbf{L}^*G(\boldsymbol{\xi}, \boldsymbol{x})\rangle = \langle \mathbf{L}\phi(\boldsymbol{\xi}), G(\boldsymbol{\xi}, \boldsymbol{x})\rangle = \phi(\mathbf{x}).$$ [118]

So we want

$$\mathbf{L}^*G(\boldsymbol{\xi}, \boldsymbol{x}) = \delta(\boldsymbol{\xi} - \mathbf{x}),$$

not $\mathbf{L}G(\boldsymbol{\xi}, \boldsymbol{x}) = \delta(\boldsymbol{\xi} - \mathbf{x})$. This isn't a problem with the one-dimensional **Laplacian** $\partial_x^{\,2}$, for in this case the operator is at least *formally* **self-adjoint**, meaning that **L*** = **L**, so assuming the **scalar product** is over an interval $[a, b]$:

$$\langle \partial_x^{\,2}\phi(\xi), G(\xi, x)\rangle = \partial_x^{\,2}\phi(\xi).G(\xi, x).\mathbf{d}\xi \,|\, [a, b] = \mathbf{d}(\partial_x\phi(\xi).G(\xi, x)) \,|\, [a, b] - \partial_x\phi(\xi).\partial_x G(\xi, x).\mathbf{d}\xi \,|\, [a, b]$$

$$= \mathbf{d}(\partial_x\phi(\xi).G(\xi, x)) \,|\, [a, b] - \mathbf{d}(\phi(\xi).\partial_x G(\xi, x)) \,|\, [a, b] + \phi(\xi).\partial_x^{\,2}G(\xi, x).\mathbf{d}\xi \,|\, [a, b]$$

$$= (\partial_x\phi(\xi).G(\xi, x)) \,|\, \mathbf{b}[a, b] - (\phi(\xi).\partial_x G(\xi, x)) \,|\, \mathbf{b}[a, b] + \langle \phi(\xi), \partial_x^{\,2}G(\xi, x)\rangle.$$

So if we can assume that $\phi(a) = \phi(b) = \partial_x\phi(a) = \partial_x\phi(b) = 0$, then we can look for $G(\xi, x)$ such that:

$$\partial_x^{\,2}G(\xi, x) = \delta(x - \xi).$$

This is easy to find. Integrate once:

$$\partial_x G(\xi, x) = H(x - \xi) + \alpha(\xi),$$

where the $\alpha(\xi)$ represents possible terms independent of x, and corresponds to a "**constant of integration**" in this two-variable context ($\beta(\xi)$ below is similar), and integrate again:

$$G(\xi, x) = (x - \xi).H(x - \xi) + x.\alpha(\xi) + \beta(\xi).$$

But even the simplest possible differential operator ∂_x *isn't* **self-adjoint**; its **adjoint** is $-\partial_x$. Does it have a **Green's function**? Well,

$$\langle \partial_x\phi(\xi), G(\xi, x)\rangle = \partial_x\phi(\xi).G(\xi, x).\mathbf{d}\xi \,|\, [a, b] = \mathbf{d}(\phi(\xi).G(\xi, x)) \,|\, [a, b] - \phi(\xi).\partial_x G(\xi, x).\mathbf{d}\xi \,|\, [a, b].$$

---

[118] Here I'm talking of the **adjoint** in the sense of the **Green's formulae** of Section 7.5, whereby for example **δ** would be the **adjoint** of **d**. In the expression above, I've used the **functional derivative** concept where **L*** is defined in such a way that the **boundary terms** of the switch are "implied", as we shall see.

So we'd need $\partial_x G(\xi, x) = -\delta(\xi - x)$, so this time:

$$G(\xi, x) = -H(\xi - x) + \alpha(\xi),$$

giving

$$-\phi(\xi).\partial_x G(\xi, x).\mathbf{d}\xi \,|\, [a, b] = +\phi(\xi).\partial_x H(\xi - x).\mathbf{d}\xi \,|\, [a, b] - \phi(\xi).\partial_x \alpha(\xi).\mathbf{d}\xi \,|\, [a, b].$$

Assume $a < x < b$ and $\alpha(\xi) = 0$ throughout, and we have $\phi(b) = \phi(a) = 0$. Then the second term drops out, and we have:

$$\langle \partial_x \phi(\xi), G(\xi, x) \rangle = (\phi(\xi).G(\xi, x)) \,|\, \mathbf{b}[a, b] + \phi(\xi).\partial_x H(\xi - x).\mathbf{d}\xi \,|\, [a, b]$$

$$= 0.G(b, x) - 0.G(a, x) + \phi(\xi).\delta(\xi - x).\mathbf{d}\xi \,|\, [a, b] = \phi(x),$$

using the rule that $\partial_x H(\xi - x) = \delta(\xi - x)$. So, yes it does!

Going back to the **Green's function** for ∂_x^2, the **boundary** terms at a and b will be zero for $\phi(b) = \phi(a) = 0$ if we require that also $G(b, x) = G(a, x) = 0$. We then have:

$$\langle \partial_x^2 \phi(\xi), G(\xi, x) \rangle = (\partial_x \phi(\xi).G(\xi, x)) \,|\, \mathbf{b}[a, b] - (\phi(\xi).\partial_x G(\xi, x)) \,|\, \mathbf{b}[a, b] + \langle \phi(\xi), \partial_x^2 G(\xi, x) \rangle$$

$$= (\partial_x \phi(\xi).0) \,|\, \mathbf{b}[a, b] - (0.\partial_x G(\xi, x)) \,|\, \mathbf{b}[a, b] + \langle \phi(\xi), \partial_x^2 G(\xi, x) \rangle$$

$$= \langle \phi(\xi), \partial_x^2 G(\xi, x) \rangle$$

but $\partial_x^2 G(\xi, x) = \delta(x - \xi)$, so this is $\langle \phi(\xi), \delta(\xi - x) \rangle = \phi(x)$.

From these two most simple examples, we can begin to see a pattern:

- **Green's functions** have a **discontinuity** one order below the order of the operator **L**, so if **L** is first-order (a ∂ form), $G(\xi, x)$ itself has a discontinuity, if **L** is a second-order operator (i.e. a ∂^2 form), $\partial_x G(\xi, x)$ will have a discontinuity.
- $G(\xi, x)$ will commonly obey the same **boundary conditions** as the set of underlying functions $\phi(x)$.
- Everywhere *except* at the **discontinuity**, $G(\xi, x)$ obeys the simple homogeneous equation: $\mathbf{L}G(\xi, x) = 0$.
- Amongst these properties I'll add that when **L** is **self-adjoint** or **L*** = **L**, G tends to be **symmetric** in ξ and x: $G(\xi, x) = G(x, \xi)$, although this isn't immediately apparent in the above.

Bernard Friedman[119] introduces a useful term for the odd **boundary terms** which we've seen above:

---

[119] B. Friedman, *Principles and Techniques of Applied Mathematics*, Dover Wiley 1956, Dover 1990.

In $\langle \mathbf{L}\phi(\xi), \psi(\xi)\rangle - \langle \phi(\xi), \mathbf{L}^*\psi(\xi)\rangle$

$= \psi(\xi).\mathbf{L}\phi(\xi).\mathbf{d}\xi \,|\, [a, b] - \phi(\xi).\mathbf{L}^*\psi(\xi)).\mathbf{d}\xi \,|\, [a, b] = J(\psi, \phi) \,|\, \mathbf{b}[a, b]$,

the function $J(\psi, \phi)$ is called the **conjunct** of ψ and ϕ.

Note that although it may be hidden in the **δ-function** notation, it's still actually the **conjunct** terms that really "give the answer". So in the ∂_x case, where $G(\xi, x) = -H(\xi - x)$, so $\mathbf{L}^*G(\xi, x) = (-\partial_x)(-H(\xi - x)) = +\,\partial_x H(\xi - x)$, the **conjunct** over $[a, b]$ is given by:

$$-\partial_x\phi(\xi).H(\xi - x).\mathbf{d}\xi \,|\, [a, b] - (+\phi(\xi).\partial_x H(\xi - x)).\mathbf{d}\xi \,|\, [a, b]$$

$$= -\partial_x\phi(\xi).G(\xi, x).\mathbf{d}\xi \,|\, [a, b] - \phi(\xi).\partial_x H(\xi - x).\mathbf{d}\xi \,|\, [a, b]$$

$$= -\mathbf{d}(\phi(\xi).H(\xi - x)) \,|\, [a, b]$$

$$= -\mathbf{d}(\phi(\xi).H(\xi - x)) \,|\, [a, x-\varepsilon] - \mathbf{d}(\phi(\xi).H(\xi - x)) \,|\, [x+\varepsilon, b]$$

$$= -\phi(\xi).H(\xi - x) \,|\, \mathbf{b}[a, x-\varepsilon] - \partial_x\phi(\xi).1.\mathbf{d}\xi \,|\, [x+\varepsilon, b]$$

$$= -\phi(x-\varepsilon).H(x - \varepsilon - x) + \phi(a).H(a - x) - \mathbf{d}\phi(\xi) \,|\, [x+\varepsilon, b]$$

$$= 0 + 0 - \phi(\xi) \,|\, \mathbf{b}[x+\varepsilon, b] = -\phi(b) + \phi(x + \varepsilon) = -\,0 + \phi(x).$$

Returning to the higher-dimensional general case, we can extend the observations above by expecting that **Green's functions** will exist subject to two conditions:

- The **function** (or **form**) $\phi(\boldsymbol{\xi})$ here has to be of **bounded support**. So what does that mean? Well, the **support** of a **function** or of a **form** is the **domain** over which it has non-zero values, so a **function** or **form** of **bounded support** is zero outside a finite or **bounded domain**. The **domain of integration** Σ of the **scalar product** is therefore chosen so that $\phi(\mathbf{x}) = 0$ on $\mathbf{b}\Sigma$. **Bounded support** knocks out the **conjunct** terms. This condition can be relaxed, however.
- G must have a **singular point** at $\mathbf{x} = \boldsymbol{\xi}$, and so the **domain of integration** chosen for the **scalar product** $\langle\rangle$ must be the **domain** Σ, but modified by having an ε-**sphere** around this **singular point**, and so is $\Sigma - S_\varepsilon$, where S_ε is the ε-**sphere** in question.

I proceed for the case where $\mathbf{L} = \boldsymbol{\Delta}$. Then by the **Green's formula** for the **Laplacian** (formula V) given in Section 7.5, we have:

$$\langle \boldsymbol{\Delta}_\xi\phi, G\rangle = \langle \phi, \boldsymbol{\Delta}_\xi G\rangle + (\boldsymbol{\delta}_\xi\phi \wedge *G - \boldsymbol{\delta}_\xi G \wedge *\phi + \phi \wedge *\mathbf{d}_\xi G - G \wedge *\mathbf{d}_\xi\phi) \,|\, \mathbf{b}(\Sigma - S_\varepsilon).$$

Here I've subscripted the **operators** with $\boldsymbol{\xi}$ to show that we're using $\mathbf{d}_\xi$ rather than $\mathbf{d}_x$ as $G(\boldsymbol{\xi}, \boldsymbol{x})$ can be **differentiated** with respect to *either* $\boldsymbol{x}$ or $\boldsymbol{\xi}$ – in either $\boldsymbol{x}$-space or in $\boldsymbol{\xi}$-space. To put it another way, both $\mathbf{d}_x G(\boldsymbol{\xi}, \boldsymbol{x})$ and $\mathbf{d}_\xi G(\boldsymbol{\xi}, \boldsymbol{x})$ are well-defined.

We choose G so that $\boldsymbol{\Delta}_\xi G(\boldsymbol{\xi}, \boldsymbol{x}) = 0$ everywhere in Σ *outside* the **singular point** $\boldsymbol{\xi} = \boldsymbol{x}$. So the first term on the RHS (right-hand side) above disappears: $\langle \phi, \boldsymbol{\Delta}_\xi G\rangle = 0$. We now need the **boundary integral** $(\cdot)\big|\mathbf{b}(\Sigma - S_\varepsilon)$ to reduce to $\phi(\boldsymbol{x})$. The condition of **bounded support** on ϕ ensures that $(\cdot)\big|\mathbf{b}\Sigma = 0$. To get the final result, G needs to be such that the $\boldsymbol{\delta}_\xi\phi$ and $\mathbf{d}_\xi\phi$ terms go to zero on $\mathbf{b}S_\varepsilon$, because we want to abstract $\phi(\boldsymbol{x})$, not its **derivatives**:

$$(\boldsymbol{\delta}_\xi\phi \wedge *G - G \wedge *\mathbf{d}_\xi\phi)\big|\mathbf{b}S_\varepsilon = 0.$$

So we get the result that $\phi(\boldsymbol{x}) = \langle \boldsymbol{\Delta}_\xi\phi, G\rangle$ if:

$$(\phi \wedge *\mathbf{d}_\xi G - \boldsymbol{\delta}_\xi G \wedge *\phi)\big|\mathbf{b}S_\varepsilon = \phi(\boldsymbol{x}).$$

It may seem hard to believe that any **function** or **form** could simultaneously satisfy all these requirements, but there is a general solution here. It differs slightly for 2-space, for which we use:

$$G(\boldsymbol{\xi}, \boldsymbol{x}) = (1/2\pi)\mathbf{ln}(1/|\boldsymbol{x} - \boldsymbol{\xi}|).$$

For $n > 2$, the solution is:

$$G(\boldsymbol{\xi}, \boldsymbol{x}) = \frac{(1/|x - \xi|^{n-2})}{[(n-2)S_n]}.$$

Here S_n is the surface area of the unit sphere (that is, of radius = 1) of dimension n. So for $n = 3$, $S_3 = 4\pi(1^2) = 4\pi$. Note that both these **functions** have a **singularity** at $\boldsymbol{x} = \boldsymbol{\xi}$ in accordance with the remarks above, and indeed with the entire drift of this Chapter. Their form admits too of their being written, if we set $r = |\boldsymbol{x} - \boldsymbol{\xi}|$, as:

$$G_2(r) = (1/2\pi)\mathbf{ln}(1/r) \qquad \text{and} \qquad G_n(r) = 1/((n-2)S_n r^{n-2}).$$

So $G_3(r) = 1/4\pi r$. Actually showing that these **functions** satisfy the **Green's function** properties above is quite fiddly. I have a stab at the $n = 2$ case in Appendix B.

If we follow up the device used in Section 7.5, where we defined an **adjoint exterior derivative** $\mathbf{d}^\dagger$ by:

$$\mathbf{d}^\dagger\phi(\psi) := \langle \phi, \boldsymbol{\delta}\psi\rangle = \langle \mathbf{d}\phi, \psi\rangle - (\phi \wedge *\psi)\big|\mathbf{b}\Sigma,$$

and define a "strictly" **adjoint Laplacian** by:

$$\boldsymbol{\Delta}^\dagger G(\phi) := \langle G, \boldsymbol{\Delta}\phi\rangle = \langle \boldsymbol{\Delta}\phi, G\rangle$$

$$= \langle \phi, \boldsymbol{\Delta}_\xi G\rangle + (\boldsymbol{\delta}_\xi\phi \wedge *G - \boldsymbol{\delta}_\xi G \wedge *\phi + \phi \wedge *\mathbf{d}_\xi G - G \wedge *\mathbf{d}_\xi\phi) \,\big|\, \mathbf{b}(\Sigma - S_\varepsilon)$$

$$= \phi(\boldsymbol{x}),$$

then we could use this to "legitimize" the form $\boldsymbol{\Delta}^{\dagger}G(\boldsymbol{x} - \boldsymbol{\xi}) = \delta(\boldsymbol{x} - \boldsymbol{\xi})$. But I woudn't recommend this: it's all too easy to slip into mistaking $\boldsymbol{\Delta}^{\dagger}$ for $\boldsymbol{\Delta}$. As we've seen above with $\partial_x^{\,2}$, there's no obvious problem with keeping within the simple direct model in which $\boldsymbol{\Delta}$ is **self-adjoint** but the switch throws up a large number of **boundary** or as we might now call them, **conjunct**, terms, which in turn give the δ-**functional** property around the **singularity** of G.

7.A Appendix A: The Cauchy Integral in Simplex Form

You may wish to be reassured, as I was, that the **integral** of a simple **pole** in the **complex plane** still gives the $2\pi j$ result even for a **simplex** rather than a little circle. The **differential form** here is $(z - z_0)^{-1}.\mathbf{d}z$, which we can trivially transform to the **origin** by setting $\zeta := z - z_0$, with $\xi = x - x_0$ and $\psi = y - y_0$. Then $\mathbf{d}z = \mathbf{d}\zeta = \mathbf{d}\xi + j.\mathbf{d}\psi$, so we have the form $\zeta^{-1}.\mathbf{d}\zeta$ which I will take over the enclosing **simplex** $\mathbf{p}_0\mathbf{p}_1\mathbf{p}_2$ with $\mathbf{p}_0 = (-2, -1)$, $\mathbf{p}_1 = (1, -1)$ and $\mathbf{p}_2 = (1, 2)$ in the (ξ, ψ)-plane, which will indeed enclose $(\xi, \psi) = (0, 0) \equiv z = z_0$.

So we need to evaluate:

$$\zeta^{-1}.\mathbf{d}\zeta \,\big|\, (\mathbf{p}_0\mathbf{p}_1 + \mathbf{p}_1\mathbf{p}_2 + \mathbf{p}_2\mathbf{p}_0).$$

Now from Section 4.5, $\zeta^{-1} = (\xi - j\psi)/(\xi^2 + \psi^2)$, so the **form** splits into the two parts:

$$\frac{\xi\mathbf{d}\xi + \psi\mathbf{d}\psi}{(\xi^2 + \psi^2)} + \frac{j(\xi\mathbf{d}\psi - \psi\mathbf{d}\xi)}{(\xi^2 + \psi^2)}.$$

Over $\mathbf{p}_0\mathbf{p}_1$, $\psi = -1$ and so $\mathbf{d}\psi = 0$, giving:

$$(\xi^2 + 1)^{-1}.\xi\mathbf{d}\xi \,\big|\, [-2, 1] + j.(\xi^2 + 1)^{-1}.\mathbf{d}\xi \,\big|\, [-2, 1]\,.$$

Over $\mathbf{p}_1\mathbf{p}_2$, $\xi = 1$ and $\mathbf{d}\xi = 0$, giving:

$$(1 + \psi^2)^{-1}.\psi\mathbf{d}\psi \,\big|\, [-1, 2] + j.(1 + \psi^2)^{-1}.\mathbf{d}\psi \,\big|\, [-1, 2]\,.$$

Over $\mathbf{p}_2\mathbf{p}_0$, $\psi = 1 + \xi$, so at $\mathbf{p}_2$, $\xi = 1 \rightarrow \psi = 2$ and at $\mathbf{p}_0$, $\xi = -2 \rightarrow \psi = -1$, and $\mathbf{d}\psi = \mathbf{d}\xi$ here. so we have:

$$\frac{\xi\mathbf{d}\xi + (1 + \xi)\mathbf{d}\xi}{(\xi^2 + (1 + \xi)^2)} \,\Big|\, [1, -2] + j.\frac{(\xi\mathbf{d}\xi - \mathbf{d}\xi - \xi\mathbf{d}\xi)}{(\xi^2 + (1 + \xi)^2)} \,\Big|\, [1, -2]$$

$$= \frac{(1 + 2\xi)\mathbf{d}\xi}{(\xi^2 + (1 + \xi)^2)} \,\Big|\, [1, -2] - j.\frac{\mathbf{d}\xi}{(\xi^2 + (1 + \xi)^2)} \,\Big|\, [1, -2].$$

Now the $(x^2 + 1)^{-1}.x\mathbf{d}x$ forms, which occur in ξ and in ψ, can be evaluated by putting $u = (1 + x^2)$, so $\mathbf{d}u = 2x\mathbf{d}x$, making these $½u^{-1}\mathbf{d}u = ½\mathbf{d}{\downarrow}u = ½\mathbf{d}(\mathbf{ln}u)$. For $x = -2 \rightarrow u = 5$, $x = 1 \rightarrow u = 2$, so
$(\xi^2 + 1)^{-1}.\xi\mathbf{d}\xi \,|\, [-2, 1] = ½(\mathbf{ln}(2) - \mathbf{ln}(5))$, and $(1 + \psi^2)^{-1}.\psi\mathbf{d}\psi \,|\, [-1, 2] = ½(\mathbf{ln}(5) - \mathbf{ln}(2))$, so these cancel.
The $(x^2 + 1)^{-1}.\mathbf{d}x$ forms, which again occur in ξ and in ψ, need a little more subtlety. On the **unit circle**, we define **tan := sin/cos** = y/x. Call it t, and use $\mathbf{d}y = x\mathbf{d}\theta$, $\mathbf{d}x = -\, y\mathbf{d}\theta$:

$$\mathbf{d}t = \mathbf{d}(y/x) = (x\mathbf{d}y - y\mathbf{d}x)/x^2 = (x.x\mathbf{d}\theta + y.y\mathbf{d}\theta)/x^2 = (x^2 + y^2)\mathbf{d}\theta/x^2 = \mathbf{d}\theta/x^2.$$

So $\mathbf{d}\theta = x^2\mathbf{d}t = (1 + t^2)^{-1}.\mathbf{d}t$, where $t = y/x = \surd(1 - x^2)/x \rightarrow t^2 = (1 - x^2)/x^2$ gives the last result. So this **integral** is the **inverse** of **tan**, called in the texts **arctan**:

$$\mathbf{d}(\mathbf{arctan}(x)) = (1 + x^2)^{-1}.\mathbf{d}x$$

and we have:

$$j.(\xi^2 + 1)^{-1}.\mathbf{d}\xi \,|\, [-2, 1] = j(\mathbf{arctan}(1) - \mathbf{arctan}(-2)) = j.(\pi/4 + \mathbf{arctan}(2))$$
$$j.(1 + \psi^2)^{-1}.\mathbf{d}\psi \,|\, [-1, 2] = j(\mathbf{arctan}(2) - \mathbf{arctan}(-1)) = j.(\mathbf{arctan}(2) + \pi/4)).$$

So these together give $2j.(\mathbf{arctan}(1) + \mathbf{arctan}(2))$. Of the last two **integrals** one is:

$$\frac{(1 + 2\xi)\mathbf{d}\xi}{(\xi^2 + (1 + \xi)^2)} = \mathbf{d}(½\mathbf{ln}(1 - 2(1 + \xi) + 2(1 + \xi)^2))$$

and over [1, −2], this gives: $½(\mathbf{ln}(5) - \mathbf{ln}(5)) = 0$. So all the **real** parts indeed sum to zero.

The other is:

$$-\; j.(\xi^2 + (1 + \xi)^2)^{-1}\,\mathbf{d}\xi \;=\; -\; j.\mathbf{d}(-\mathbf{arctan}(1 - 2(1+\xi))).$$

Over [1, −2], this gives $j.(\mathbf{arctan}(3) - \mathbf{arctan}(-3))$. Now **arctan** is odd: i.e. $\mathbf{arctan}(-x) = -\mathbf{arctan}(x)$, so this is $2j.\mathbf{arctan}(3)$. So all told we should have:

$$2j.(\mathbf{arctan}(1) + \mathbf{arctan}(2) + \mathbf{arctan}(3)) = 2\pi j$$

because **arctan**(1) + **arctan**(2) + **arctan**(3) *does equal* π. *This is the right answer.*

7.B Appendix B: The Laplacian Green's Function for n = 2

The **Newtonian potential** of a charge distribution $\rho(\boldsymbol{x})$, where $\boldsymbol{x}$ is a point in 3-space, is given by:

$$\mathrm{U}(\boldsymbol{x}) = (\rho(\boldsymbol{\xi})/|\boldsymbol{\xi} - \boldsymbol{t}|).\mathbf{d}\xi\mathbf{d}\psi\mathbf{d}\zeta \,\big|\, \mathbf{V}$$

where **V** is all of 3-space. This can be written as a **convolution**, which is an **integral** of the form $\phi(\xi).\theta(x - \xi).\mathbf{d}\xi \mid \mathbf{V}$, and commonly this is put as $\phi * \theta$. Here this would simply take the form:

$$U = \rho * 1/|\mathbf{x}| = \rho * 1/r, \text{ where } r := \sqrt{(x^2 + y^2 + z^2)}.$$

Here $\rho(\mathbf{x})$ is a **function** over 3-space, but the definition of **potential** could now be extended to any **distribution** as:

$$U_T(\mathbf{x}) = T * 1/r.$$

Applying the **Laplacian**, $\Delta U = \Delta(T * 1/r)$, would here give:

$$\Delta U = T * \Delta(1/r).$$

Now $\Delta(1/r)$ can be evaluated from $r = \sqrt{(x^2 + y^2 + z^2)} = (r^2)^{½}$, which gives:

$$\mathbf{d}r = ½(r^2)^{½-1}.\mathbf{d}(r^2) = ½(r^2)^{-½}.\mathbf{d}(x^2 + y^2 + z^2) = (1/r).(x\mathbf{d}x + y\mathbf{d}y + z\mathbf{d}z)$$

so $\quad \mathbf{d}(1/r) = -r^{-2}.\mathbf{d}r = -(1/r^3).(x\mathbf{d}x + y\mathbf{d}y + z\mathbf{d}z)$

or $\quad \partial_x(1/r) = -x/r^3$, and similarly $\partial_y(1/r)$ and $\partial_z(1/r)$.

So $\quad \partial_x{}^2(1/r) = \partial_x(-x/r^3) = +3r^{-4}.x.\partial_x r - r^{-3} = 3r^{-5}x^2 - r^{-3}$

and ditto y, ditto z. So adding all three, we get:

$$3r^{-5}(x^2 + y^2 + z^2) - 3r^{-3} = 3r^{-5}r^2 - 3r^{-3} = 3r^{-3} - 3r^{-3} = 0.$$

So this is a function which is zero everywhere away from the **origin**. But at $r = 0$, we have a **singular** term involving **division by zero**.

In 2-space we will need to show, from the formulae given in Section 7.7, that, in effect:

$$(1/2\pi)\Delta\mathbf{log}(1/|z - z_0|) = -\delta(z - z_0).$$

To do this, we will need to evaluate those awkward products in 2-space that appeared in Section 7.5. First, remember that in 2-space the **Hodge duals** are:

$\quad *\mathbf{d}x\mathbf{d}y = 1 \quad$ and $\quad *1 = \mathbf{d}x\mathbf{d}y$

and $\quad *\mathbf{d}x = \mathbf{d}y \quad$ and $\quad *\mathbf{d}y = -\mathbf{d}x$.

Now $\Delta = \mathbf{d\delta} + \mathbf{\delta d}$, acting on a **0-form** $\phi(z)$, since $np + p + 1$, with $n = 2$ and $p = 0$, is 1, so we have $\boldsymbol{\delta}\phi(z) = -*\mathbf{d}(*\phi(z)) = *\mathbf{d}(\phi(z)\mathbf{d}x\mathbf{d}y) = 0$, because this is **d** acting on the **pseudoscalar**. On the other hand, $\mathbf{d}\phi(z) = \partial_x\phi(z).\mathbf{d}x + \partial_y\phi(z).\mathbf{d}y$, so $*$ on this gives:

$$*\mathbf{d}\phi(z) = \partial_x\phi(z).\mathbf{d}y - \partial_y\phi(z).\mathbf{d}x,$$

and since now p is 1, $np + p + 1 = 4$, so there's no sign change. **d** on the $*\mathbf{d}\phi(z)$ form above gives:

$$\partial_x{}^2\phi(z).\mathbf{dxdy} - \partial_y{}^2\phi(z).\mathbf{dydx} = [\partial_x{}^2\phi(z) + \partial_y{}^2\phi(z)].\mathbf{dxdy}$$

and $*$ again gives just: $\partial_x{}^2\phi(z) + \partial_y{}^2\phi(z)$. So $\Delta\phi(z) = \mathbf{d\delta}\phi(z) + \mathbf{\delta d}\phi(z) = \mathbf{\delta d}\phi(z) = \partial_x{}^2\phi(z) + \partial_y{}^2\phi(z)$, as expected.

Now, from Section 7.5, we know we have to show that:

$$\mathbf{lim}_{(\varepsilon\to 0)}(\theta \mid S_\varepsilon) = \mathbf{lim}_{(\varepsilon\to 0)}((\mathbf{\delta}\phi \wedge *\gamma - \mathbf{\delta}\gamma \wedge *\phi + \phi \wedge *\mathbf{d}\gamma - \gamma \wedge *\mathbf{d}\phi) \mid S_\varepsilon) = -\phi[z_0].$$

where ϕ is as we've just been using it, and γ is the **Green's function**, which I'll write as $G(z, z_0)$. Now $\mathbf{\delta}\phi(z) = 0$, so the first term goes. But similarly, $\mathbf{\delta}\gamma = -*\mathbf{d}(*G(z, z_0)) = -*\mathbf{d}(G(z, z_0)\mathbf{dxdy}) = 0$, so the second term goes too. So, allowing that $\wedge$ goes to ordinary multiplication when one **operand** is a **0-form**, we have:

$$\phi(z).(\partial_x G(z, z_0).\mathbf{dy} - \partial_y G(z, z_0).\mathbf{dx}) - G(z, z_0).(\partial_x\phi(z).\mathbf{dy} - \partial_y\phi(z).\mathbf{dx}).$$

From Section 7.7, we expect $G(z, z_0) = (1/(2\pi).\mathbf{log}[1/(|z - z_0|)]$. If we integrate the **1-form** just given over the **boundary** of a small S_ε circle about z_0, we *should* get $-\phi(z_0)$.

First, $|z - z_0| = \surd((x - x_0)^2 + (y - y_0)^2)$, so:

$$G(z, z_0) = (1/(2\pi)).\mathbf{log}[1/\surd((x - x_0)^2 + (y - y_0)^2)] = -(1/4\pi).\mathbf{log}((x - x_0)^2 + (y - y_0)^2)$$

by $\mathbf{log}(x^{-1/2}) = -\tfrac{1}{2}.\mathbf{log}(x)$. Also $(x - x_0)^2 + (y - y_0)^2 = \varepsilon^2$, which is constant during the **integration** about S_ε. Also $\partial_x\mathbf{log}(\varepsilon^2) = (1/\varepsilon^2).2(x - x_0)$ and $\partial_y\mathbf{log}(\varepsilon^2) = (1/\varepsilon^2).2(y - y_0)$.

But we also have $\mathbf{dx} = -(y - y_0).\mathbf{d}\theta$ and $\mathbf{dy} = (x - x_0).\mathbf{d}\theta$,[120] where $\mathbf{b}S_\varepsilon \equiv \mathbf{b}S_\varepsilon(\theta) = \varepsilon.\mathbf{e}^{j\theta}$ over θ:[0, 2π], with ε constant. So our **integral** over S_ε is:

$$\begin{aligned}
&\phi(z).(\partial_x G(z, z_0).\mathbf{dy} - \partial_y G(z, z_0).\mathbf{dx}) - G(z, z_0).(\partial_x\phi(z).\mathbf{dy} - \partial_y\phi(z).\mathbf{dx}) \mid \mathbf{b}S_\varepsilon\\
&= \phi(z)(-(1/4\pi).[(1/\varepsilon^2).2(x - x_0).(x - x_0).\mathbf{d}\theta - (1/\varepsilon^2).2(y - y_0).(-(y - y_0).\mathbf{d}\theta)]\\
&\qquad + (1/4\pi)\mathbf{log}(\varepsilon^2)[(x - x_0).\partial_x\phi(z).\mathbf{d}\theta - (-(y - y_0))\partial_y\phi(z).\mathbf{d}\theta] \mid [0, 2\pi]
\end{aligned}$$

$$\begin{aligned}
&= \phi(z)(-(1/2\pi\varepsilon^2).((x - x_0)^2 + (y - y_0)^2).\mathbf{d}\theta \mid [0, 2\pi]\\
&\qquad + (1/4\pi)\mathbf{log}(\varepsilon^2)[(x - x_0).\partial_x\phi(z) + (y - y_0).\partial_y\phi(z)].\mathbf{d}\theta \mid [0, 2\pi]
\end{aligned}$$

$$= \phi(z)(-\varepsilon^2/2\pi\varepsilon^2).\mathbf{d}\theta \mid [0, 2\pi] + (1/4\pi)\mathbf{log}(\varepsilon^2)[(x - x_0).\partial_x\phi(z) + (y - y_0).\partial_y\phi(z)].\mathbf{d}\theta \mid [0, 2\pi].$$

The first term is clearly our answer $-\phi(z_0)$, because, taking $\mathbf{lim}_{(\varepsilon\to 0)}\phi(z).\mathbf{d}\theta \mid [0, 2\pi] \approx$
$\phi(z_0).(\mathbf{d}\theta \mid [0, 2\pi]) = 2\pi\phi(z_0)$. So how do we get rid of the second term? Well, the offending term is:

$$G(z, z_0).(\partial_x\phi(z).\mathbf{dy} - \partial_y\phi(z).\mathbf{dx}).$$

---

[120] From $x - x_0 = \varepsilon.\mathbf{cos}\theta$ and $y - y_0 = \varepsilon.\mathbf{sin}\theta$, so $\mathbf{dx} = -\varepsilon.\mathbf{sin}\theta.\mathbf{d}\theta = (y - y_0).\mathbf{d}\theta$, and $\mathbf{dy} = \varepsilon.\mathbf{cos}\theta.\mathbf{d}\theta = (x - x_0).\mathbf{d}\theta$.

Now $G(z, z_0) = (2\pi)^{-1}.\mathbf{log}(\varepsilon^{-1}) = -(2\pi)^{-1}.\mathbf{log}(\varepsilon)$, and

$$\partial_x\phi(z).\mathbf{d}y - \partial_y\phi(z).\mathbf{d}x = \partial_x\phi(z).\varepsilon.\mathbf{cos}\theta.\mathbf{d}\theta + \partial_y\phi(z).\varepsilon.\mathbf{sin}\theta.\mathbf{d}\theta.$$

So all told, the term is:

$$-(2\pi)^{-1}.\varepsilon\mathbf{log}(\varepsilon).[\partial_x\phi(z).\mathbf{cos}\theta + \partial_y\phi(z).\mathbf{sin}\theta].\mathbf{d}\theta \,\big|\, \mathbf{bS}_\varepsilon$$

$$\approx \mathbf{lim}_{(\varepsilon\to 0)}\{-(2\pi)^{-1}.\varepsilon\mathbf{log}(\varepsilon).\partial_x\phi(z_0).[\mathbf{cos}\theta.\mathbf{d}\theta] \,\big|\, [0, 2\pi]\} + \text{similar in } (\mathbf{sin}\theta.\mathbf{d}\theta).$$

$$= \partial_x\phi(z_0).\mathbf{lim}_{(\varepsilon\to 0)}\{-(2\pi)^{-1}.\varepsilon\mathbf{log}(\varepsilon).[\mathbf{cos}\theta.\mathbf{d}\theta] \,\big|\, [0, 2\pi]\} + \text{similar in } (\mathbf{sin}\theta.\mathbf{d}\theta).$$

We can expand $\mathbf{log}(\varepsilon) = (\varepsilon-1) - \tfrac{1}{2}(\varepsilon-1)^2 - \tfrac{1}{3}(\varepsilon-1)^3 \ldots$[121]
so $\varepsilon.\mathbf{log}(\varepsilon) = \varepsilon^2 - \varepsilon - \tfrac{1}{2}\varepsilon^3 + \tfrac{1}{2}\times 2\varepsilon^2 - \tfrac{1}{2}\varepsilon \ldots$ and all these terms will $\to 0$ as $\varepsilon \to 0$. Furthermore, $\mathbf{cos}\theta.\mathbf{d}\theta \,|\, [0, 2\pi]$ $= \mathbf{d}(\mathbf{sin}\theta) \,|\, [0, 2\pi] = \mathbf{sin}(2\pi) - \mathbf{sin}(0) = 0$ anyway. So the offending term will go to zero!

So this second **integral** is indeed zero, and we *do* have:

$$\mathbf{lim}_{(\varepsilon\to 0)}[\phi(z).(\partial_x G(z, z_0).\mathbf{d}y - \partial_y G(z, z_0).\mathbf{d}x) - G(z, z_0).(\partial_x\phi(z).\mathbf{d}y - \partial_y\phi(z).\mathbf{d}x)] \,\big|\, \mathbf{bS}_\varepsilon$$

$$= -\phi(z_0).$$

So the method does work!

7.C Appendix C: Distributions on the Unit Circle

> This Appendix was largely written before the bulk of this Chapter, so it goes over some ground already covered. I was planning to drop it but as it has some ideas highly relevant to **Fourier series** in the next Chapter, and I still like its approach, I've decided to leave it in.

The simplest example of the theory of **distributions** is that where the **domain** of the **vector space of functions** is the **unit circle** that dominated Chapter 4.

This Appendix discusses some of the issues relating to this special case.

I will formally define the **vector space** D(Γ), where Γ is the **unit circle** in the plane, as the space of **functions** whose **domain** is Γ. Such a **function** would be $\phi(x, y)$ where $x^2 + y^2 = 1$, or a **function** of an **angle** θ: $\phi(\theta)$.

It will turn out to be useful to bring in the **arc length parameterization** with θ introduced in Section 4.4 by:

$$\mathbf{d}x = -y.\mathbf{d}\theta$$
$$\mathbf{d}y = x.\mathbf{d}\theta$$

[121] Abramowitz, M and Stegun, I.A., *Handbook of Mathematical Functions*, Dover 1965

which from Section 4.6, gives the special property that:

$$\mathbf{d}(x + jy) = j(x + jy).\mathbf{d}\theta$$

or

$$\mathbf{d}z(\theta) = j.z(\theta).\mathbf{d}\theta$$

and so

$$z(\theta) = \mathbf{e}^{j\theta}.$$

Similarly, we obtained a **complex conjugate** form:

$$\mathbf{d}(x - jy) = -j(x - jy).\mathbf{d}\theta$$

or

$$\mathbf{d}z*(\theta) = -j.z*(\theta).\mathbf{d}\theta$$

and so

$$z*(\theta) = \mathbf{e}^{-j\theta}.$$

As θ ranges from 0 to 2π, $\mathbf{e}^{j\theta}$ travels around the **unit circle** once *anticlockwise*, and $\mathbf{e}^{-j\theta}$ travels round once *clockwise*. If we make the change of **parameter** $t := \theta/2\pi$, so that $\mathbf{d}\theta = 2\pi.\mathbf{d}t$, these equations take the form:

$$\mathbf{d}x = -2\pi y.\mathbf{d}t$$
$$\mathbf{d}y = 2\pi x.\mathbf{d}t$$

$$\mathbf{d}(x + jy) = 2\pi j(x + jy).\mathbf{d}t$$
$$\mathbf{d}z(t) = 2\pi j.z(t).\mathbf{d}t$$
$$z(t) = \mathbf{e}^{2\pi jt}.$$

The **complex conjugate** form now becomes:

$$\mathbf{d}(x - jy) = -2\pi j(x - jy).\mathbf{d}t$$
$$\mathbf{d}z*(t) = -2\pi j.z^*(t).\mathbf{d}t$$
$$z^*(t) = \mathbf{e}^{-2\pi jt}.$$

Now t runs from 0 to 1, so the **parametric interval** is the simpler [0, 1]. It is in this form that Laurent Schwartz recommends using them in what may well be regarded as the definitive work on the use of **distribution theory**, his *Mathematics for the Physical Sciences*, an ambitious title which is not entirely unjustified.[122] Note the identification:

$$\Gamma \equiv \mathbf{e}^{j\theta}:[0, 2\pi] \equiv \mathbf{e}^{2\pi jt}:[0, 1].$$

Here I use a *colon* to indicate the **domain** of the **function**. These two **functions** actually describe Γ in θ and in t respectively. Do not confuse this with the **integral operator** "|".

The equations above should be familiar from Section 4.9 as the ε-notation forms:

$$\mathbf{d}\varepsilon_x(\theta) = -2\pi.\varepsilon_y(\theta).\mathbf{d}\theta$$

$$\mathbf{d}\varepsilon_y(\theta) = 2\pi.\varepsilon_x(\theta).\mathbf{d}\theta.$$

---

[122] His page 152.

Also note that, for example, $\mathbf{e}^{2j\theta}$ or $\mathbf{e}^{4\pi jt}$ *also* describes Γ, but these two **functions** run round twice as fast: $\mathbf{e}^{2j(\pi/2)}$ = $\mathbf{e}^{4\pi j(0.25)}$ already equals $\mathbf{e}^{j\pi}$ = (x = −1, y = 0), the due West point, so is already half-way round the circle at only a quarter of the way through the former **domain**. Similarly, $\mathbf{e}^{-2j\theta}$ or $\mathbf{e}^{-4\pi jt}$ *also* describes Γ again, but this time going round twice as fast *clockwise*.

So any **function** $\mathbf{e}^{\pm jk\theta} = \mathbf{e}^{\pm 2\pi jkt} = \mathbf{e}\!\uparrow\!(\pm jk\theta) = \mathbf{e}\!\uparrow\!(\pm 2\pi jkt)$ describes the **unit circle** Γ.

Other **functions** on Γ that appear a lot are x and y themselves, the **cosine** and **sine** of any angle $\boldsymbol{\theta}$ in Γ. Now, writing $\boldsymbol{\theta}$ as $\theta_x + j\theta_y$, since

$$\mathbf{e}^{j\theta} = \theta_x + j\theta_y \qquad \text{and} \qquad \mathbf{e}^{-j\theta} = \theta_x - j\theta_y$$

$\theta_x = \mathbf{cos}(\theta) = ½(\mathbf{e}^{j\theta} + \mathbf{e}^{-j\theta})$ and $\theta_y = \mathbf{sin}(\theta) = ½j(\mathbf{e}^{j\theta} - \mathbf{e}^{-j\theta})$, two very useful formulae.

To obtain our first examples of **distributions**, we can take the idea from Section 5.4 that elements of the basic **vector space** can *themselves* act as members of the **dual space** (which is here the space of **distributions** on the elements of the basic **function space**) by a **scalar product**. The **scalar product** we use here is the **integral** introduced in Section 7.1, which here takes the form:

$$\mathrm{T}(\boldsymbol{\theta}).\phi(\boldsymbol{\theta}).\mathbf{d}\boldsymbol{\theta} \,\big|\, \Gamma$$

where $\boldsymbol{\theta}$, written bold, equals $\theta_x + j\theta_y$, so this is a **complex line integral** like those discussed in Section 7.2. Here T($\boldsymbol{\theta}$) is a **function** on $\boldsymbol{\theta}$ acting as a **distribution**, because through this **scalar product**, T($\boldsymbol{\theta}$) defines a **function** which takes elements of the **function space** D(Γ) as **arguments**, and, **integrated** over Γ, returns a *number*, which here may be **real** or **complex**. It thus defines a member of the **dual space** of D(Γ), which space I will label $\mathrm{D}^{\dagger}(\Gamma)$.[123]

On the space D(Γ), it is usually easiest to evaluate the **integral** through the use of the **real parameters** θ or t. So, rather than the **scalar product** above, we use the simpler version **pulled back** to the **parameter** space, given by:

$$\mathrm{T}(\theta).\phi(\theta).\mathbf{d}\theta \,\big|\, [0, 2\pi].$$

For example, **cos(**$m\boldsymbol{\theta}$**)** or **cos**($m\theta$) defines a **distribution** which acts on $\phi(\theta) = \uparrow(jk\theta)$ to give the value:

$$\mathbf{cos}(m\theta).\mathbf{e}\!\uparrow\!(jk\theta).\mathbf{d}\theta \,\big|\, [0, 2\pi].$$

But $\mathbf{cos}(\theta) = ½(\uparrow(j\theta) + \uparrow(-j\theta))$, so this becomes:

$$½[\uparrow(jm\theta).\uparrow(jk\theta) + \uparrow(-jm\theta).\uparrow(jk\theta)].\mathbf{d}\theta \,\big|\, [0, 2\pi]$$

$$= ½[\uparrow j\theta(m + k) + \uparrow j\theta(m - k)].\mathbf{d}\theta \,\big|\, [0, 2\pi].$$

[123] The possibility that the **line integral** in the plane may give a **complex** result can easily be reconciled with the original definition of Section 5.4 if we suggest that *either* the **real** (x) or the **imaginary** (y) part of the **integral** value can be taken as a **real number** result. In this sense, *two* **distributions** are defined. Alternatively, we could extend the definition of the **dual space** to allow for **complex** values.

Now if $p \neq 0$, $\mathbf{e}{\uparrow}(jp\theta).\mathbf{d}\theta = \mathbf{d}((1/jp).\mathbf{e}{\uparrow}(jp\theta))$, so if $m \neq k$ and $m \neq -k$, we get two terms of the form:

$$\mathbf{d}((1/jp).\mathbf{e}{\uparrow}(jp\theta)) \,\big|\, [0, 2\pi] = (1/jp).\mathbf{e}{\uparrow}(jp\theta) \,\big|\, \mathbf{b}[0, 2\pi] = (1/jp).\mathbf{e}{\uparrow}(jp2\pi) - (1/jp).\mathbf{e}{\uparrow}(jp0).$$

But for p equal to any non-zero integer, $\mathbf{e}{\uparrow}(jp2\pi) = (\mathbf{e}{\uparrow}(j2\pi)) \uparrow p = (1 + j.0) \uparrow p = 1^p = 1$, and $\mathbf{e}{\uparrow}(jp0) = (\mathbf{e}{\uparrow}0) \uparrow p = (1) \uparrow p = 1^p = 1$, because both $\theta = 0$ and $\theta = 2\pi$ **map** into the same point (1, 0) or $1 + j.0 = 1$, so our **integral** is $(1/jp)[1 - 1] = 0$.

But if $m = -k$, the *first* term becomes:

$$(\mathbf{e}{\uparrow}j\theta(m + k)).\mathbf{d}\theta \,\big|\, [0, 2\pi] = (\mathbf{e}{\uparrow}(j\theta{\times}0)).\mathbf{d}\theta \,\big|\, [0, 2\pi] = (\mathbf{e}{\uparrow}0).\mathbf{d}\theta \,\big|\, [0, 2\pi] = 1.\mathbf{d}\theta \,\big|\, [0, 2\pi] = 2\pi - 0$$

and if $m = k$, the *second* term gives:

$$(\mathbf{e}{\uparrow}j\theta(m - k)).\mathbf{d}\theta \,\big|\, [0, 2\pi] = (\mathbf{e}{\uparrow}0).\mathbf{d}\theta \,\big|\, [0, 2\pi] = 1.\mathbf{d}\theta \,\big|\, [0, 2\pi] = 2\pi.$$

So remembering the ½ in front, we have all told:

$$\mathbf{cos}(m\theta).\mathbf{e}{\uparrow}(jk\theta).\mathbf{d}\theta \,\big|\, [0, 2\pi] = \{\pi \text{ if } m = -k \text{ or if } m = k, \text{ otherwise } 0\}.$$

The two **scalar products** given above differ slightly, because, since $\boldsymbol{\theta} = \theta_x + j\theta_y$:

$$\mathbf{d}\boldsymbol{\theta} = \mathbf{d}(\theta_x + j\theta_y) = \mathbf{d}(\mathbf{e}{\uparrow}j\theta) = j(\mathbf{e}{\uparrow}j\theta).\mathbf{d}\theta = j\boldsymbol{\theta}.\mathbf{d}\boldsymbol{\theta},$$

and here $\mathbf{cos}(m\theta).\mathbf{e}{\uparrow}(jk\theta).\mathbf{d}\theta$, in terms of $\boldsymbol{\theta}$, would be $\mathbf{cos}(m\boldsymbol{\theta}).\boldsymbol{\theta}^k.\mathbf{d}\theta = \mathbf{cos}(m\boldsymbol{\theta}).\boldsymbol{\theta}^k.(j\boldsymbol{\theta})^{-1}.\mathbf{d}\boldsymbol{\theta}$. Indeed, $\mathbf{cos}(m\theta) = ½(\mathbf{e}{\uparrow}(jm\theta) + \mathbf{e}{\uparrow}(-jm\theta)) = ½((\mathbf{e}{\uparrow}j\theta)^m + (\mathbf{e}{\uparrow}j\theta)^{-m}) = ½(\boldsymbol{\theta}^m + \boldsymbol{\theta}^{-m})$, so we should have:

$$\mathbf{cos}(m\theta).\mathbf{e}{\uparrow}(jk\theta).\mathbf{d}\theta = ½(\boldsymbol{\theta}^m + \boldsymbol{\theta}^{-m}).j^{-1}\boldsymbol{\theta}^{k-1}.\mathbf{d}\boldsymbol{\theta}.$$

So now we have two **integrals** of the form:

$$\boldsymbol{\theta}^{m+k-1}.\mathbf{d}\boldsymbol{\theta} \,\big|\, \Gamma \qquad \text{and} \qquad \boldsymbol{\theta}^{-m+k-1}.\mathbf{d}\boldsymbol{\theta} \,\big|\, \Gamma.$$

Writing the **complex number** $\boldsymbol{\theta}$ as z, these are:

$$z^{m+k-1}.\mathbf{d}z \,\big|\, \Gamma \qquad \text{and} \qquad z^{-m+k-1}.\mathbf{d}z \,\big|\, \Gamma.$$

If $m = -k$ or $m = k$¸ one or other of these takes the form:

$$\mathbf{d}z/z \,\big|\, \Gamma = \mathbf{d}z/(z - 0) \,\big|\, \Gamma.$$

Here we are integrating over a circle Γ. But from Section 7.2, we know these to evaluate to $2\pi j$. So now bring in the ½ and the j^{-1} from the full expression $½(\boldsymbol{\theta}^m + \boldsymbol{\theta}^{-m}).j^{-1}\boldsymbol{\theta}^{k-1}.\mathbf{d}\boldsymbol{\theta}$ and we recover our result above that when $m = -k$ or $m = k$¸

$$½(\boldsymbol{\theta}^m + \boldsymbol{\theta}^{-m}).j^{-1}\boldsymbol{\theta}^{k-1}.\mathbf{d}\boldsymbol{\theta} \,\big|\, \Gamma = 2\pi j(½j^{-1}) = \pi.$$

But if $m \neq k$ and $m \neq -k$, then from the same **integration** in Section 7.2, prefacing **Cauchy's integral formula**, we have:

$$\mathbf{d}z.z^m \,\big|\, \Gamma = \mathbf{d}z.(z-0)^m \,\big|\, \Gamma = 0, \qquad \text{for } m \neq 0.$$

So with the appropriate substitutions, both methods, and both **scalar products**, do in fact give the same result!

Obviously the **Dirac delta distribution** returns a number, which may be **complex** if the **range** of ϕ is **complex**, and it is linear by:

$$\langle \delta_{(a)}, p\phi + q\psi \rangle = p\phi(a) + q\psi(a)$$

where a, p and q are **scalars** (i.e. numbers), and ϕ and ψ are **functions**. So it *is* an acceptable element of the **dual space**. In $D^{\dagger}(\Gamma)$, the appropriate form would be:

$$\langle \delta_{(\theta 0)}, \phi \rangle := \phi(\theta_0).$$

We can define **derivatives** of δ, following the ideas outlined in Section 7.5. Again, we look at the behaviour of **derivatives** for **distributions** that are **functions**, and whose action can be defined by the usual **integral scalar product**. Take a **function** $g(\theta) = g(\mathbf{e}^{j\theta})$ and use the $\mathbf{d}\theta \,|\, [0, 2\pi]$ **scalar product**:

$$\langle g, \phi \rangle = g(\mathbf{e}^{j\theta}).\phi(\mathbf{e}^{j\theta}).\mathbf{d}\theta \,\big|\, [0, 2\pi].$$

Now the **derivative** of g, $\partial_\theta\, g$, gives under **integration by parts** (using **axiom II** of the **differential operator** and defined in Section 2.4):

$$\partial_\theta[g(\mathbf{e}^{j\theta})].\phi(\mathbf{e}^{j\theta}).\mathbf{d}\theta = \mathbf{d}[g(\mathbf{e}^{j\theta}).\phi(\mathbf{e}^{j\theta})] - g(\mathbf{e}^{j\theta}).\partial_\theta\phi(\mathbf{e}^{j\theta}).\mathbf{d}\theta.$$

This is simply **axiom II** with one of the right-hand terms isolated on the left:

$$\mathbf{d}(xy) = x.\mathbf{d}y + y.\mathbf{d}x \longrightarrow y.\mathbf{d}x = \mathbf{d}(xy) - x.\mathbf{d}y$$

where here we set $x := g(\mathbf{e}^{j\theta})$ and $y := \phi(\mathbf{e}^{j\theta})$. Now $\mathbf{d}[g(\mathbf{e}^{j\theta}).\phi(\mathbf{e}^{j\theta})] \,|\, [0, 2\pi] = g(\mathbf{e}^{j\theta}).\phi(\mathbf{e}^{j\theta}) \,|\, \mathbf{b}[0, 2\pi]$ and this is:

$$g(\mathbf{e}^{j2\pi}).\phi(\mathbf{e}^{j2\pi}) - g(\mathbf{e}^{j0}).\phi(\mathbf{e}^{j0}) = g(1+j.0).\phi(1+j.0) - g(1+j.0).\phi(1+j.0) = 0.$$

So we have all told:

$$\partial_\theta[g(\mathbf{e}^{j\theta})].\phi(\mathbf{e}^{j\theta}).\mathbf{d}\theta \,\big|\, [0, 2\pi] = -\, g(\mathbf{e}^{j\theta}).\partial_\theta\phi(\mathbf{e}^{j\theta}).\mathbf{d}\theta \,\big|\, [0, 2\pi].$$

In a general context, this takes the form:

$$\partial_x g(x).\phi(x) \,\big|\, [-\infty, \infty] = -g(x).\partial_x\phi(x) \,\big|\, [-\infty, \infty]$$

or $$\langle \partial_x g, \phi \rangle = -\langle g, \partial_x\phi \rangle.$$

Since $\partial_x\phi$ is always well-defined for a member of the **function space,** this **adjoint** switch can be used as the *definition* of the **derivative** for a **distribution** which will not otherwise have one. *This is a hugely important definition.*[124]

This definition can be written in a strict **adjoint** form if we put it as $\langle \partial_x g, \phi\rangle = \langle g, -\partial_x\phi\rangle$, so that ∂_x and $-\partial_x$ are **adjoint operators** for the **distribution product**. So now, for the **delta distribution** on Γ, we can write:

$$\langle \partial_\theta(\delta), \phi\rangle := -\langle \delta, \partial_\theta\phi\rangle := -\partial_\theta\phi(0)$$

and

$$\langle \partial_\theta(\delta_{(\theta 0)}), \phi\rangle := -\langle \delta_{(\theta 0)}, \partial_\theta\phi\rangle := -\partial_\theta\phi(\theta_0),$$

where I write $\phi(\mathbf{e}^{j\theta}) = \phi(x(\theta) + jy(\theta))$ as just $\phi(\theta)$, hopefully an acceptable shorthand!

On the **unit circle** Γ, we couldn't have a **Heaviside function**, because the lack of **ordering** (Section 4.3) means there's no set of θ such that $\theta > \theta_0$, but we *do* have "betweenness", using the **minor arc**, or we can talk of the arc running *anticlockwise* from θ_a to θ_b. So we can have a **pulse**:

$\Pi(\theta, \theta_a, \theta_b) := 1$ in the arc running *anticlockwise* from θ_a to θ_b, else 0.

In developing the **derivative** rule here, we need to be careful if we use the ordinary **parameterization**, because if $\theta_a = \sqrt{½} - j\sqrt{½}$ (South-East) and $\theta_b = \sqrt{½} + j\sqrt{½}$ (North-East), we'll cross the $\theta = 0$ point, so there's a **discontinuity** in the **parameterization** in this case. To circumvent this, we make a change of variable to shift the point of discontinuity to θ_a, so using $\theta^\dagger = \theta - \theta_a$, we have:

$$\langle\, \partial_\theta\Pi(\theta, \theta_a, \theta_b), \phi\,\rangle = -\, \partial_\theta\phi(\mathbf{e}^{j\theta}).\mathbf{d}\theta^\dagger \,\big|\, [0, \theta_b - \theta_a] = -\, \partial_\theta\phi(\mathbf{e}{\uparrow}j(\theta^\dagger + \theta_a)).\mathbf{d}\theta^\dagger \,\big|\, [0, \theta_b - \theta_a]$$

$$= -\mathbf{d}(\phi(\mathbf{e}{\uparrow}j(\theta^\dagger + \theta_a))) \,\big|\, [0, \theta_b - \theta_a]$$

$$= -\, \phi(\mathbf{e}{\uparrow}j(\theta^\dagger + \theta_a))) \,\big|\, \mathbf{b}[0, \theta_b - \theta_a] = -\, (\phi(\mathbf{e}{\uparrow}j(\theta_b - \theta_a + \theta_a)) - \phi(\mathbf{e}{\uparrow}j(0 + \theta_a)))$$

$$= \phi(\mathbf{e}{\uparrow}j\theta_a) - \phi(\mathbf{e}{\uparrow}j\theta_b) = \phi(\theta_a) - \phi(\theta_b).$$

So again we have:

$$\partial_\theta\Pi(\theta, \theta_a, \theta_b) \equiv \delta(\theta_a) - \delta(\theta_b)$$

much as we would expect.

[124] I might also add that Schwartz's heavy use of such **adjoint definitions** was my main inspiration to do so for basic **integration.**

7.D Appendix D: Delta Functionals of the Form δ(φ(*x*)).

It might seem reasonable to define the action of a **distribution** or **functional** like $\delta(\phi(x))$ as simply abstracting the **zeroes** of ϕ within the **domain** in question, where the **zeroes** of any **function** are the points where it has the value 0. So we might define $\delta(x^2 - 4)$ simply as $\delta(x + 2) + \delta(x - 2)$ because $x^2 - 4 = 0$ at $x = \pm 2$.

But to remain consistent with the general emphasis on **integration by parts**, that is not what is done, although what *is* done might itself be called into question, as it uses the principle that **Jacobians** should always be used as **absolute values** even if a change of **coordinates** is **chiral** (that is, its **Jacobian determinant** involves a change of sign).

So in an **integral** like $g(x)\delta(x^2 - 4)\mathbf{d}x \mid [-\infty, \infty]$, we'd make the change of variable $u := x^2 - 4$, so that $\mathbf{d}u = 2x\mathbf{d}x$, so $\mathbf{d}x = \mathbf{d}u/2x$ and $x = \pm\sqrt{(u + 4)}$ to give:

$$g(\pm\sqrt{(u + 4)})\delta(u)\mathbf{d}u/(\pm 2\sqrt{(u + 4)}) \mid [-\infty, \infty]$$

giving, by *my* reckoning:

$$g(+\sqrt{4})/(+2\sqrt{4}) + g(-\sqrt{4})/(-2\sqrt{4}) = g(2)/4 - g(-2)/4.$$

But if we insist that $\delta(u(x))\mathbf{d}x$ must equate to $\delta(u)\mathbf{d}u/|\partial_x u(x)| = \delta(u)\mathbf{d}u/|\partial_x u(x(u = 0))|$, then since here $|\partial_x u(x)| = 2|x| = 2|\pm\sqrt{4}| = 4$ at both points where u is zero, we get:

$$g(+\sqrt{4})/(+2\sqrt{4}) + g(-\sqrt{4})/(+2\sqrt{4}) = g(2)/4 + g(-2)/4.$$

As it happens, this is closer to my original suggestion that $\delta(x^2 - 4) = \delta(x + 2) + \delta(x - 2)$.
Roy Hoskins[125] gives another interesting example:

$$(\mathbf{cos}(t) + \mathbf{sin}(t)).\delta'(t^3 + t^2 + t)\mathbf{d}t \mid [-\infty, \infty],$$

where $\delta'(x)$ is the **OD** $\partial_x[\delta(x)]$. We've already defined $\partial_t[\delta(t)]$ in Section 7.5 by:

$$\partial_t \delta(t - a)(f) := -\partial_t f(a).$$

Now $t^3 + t^2 + t = t(t^2 + t + 1) = t(t - ½(1 + j\sqrt{3}))(t - ½(1 + j\sqrt{3}))$, so only one **zero** of this, at $t = 0$, is in $R \equiv [-\infty, \infty]$. So putting $u = t^3 + t^2 + t$, $\mathbf{d}u = (3t^2 + 2t + 1)\mathbf{d}t$, we must have:

$$(\mathbf{cos}(t(u)) + \mathbf{sin}(t(u)))(3t^2(u) + 2t(u) + 1)^{-1}\partial_u\delta(u)\mathbf{d}u \mid [-\infty, \infty]$$

where the **domain of integration** remains the same $[-\infty, \infty]$ because $u \rightarrow \pm\infty$ as $t \rightarrow \pm\infty$, and $u > 0$ when $t > 0$, and $u < 0$ when $t < 0$. By our definition, $\partial_u \delta(u)(f) := -\partial_u f(0)$, so this must give the result:

$$-\partial_u[(\mathbf{cos}(t(u)) + \mathbf{sin}(t(u)))(3t^2(u) + 2t(u) + 1)^{-1}](u = 0)$$

---

[125] R.F. Hoskins, *The Delta Function*, Ellis Horwood Press 2nd Edition 2009.

$= -\ [(-\mathbf{sin}(t(u)) + \mathbf{cos}(t(u)))(3t^2(u) + 2t(u) + 1)^{-1}$

$- (\mathbf{cos}(t(u)) + \mathbf{sin}(t(u)))(3t^2(u) + 2t(u) + 1)^{-2}(6t(u) + 2)](u = 0).$

But because $t(u) = 0$ when $u = 0$, this is:

$- [(-\mathbf{sin}(0) + \mathbf{cos}(0))(3 \times 0^2 + 2 \times 0 + 1)^{-1}$

$- (\mathbf{cos}(0) + \mathbf{sin}(0))(3 \times 0^2 + 2 \times 0 + 1)^{-2}(6 \times 0 + 2)]$

$= -\ (-\mathbf{sin}(0) + \mathbf{cos}(0)) + (\mathbf{cos}(0) + \mathbf{sin}(0)) \times 2 = 2 - 1 = 1.$

An interesting case is $\delta(\mathbf{sin}(\theta))$, which under the simple idea that the expansion of any $\delta(\phi)$ form should be as Σ_k $\delta(\mathbf{zero}_k(\phi))$ should give on the **unit circle** $\delta(\theta) + \delta(\theta - \pi)$. That is indeed the correct answer, but because here ∂_θ $(\phi(\theta)) = \partial_\theta(\mathbf{sin}(\theta)) = \mathbf{cos}(\theta)$ here, and at the two **zeroes** of $\mathbf{sin}(\theta)$, $\theta = 0$ and $\theta = \pi$, $1/|\mathbf{cos}(\theta)| = 1$.

It is important to realize that the conventional version of $\delta(\mathbf{sin}(\theta))$ is somewhat nastier than this, because it is presented *not* for θ on the **unit circle**, but with θ **parameterizing** a **periodic function** along the **real line** with θ running from $-\infty$ to $+\infty$. Because *this* θ corresponds to endless sweeps around the **unit circle**, the two **zeroes** of $\mathbf{sin}(\theta)$ are now picked up again and again, and we obtain the form:

$\delta(\mathbf{sin}(\theta)) = \Sigma_{k=-\infty}^{\infty} \delta(\theta - k\pi)$

$= \ldots \delta(\theta - (-3\pi)) + \delta(\theta - (-2\pi)) + \delta(\theta - (-\pi)) + \delta(\theta - 0)$

$+ \delta(\theta - \pi) + \delta(\theta - 2\pi) + \delta(\theta - 3\pi) + \ldots$

$= \ldots \delta(\theta + 3\pi) + \delta(\theta + 2\pi) + \delta(\theta + \pi) + \delta(\theta) + \delta(\theta - \pi) + \delta(\theta - 2\pi) + \delta(\theta - 3\pi) + \ldots$

7.X Exercises

1. Evaluate $[(2z - 1)/(z^2 - z)].\mathbf{d}z$ over the circle centre $z = ½$, radius = 1, by expanding using **partial fractions** and then using the **Cauchy integral formula**.
2. Expanding $\mathbf{sin}(z)$ in a **Taylor expansion** about $z = 0$, evaluate $\mathbf{sin}(z).\mathbf{d}z/z^4$ around the **unit circle** in the **complex plane**. (Be very careful here – note the $m = -1$ in the derivation of **Cauchy's integral formula**.) (E. Kreyszig, *Advanced Engineering Mathematics*, Wiley 1967.)
3. Using a technique similar to that in the last question, show that if $\varphi(z)$ has a **pole of order** m at a, and so can be written as a partial fractions expansion of the form $\varphi(z) = c_m/(z - a)^m +$

$c_{m-1}/(z-a)^{m-1} + \ldots + c_2/(z-a)^2 + c_1/(z-a)$ + {positive powers of $(z-a)$}, the **residue** of $\varphi(z)$ at $z = a$ is $(1/(m-1)!).\mathbf{lim}_{z\to a}\partial_z^{m-1}((z-a)^m\varphi(z))$.

4. Using **integration by parts** and a suitable change of variable, show that $\delta(ax + b) \equiv (1/|a|).\delta(x + b/a)$. Similarly show that if $\varphi(x)$ has **zeroes** $\{x_1, x_2, \ldots x_n, \ldots\}$, $\delta(\varphi(x)) = \Sigma_n(1/|\partial_x\varphi(x_n)|).\delta(x - x_n)$.

5. Can you justify $\delta(xy) = [\delta(x) + \delta(y)]/(x^2 + y^2)^{½}$? (R.N.Bracewell, McGraw-Hill 2000.)

6. Evaluate the **Green's formula** $\langle \mathbf{d}\phi \cdot \psi\rangle - \langle \phi \cdot \boldsymbol{\delta}\psi\rangle = (\phi \wedge *\psi) \mid \mathbf{b}\Sigma$ for $\phi = x^2y$ and $\psi = \mathbf{d}(y^2x)$ over the unit disc $(x^2 + y^2 \leq 1)$ as Σ.

7. **Green's functions** for a **differential operator** L are often represented by the equation $LG(x, \xi) = \delta(x - \xi)$ using the **Dirac delta function**. Justify this notation as far as possible. It indicates the **Green's function** is the "response" to a unit pulse "source". Show also that if L is **self-adjoint**, then its **Green's function** is symmetric or for any ξ and x, $G(x, \xi) = G(\xi, x)$.

8. If $L = a(x).\partial_x^2 + b(x).\partial_x + c(x)$, show that its **adjoint** L^* is $L^*v = \partial_x^2(av) - \partial_x(bv) + cv$, and the **conjunct** $\langle v \cdot Lu\rangle - \langle u \cdot L^*v\rangle = (vLu - uL^*v).\mathbf{d}x \mid [\alpha, \beta] = J(u, v) \mid [\alpha, \beta]$ is given as: $J(u, v) = av\partial_x u - u\partial_x(av) + buv$. Hence for $L = \partial_x^2$, $J(u, v) = v\partial_x u - u\partial_x v$.

9. Hence show that if u obeys $Lu = 0$, and v obeys $L^*v = 0$, $J(u, v)$ is a constant.

10. Show that the operator above, $Lu = a(x).\partial_x^2u + b(x).\partial_x u + c(x).u$, if $a(x)$ is non-zero over $[\alpha, \beta]$, can be put in the form $Lu = \partial_x(p(x).\partial_x u) + q(x).u$ by defining $p(x) = \mathbf{e}\uparrow((b(\xi)/a(\xi)).\mathbf{d}\xi \mid [\alpha, x])$, $q(x) = (c(x)/a(x)).p(x)$, and multiplying Lu through by $p(x)/a(x)$.

11. Show that $Lu = \partial_x(p(x).\partial_x u) + q(x).u$ is formally **self-adjoint**.

12. Consider the general class of **ordinary differential equation** $\partial_x(p(x).\partial_x y(x)) + q(x).y(x) = -f(x)$. A **Green's function** to solve this problem as $y(x) = G(x, \xi).f(\xi).\mathbf{d}\xi \mid [a, b]$ where we have boundary conditions $y(b) = y(a) = 0$ must obey:

    a. $\partial_x(p(x).\partial_x G(x, \xi)) + q(x).G(x, \xi) = 0$ for $x \neq \xi$

    b. $G(x = b) = G(x = a) = 0$

    c. $G(x = \xi + \varepsilon) - G(x = \xi - \varepsilon) \to 0$ as $\varepsilon \to 0$

    d. $\partial_x G(x = \xi + \varepsilon) - \partial_x G(x = \xi - \varepsilon) \to -1/p(\xi)$ as $\varepsilon \to 0$

    e. $G(x, \xi) = G(\xi, x)$

    So **Green's functions** are not easy things to come by! Note the "step **discontinuity**" in the first derivative (d) which gives the delta functional behaviour. Now take the simplest case of $p(x) = 1$, $q(x) = 0$, so $\partial_x^2y(x) = -f(x)$, and assume $y(0) = y(1) = 0$. Show that setting $G(x,\xi) = (1 - \xi)x$ if $x \leq \xi$ and $G(x, \xi) = \xi(1 - x)$ if $x \geq \xi$ [i.e. $G(x,\xi) = x(1 - \xi).H(\xi - x) + \xi(1 - x).H(x - \xi)$] fits the listed criteria and so solves this problem as the pair of integrals: $y(x) = (1 - x)\xi.f(\xi).\mathbf{d}\xi \mid [0, x] + (1 - \xi)x.f(\xi).\mathbf{d}\xi \mid [x, 1]$. Try it for $f(x) = 1$.

Chapter 8

FOURIER-LAPLACE TRANSFORMS

8.1 Discrete Parallels

Much of the theory of discrete **vector spaces** developed in Chapter 5 carries over to **functional spaces** where the "**vectors**" are **functions**. A couple of these parallels are highly relevant to this Chapter.

Under the **integral scalar product** $\langle \omega, \phi \rangle := \omega \wedge *\phi \,\big|\, \mathbf{D}$ that I introduced in Section 7.4, we can say that two **differential forms** of the same **grade** ω and ϕ are **orthogonal** if:

$$\langle \omega, \phi \rangle = 0.$$

It may seem a rather quaint idea to talk of such abstract constructs as being, apparently, "at right angles" to one another, the ordinary meaning of this term in the spaces of **standard vectors** discussed in Chapter 5. But the concept has its uses.

In particular, suppose we have a **vector x** in ordinary 3-space, expressed in terms of an **orthonormal** set $\{\mathbf{e}_i\}$ of three vectors. Then, by **orthonormality**,

$$\mathbf{e}_n \cdot \mathbf{e}_m = \delta_{nm}$$

where δ_{nm} is the **Kronecker delta** symbol introduced in Section 5.4 (and in Section 1.8), which has the value 1 if $n = m$, but is otherwise zero.

Now if $\mathbf{x} = a_1\mathbf{e}_1 + a_2\mathbf{e}_2 + a_3\mathbf{e}_3$, then:

$$\mathbf{x} \cdot \mathbf{e}_2 = a_1(\mathbf{e}_1 \cdot \mathbf{e}_2) + a_2(\mathbf{e}_2 \cdot \mathbf{e}_2) + a_3(\mathbf{e}_3 \cdot \mathbf{e}_2) = a_1.0 + a_2.1 + a_3.0 = a_2.$$

In the same way, the **scalar product** of **x** with any of $\mathbf{e}_1$ or $\mathbf{e}_2$ or $\mathbf{e}_3$ will isolate the corresponding **component** in the expansion of **x**.

So $\quad \mathbf{x} = (\mathbf{x} \cdot \mathbf{e}_1)\mathbf{e}_1 + (\mathbf{x} \cdot \mathbf{e}_2)\mathbf{e}_2 + (\mathbf{x} \cdot \mathbf{e}_3)\mathbf{e}_3.$

We will encounter a very similar situation in Section 8.2, in the theory of **Fourier Series**. Because of this parallel, the **components** a_i = (**x** · **e**$_i$) of **x** are called its **Fourier coefficients**. But in the **Fourier series** case, the number of **basis** elements **e**$_n$ is *infinite*! This is an infinite dimensional space.

Particularly important too is the extension of the theory of **eigenvalues** and **eigenvectors** to **functional spaces.** For this we need to have an analogous concept of a **linear operator**, which for generality I will define as operating between **differential forms**. A **linear operator** L on a **form** ω is simply a **function** whose result is also a **form**, and that obeys:

$$L(a\omega + b\phi) = a.L\omega + b.L\phi.$$

Here ω and ϕ are **forms** and a and b are simply numbers. The **differential operator** or **exterior derivative** is a **linear operator** on **forms**, as are the **co-differential operator** and the **Laplace-Beltrami operator** or **Laplacian** introduced in Section 7.5. Keep in mind that the **forms** involved in this definition may be a **0-form** or a **scalar function**, and historically, the vast bulk of research on these **linear operators** has been devoted to this case. One of the simplest **linear operators** is what I will call the **Fourier operator** $-\partial^2$ which operates from a **0-form** to generate another **0-form**:

$$-\partial^2 g(x) = -\partial_x{}^2 g(x)$$

$$-\partial^2(2x^5) = -\partial(10x^4) = -\ 40x^3.$$

Strictly speaking there is a **Fourier operator** for every **variable**, so we'd have $-\partial_x{}^2$, $-\partial_y{}^2$ $-\partial_z{}^2$, and $-\partial_t{}^2$ in *xyzt*-space, the normal space of physics with three space dimensions and one time t. It's worth emphasizing this here because the vast majority of the theory of **Fourier-Laplace transforms**, on which this Chapter makes but a beginning, involves the *time* case $-\partial_t{}^2$.

An **eigenfunction** for a **linear operator** is defined analogously to the ordinary **vector** case. So $\phi(x)$ is an **eigenfunction** for L if:

$$L\phi(x) = \lambda\phi(x)$$

where λ is a **real** or **complex number**, the corresponding **eigenvalue**. This Chapter is all about the **eigenfunctions** of the Fourier **operator** $-\partial^2$, which for $-\partial_x{}^2$ are simply the **functions** $\mathbf{e}^{j\lambda x}$ or **e**↑$j\lambda x$:

$$-\partial_x{}^2\mathbf{e}^{j\lambda x} = -\ \partial_x j\lambda.\mathbf{e}^{j\lambda x} = -\ j^2\lambda^2\mathbf{e}^{j\lambda x} = \lambda^2\mathbf{e}^{j\lambda x}.$$

So $\mathbf{e}^{j\lambda x}$ is an **eigenfunction** for $-\partial_x{}^2$ with **real eigenvalue** λ^2. The **eigenvalues** are **real** because the **operator** is **Hermitian**, a concept touched on in the context of **matrices** in Exercise 5.X.9, which for an **operator** on **forms** is defined as follows.

A **Linear Operator** H is **Hermitian** if:

$$\langle\omega^*, H\phi\rangle = \langle\phi^*, H\omega\rangle^*$$

where * indicates the **complex conjugate**, ω and ϕ are **forms** and $\langle\ ,\ \rangle$ is the **scalar product** for **forms** introduced in Section 7.4. This looks pretty nasty, but it's quite easy for the **Fourier operator** where the **domain of integration** for the **scalar product** is either the full range of x:[$-\infty$, ∞], which is defined as a **limiting** value $\mathbf{lim}_{a\to\infty}\ \phi(x)\mathbf{d}x \mid [-a, a]$, or else it is the **unit circle parameterized** with t:[0, 2π].

The **unit circle** case is the easiest to demonstrate. For **complex 0-forms** $f(t)$ and $g(t)$[126] on the **unit circle** *obeying the* **periodicity** *conditions*:

$$f(2\pi) = f(0) \text{ and } g(2\pi) = g(0), \text{ and } \partial_t f(2\pi) = \partial_t f(0) \text{ and } \partial_t g(2\pi) = \partial_t g(0),$$

then

$$\langle f^*, -\partial_t^2 g\rangle = -f^*(t).\partial_t^2 g(t).\mathbf{d}t \mid [0, 2\pi].$$

Now $\mathbf{d}(f^*(t).\partial_t g(t)) = \mathbf{d}f^*(t).\partial_t g(t)) + f^*(t).\partial_t^2 g(t).\mathbf{d}t$ by our usual axiom II of the **calculus**, so "**integrating by parts**":

$$-f^*(t).\partial_t^2 g(t).\mathbf{d}t \mid [0, 2\pi] = -\mathbf{d}(f^*(t).\partial_t g(t)) \mid [0, 2\pi] + \mathbf{d}f^*(t).\partial_t g(t)) \mid [0, 2\pi].$$

$\mathbf{d}(f^*(t).\partial_t g(t)) \mid [0, 2\pi] = f^*(t).\partial_t g(t) \mid \mathbf{b}[0, 2\pi] = f^*(2\pi).\partial_t g(2\pi) - f^*(0).\partial_t g(0) = 0$ by our **periodicity** premises on f and g, so, expanding the rightmost term as $\partial_t f^*.\partial_t g.\mathbf{d}t$ and re-applying axiom II in the form:

$$\mathbf{d}(\partial_t f^*.g) = \mathbf{d}\partial_t f^*.g + \partial_t f^*.\mathbf{d}g,$$

we obtain

$$\partial_t f^*.\partial_t g.\mathbf{d}t \mid [0, 2\pi] = \mathbf{d}(\partial_t f^*.g) \mid [0, 2\pi] - \mathbf{d}(\partial_t f^*).g \mid [0, 2\pi].$$

Again $\mathbf{d}(\partial_t f^*.g) \mid [0, 2\pi] = \partial_t f^*(2\pi).g(2\pi) - \partial_t f^*(0).g(0) = 0$, so all told:

$$\langle f^*, -\partial_t^2 g\rangle = -f^*(t).\partial_t^2 g(t).\mathbf{d}t \mid [0, 2\pi] = -\mathbf{d}(\partial_t f^*).g \mid [0, 2\pi]$$

$$= -\partial_t^2 f^*(t).g(t)\mathbf{d}t \mid [0, 2\pi].$$

But by the **complex conjugate** property $(z_1z_2)^* = z_1^*.z_2^*$, $(\partial_t^2 f^*(t).g(t))^* = \partial_t^2 f(t).g^*(t)$, so this last expression is:

$$-g(t).\partial_t^2 f^*(t).\mathbf{d}t \mid [0, 2\pi] = \langle g^*, -\partial_t^2 f\rangle^*,$$

or

$$\langle f^*, -\partial_t^2 g\rangle = \langle g^*, -\partial_t^2 f\rangle^*$$

and $-\partial_t^2$ is **Hermitian**.

---

[126] i.e. f and g take **complex values** although t is **real**.

Observe also that $\mathbf{e}^{-j\lambda x}$ is also an **eigenfunction** of $-\partial_x^2$ with the same **eigenvalue**:

$$-\partial_x^2(\mathbf{e}^{-j\lambda x}) = +j\lambda.\partial_x(\mathbf{e}^{-j\lambda x}) = -j^2\lambda^2.(\mathbf{e}^{-j\lambda x}) = \lambda^2(\mathbf{e}^{-j\lambda x}).$$

This phenomenon, where two **eigenfunction** share the same **eigenvalue**, is known as **degeneracy** and the poor **eigenvalue** is called **degenerate**.

Before leaving this introduction, I'd like to emphasize again that you should keep in mind that often the parallels with discrete spaces may be downright misleading in the **functional** context.

8.2 The Fourier Series

In Section 4.9 I introduced the ε() function, $\varepsilon(\xi) = \mathbf{e}\uparrow(2\pi j\xi)$. Being derived from the **exponential function**, ε() shares the **exponential property**:

$$\varepsilon(\xi + \psi) = \mathbf{e}\uparrow(2\pi j(\xi + \psi)) = \mathbf{e}\uparrow(2\pi j\xi + 2\pi j\psi) = \mathbf{e}\uparrow(2\pi j\xi).\mathbf{e}\uparrow(2\pi j\psi) = \varepsilon(\xi).\varepsilon(\psi).$$

So $\varepsilon(2\xi) = \varepsilon(\xi + \xi) = \varepsilon(\xi).\varepsilon(\xi) = (\varepsilon(\xi))^2$. Let's just for a moment call this $\varepsilon^2(\xi)$.

Now $\varepsilon^2(½) = \varepsilon(2.½) = \varepsilon(1) = (1, 0)$ the **unit** of **complex multiplication**. Similarly, $\varepsilon^2(¼) = \varepsilon(½) = \mathbf{e}^{2\pi j ½} = \mathbf{e}^{\pi j}$ $= \mathbf{cos}(\pi) + j.\mathbf{sin}(\pi) = -1$, and $\varepsilon^2(1/8) = \varepsilon(¼) = \mathbf{e}^{½\pi j} = \mathbf{cos}(\pi/2) + j.\mathbf{sin}(\pi/2) = j$.

In general $\varepsilon^2(\xi)$ is the *sum* of the **angles** represented by ε(ξ) and ε(ξ), and so is *twice as far round the* **unit circle** as ε(ξ) itself.

Similarly, ε(3ξ) is the sum of the **angles** represented by ε(ξ) and ε(ξ) and ε(ξ), and is *three times* as far around the **unit circle** as ε(ξ) itself. If we call this $\varepsilon^3(\xi)$, then $\varepsilon^3(⅓) = \varepsilon(3.⅓) = \varepsilon(1) = (1, 0)$ and has already got back to the starting point ε(0) = (1, 0).

> In general, ε(*n*ξ) travels *n* times around the **unit circle** as ξ increases by 1, and so has a **period** of 1/*n*.

But if we regard ε(ξ) as $\varepsilon_x(\xi) + j.\varepsilon_y(\xi) = \mathbf{cos}(2\pi\xi) + j.\mathbf{sin}(2\pi\xi)$, *each of these component functions for* ε(*n*ξ) *must also have* **period** 1/*n*.

Figure 8.2.1 illustrates the first three "*x*" **components**, $\varepsilon_x(\xi) = \mathbf{cos}(2\pi\xi)$ in black, $\varepsilon_x(2\xi) = \mathbf{cos}(4\pi\xi)$ in red, and $\varepsilon_x(3\xi) = \mathbf{cos}(6\pi\xi)$ in blue.[127]

[127] You can download the full colour version of this image from the website www.algebraiccalculus.com.

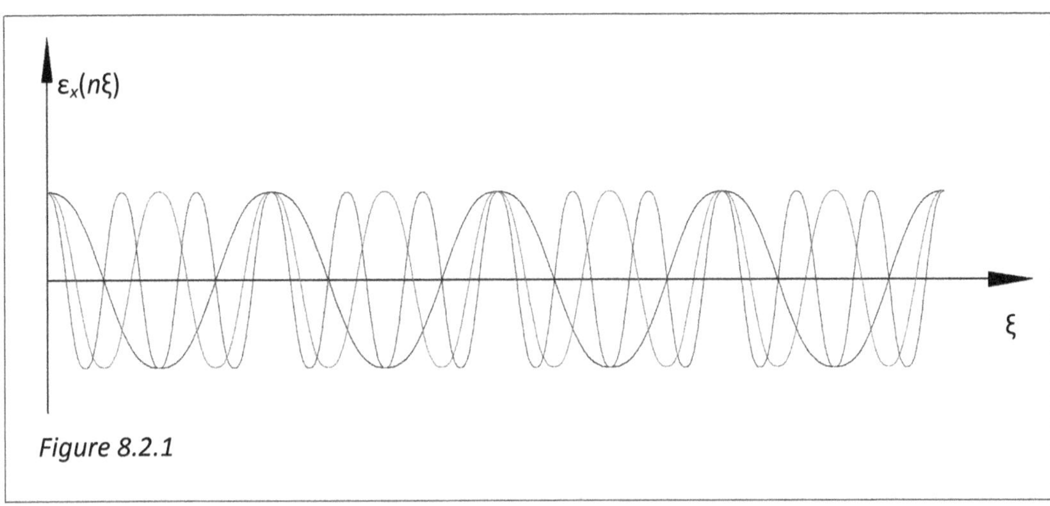

Figure 8.2.1

As well as being **periodic** in $1/n$, each of the $\varepsilon(n\xi)$ functions is also, as we can see in the Figure, **periodic** in 1 because

$$\varepsilon(n(\xi + 1)) = \varepsilon(n\xi + n) = \varepsilon(n\xi).\varepsilon(n) = \varepsilon(n\xi).$$

So again, each of the **components** $\varepsilon_x(n\xi)$ and $\varepsilon_y(n\xi)$ must also be **periodic** in 1. Now $\varepsilon_x(\xi)$ is **even** and $\varepsilon_y(\xi)$ is **odd**:

$$\varepsilon_x(-\xi) = \varepsilon_x(\xi)$$
$$\varepsilon_y(-\xi) = -\varepsilon_y(\xi),$$

familiar properties of **cos** and **sin**. But this property goes over to the products:

$$\varepsilon(2\xi) = \varepsilon(\xi).\varepsilon(\xi) = (\varepsilon_x(\xi) + j.\varepsilon_y(\xi)).(\varepsilon_x(\xi) + j.\varepsilon_y(\xi)) = (\varepsilon_x^2(\xi) - \varepsilon_y^2(\xi)) + 2j(\varepsilon_x(\xi).\varepsilon_y(\xi))$$

so

$$\varepsilon(-2\xi) = \varepsilon(-\xi-\xi) = \varepsilon(-\xi).\varepsilon(-\xi) = (\varepsilon_x^2(-\xi) - \varepsilon_y^2(-\xi)) + 2j(\varepsilon_x(-\xi).\varepsilon_y(-\xi)),$$

but $\varepsilon_x^2(-\xi) = \varepsilon_x^2(\xi)$, $\varepsilon_y^2(-\xi) = (-1.\varepsilon_y(\xi)).(-1.\varepsilon_y(\xi)) = \varepsilon_y^2(\xi)$, and $\varepsilon_x(-\xi).\varepsilon_y(-\xi) = \varepsilon_x(\xi).(-\varepsilon_y(\xi)) = -\,\varepsilon_x(\xi).\varepsilon_y(\xi)$, so

$$\varepsilon_x(-2\xi) = (\varepsilon_x^2(-\xi) - \varepsilon_y^2(-\xi)) = (\varepsilon_x^2(\xi) - \varepsilon_y^2(\xi)) = \varepsilon_x(2\xi)$$

and

$$\varepsilon_y(-2\xi) = 2j(\varepsilon_x(-\xi).\varepsilon_y(-\xi)) = -2\,j(\varepsilon_x(\xi).\varepsilon_y(\xi)) = -\varepsilon_y(2\xi).$$

So $\varepsilon_x(n\xi)$ has a **period** of 1, and is **even**, while $\varepsilon_y(n\xi)$ also has a **period** of 1, but is **odd**.

In perhaps a flash of inspiration, it occurred to Jean-Baptiste-Joseph Fourier (1768 – 1830) that maybe *all* **even periodic** functions could be expressed as sums of the $\varepsilon_x(n\xi)$ functions, and *all* **odd periodic** functions could be expressed as sums of the $\varepsilon_y(n\xi)$ functions, and *all* **periodic** functions altogether could be expressed as sums of the two.

> This hypothesis is therefore that the total set of $\{\varepsilon_x(n\xi), \varepsilon_y(n\xi)\}$ functions for $n = -\infty$ to $n = +\infty$ thereby constitutes a **vector basis** for all **(unit) periodic** functions.
> The expression of a function in terms of this **basis** is called its **Fourier series.**

If we can interpret the $\varepsilon(n\xi)$ functions as a **vector basis**, we can also interpret them as **eigenfunctions**, in this case the **eigenfunctions** for $-\partial^2$ on $[0, 1] \rightarrow \mathbf{C}_1$ (the **unit circle**). First note that they give the correct **periodic boundary conditions** such that:

$$\uparrow(2\pi jn.1) = \uparrow(2\pi jn.0) = 1.$$

Then clearly

$$-\partial_\xi^{\,2}\varepsilon(n\xi) = -\partial_\xi^{\,2}\mathbf{e}^{2\pi jn\xi} = -2\pi.jn.\partial_\xi\mathbf{e}^{2\pi jn\xi} = -4\pi^2j^2n^2.\mathbf{e}^{2\pi jn\xi} = 4\pi^2n^2\mathbf{e}^{2\pi jn\xi}$$

giving the **eigenvalues** $4\pi^2n^2$ which are shared by $\mathbf{e}^{2\pi jn\xi}$ and $\mathbf{e}^{-2\pi jn\xi}$.

The idea can be extended beyond functions with just the **unit period** (= 1).

If the **period** were T rather than 1, we might simply pre-multiply the argument by $1/T$, because

$$\varepsilon(1 + \xi) = \varepsilon(\xi) \longrightarrow \varepsilon((1/T)(T + \psi)) = \varepsilon(1 + \psi/T) = \varepsilon(\psi/T),$$

so $\varepsilon(\psi/T)$ is **periodic** in T, or defining $\varepsilon_{1/T}(\psi) := \varepsilon(\psi/T)$,

$$\varepsilon_{1/T}(T + \psi) = \varepsilon_{1/T}(\psi).$$

For simplicity I'll develop the theory using the standard **period** of 1.

I'll treat the expansion in the **plane** using the natural mapping of the $\varepsilon(n\xi)$ functions into **complex numbers**. This clearly subsumes the **real** cases as the **projections** onto the **real** and **imaginary axes** in the **complex plane**, which are plotted as graphs of ξ over [0, 1] for each of the three loops shown in *Figure 8.2.2* below, which shows the general pattern.

A **periodic** function is defined by a **mapping** λ from [0, 1] on the **real line** onto a **loop** in the **complex plane** as shown by, for example, Λ_1 in the *Figure*. The **loop** doesn't have to include the **origin** in its interior, nor does it even need to be **continuous**. The fragmented "**loop**" Λ_2 in the *Figure* can still, as it turns out, be expressed by a **Fourier series** but the breaks or "jumps" will be assigned the values of their mid-points.

It doesn't even need to be a **loop** defined with unique values at each angle θ around the **origin** like Λ_2: in the problems there is an example where you are asked to find the **Fourier series** of two crossed lines intersecting at the **origin**.

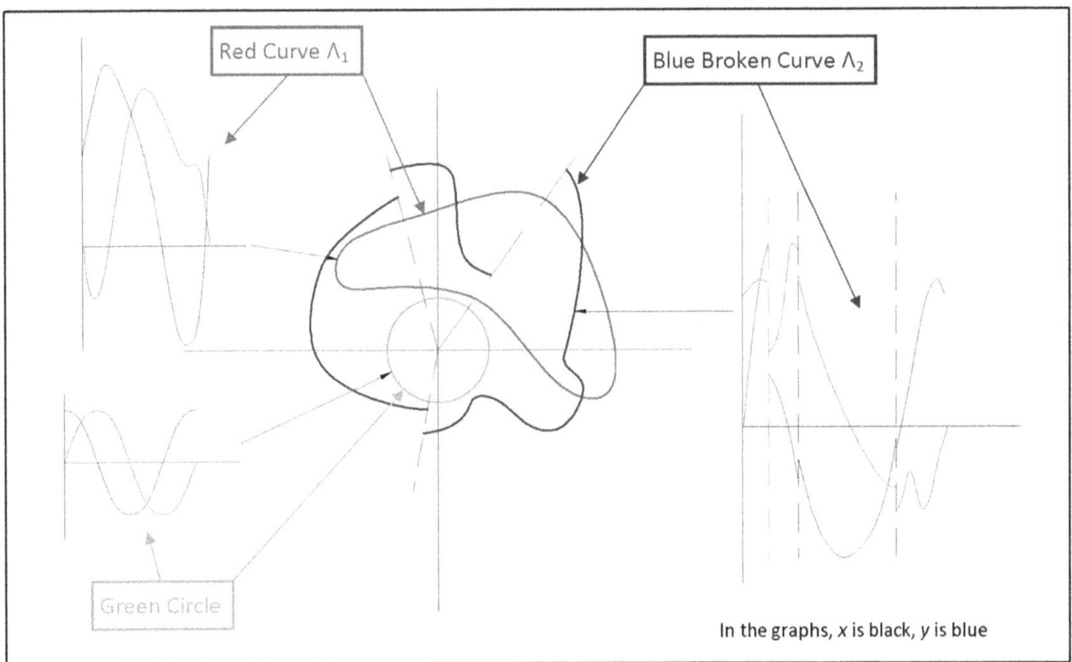

Figure 8.2.2

Because the function is **periodic** with **period** 1, λ describes the *same* **loop** *again* as its argument ξ runs from 1 to 2, and yet again from 2 to 3, and so on, and this behaviour is precisely reflected in the ε(*n*ξ) functions.

The **projections** onto the **axes** are also sketched in the *Figure* as separate graphs of the *x* and *y* components for each **loop** plotted against ξ. [128]

The **real axis** or *x* values correspond to the plot of $\Pi_x\Lambda(\xi) = (\Lambda_x(\xi), 0)$, shown in the graphs in black, and the *y* values on the **imaginary axis** to $\Pi_y\Lambda(\xi) = (0, \Lambda_y(\xi))$, here plotted in blue. For all three **loops**, the graphs begin at where the **loop** meets the *x* axis in the plane. So for the circle **loop** the black graph is the **cos** or $\varepsilon_x()$ part, the blue the **sin** or $\varepsilon_y()$.

$\Pi_x\Lambda(\xi)$ may not necessarily correspond to pure sums of $\varepsilon_x(n\xi)$ functions, nor $\Pi_y\Lambda(\xi)$ to sums of $\varepsilon_y(n\xi)$ functions, because we will assume the **coefficients** - the **scalars** - in the **Fourier series** can themselves be **complex**, so it needn't be that the **real** function $\Pi_x\Lambda(\xi)$ is **even**, or that $\Pi_y\Lambda(\xi)$ is **odd**.

[128] You can download the full colour version of this image from the website www.algebraiccalculus.com.

To analyse the **Fourier series**, we express it as:

$$\lambda(\xi) = \sum_{-\infty}^{\infty} c_n \varepsilon(n\xi)$$

and try to find its **Fourier coefficients** $\{c_n\}$ using techniques analogous to those of Chapter 5.

The difference is that we now have to use an **integral scalar product** as discussed in Section 7.4.

Note that this series contains a term $c_0.\varepsilon(0\xi) = c_0.\varepsilon(0) = c_0.1 = c_0$. So, for example,

$$\lambda_c(\xi) = c_0 + r.\varepsilon(\xi)$$

would describe a circle of radius r about $c_0 = (x_0, y_0)$ in the **complex plane**.

To get the **Fourier coefficients** we need to evaluate the **scalar products**:

$$\mathbf{e}_m \cdot \mathbf{e}_n = \langle \varepsilon(m\xi), \varepsilon(n\xi)\rangle = \varepsilon(m\xi) \wedge *\varepsilon(n\xi) \,|\, [0, 1] = \varepsilon(m\xi).\varepsilon(n\xi).\mathbf{d}\xi \,|\, [0, 1].$$

Here I've used the notation $\mathbf{e}_n \equiv \varepsilon(n\xi)$ to emphasize the parallel between these **basis functions** and their more familiar **vector** analogues.

Now

$$\varepsilon(m\xi).\varepsilon(n\xi) = \varepsilon((m + n)\xi) = \mathbf{e}\uparrow(2\pi j(m + n)\xi),$$

so

$$\mathbf{d}(\varepsilon(m\xi).\varepsilon(n\xi)) = \mathbf{d}(\varepsilon((m + n)\xi)) = 2\pi j(m + n).\mathbf{e}\uparrow(2\pi j(m + n)\xi).\mathbf{d}\xi$$

$$= 2\pi j(m + n).\varepsilon((m + n)\xi).\mathbf{d}\xi,$$

so

$$\varepsilon(m\xi).\varepsilon(n\xi).\mathbf{d}\xi = (1/(2\pi j(m + n))).\mathbf{d}(\varepsilon((m + n)\xi)).$$

This means that

$$\langle \varepsilon(m\xi), \varepsilon(n\xi)\rangle = \varepsilon(m\xi).\varepsilon(n\xi).\mathbf{d}\xi \,|\, [0, 1] = (1/(2\pi j(m + n))).\mathbf{d}(\varepsilon((m + n)\xi)) \,|\, [0, 1]$$

$$= (1/(2\pi j(m + n))).(\varepsilon((m + n)\xi)) \,|\, \mathbf{b}[0, 1]$$

$$= (1/(2\pi j(m + n))).[\varepsilon((m + n).1) - \varepsilon((m + n).0)]$$

$$= (1/(2\pi j(m + n))).[\varepsilon(m + n) - \varepsilon(0)].$$

Now $\varepsilon(0) = 1$ (strictly (1, 0) in the **complex plane**) and for the sum of any whole numbers like $m + n$, $\varepsilon(m + n) = 1$ because ε is **periodic** with **period** 1.

So it looks like

$\langle \varepsilon(m\xi), \varepsilon(n\xi) \rangle = (1/(2\pi j(m + n))).[1 - 1] = 0.$

So the **scalar product** between any two of these **eigenfunctions** is zero. In other words, they are all, apparently, **orthogonal**.

There are two exceptional cases, and they both reduce to the same form. If $n = m = 0$ or if $m = -n$, then the **denominator** above $2\pi j(n + m) = 0$ is also 0, and the above evaluation gives 0/0 which is *undefined*. For example, if $m = -n$, when $m + n = 0$, we get:

$\langle \varepsilon(-n\xi), \varepsilon(n\xi) \rangle = (1/(2\pi j(0))).0 = 0/0.$

This is actually a good example of an apparently sound and rigorous derivation that breaks down unexpectedly. But these two cases can be trivially evaluated another way.

Taking $m = -n$,

$\langle \varepsilon(-n\xi), \varepsilon(n\xi) \rangle = \mathbf{e}\uparrow(2\pi j(-n + n)\xi).\mathbf{d}\xi \mid [0, 1] \; = \mathbf{e}\uparrow(2\pi j(0)\xi).\mathbf{d}\xi \mid [0, 1]$

$= \mathbf{e}\uparrow 0.\mathbf{d}\xi \mid [0, 1]$

$= \mathbf{d}\xi \mid [0, 1] = 1 - 0 = 1.$

So:

$\langle \varepsilon(n\xi), \varepsilon(-n\xi) \rangle = \langle \mathbf{e}\uparrow(2\pi jn\xi), \mathbf{e}\uparrow(-2\pi jn\xi) \rangle = 1$

and

$\langle \varepsilon(0\xi), \varepsilon(0\xi) \rangle = \langle \mathbf{e}\uparrow(2\pi j0\xi), \mathbf{e}\uparrow(-2\pi j0\xi) \rangle = \langle \mathbf{e}\uparrow 0, \mathbf{e}\uparrow 0 \rangle = \langle 1, 1 \rangle = 1.$

So the full set of **eigenfunctions** is actually (quasi-)**orthonormal**, where I say "quasi-" because any **eigenfunction** with $n \neq 0$ has a non-zero **scalar product** only with its "partner" **eigenfunction** sharing the same **eigenvalue** $4\pi^2n^2$ but with $-n$ instead of n.

These (quasi-)**orthonormality** relations are, however, quite adequate to evaluate the **Fourier coefficients** of any **Fourier series** in the **eigenfunctions** $\varepsilon(n\xi)$.

So given

$\lambda(\xi) = \sum_{-\infty}^{\infty} c_n \varepsilon(n\xi)$

we can find c_0 by:

$\mathbf{e}_0 \cdot \lambda = \langle \varepsilon(0.\xi), \sum_{-\infty}^{\infty} c_n \varepsilon(n\xi) \rangle = \sum_{-\infty}^{\infty} c_n \langle \varepsilon(0.\xi), \varepsilon(n\xi) \rangle$

$= c_0.\langle \varepsilon(0.\xi), \varepsilon(0\xi) \rangle + \sum_{-\infty}^{\infty} c_{n\neq 0} \langle \varepsilon(0.\xi), \varepsilon(n\xi) \rangle$

$= c_0.1 + \sum_{-\infty}^{\infty} c_{n\neq 0}.0 = c_0,$

and c_n by:

$$\mathbf{e}_{-n} \cdot \lambda = \langle \varepsilon(-n\xi), \textstyle\sum_{-\infty}^{\infty} c_m \varepsilon(m\xi)\rangle = \sum_{-\infty}^{\infty} c_m \langle \varepsilon(-n.\xi), \varepsilon(m\xi)\rangle$$

$$= c_n.\langle \varepsilon(-n.\xi), \varepsilon(n\xi)\rangle + \textstyle\sum_{-\infty}^{\infty} c_{m\neq -n} \langle \varepsilon(-n.\xi), \varepsilon(m\xi)\rangle$$

$$= c_n.1 + \textstyle\sum_{-\infty}^{\infty} c_{m\neq -n}.0 = c_n.$$

As an example, consider the square pulse train shown in *Figure 8.2.3* below. As drawn, we assume the pulse has value 1 over the **interval** [−¼, ¼] but 0 over [¼, ¾] then back to 1. To make things easier, we'll evaluate $\Pi(t - ¼)$ – shifted *right*, think of $H(t - ¼)$ which has the step at $t = +¼$ – which has the value 1 over [0, ½] and 0 over [½, 1].

Then $\langle \varepsilon(-nt), \Pi(t - ¼)\rangle = \varepsilon(-nt).\mathbf{d}t\,|\,[0, ½] = \mathbf{e}^{-2\pi jnt}\mathbf{d}t\,|\,[0, ½] = \mathbf{d}\{(1/(-2\pi jn)).\mathbf{e}^{-2\pi jnt}\}\,|\,[0, ½]$

$$= \{(1/(-2\pi jn)).\mathbf{e}^{-2\pi jn½}\} - \{(1/(-2\pi jn)).\mathbf{e}^{-2\pi jn0}\} = \{(1/(-2\pi jn)).\mathbf{e}^{-\pi jn}\} - \{(1/(-2\pi jn)).1\}$$

$$= (j/(2\pi n)).(\mathbf{e}\uparrow(-\pi jn) - 1), \text{ using } 1/(-j) = j.$$

This is c_n for $\Pi(t - ¼)$. So to recover our original form we use:

$$\Pi(t) = \textstyle\sum_{-\infty}^{\infty} c_m\, \varepsilon(m(t+¼)) = \sum_{-\infty}^{\infty} (j/(2\pi m)).(\mathbf{e}\uparrow(-\pi jm) - 1).\varepsilon(m(t+¼))$$

$$= \textstyle\sum_{-\infty}^{\infty} (j/(2\pi m)).\varepsilon(¼m).(\mathbf{e}\uparrow(-\pi jm) - 1).\varepsilon(mt).$$

But $(\mathbf{e}\uparrow(\pm 2\pi j) - 1) = (\mathbf{e}\uparrow(\pm 4\pi j) - 1) = (\mathbf{e}\uparrow(\pm 6\pi j) - 1) = \ldots (\mathbf{e}\uparrow 0 - 1) = 1 - 1 = 0$ for all **even** m. So the terms in **even** m drop out. This is as we would expect since the *shifted* function $\Pi(t - ¼)$, or more precisely $\Pi(t - ¼) - ½$, is **odd**. The $\varepsilon(¼m)$ term pushes it over to being **even** as depicted in the *Figure*, with or without the − ½.

Now for *odd m*, $\varepsilon(¼m) = \mathbf{e}\uparrow(2\pi¼jm) = \mathbf{e}\uparrow(½\pi jm)$ which gives, for m = 1, 3, 5, . . .

$$\mathbf{e}\uparrow(½\pi j), \mathbf{e}\uparrow((3/2)\pi j), \mathbf{e}\uparrow((5/2)\pi j) \ldots = j, -j, j, -j, \ldots$$

So our **series** is

$$(j/2\pi).j.(\mathbf{e}\uparrow(-\pi j) - 1).\varepsilon(t) - (j/6\pi).j.(\mathbf{e}\uparrow(-3\pi j) - 1).\varepsilon(3t) + (j/10\pi).j.(\mathbf{e}\uparrow(-5\pi j) - 1).\varepsilon(5t) \ldots$$

but again,

$$\mathbf{e}\uparrow(-\pi j) = \mathbf{e}\uparrow(\pi j).\mathbf{e}\uparrow(-2\pi j) = \mathbf{e}\uparrow(\pi j).1 = -1$$
$$\mathbf{e}\uparrow(-3\pi j) = \mathbf{e}\uparrow(\pi j).\mathbf{e}\uparrow(-4\pi j) = \mathbf{e}\uparrow(\pi j).1 = -1, \text{ etc.}$$

so the series so far becomes:

$(j/2\pi).j.(-2).\varepsilon(t) - j(j/6\pi).(-2).\varepsilon(3t) + j(j/10\pi)(-2).\varepsilon(5t) \ldots$

$= (1/\pi)\varepsilon(t) - (1/3\pi).\varepsilon(3t) + (1/5\pi)\varepsilon(5t) - \ldots$

But we also have a **series** for $m \rightarrow -m$. The reversed signs in the $j/(2\pi m)$ and the $\varepsilon(-¼m)$ cancel, and in the $\mathbf{e}\uparrow(-(-m)\pi j)$ terms we now have e.g. $\mathbf{e}\uparrow 3\pi j = \mathbf{e}\uparrow(2\pi j + \pi j) = \mathbf{e}\uparrow 2\pi j.\ \mathbf{e}\uparrow\pi j = 1.\ \mathbf{e}\uparrow\pi j = 1.(-1) = -1$ again, so the complementary **series** is:

$(1/\pi)\varepsilon(-t) - (1/3\pi).\varepsilon(-3t) + (1/5\pi)\varepsilon(-5t) - \ldots$

Adding the two together, we get terms like:

$\varepsilon(t) + \varepsilon(-t) = \varepsilon_x(t) + j.\varepsilon_y(t) + \varepsilon_x(t) - j.\varepsilon_y(t) = 2.\varepsilon_x(t) = 2.\mathbf{cos}(2\pi t)$

so we finally end up with

$2\{(1/\pi)\mathbf{cos}(2\pi t) - (1/3\pi).\mathbf{cos}(6\pi t) + (1/5\pi)\mathbf{cos}(10\pi t) - \ldots\}$

and as this is an expansion strictly in terms of **cosines**, it is **even** as required in the *Figure 8.2.3* below.

There's still one thing wrong: *this* series would be centred vertically about the $y = 0$ line. We need to add in the term for $\mathbf{e}_0$ or ε_0:

$\langle\varepsilon(-0t), \Pi(t - ¼)\rangle = \varepsilon(-0t).\mathbf{d}t\,|\,[0, ½] = 1.\mathbf{d}t\,|\,[0, ½] = ½.$

This is a constant, so the $t \pm ¼$ shift doesn't affect it, so the final result is:

$½ + (2/\pi)\{\mathbf{cos}(2\pi t) - (1/3).\mathbf{cos}(6\pi t) + (1/5).\mathbf{cos}(10\pi t) - \ldots\}.$[129]

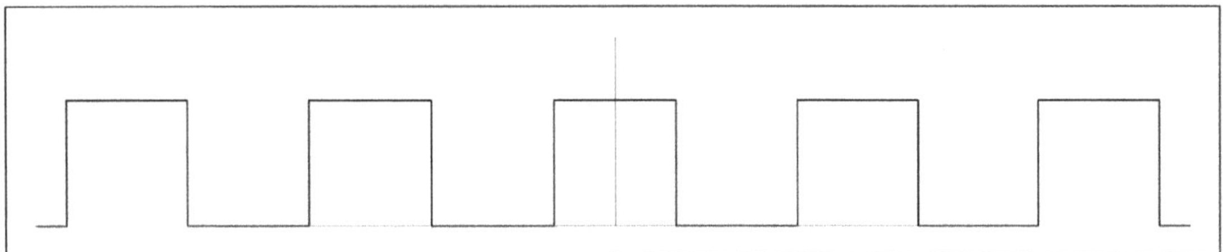

Figure 8.2.3

Note how the **Fourier series**, although developed in terms of **complex** functions, returns a **real series** for a **real function**. It always correctly puts the **series** back in the right part of the **complex plane**.

[129] To be doubly sure I put together a C# program to check this is right! It took about three minutes. Always do this!

Given that we can evaluate the **coefficients** this way, we can approach the more difficult question of establishing **completeness** – that for an arbitrary function λ(ξ) the **Fourier series** so defined does indeed give the correct values of λ(ξ) for all values of ξ between 0 and 1.

At this point we have to introduce **limit** concepts, because the proof proceeds in two stages. First we have to show that the *infinite sum* $\sum_{-\infty}^{\infty} c_n \varepsilon(n\xi)$ does indeed **converge** to a meaningful value for each ξ in the **interval** [0, 1], and then we have to show that that value, the **limit** of the sum, *does* equal the required value of λ(ξ).

What this boils down to is showing that for any given positive number ε,[130] *however small*, there is an **integer** (whole number) *N* such that:

$$| \lambda(\xi) - \sum_{-N}^{N} c_n \varepsilon(n\xi) | < \varepsilon.$$

Here the bars indicate **absolute value**.

In other words, the sum over all terms from −*N* to *N* comes within a distance ε of the correct value and we can ignore terms with $n < -N$ or $n > N$ as negligibly small. Note that *N* may be a function of ξ itself – in **limit**-speak, the **series** may **converge** *faster* at some points than at others.

As always these **limit** arguments rather rapidly descend into deep waters (and this is a very simple case) so I'll only sketch how part of the argument proceeds for λ functions with the property that λ has **continuous** first and second **derivatives** $\partial_\xi\lambda(\xi)$ and $\partial_\xi^2\lambda(\xi)$, and again we start with:[131]

$$\lambda(\xi) = \sum_{-\infty}^{\infty} c_n \varepsilon(n\xi)$$

which, by the way, we can also write as:

$$\lambda(\xi) = c_0 + \Sigma_1^{\infty}(a_n.\mathbf{cos}(2\pi n\xi) + b_n.\mathbf{sin}(2\pi n\xi)).$$

Here, by analogy with our earlier formulae,

$$c_n = \langle \varepsilon(-n\xi), \lambda(\xi) \rangle = \varepsilon(-n\xi).\lambda(\xi).\mathbf{d}\xi \,|\, [0, 1] = \mathbf{e}{\uparrow}(2\pi jn\xi).\lambda(\xi).\mathbf{d}\xi \,|\, [0, 1].$$

But $\mathbf{d}(\mathbf{e}{\uparrow}(2\pi jn\xi).\lambda(\xi)) = 2\pi jn.\mathbf{e}{\uparrow}(2\pi jn\xi).\mathbf{d}\xi.\lambda(\xi) + \mathbf{e}{\uparrow}(2\pi jn\xi).\partial_\xi\lambda(\xi).\mathbf{d}\xi$, so **integrating by parts**,

$$c_n = \mathbf{e}{\uparrow}(2\pi jn\xi).\lambda(\xi).\mathbf{d}\xi \,|\, [0, 1]$$

$$= (1/2\pi jn).\{\mathbf{d}(\mathbf{e}{\uparrow}(2\pi jn\xi).\lambda(\xi)) - \mathbf{e}{\uparrow}(2\pi jn\xi).\partial_\xi\lambda(\xi).\mathbf{d}\xi\} \,|\, [0, 1].$$

The first term, $\mathbf{d}(\mathbf{e}{\uparrow}(2\pi jn\xi).\lambda(\xi)) \,|\, [0, 1] = \mathbf{e}{\uparrow}(2\pi jn1).\lambda(1) - \mathbf{e}{\uparrow}(2\pi jn0).\lambda(0) = 0$.

---

[130] Nothing to do with our ε() functions – this is a distinct usage.

[131] This method is from E. Kreyszig, *Advanced Engineering Mathematics*, 2nd Ed, Wiley 1968. Interpreted in terms of **cos** *and* **sin** we need only sum from 1 to *n* by $\mathbf{cos}(x) = ½(\mathbf{e}^{jx} + \mathbf{e}^{-jx})$ and $\mathbf{sin}(x) = -½j(\mathbf{e}^{jx} - \mathbf{e}^{-jx})$.

So $c_n = -\,(1/2\pi jn).\mathbf{e}\!\uparrow\!(2\pi jn\xi).\partial_\xi\lambda(\xi).\mathbf{d}\xi\}\,|\,[0, 1]$.

a second **integration by parts** gives

$$c_n = -(1/2\pi jn)^2\{\mathbf{d}(\mathbf{e}\!\uparrow\!(2\pi jn\xi).\partial_\xi\lambda(\xi)) - \mathbf{e}\!\uparrow\!(2\pi jn\xi).\partial_\xi^{\,2}\lambda(\xi).\mathbf{d}\xi\,\}\,|\,[0, 1].$$

Again the first term goes to zero, and since $\partial_\xi^{\,2}\lambda(\xi)$ is **continuous** over [0, 1], we must have:

$$|\partial_\xi^{\,2}\lambda(\xi)| < M$$

for some positive number M (**complex numbers** have **real magnitudes** too, remember),

$$|c_n| = |\,(1/2\pi jn)^2.\mathbf{e}\!\uparrow\!(2\pi jn\xi).\partial_\xi^{\,2}\lambda(\xi).\mathbf{d}\xi\,|\,[0, 1]| < (1/2\pi n)^2 M,$$

a conclusion which is easily justified by regarding the **integral** as what is called a Riemann sum, but which is not so obvious in the context of my presentation,

so the value of the **series** must be at most equal to:

$$|c_0| + (1/4\pi^2)M.(1 + (1/2^2) + (1/3^2) + (1/4^2) + \ldots)$$

and this **series** does happen to be **convergent**. Note that we have yet to prove that it **converges** to $\lambda(\xi)$!

For full details it's best to consult Schwartz, *op.cit*. The full development is quite long-winded.

8.3 The Fourier Transform

The **Fourier transform** comes about by using functions of the form $\varepsilon(y\xi)$ rather than $\varepsilon(n\xi)$. In this Section I'll switch to using x rather than ξ as the primary argument.

Given a function $\phi(x)$ of **bounded support**, as defined in Section 7.7, we define its **Fourier transform** as:

$$\mathfrak{F}\phi(y) = \langle\varepsilon(-yx), \phi(x)\rangle = \varepsilon(-yx).\phi(x).\mathbf{d}x\,|\,\mathbf{I}_\phi$$

where $\mathbf{I}_\phi$ is an **interval** large enough to contain the **support** of ϕ.

> Here $\mathfrak{F}\phi$ is a new function, taking y as its argument. Note that it's the *function* that's **transformed**.

There are actually *two* possible **transforms**; this one we could write as $\mathfrak{F}_-$. There is also the form:

$\mathfrak{F}_+\phi(y) = \langle \varepsilon(yx), \phi(x) \rangle = \varepsilon(yx).\phi(x).\mathbf{d}x \mid \mathbf{I}_\phi$.

They differ simply by the sign of the argument of ε().

First note that the definition works straight away for **distributions**:

$\mathfrak{F}\delta_{x0}(y) = \langle \varepsilon(-yx), \delta(x - x_0) \rangle = \varepsilon(-yx_0)$,

and we also have $\mathfrak{F}_+\delta_{x0}(y) = \langle \varepsilon(yx), \delta(x - x_0) \rangle = \varepsilon(yx_0)$.

Now $\varepsilon(yx_0)$ and $\varepsilon(-yx_0)$ themselves, i.e. $\mathbf{e}\uparrow(\pm 2\pi jyx_0) = \mathbf{cos}(2\pi jyx_0) \pm j.\mathbf{sin}(2\pi jyx_0)$, are clearly *not* of **bounded support**, so they *don't have* **Fourier transforms**.

Indeed if we try to take the analogous **integral** (now in *y*) of, say $\mathfrak{F}_+\varepsilon(-yx_0)$, we get:

$\langle \varepsilon(yx), \varepsilon(-yx_0) \rangle = \mathbf{e}\uparrow(2\pi jyx).\mathbf{e}\uparrow(-2\pi jyx_0).\mathbf{d}y \mid [-a, a] = \mathbf{e}\uparrow(2\pi jy(x - x_0)).\mathbf{d}y \mid [-a, a]$

$= (1/(2\pi j(x - x_0))).\mathbf{d}(\mathbf{e}\uparrow(2\pi jy(x - x_0))) \mid [-a, a]$

$= (1/(2\pi j(x - x_0))).\mathbf{e}\uparrow(2\pi jy(x - x_0)) \mid \mathbf{b}[-a, a]$

$= (1/(2\pi j(x - x_0))).[\mathbf{e}\uparrow(2\pi ja(x - x_0)) - \mathbf{e}\uparrow(-2\pi ja(x - x_0))]$.

But $(1/(2j)).[\mathbf{e}^{jx} - \mathbf{e}^{-jx}] = \mathbf{sin}(x)$, so this is:

$\mathbf{sin}(2\pi a(x - x_0))/(\pi(x - x_0))$

and depends on a. So even treated as an **improper integral**, taking the **limit** as $a \longrightarrow \infty$, the value of this integral, being of the form $\alpha.\mathbf{sin}(\beta a)$, will just keep on oscillating.

The property that **Fourier transforms** of functions of **bounded support** are themselves unbounded turns out to be a general rule:

> If a function ϕ is of **bounded support** and so has a well-defined **Fourier transform** $\mathfrak{F}\phi$, then that **transform** $\mathfrak{F}\phi$ *itself* will *not* be of **bounded support** and so will *not* have a defined **Fourier transform**.

I'll give the reference to a proof of this a little later. It's actually a special but very important case of the **uncertainty principle** so beloved of quantum physicists, which is arguably the most misinterpreted result in physics.

As it stands here, this result is a godsend, because from, for example,

$$\mathfrak{F}\delta_{x0}(y) = \langle \varepsilon(-yx), \delta(x - x_0)\rangle = \varepsilon(-yx_0),$$

we can clearly refer to the **inverse Fourier transform** $\mathfrak{F}^{-1}$ of $\varepsilon(-yx_0)$ as $\delta_{x0}(y)$:

$$\mathfrak{F}^{-1}\varepsilon(-yx_0) = \delta_{x0}(y) = \delta(x - x_0)$$

and from $\mathfrak{F}_+\delta_{x0}(y) = \langle \varepsilon(yx), \delta(x - x_0)\rangle = \varepsilon(yx_0)$, we may similarly refer to

$$\mathfrak{F}_+^{-1}\varepsilon(yx_0) = \delta_{x0}(y) = \delta(x - x_0)$$

as the **inverse** of $\mathfrak{F}_+$ on $\delta_{x0}(y) = \delta(x - x_0)$.

Now here's the trick: whenever we have $\mathfrak{F}\phi(y) = \langle \varepsilon(-yx), \phi(x)\rangle$, although we *do* have

$$\mathfrak{F}^{-1}(\mathfrak{F}\phi(y)) = \phi(x)$$

or

$$\mathfrak{F}^{-1}(\mathfrak{F}\phi) = \phi,$$

written strictly as a *functional* **transform**, *inevitably*:

$\mathfrak{F}_\pm(\mathfrak{F}\phi) = \langle \varepsilon(\pm yx), \mathfrak{F}\phi(y)\rangle$ is *undefined*.

So we're free to define $\mathfrak{F}_\pm(\mathfrak{F}\phi)$ however we want.

The device we choose is always to define:

$$\mathfrak{F}_+(\mathfrak{F}_-\phi(x)) := \mathfrak{F}_-^{-1}(\mathfrak{F}_-\phi(x)) = \phi(x),$$
$$\mathfrak{F}_-(\mathfrak{F}_+\phi(x)) := \mathfrak{F}_+^{-1}(\mathfrak{F}_+\phi(x)) = \phi(x).$$

In other words, we *define* $\mathfrak{F}_-^{-1} \equiv \mathfrak{F}_+$ and $\mathfrak{F}_+^{-1} \equiv \mathfrak{F}_-$.

As with our earlier results on the **calculus**, this is normally presented as a *theorem*, the **Fourier Inversion Theorem**, and one that applies only to a very restricted set of functions, largely excluding most of those we need all the time in **Fourier analysis**, and also involving a quite extraordinarily tortuous proof.

The proof also involves the difficult concept of **improper integrals** touched on in the last Chapter, a concept which is unstable because taking the **limits** according to different prescriptions can lead to different results. But it's the **Fourier transform** that is important, not the idea of **improper integrals**, and the **transform** should be made the prior concept and defined independently.

Observe that this definition also gives the result:

$$\langle \varepsilon(yx), \varepsilon(-yx_0)\rangle := \delta(x - x_0)$$

unambiguously, because $\varepsilon(-yx_0) = \mathfrak{F}\delta_{x0}(y)$, so by *definition* now we extend the $\mathfrak{F}_+$ product to $\varepsilon(-yx_0)$ by:

$$\mathfrak{F}_+(\varepsilon(-yx_0)) := \langle \varepsilon(yx), \varepsilon(-yx_0)\rangle := \delta(x - x_0).$$

As an **integral** this product was ambiguous.[132] This result, which we may put in the form

$$\langle \varepsilon(yx), \varepsilon(-zx)\rangle := \delta(y - z),$$

is a commonly assumed alternative statement of the **inversion theorem**. It gives, by associativity of the products $\langle,\langle\rangle\rangle \to \langle\langle\rangle,\rangle$:

$$\langle \varepsilon(yx), \mathfrak{F}_-(\phi(z))\rangle = \langle \varepsilon(yx), \langle \varepsilon(-zx), \phi(z)\rangle\rangle$$

$$= \langle\langle \varepsilon(yx), \varepsilon(-zx)\rangle, \phi(z)\rangle = \langle \delta(y - z), \phi(z)\rangle = \phi(y)$$

and

$$\langle \varepsilon(-yx), \mathfrak{F}_+(\phi(z))\rangle = \langle \varepsilon(-yx), \langle \varepsilon(zx), \phi(z)\rangle\rangle$$

$$= \langle\langle \varepsilon(-yx), \varepsilon(zx)\rangle, \phi(z)\rangle = \langle \delta(y - z), \phi(z)\rangle = \phi(y).$$

Since these correspond to

$$\mathfrak{F}_+(\mathfrak{F}_-(\phi(z)) = \phi(z) \qquad \text{and} \qquad \mathfrak{F}_-(\mathfrak{F}_+(\phi(z)) = \phi(z)$$

they clearly correspond to statements that $\mathfrak{F}_+$ and $\mathfrak{F}_-$ are **inverses**, the essence of the **inversion theorem**.

These ideas are so important that I'll recapitulate them, putting them now into an axiomatic form. This treatment is tantamount to *identifying* $\mathfrak{F}_-$ as $\mathfrak{F}$ and $\mathfrak{F}_+$ as $\mathfrak{F}^{-1}$ *from the start*:

Axiom I: If f(x) is of **bounded support**, with **support** supp(f),

[132] Even treated as an **improper integral**, taking $a \to \infty$ in our earlier result, it's **divergent**.

$$\mathfrak{F}(\mathrm{f}(x)) := \langle \mathrm{f}(x), \varepsilon(-xy)\rangle := \mathrm{f}(x).\varepsilon(-xy).\mathbf{d}x \,|\, \mathrm{supp(f)}.$$

This is to be understood to include defined **distributions** so that e.g.

$$\mathfrak{F}(\delta(x-z)) = \langle \delta(x-z), \varepsilon(-xy)\rangle = \varepsilon(-zy).$$

Axiom II: $\mathfrak{F}(\varepsilon(xz)) = \langle \varepsilon(xz), \varepsilon(-xy)\rangle := \delta(y-z).$ **(Inversion Theorem)**

Axiom III: The product is **associative**.

For f of **bounded support** we define $\mathfrak{F}^{-1}$ by **inversion** so that the **transform** below:

$$\mathfrak{F}(\mathrm{f}(x)) = \mathrm{f}(x).\varepsilon(-xy).\mathbf{d}x \,|\, \mathrm{supp(f)} =: \phi(y)$$

defines

$$\mathfrak{F}^{-1}(\phi(y)) := \mathrm{f}(x).$$

So $\mathfrak{F}^{-1}(\delta(y-z)) := \varepsilon(xz).$

This gives that $\mathfrak{F}^{-1}(\delta(y-z)) = \langle \delta(y-z), \varepsilon(xy)\rangle = \varepsilon(xz)$ and this gives the formula for $\mathfrak{F}^{-1}()$ which can be taken to obey a rule identical to Axiom I, giving:

Axiom IIb: If $\phi(y)$ is of **bounded support**,

$$\mathfrak{F}^{-1}(\phi(y)) := \langle \phi(y), \varepsilon(xy)\rangle := \phi(y).\varepsilon(xy).\mathbf{d}x \,|\, \mathrm{supp}(\phi).$$

This is to be understood to include defined **distributions** so that e.g.

$$\mathfrak{F}^{-1}(\delta(y-z)) = \langle \delta(y-z), \varepsilon(xy)\rangle = \varepsilon(xz).$$

We again now *define* $\mathfrak{F}(\mathrm{f}(x)) = \phi(y)$ if $\mathfrak{F}^{-1}(\phi(y)) = \phi(y).\varepsilon(xy).\mathbf{d}x \,|\, \mathrm{supp}(\phi) =: \mathrm{f}(x)$.

The two will never conflict because if $\mathfrak{F}(\mathrm{f}(x))$ is well-defined – the **integral** or **distribution** can be evaluated – then $\mathfrak{F}^{-1}(\phi(y))$ won't be and *vice versa*.

So from axiom IIb above, we can recover the **inversion formula** by:

$$\mathfrak{F}^{-1}(\delta(y-z)) = \langle \delta(y-z), \varepsilon(xy)\rangle = \varepsilon(xz)$$

$$\rightarrow \quad \mathfrak{F}(\varepsilon(xz)) = \langle \varepsilon(xz), \varepsilon(-xy)\rangle = \delta(y-z).$$

So axioms II and IIb are interchangeable alternatives.

Let's look at another example again, the **unit pulse** Π shown in *Figure 8.3.1* below, which has the value zero everywhere outside the **interval** [−½, ½], wherein it's 1.

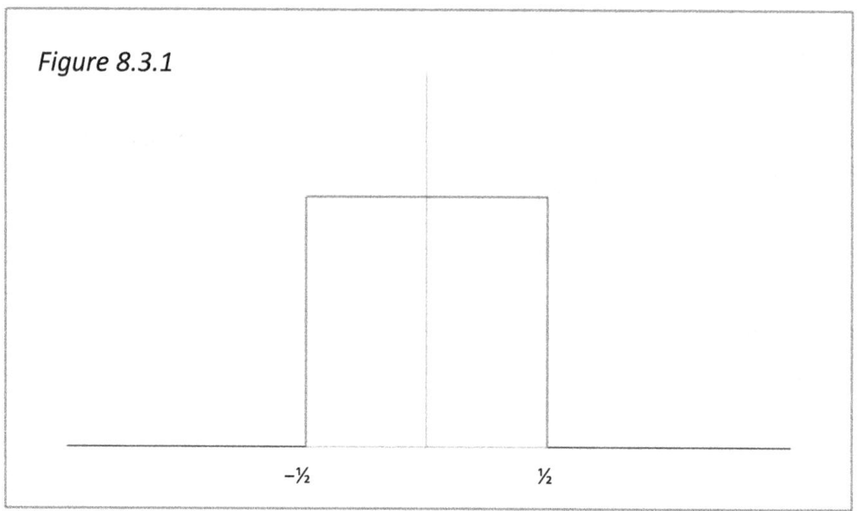

Its **Fourier transform** is ⟨ε(−*yx*), Π(*x*)⟩ = **e**↑(−2π*jyx*).**d***y* | [−½, ½]. But we've seen this integral above with $x - x_0$ instead of *x*, and *a* instead of ½. In that form it evaluated to:

$$\mathbf{sin}(2\pi a(x - x_0))/(\pi(x - x_0)).$$

So its value here will be:

$$\mathbf{sin}(2\pi\tfrac{1}{2}(x))/(\pi(x)) = \mathbf{sin}(\pi x)/(\pi x).$$

This is a function which is much used in signal theory precisely because it's the **Fourier transform** of Π. It was given the name of the **sinc** function by Phillip M. Woodward in 1952 (Wikipedia).

$$\mathbf{sinc}(x) := \mathbf{sin}(\pi x)/(\pi x).$$

The function has a **singularity** at $x = 0$, where its value runs off to infinity. By convention, we *define* **sinc**(0) := 1. The **sinc** function is illustrated below in *Figure 8.3.2*.

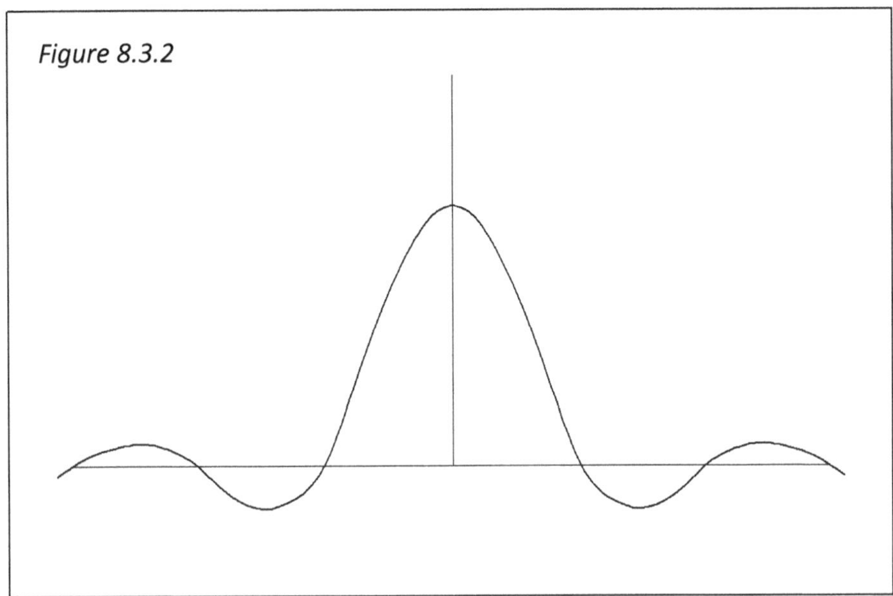
Figure 8.3.2

As it happens, the **sinc** function, with the $a \to \infty$ as suggested above, *can* be construed as a **delta function** because it can be shown that:

$$\mathbf{lim}_{a\to 0}\{(\mathbf{sin}(\pi x/a)/\pi x).\phi(x).\mathbf{d}x \,|\, [-\infty, \infty]\}$$
$$= \mathbf{lim}_{a\to 0}\{(1/a)(\mathbf{sinc}(x/a).\phi(x).\mathbf{d}x \,|\, [-\infty, \infty]\} = \phi(0)$$

which is tantamount to regarding $\mathbf{lim}_{a\to 0}(1/a)(\mathbf{sinc}(x/a) = \delta(x)$. But as we saw above this simply *isn't so*. It's only in the context of the containing **integral** that the **delta function** behaviour appears (and in this $\phi(x)$ needs to be "**compact**", a slightly more general concept comprehending the notion of **bounded support**).

So it seems to me that simply to *define* the product $\langle\varepsilon(yx), \varepsilon(-zx)\rangle$ as $\delta(y - z)$ puts us on firmer ground!

We run into similar difficulties if we try to **invert** the **sinc** function as an **improper integral** to recover the Π **pulse**.[133] I'll develop this case briefly as it's actually a good example of the complexities brought about by the traditional approach of trying to define **Fourier transforms** or their **inverse** by means of **improper integrals**.

Here we want

$$\langle\varepsilon(yx), \mathbf{sinc}(x)\rangle = \mathbf{e}\uparrow(2\pi jyx).\mathbf{sin}(\pi x)/(\pi x).\mathbf{d}x \,|\, [-\infty, \infty]$$

where I put the "**support**" of **sinc** as $[-\infty, \infty]$ because **sinc** is non-zero *everywhere*. We evaluate this as an **improper integral** in the form:

$$\mathbf{lim}_{a\to\infty}\{\mathbf{e}\uparrow(2\pi jyx).\mathbf{sin}(\pi x)/(\pi x).\mathbf{d}x \,|\, [-a, a]\}.$$

133 I take this evaluation from http://math.stackexchange.com/questions/25903/derive-fourier-transform-of-sinc-function?rq=1, somewhat modified.

Now $\mathbf{sin}(\pi x) = (1/2j)(\mathbf{e}\uparrow(\pi jx) - \mathbf{e}\uparrow(-\pi jx))$, so the integral ends up as:

$$(1/2j)\{\mathbf{e}\uparrow(2\pi jyx).\mathbf{e}\uparrow(\pi jx) - \mathbf{e}\uparrow(2\pi jyx).\mathbf{e}\uparrow(-\pi jx)\}/(\pi x).\mathbf{d}x \,\big|\, [-\infty, \infty]$$

$$= (1/2j)\{\mathbf{e}\uparrow(2\pi jyx + \pi jx) - \mathbf{e}\uparrow(2\pi jyx - \pi jx)\}/(\pi x).\mathbf{d}x \,\big|\, [-\infty, \infty]$$

$$= (1/2j\pi)\{\mathbf{e}\uparrow(\pi j(2y + 1)x)/x - \mathbf{e}\uparrow(\pi j(2y - 1)x)/x\}.\mathbf{d}x \,\big|\, [-\infty, \infty].$$

Now these are two integrals of the **Cauchy integral formula** type, where we have:

$$[\phi(z)/(z - z_0)].\mathbf{d}z \,\big|\, \mathbf{C} = 2\pi j.\phi(z_0).$$

To understand the logic here, you'll need to look in the Appendix at the end of the book at Section A.6 on **contour integration**. It's explained there that to evaluate such integrals we actually take a more subtle **contour** by-passing the **singular pole** but using an evaluation method directly analogous to that in solving for the **Cauchy integral formula**, but which gives a value of j as the **imaginary part** for the loop C_ε over the **pole**.

Here $z_0 = 0$, and $\phi(z) = \mathbf{e}\uparrow(\pi j(2y \pm 1)x)$. We take a **contour C** that slightly bypasses the two **singularities** on the **real line** at $2y = \pm 1$ by the device of a little semicircle around each C_ε, and we close the path by a semi-circle at infinity *either* in the lower half-plane, or in the upper half-plane.

We make a change of variables for the two integrals, so that each entails just one pole, the first catching the $y = -½$ pole and the second catching the $y = ½$ one, as follows.

If we put $z_1 := (2y + 1)x$ in the first and $z_2 := (2y - 1)x$ in the second, we get:

$$\{\mathbf{e}\uparrow(\pi j(2y + 1)x)/x\}.\mathbf{d}x \longrightarrow \{(2y + 1).\mathbf{e}\uparrow(\pi jz_1)/z_1\}.\mathbf{d}z_1/(2y + 1) = \{\mathbf{e}\uparrow(\pi jz_1)/z_1\}.\mathbf{d}z_1$$

and

$$\{\mathbf{e}\uparrow(\pi j(2y - 1)x)/x\}.\mathbf{d}x \longrightarrow \{(2y - 1).\mathbf{e}\uparrow(\pi jz_2)/z_2\}.\mathbf{d}z_2/(2y - 1) = \{\mathbf{e}\uparrow(\pi jz_2)/z_2\}.\mathbf{d}z_2$$

and the **singular points** now are where $z_1 = (2y + 1)x = 0$, which is when $y = -½$ and $z_2 = 0$ at $y = ½$.

Suppose $y = -1$, *outside* the $[-½, ½]$ interval.
Then $(2y + 1)x = (-2 + 1)x = -x$, and $(2y - 1)x = (-2 - 1)x = -3x$. *Both are negative factors* on x. So to close the **contour C** in such a way that $\mathbf{e}\uparrow(\pi jz_1) \longrightarrow 0$ and $\mathbf{e}\uparrow(\pi jz_2) \longrightarrow 0$ at large magnitudes of z_1 and z_2 we need to close it for *both* integrals in the *lower* half-plane.

For example, putting $-x$ as $-\zeta$ in the z_1 case to emphasize its extension through the **complex plane**,

$$\mathbf{e}\uparrow(\pi jz_1)/z_1\}.\mathbf{d}z_1 \longrightarrow \{\mathbf{e}\uparrow(-\pi j\zeta)/(-\zeta)\}.\mathbf{d}(-\zeta) = \{\mathbf{e}\uparrow(-\pi j\zeta)/\zeta\}.\mathbf{d}\zeta.$$

Setting $\zeta = a(\mathbf{cos}(\theta) + j.\mathbf{sin}(\theta)) \longrightarrow \{\mathbf{e}\uparrow(-\pi ja.\mathbf{cos}(\theta)).\mathbf{e}\uparrow(+\pi a.\mathbf{sin}(\theta))/\zeta\}.\mathbf{d}\zeta$

This is actually the *opposite* case to that in the Appendix A.6, so we need to close the **contour**, and also to take the little C_ε loop around the **singularity**, in the *lower* half-plane.

Now suppose $y = +1$, also outside the [−½, ½] interval.
Then $(2y + 1)x = (2 + 1)x = +3x$, and $(2y - 1)x = (2 - 1)x = +x$. *Both are positive factors* on x. So to close the **contour C** in such a way that **e**↑$(\pi j z_1) \to 0$ and **e**↑$(\pi j z_2) \to 0$ at large magnitudes of z_1 and z_2 we need to close it for both in the *upper* half-plane.

But if $y = 0$, *within* the [−½, ½] interval, then $(2y + 1)x = x$, and $(2y - 1)x = -x$. So *now* in the *first* integral, in z_1, we need to close the **contour C** in the upper half-plane, and in the *second* integral, in z_2, we need to close it in the *lower*.

The upshot is that in the first two cases, both integrals give the *same* value because they're both respectively $(1/2j\pi)(\pi j)$ = ½, and ½ − ½ = 0, so the two integrals cancel.

But in the $y = 0$ case, the two integrals have opposite values because the C_ε loop around the singularity, although always running left to right, in the first the loop passes *over* the **pole** and so *clockwise* and in the second *under* and *anticlockwise* so we get the result ½ − (−½) = ½ + ½ = 1.

"Recondite" is hardly the word, yet this is not an exceptionally complicated example. My "definitive" approach is far cleaner.

There's a point of principle here too, which I'll box below, that if we *must* use **improper integrals . . .**

> Instead of adapting our concept of an acceptable **Fourier transform** to fit a pre-defined notion of an **improper integral**, our rule now is that we adapt the acceptable definitions of **improper integrals** to maintain the definition of the **Fourier transform**.

This is far from being an idle conjecture. Strictly, within the framework of this book, **integrals** over the **real line *R*** are not defined because ***R*** <u>doesn't have a</u> **<u>boundary</u>**. **b*R*** is not defined. So for a function that is *not* of **bounded support**, like ε(*xz*), we're effectively *defining* the extension of **integration** to ***R*** by:

$$\varepsilon(xz).\varepsilon(-xy).\mathbf{d}x \,\big|\, \boldsymbol{R} := \langle \varepsilon(xz), \varepsilon(-xy) \rangle = \delta(z - y).$$

Sometimes we do have a function that *doesn't* have **bounded support**, and doesn't appear as the **Fourier transform** of any function that does, but where it really does seem reasonable to give it a **Fourier transform**. The classic case is that of the unique function that is *its own* **Fourier transform**, the **Gaussian function e**↑$(-\pi x^2)$.

If there exists a unique function Γ(x) that is its own **transform** or such that:[134]

$$\mathfrak{F}(\Gamma(x)) = \langle\Gamma(x), \varepsilon(-xy)\rangle = \Gamma(y)$$

then

$$\mathfrak{F}^{-1}(\Gamma(y)) = \langle\Gamma(y), \varepsilon(xy)\rangle = \Gamma(x).$$

Straight away this suggests that $\Gamma(x) = \Gamma(-x)$, something we know to be true about the **Gaussian** $\mathbf{e}\uparrow(-\pi x^2)$.

Note that by our fundamental thesis, Γ(x) cannot be of **bounded support** because if Γ(x) is of **bounded support**, $\Gamma(y) = \mathfrak{F}(\Gamma(x))$ cannot be, so we have a contradiction.

To evaluate the **transform** of the **Gaussian**, we proceed by "completing the square" along the lines of Section 1.6. The **transform** is:

$$\langle\, \varepsilon(-yx), \mathbf{e}\uparrow(-\pi x^2)\rangle$$

and as the **Gaussian** is non-zero everywhere we do have to treat the result as an **improper integral** over $[-\infty, \infty]$.

So we try to find:

$$\varepsilon(-yx).\mathbf{e}\uparrow(-\pi x^2).\mathbf{d}x \,|\, [-\infty, \infty] = \mathbf{e}\uparrow(-2\pi jyx).\mathbf{e}\uparrow(-\pi x^2).\mathbf{d}x \,|\, [-\infty, \infty]$$

$$= \mathbf{e}\uparrow(-\pi(x^2 + 2jyx)).\mathbf{d}x \,|\, [-\infty, \infty].$$

Now $(x + jy)^2 = x^2 + 2jyx - y^2$, so

$$\mathbf{e}\uparrow(-\pi(x^2 + 2jyx)) = \mathbf{e}\uparrow(-\pi(x + jy)^2 - \pi y^2) = \mathbf{e}\uparrow(-\pi(x + jy)^2).\mathbf{e}\uparrow(-\pi y^2)$$

and our integral is:

$$\mathbf{e}\uparrow(-\pi y^2).\mathbf{e}\uparrow(-\pi(x + jy)^2).\mathbf{d}x \,|\, [-\infty, \infty].$$

Put $z := x + jy$ and here $\mathbf{d}z = \mathbf{d}x$, so this becomes:

$$\mathbf{e}\uparrow(-\pi y^2).\mathbf{e}\uparrow(-\pi z^2).\mathbf{d}z \,|\, [-\infty, \infty].$$

[134] Note that I'm just using Γ here as an arbitrary function variable; this is *not* the **gamma function**, which is the extension of **factorials** to continuous variables.

Unfortunately, the integral $\mathbf{e}{\uparrow}(-\pi z^2).\mathbf{d}z\,|\,[-\infty, \infty]$ is yet again not an elementary result, but this time it is a well-established one (see https://en.wikipedia.org/wiki/Gaussian_integral). It evaluates to 1.

So we can reasonably say: $\mathfrak{F}(\mathbf{e}{\uparrow}(-\pi x^2)) = \mathbf{e}{\uparrow}(-\pi y^2)$.

Quite often authors talk as if the original function and its **Fourier transform** were on different **domains**; for example, Bracewell actually calls a chapter in his book "The Two Domains". This is natural in many applications of **Fourier transforms** – Bracewell's text emphasizes the conversion of signals from functions of time to functions of frequency, and in quantum mechanics the **Fourier transform** takes one from the spatial to the momentum domain. But there's nothing to that effect in the *maths*: if we had $\phi(x)$, for example, its **transform** is

$$\langle \varepsilon(-yx), \phi(x) \rangle$$

but the x here is just a "dummy variable"[135] which drops out in the integration, so we could easily rename it to ξ and rename y to x to get:

$$\mathfrak{F}\phi(x) = \langle \varepsilon(-x\xi), \phi(\xi) \rangle$$

making the **transform** itself, $\mathfrak{F}\phi$, also a function over x. It's the *function* that is transformed, not the **domain**.

On this basis, we could write the **transform** of the **Gaussian**, which we might call **G**, (not Γ, as that's used for another rather similar function, the **real** analogue of the **factorial function**), as:

$$\mathfrak{F}(\mathbf{G}) = \mathbf{G}.$$

Finally you'll want the reference for that proof that if ϕ is of **bounded support**, its **transform** isn't. I sketch this from Schwartz, p. 190, Theorem 5. The basic idea is that if we have a **distribution** with **bounded support** U_x and a function $\phi(\lambda, x)$, then the only way that $\langle U_x, \phi(\lambda, x) \rangle$ can be zero for all values of λ outside a bounded area is if either $U_x \equiv 0$ identically or $\phi(\lambda, x)$ is identically zero outside the bounded area. This result he proves from basic principles of **distributions** developed at the start of his chapter III. Since $\varepsilon(\lambda x)$ is non-zero everywhere for all values of λ, $\langle U_x, \varepsilon(\lambda x) \rangle$ **cannot be of** bounded **support.**

Interestingly, as a by-product of this proof, $\langle U_x, \varepsilon(\lambda x) \rangle$ must be an **analytic** function in the sense of **complex analysis** – i.e. it's **holomorphic**. (A **holomorphic** function is a complex-valued function of one

[135] Variables that serve only to indicate the form of an **integral** and which do not appear in the final result of the **integration** are often called "**dummy variables**", although a more correct term is a **bound variable**. So in the **integral** $x^2 y_0.\mathbf{d}x\,|\,[0, 1] = \mathbf{d}(\tfrac{1}{3}x^3 y_0)\,|\,[0, 1] = \tfrac{1}{3}y_0$, where y_0 is treated as a **constant**, x is a **dummy variable**, and it could easily be replaced by, say, ξ as in $\xi^2 y_0\mathbf{d}\xi\,|\,[0, 1] = \tfrac{1}{3}y_0$. The index of a **summation** is another example of a **dummy variable**. In $\Sigma_i x_{ik}$, we could replace i by m as $\Sigma_m x_{mk}$ would represent the same result. But we couldn't replace x by y_0 in the **integral** or i by k in the sum without changing the sense.

or more complex variables that is complex differentiable in a neighborhood of every point in its domain.)

Note that in the development of this proof, Schwartz uses as the definition of the **Fourier transform** for a **distribution** U, given that the **transform** of a function ϕ is well-defined:

$$\langle \mathfrak{F}U, \phi\rangle := \langle U, \mathfrak{F}\phi\rangle.$$

I may refer to this as the **Schwartz rule**. In my presentation, it's simply a statement of the **associativity** principle (**Axiom III** above):

$$\langle\langle U_y, \varepsilon(-yx)\rangle, \phi(x)\rangle = \langle U_y, \langle\varepsilon(-yx), \phi(x)\rangle\rangle$$

and I would tend to regard the question of whether U_y is a function or a **distribution** as largely immaterial.

Also note that the **associativity principle** is essentially the extension of **Fubini's theorem** of double integration to the product involved in the **Fourier transform**. The entire problem with the conventional presentation is that, seen as **improper integrals**, **associativity** breaks down in the all-important case:

$$\langle\varepsilon(zx), \langle\varepsilon(-yx), \phi(x)\rangle\rangle \equiv \langle\langle\varepsilon(zx), \varepsilon(-yx)\rangle, \phi(x)\rangle = \phi(z).$$

The trouble here is that *on its own*, $\langle\varepsilon(zx), \varepsilon(-yx)\rangle \neq \delta(z - y)$ as **associativity** would require. By defining **associativity** as an *axiomatic requirement* of the ⟨⟩ *product*, and not trying to define it in terms of "**improper integrals**", my approach circumvents this difficulty.

The **Schwartz rule** is also a statement that the **Fourier transform** is, or is required to be, **self-adjoint** under the operation ⟨⟩ between elements of the **basic space** and the **dual space**.

Now let $U = \mathfrak{F}^{-1}(\psi)$, so $\mathfrak{F}(U) = \psi$, and

$$\langle\mathfrak{F}\mathfrak{F}^{-1}(\psi), \phi\rangle = \langle\psi, \phi\rangle = \langle\mathfrak{F}^{-1}(\psi), \mathfrak{F}(\phi)\rangle.$$

But $\phi = \mathfrak{F}^{-1}(\mathfrak{F}(\phi))$, so $\langle\psi, \mathfrak{F}^{-1}(\mathfrak{F}(\phi))\rangle = \langle\mathfrak{F}^{-1}(\psi), \mathfrak{F}(\phi)\rangle$.

So $\mathfrak{F}^{-1}$ is also **self-adjoint**.

Note that for $z = 0$ the formulae above for the **transform** of $\varepsilon(zx)$ reduce to the simple form:

$$\mathfrak{F}(\varepsilon(zx)) = \mathfrak{F}(\varepsilon(0)) = \mathfrak{F}(1) = \delta(y), \text{ and so } \mathfrak{F}^{-1}(\delta(y)) = 1.$$

It's also not too difficult to develop all the basic properties of the **Fourier transform** such as:

- $\mathfrak{F}(f(x)) = \phi(y) \rightarrow \mathfrak{F}(f(ax)) = |a|^{-1}\phi(y/a)$
- $\mathfrak{F}(f(x)) = \phi(y) \rightarrow \mathfrak{F}(f(x - a)) = \varepsilon(-ay).\phi(y)$
- $\mathfrak{F}(f(x)) = \phi(y)$ and $\mathfrak{F}(g(x)) = \psi(y) \rightarrow \mathfrak{F}(f(u).g(x - u).\mathbf{dx}\,|\,[-\infty, \infty]) = \phi(y).\psi(y)$ *but this does depend on some prior definition of the* **improper integral**, at least as stated. But *this is no problem if* f *and g are of* **bounded support**.
- $\mathfrak{F}(f(x)) = \phi(y) \rightarrow \mathfrak{F}(\partial_x f(x)) = 2\pi jy.\phi(y)$

For example,

$$\langle\phi(ax), \varepsilon(-yx)\rangle = \phi(ax).\varepsilon(-yx).\mathbf{dx}\,|\,\mathbf{I}_\phi = (1/|a|).\phi(ax).\mathbf{e}\uparrow(-2\pi j(ax)(y/a)).\mathbf{d}(ax)\,|\,\mathbf{I}_{\phi*}$$

$$= (1/|a|).\mathfrak{F}\phi(y/a).$$

Here I denote the **support** in the modified variable *ax* by $\mathbf{I}_{\phi*}$.

Again,

$$\langle\phi(x - a), \varepsilon(-yx)\rangle = \phi(x - a).\varepsilon(-yx).\mathbf{dx}\,|\,\mathbf{I}_\phi$$

$$= \phi(x - a).\varepsilon(-y(x - a)).\varepsilon(-ya).\mathbf{d}(x-a)\,|\,\mathbf{I}_{\phi*} = \phi(u).\varepsilon(-y(u)).\varepsilon(-ya).\mathbf{d}(u)\,|\,\mathbf{I}_\phi$$

$$= \varepsilon(-ya).\mathfrak{F}(\phi(x)).$$

The most important is:

$$\langle\partial_x\phi(x), \varepsilon(-yx)\rangle = \partial_x\phi(x).\varepsilon(-yx).\mathbf{dx}\,|\,\mathbf{I}_\phi$$

$$= \mathbf{d}(\phi(x).\varepsilon(-yx))\,|\,\mathbf{I}_\phi - \phi(x).\partial_x\varepsilon(-yx).\mathbf{dx}\,|\,\mathbf{I}_\phi \text{ by parts.}$$

If $\phi(x)$ is of **bounded support**, it must be zero at the end points x_1 and x_2 of $\mathbf{I}_\varphi$, so

$$\mathbf{d}(\phi(x).\varepsilon(-yx))\,|\,\mathbf{I}_\phi = \mathbf{d}(\phi(x).\varepsilon(-yx))\,|\,[x_1, x_2] = \phi(x_2).\varepsilon(-yx_2) - \phi(x_1).\varepsilon(-yx_1)$$

$$= 0 - 0 = 0$$

So

$$\langle\partial_x\phi(x), \varepsilon(-yx)\rangle = -\,\phi(x).\partial_x\varepsilon(-yx).\mathbf{dx}\,|\,\mathbf{I}_\phi = -\,\phi(x).\partial_x\mathbf{e}\uparrow(-2\pi jyx).\mathbf{dx}\,|\,\mathbf{I}_\phi$$

$$= -\,(-2\pi jy).\phi(x).\mathbf{e}\uparrow(-2\pi jyx).\mathbf{dx}\,|\,\mathbf{I}_\phi = +2\pi jy.\mathfrak{F}(\phi(x)).$$

I'll deal with the third formula above in the Section on convolution.

Of course we have directly that:

$$\mathfrak{F}(½(\delta(x+z)+\delta(x-z))=½(\langle\delta(x+z),\varepsilon(-xy)\rangle+\langle\delta(x-z),\varepsilon(-xy)\rangle)$$

$$=½(\varepsilon(yz)+\varepsilon(-yz))=½(2\mathbf{cos}(2\pi.yz))=\mathbf{cos}(2\pi.yz)$$

and

$$\mathfrak{F}(-½\,j(\delta(x+z)-\delta(x-z))=-½\,j(\langle\delta(x+z),\varepsilon(-xy)\rangle-\langle\delta(x-z),\varepsilon(-xy)\rangle)$$

$$=-½\,j(\varepsilon(yz)-\varepsilon(-yz))=-½\,j(\mathbf{cos}(2\pi.yz)+j.\mathbf{sin}(2\pi yz)-\mathbf{cos}(2\pi.yz)+j.\mathbf{sin}(2\pi yz))$$

$$=-j^2.\mathbf{sin}(2\pi yz)=\mathbf{sin}(2\pi.yz).$$

I'll finish this long Section with a note on the **orthogonality** relations of **Fourier series** and **transforms.**

If we write: $\mathbf{e}_n := \varepsilon(n\xi)$,

then our **orthogonality integrals** over [0, 1] for the **Fourier series,** $\varepsilon(n\xi)\wedge *\varepsilon(m\xi)\,|\,[0,1]$, can now be expressed very simply as obeying:

$$\mathbf{e}_n\cdot\mathbf{e}_{-m}=\delta_{nm}$$

using the **Kronecker delta** symbol introduced in Chapter 5. But beware that this has $-m$ on the left but m on the right. This formulation covers the zero case too because for $m = 0$, we have $-0 = 0$.

It is important to grasp that this neat form only comes out because our **integral scalar product** is here over a *finite* **interval** [0, 1].

The analogous "result" for the **scalar product** over $[-\infty, \infty]$ appearing in the **Fourier transform** is the very tentative assertion:

$$\mathfrak{F}_x(\mathbf{e}\!\uparrow\!(2\pi jxy_0))=\langle\mathbf{e}\!\uparrow\!(2\pi jxy_0),\mathbf{e}\!\uparrow\!(-2\pi jxy)\rangle_x=\mathbf{e}\!\uparrow\!(2\pi jx(y_0-y)).\mathbf{dx}\,|\,[-\infty,\infty]$$

$$=\delta(y-y_0).$$

Defining $\mathfrak{E}_y := \varepsilon(xy) = \mathbf{e}\!\uparrow\!(2\pi jxy)$, and writing *this* **scalar product** as:

$$\mathfrak{E}_y\cdot\mathfrak{E}_z:=\langle\mathbf{e}\!\uparrow\!(2\pi jyx),\mathbf{e}\!\uparrow\!(2\pi jzx)\rangle_x$$

we seem to have:

$$\mathfrak{E}_y\cdot\mathfrak{E}_{-z}=\delta(y-z).$$

This of course gives the **inversion theorem** just the same way as we derived the **Fourier coefficients** above.

But be very careful of this: the **integral** in the **scalar product** is well-defined, but for this case, $\mathbf{e}\uparrow(2\pi jx(y_0 - y)).\mathbf{d}x \mid [-\infty, \infty]$, it is very definitely *indeterminate*.

8.4 The Laplace Transform

The **Laplace Transform** can be seen as a way to get a **transform** out even for **functions** that don't have **bounded support** and equally fail the strict **convergence** requirements needed at $x = \pm\infty$ for evaluating the **Fourier transform** as an **improper integral**. It forms the foundation of almost the entire subject of Control Engineering.

To bring the divergence towards $+\infty$ under control, we multiply the given **function** $f(t)$ by a "tempering factor" $\mathbf{e}^{-\alpha t}$ where α is a positive **real number** $\alpha > 0$, and this function, being equal to $1/\mathbf{e}^{\alpha t}$, diminishes rapidly towards zero as $t \rightarrow \infty$.

To bring the divergence towards $-\infty$ under control, we simply impose the **Heaviside function** $H(t - d)$ which just zeroes the **function** for values less than d.

We now take the slightly different **Fourier transform** $\mathfrak{F}^{+}$ of the "well-behaved" composite **function**:

$$H(t - d).f(t).\mathbf{e}^{-\alpha t},$$

which is:

$$H(t - d).f(t).\mathbf{e}^{-\alpha t}.\mathbf{e}^{-j\omega t}.\mathbf{d}t \mid [-\infty, \infty]. \qquad (8.4.1)$$

This is the **Laplace Transform** of $f(t)$.

$\mathfrak{F}^{+}$ differs from the **transform** I used before because it involves the product:

$$\langle \mathbf{e}\uparrow(-jyx), \varphi(x) \rangle_x$$

instead of

$$\langle \varepsilon(-yx), \varphi(x) \rangle_x = \langle \mathbf{e}\uparrow(-2\pi jyx), \varphi(x) \rangle_x.$$

This gives a slightly less clean form for the **inversion theorem** as:

$$\varphi(z) = (1/2\pi).\langle \mathbf{e}\uparrow(jyz), \langle \mathbf{e}\uparrow(-jyx), \varphi(x) \rangle_x \rangle_y.$$

Note the leading $(1/2\pi)$.[136] This can be written as

[136] The various conventions in use are well described in Bracewell, p.6.

$$2\pi.\varphi(z) = \langle \mathbf{e}\uparrow(jyz), \langle \mathbf{e}\uparrow(-jyx), \varphi(x)\rangle_x \rangle_y.$$

and it's in this form that I use it below.

It's customary to choose t as the **argument** of f in discussion of the **Laplace transform** because these **transforms** are very often taken in the time domain.

Since $H(t - d) = 0$ for $t < d$, the **integral** in (8.4.1) splits, using the additive property of the **simplex argument** that $[-\infty, \infty] = [-\infty, d] + [d, \infty]$, into:

$$0.f(t).\mathbf{e}^{-\alpha t}.\mathbf{e}^{-j\omega t}.\mathbf{d}t \,\big|\, [-\infty, d] + 1.f(t).\mathbf{e}^{-\alpha t}.\mathbf{e}^{-j\omega t}.\mathbf{d}t \,\big|\, [d, \infty]$$

$$= f(t).\mathbf{e}^{-\alpha t}.\mathbf{e}^{-j\omega t}.\mathbf{d}t \,\big|\, [d, \infty].$$

Now $\mathbf{e}^{-\alpha t}\mathbf{e}^{-j\omega t} = \mathbf{e}^{-(\alpha + j\omega)t}$, and so if we write $p := \alpha + j\omega$, the **integral** now reduces to:

$$f(t).\mathbf{e}^{-pt}.\mathbf{d}t \,\big|\, [d, \infty]$$

and finally, although it is not essential to the logic, d is always taken as zero, so the final, standard, form for the **Laplace transform** is:

$$\lambda(p) = \mathfrak{L}(f(t)) := f(t).\mathbf{e}^{-pt}.\mathbf{d}t \,\big|\, [0, \infty].$$

Because α has to be larger than a certain value a in order to "dominate" the original $f(t)$, *all these formulae only hold for* $\alpha > a$.

Because of its derivation from the **Fourier transform**, the **Laplace transform** already has its **inversion** well-defined. Regarding the **Fourier Inversion Theorem** above as a double integral over the plane of t and ω, we will have (taking $d = 0$, as always hereafter and calling the plane over which the theorem operates the ωt-**plane**):

$$2\pi.H(t_0).f(t_0).\mathbf{e}\uparrow(-\alpha t_0) = [H(t).f(t).\mathbf{e}^{-\alpha t}].\mathbf{e}\uparrow(-j\omega(t - t_0)).\mathbf{d}\omega\mathbf{d}t \,\big|\, \omega t\text{-}\mathbf{plane}.$$

Since this LHS is only non-zero for $t_0 > 0$, as it should be, the $H(t_0)$ on the LHS is often omitted, when we get the **Laplace inversion formula** as:

$$2\pi.f(t_0).\mathbf{e}\uparrow(-\alpha t_0) = [\mathfrak{L}(f(t))].\mathbf{e}\uparrow(j\omega t_0).\mathbf{d}\omega \,\big|\, [-\infty, \infty].$$

Multiplying through by $\mathbf{e}\uparrow(\alpha t_0)/2\pi$, this gives:

$$f(t_0) = (1/2\pi).[\mathfrak{L}(f(t))].\mathbf{e}\uparrow((\alpha + j\omega).t_0).\mathbf{d}\omega \,\big|\, [-\infty, \infty]$$

$$= (1/2\pi j).[\lambda(p)].\mathbf{e}\uparrow(pt_0).\mathbf{d}p \,\big|\, [\alpha - j\infty, \alpha + j\infty]$$

where I have used the change of variables $p = \alpha + j\omega$, which gives $\mathbf{d}\omega = \mathbf{d}p/j$, since α is a constant for this **integration**, and where $\omega = \pm\infty$, we get $p = \alpha + j(\pm\infty)$.

Do keep in mind that although p is **complex**, ω, t, and α are all **real**.

The **line segment** $[\alpha - j\infty, \alpha + j\infty]$ in the **complex** p plane with $\alpha > a$ along which this **integral** is defined is sometimes called the **Bromwich line**, and the **inversion integral** the **Bromwich integral**.

But with the **Laplace transform** even more than with the **Fourier transform**, the actual **inversion integral** is hardly ever attempted. Instead, in accord with my development in the last Section, tables of "known" **transforms** are accumulated, and the "inversion" of any form is almost always performed by looking up the form we've got in these known results.

As with the **Fourier transform**, the most useful properties of the **Laplace transform** in applications appear when it is applied to the **ordinary derivatives** of the original function:

$$\mathfrak{L}(\partial_t f(t)) = \partial_t f(t).\mathbf{e}^{-pt}.\mathbf{d}t \,|\, [0, \infty]$$

$$= \mathbf{d}(f(t).\mathbf{e}^{-pt}) \,|\, [0, \infty] - f(t).\mathbf{d}(\mathbf{e}^{-pt}) \,|\, [0, \infty]$$

$$= f(\infty).\mathbf{e}^{-p.\infty} - f(0).\mathbf{e}^{-p.0} + p.f(t).\mathbf{e}^{-pt}.\mathbf{d}t \,|\, [0, \infty]$$

$$= 0 - f(0) + p.\mathfrak{L}(f(t))$$

$$= p.\mathfrak{L}(f(t)) - f(0).$$

In a similar way, for the second **derivative**, we can show that:

$$\mathfrak{L}(\partial_t^2 f(t)) = p^2.\mathfrak{L}(f(t)) - p.f(0) - \partial_t f(0).$$

Rather interesting is the fact that in this case we also have a **transform formula** for an **integral**. Here the **transform** variable t appears as one of the arguments defining the **interval** over which the **integration** is defined:

$$\mathfrak{L}(f(\tau).\mathbf{d}\tau \,|\, [0, t]) = (f(\tau).\mathbf{d}\tau \,|\, [0, t]).\mathbf{e}^{-pt}.\mathbf{d}t \,|\, [0, \infty].$$

To work out this result, we define $g(t) := f(\tau).\mathbf{d}\tau \,|\, [0, t]$. Now by the definition of **integration**:

$$\mathbf{d}g(\tau) \,|\, [0, t] = g(\tau) \,|\, \mathbf{b}[0, t] = g(t) - g(0)$$

so $\quad g(t) = \mathbf{d}g(\tau) \,|\, [0, t] + g(0) = \partial_t g(\tau).\mathbf{d}\tau \,|\, [0, t] + g(0) = f(\tau).\mathbf{d}\tau \,|\, [0, t].$

But $g(0) = f(\tau).\mathbf{d}\tau \,|\, [0, 0]$, and this must equal 0.

So: $\partial_t g(\tau).\mathbf{d}\tau \,|\, [0, t] = f(\tau).\mathbf{d}\tau \,|\, [0, t]$.

Now take any **antiderivative** of $f(\tau).\mathbf{d}\tau$ which we'll call $\phi(\tau)$. Then the **integral** on the right must be:

$$f(\tau).\mathbf{d}\tau \,|\, [0, t] = \mathbf{d}(\phi(\tau)) \,|\, [0, t] = \phi(t) - \phi(0).$$

So $\phi(t) - \phi(0) = g(t) - g(0) = g(t)$. **Differentiating**, this gives:

$$\partial_t \phi(t) - \partial_t \phi(0) = \partial_t \phi(t) = \partial_t g(t).$$

But $\partial_t \phi(t) = f(t)$ by definition, so $f(t) = \partial_t g(t)$. That means that:

$$\mathfrak{L}(f(t)) = \mathfrak{L}(\partial_t g(t)) = p.\mathfrak{L}(g(t)) - g(0),$$

but $g(0) = 0$, so:

$$\mathfrak{L}(g(t)) = \mathfrak{L}(f(\tau).\mathbf{d}\tau \,|\, [0, t]) = \mathfrak{L}(f(t))/p.$$

This breaks down for the **Fourier transform** because, although we do have still:

$$\mathfrak{F}(f(x)) = \mathfrak{F}(\partial_x g(x)) = jy.\mathfrak{F}(g(x))$$

for $g(x) := f(\xi).\mathbf{d}\xi \,|\, [0, x]$, we can no longer assume that this means that:

$$\mathfrak{F}(g(x)) = \mathfrak{F}(f(x))/jy.$$

The reason is that although $p = \alpha + j\omega$ in the **Laplace transform** is always non-zero because $\alpha > 0$, *we cannot make the same assumption about y in the* **Fourier**. When y is zero, the expression on the RHS becomes undefined.

The analogous formula for the **Fourier transform** turns out to be:

$$\mathfrak{F}(g(x)) = \mathfrak{F}(f(\xi).\mathbf{d}\xi \,|\, [0, x]) = \mathfrak{F}(f(x))/jy + C.\delta(y).$$

I won't attempt to prove this.

8.5 Convolution

The **Fourier-Laplace transforms** introduced above are evidently **linear**:

e.g. $\mathfrak{F}(a.g(x)) = a.\mathfrak{F}(g(x))$
$\mathfrak{F}(f(x) + g(x)) = \mathfrak{F}(f(x)) + \mathfrak{F}(g(x))$.

So they behave like the **d operator** in Chapter 2 does in its axiom I. What about axiom II? How does a **Fourier-Laplace transform** operate over a *product* of two **functions**? This is where **convolution**[137] comes in.

For

$f(x) = \langle \phi(y), \varepsilon(xy) \rangle = \langle \phi(y), \mathbf{e}{\uparrow}(2\pi jxy) \rangle,$

and

$g(x) = \langle \gamma(y), \varepsilon(jxy) \rangle = \langle \gamma(y), \mathbf{e}{\uparrow}(2\pi jxy) \rangle,$

which we can also write as $\langle \gamma(v), \mathbf{e}{\uparrow}(2\pi jxv) \rangle$, where I've renamed the "dummy variable" y to v in this second **integral** so that I can combine factors from the two:

$$\mathfrak{F}(f(x).g(x)) = \mathfrak{F}(\{\phi(y).\mathbf{e}{\uparrow}(2\pi jxy).\mathbf{d}y \,|\, [-\infty, \infty]\}\{\gamma(v).\mathbf{e}{\uparrow}(2\pi jxv).\mathbf{d}v \,|\, [-\infty, \infty]\})$$

$$= \mathfrak{F}(\phi(y).\gamma(v).\mathbf{e}{\uparrow}(2\pi jx(y + v)).\mathbf{d}y.\mathbf{d}v \,|\, [-\infty, \infty]\times_c[-\infty, \infty]).$$

Now put $u = y + v$, so that $v = u - y$, and $\mathbf{d}y\mathbf{d}v = \mathbf{d}y\mathbf{d}u$, and we get:

$$\mathfrak{F}(f(x).g(x)) = \mathfrak{F}(\phi(y).\gamma(u - y).\mathbf{e}{\uparrow}(2\pi jxu).\mathbf{d}y.\mathbf{d}u \,|\, [-\infty, \infty]\times_c[-\infty, \infty])$$

$$= \mathfrak{F}(\{\phi(y).\gamma(u - y).\mathbf{d}y \,|\, [-\infty, \infty]\}.\mathbf{e}{\uparrow}(2\pi jxu)\ .\mathbf{d}u \,|\, [-\infty, \infty])$$

$$= \mathfrak{F}(\langle\{\phi(y).\gamma(u - y).\mathbf{d}y \,|\, [-\infty, \infty]\}, \mathbf{e}{\uparrow}(2\pi jxu)\rangle)$$

$$= \{\phi(y).\gamma(u - y).\mathbf{d}y \,|\, [-\infty, \infty]\},$$

since we have the **inversion theorem** $\mathfrak{F}(\langle\psi(z), \mathbf{e}{\uparrow}(2\pi jxz)\rangle) = \langle\langle\psi(z), \varepsilon(xz)\rangle, \varepsilon(-xu)\rangle = \psi(u)$, where I've again renamed the inner u to z.

The peculiar **integral** $\phi(y).\gamma(u - y).\mathbf{d}y \,|\, [-\infty, \infty]$ is called the **convolution** of ϕ and γ and is often written as:

$\phi * \gamma$ or $\phi(y) * \gamma(y).$

So the result above is that,

if $\phi(y) = \mathfrak{F}(f(x))$
and $\gamma(y) = \mathfrak{F}(g(x))$

then $\phi(y) * \gamma(y) = \mathfrak{F}(f(x).g(x)).$

[137] Which, curiously, Lighthill doesn't mention.

Not only is the **Fourier transform** of a product a **convolution**, but the **Fourier transform** of a **convolution** is a product:

$$\mathfrak{F}\{f(\xi).g(x-\xi).\mathbf{d}\xi \mid [-\infty, \infty]\}$$

$$= (f(\xi).g(x-\xi).\mathbf{d}\xi \mid [-\infty, \infty]).\varepsilon(-xy).\mathbf{d}x \mid [-\infty, \infty],$$

or putting $u = x - \xi$ with $x = u + \xi$ and $\mathbf{d}\xi\mathbf{d}x = \mathbf{d}\xi\mathbf{d}u$, we get:

$$(f(\xi).g(u)).\mathbf{e}\uparrow(-2\pi juy).\varepsilon(-\xi y).\mathbf{d}\xi.\mathbf{d}u \mid [-\infty, \infty] \times_c [-\infty, \infty]$$

$$= \{f(\xi).\varepsilon(-\xi y).\mathbf{d}\xi \mid [-\infty, \infty]\}.\{g(u)).\varepsilon(-uy).\mathbf{d}u \mid [-\infty, \infty]\}$$

because these are completely separable **integrals** with u not appearing in any of the factors of the $\mathbf{d}\xi$ **integration** nor ξ in any of those of the $\mathbf{d}u$, and so this comes to:

$$= \gamma(y).\phi(y) = \mathfrak{F}(f(x)).\mathfrak{F}(g(y)).$$

In case you're unhappy about these changes of variables, we are simply defining:

$$u = x - \xi$$
$$\xi = \xi$$

so that $\mathbf{d}u = \mathbf{d}x - \mathbf{d}\xi$, so $\mathbf{d}x = \mathbf{d}u + \mathbf{d}\xi$ and $\mathbf{d}\xi \wedge \mathbf{d}x = \mathbf{d}\xi \wedge (\mathbf{d}u + \mathbf{d}\xi) = \mathbf{d}\xi \wedge \mathbf{d}u + \mathbf{d}\xi \wedge \mathbf{d}\xi$, which equals $\mathbf{d}\xi \wedge \mathbf{d}u$.

The relationship between **convolution** and multiplication extends to $\mathfrak{F}^{-1}()$ too. Given a product $\phi(y).\eta(y)$

$$\mathfrak{F}^{-1}(\phi(y).\eta(y)) = \langle\phi(y).\eta(y), \varepsilon(xy)\rangle = \phi(y).\eta(y).\mathbf{e}\uparrow(2\pi jxy).\mathbf{d}y \mid [-\infty, \infty].$$

But as we've just seen, $\phi(y).\eta(y) = \mathfrak{F}\{f(\xi).g(x-\xi).\mathbf{d}\xi \mid [-\infty, \infty]\} = \mathfrak{F}(f(x) * g(x))$. So applying $\mathfrak{F}^{-1}$ to this equation:

$$\mathfrak{F}^{-1}(\phi(y).\eta(y)) = \mathfrak{F}^{-1}\mathfrak{F}(f(x)*g(x)) = f(x) * g(x).$$

Similarly, $\phi(y) * \eta(y) = \mathfrak{F}(f(x).g(x))$ gives:

$$\mathfrak{F}^{-1}(\phi(y) * \eta(y)) = \mathfrak{F}^{-1}\mathfrak{F}(f(x).g(x)) = f(x).g(x).$$

Similar results can be obtained for the **Laplace transform** but things don't go quite as smoothly. So, for example, it we have a **convolution** $f(t) * g(t)$,

$$\mathfrak{L}(f(t) * g(t)) = f(t-\tau).g(\tau).\mathbf{d}\tau \mid [-\infty, \infty].\mathbf{e}^{-pt}.\mathbf{d}t \mid [0, \infty].$$

Now we need either to make the assumption that $f(t) = 0$ for $t < 0$, and $g(t) = 0$ for $t < 0$, or we *redefine* the **convolution integral** to be:

$$f(t-\tau).g(\tau).\mathbf{d}\tau \mid [-\infty, \infty] \rightarrow f(t-\tau).g(\tau).\mathbf{d}\tau \mid [0, \infty].$$

Either way, the *double* **integral** in $\mathbf{d}\tau\mathbf{d}t$ is now *only over the quadrant of the* τt*-plane where both* τ *and* t *are positive.*

Put $u = t - \tau$, so $t = u + \tau$ and $\mathbf{d}\tau\mathbf{d}t = \mathbf{d}\tau(\mathbf{d}u + \mathbf{d}\tau) = \mathbf{d}\tau\mathbf{d}u$, but now $t = 0 \rightarrow u = -\tau$, and we get:

$$\mathfrak{L}(f(t) * g(t)) = f(u).g(\tau).\mathbf{e}^{-p(u+\tau)}.\mathbf{d}\tau \mid [0, \infty]\mathbf{d}u \mid [-\tau, \infty]$$

$$= g(\tau)\mathbf{e}^{-p\tau}.\mathbf{d}\tau(f(u).\mathbf{e}^{-pu}.\mathbf{d}u \mid [-\tau, \infty]) \mid [0, \infty].$$

Now $\tau > 0$, so $-\tau < 0$, but if $f(u) = 0$ for $u < 0$, then the inner **integral**

$$f(u).\mathbf{e}^{-pu}.\mathbf{d}u \mid [-\tau, \infty] = f(u).\mathbf{e}^{-pu}.\mathbf{d}u \mid [0, \infty]$$

and so the **integrals** are again separable, and we obtain:

$$\mathfrak{L}(f(t) * g(t)) = (f(u).\mathbf{e}^{-pu}.\mathbf{d}u \mid [0, \infty]).(g(\tau)\mathbf{e}^{-p\tau}.\mathbf{d}\tau \mid [0, \infty]) = \mathfrak{L}(f(t)).\mathfrak{L}(g(t)).$$

For $\mathfrak{L}(f(t)g(t))$, we need to substitute the Bromwich formulae for $f(t)$ and $g(t)$:

$$f(t)g(t) = \{(1/2\pi j).[\phi(p)].\mathbf{e}\uparrow(pt).\mathbf{d}p \mid [\alpha - j\infty, \alpha + j\infty]\}$$
$$\times \{(1/2\pi j).[\eta(q)].\mathbf{e}\uparrow(qt).\mathbf{d}q \mid [\alpha - j\infty, \alpha + j\infty]\}$$

choosing α large enough to "dominate" both $f(t)$ and $g(t)$, and writing $\mathfrak{L}(f(t))$ as $\phi(p)$ and $\mathfrak{L}(g(t))$ as $\eta(t)$. Again I've renamed the "dummy variable" in the second **integral**, from p to q this time, so we can mix the two **integrals**. This gives:

$$(-1/4\pi^2)\phi(p)\eta(q).\mathbf{e}^{(p+q)t}.\mathbf{d}p\mathbf{d}q \mid [\alpha - j\infty, \alpha + j\infty] \times_c [\alpha - j\infty, \alpha + j\infty].$$

Put $r = p + q$, so $q = r - p$ and $\mathbf{d}p\mathbf{d}q = \mathbf{d}p\mathbf{d}r$, and we get:

$$(-1/4\pi^2)\phi(p)\eta(r-p).\mathbf{e}^{rt}.\mathbf{d}p\mathbf{d}r \mid [\alpha - j\infty, \alpha + j\infty] \times_c [\alpha - j\infty, \alpha + j\infty].$$

$$= (-1/4\pi^2)\{\phi(p)\eta(r-p).\mathbf{d}p \mid [\alpha - j\infty, \alpha + j\infty]\}.\mathbf{e}^{rt}.\mathbf{d}r\,[\alpha - j\infty, \alpha + j\infty]$$

$$= (1/2\pi j)\mathfrak{L}^{-1}(\phi(p)\eta(r-p).\mathbf{d}p \mid [\alpha - j\infty, \alpha + j\infty]).$$

Note that the $\mathfrak{L}^{-1}$ here will "supply" the second $1/2\pi j$. So operating on both sides with $\mathfrak{L}$, since $\mathfrak{L}\mathfrak{L}^{-1} = 1$, we get:

$$\mathfrak{L}(f(t)g(t)) = (1/2\pi j)(\phi(p)\eta(r-p).\mathbf{d}p \mid [\alpha - j\infty, \alpha + j\infty]).$$

Schwartz gives a rather more formal picture of **convolution** which I will briefly outline over the remainder of this Section.

We can easily extend the idea of a **tensor product** of **dual space** elements from the version in Section 5.5. Given two **distributions** S_x and T_y, we define their **tensor product** to act on two test **functions** ϕ and ψ as:

$$\langle S_x \otimes T_y, (\phi(x), \psi(y))\rangle := \langle S_x, \phi(x)\rangle.\langle T_y, \psi(y)\rangle.$$

In words, each factor of the product operates on its "own" **function** and we multiply the results together. If the separate **arguments** $(\phi(x), \psi(y))$ are replaced by a single **function** of two variables, $\phi(x, y)$, then the definition extends easily by the ploy of regarding the "partial evaluation" on each **argument** of ϕ to give a new **function** – not a **distribution** – in its own right. So:

$$\theta(x) = \langle T_y, (\phi(x, y)\rangle \qquad \text{or} \qquad \Sigma(y) = \langle S_x, (\phi(x, y)\rangle.$$

It is taken as a basic premise of the theory that there is a generalized **Fubini theorem** so that each of these **functions**, when acted on by the "other" **distribution**, returns the *same* value, thereby defining the value of:

$$\langle S_x \otimes T_y, (\phi(x, y)\rangle := \langle S_x, \theta(x)\rangle = \langle T_y, \Sigma(y)\rangle.$$

If the **distributions** are themselves **functions**, $s(x)$ and $t(y)$, this appears in terms of the **scalar product** as:

$$s(x).\mathbf{d}x\ [t(y).\phi(x, y).\mathbf{d}y \mid [-\infty, \infty]] \mid [-\infty, \infty] = t(y).\mathbf{d}y\ [s(x).\phi(x, y).\mathbf{d}x \mid [-\infty, \infty]] \mid [-\infty, \infty].$$

I have here greyed off the outer **integration** to make the iterated **integral** structure clear. In Chapter 6 I have expressed reservations about whether the **Fubini theorem** should strictly involve a sign change or not. One way out of this dilemma might be to regard all **integrations** involving the **pseudoscalar** of **differential forms** – which we might do here, regarding ourselves as being in a 2-space with **pseudoscalar** $\mathbf{d}x\mathbf{d}y$ – as *unsigned* or *positive* because these necessarily represent **volume integrals**.

Be that as it may, the **convolution** of two **distributions** is defined as:

$$\langle S_x * T_y, \phi\rangle := \langle S_x \otimes T_y, (\phi(x + y)\rangle$$

where ϕ now has a single **argument**. In case this causes distress to the reader, you can always think of it as $\langle S_x \otimes T_y, (\phi(\Sigma(x, y))\rangle$ with $\Sigma(x, y) = x + y$, where ϕ becomes an outer **function** on an inner sum. So if the **distributions** are now **functions**, we get the form seen above by putting $\xi = x + y$:

$$s(x).dx\ [t(y).\phi(x + y).\mathbf{d}y \,|\, [-\infty, \infty]] \,|\, [-\infty, \infty]$$

$$= s(x).dx\ [t(\xi - x).\phi(\xi).\mathbf{d}\xi \,|\, [-\infty, \infty]] \,|\, [-\infty, \infty]$$

or

$$t(y).dy\ [s(x).\phi(x+y).\mathbf{d}x \,|\, [-\infty, \infty]] \,|\, [-\infty, \infty]$$

$$= t(y).dy\ [s(\xi - y).\phi(\xi).\mathbf{d}\xi \,|\, [-\infty, \infty]] \,|\, [-\infty, \infty].$$

These two expansions give the usual form for the **convolution** of two **functions**:

$$(s * t)(\xi) := s(x).t(\xi - x).\mathbf{d}x \,|\, [-\infty, \infty] = t(y).s(\xi - y).\mathbf{d}y \,|\, [-\infty, \infty]$$

or swapping round the names of the variables to a more acceptable form:

$$(s * t)(x) := s(\xi).t(x - \xi).\mathbf{d}\xi \,|\, [-\infty, \infty] = t(\xi).s(x - \xi).\mathbf{d}\xi \,|\, [-\infty, \infty].$$

The substitution $y = x - \xi$, and so $\xi = x - y$, followed by renaming y back to ξ, switches between these two forms.[138]

It is possible to develop an *algebra* of **convolutions**, and an *invertible* one too. In other words, we can *solve for things* in this algebra. In this algebra, the **Dirac delta distribution** constitutes the **unit element**, as we shall now see.

I follow Schwartz's own notation here. He rather shies away from the conventional $\delta(\xi - a)$ and prefers to write this as $\delta_{(a)\xi}$. So, for example, his $\langle \delta_{(b)\xi}, \phi(\xi + \eta)\rangle$ would normally be written as $\langle \delta(\xi - b), \phi(\xi + \eta)\rangle$and evaluates to $\phi(b + \eta)$.

The first of the key theorems of the **convolution** algebra is that $\delta(\xi)$ is the **unit element**:

(1.) $\delta * T = T$

which comes from:

$$\langle \delta * T, \phi\rangle = \langle \delta_\xi \otimes T_\eta, \phi(\xi + \eta)\rangle = \langle T_\eta, \langle \delta_\xi, \phi(\xi + \eta)\rangle\rangle = \langle T_\eta, \phi(\eta)\rangle.$$

Then (2.) $\delta_{(a)} * T = T_{x-a}$

[138] $\mathbf{d}\xi$ would actually change sign, but $[-\infty, \infty]$ would become $[\infty, -\infty]$, so swapping this back recovers the original signature.

from: $\langle T_{x-a}, \phi\rangle := \langle T(x-a), \phi(x)\rangle = \langle T(x), \phi(x+a)\rangle$
by a change of variables, and so

$$\langle \delta_{(a)} * T, \phi\rangle = \langle \delta_{(a)\xi} \otimes T_\eta, \phi(\xi+\eta)\rangle =$$
$$\langle T_\eta, \langle \delta_{(a)\xi}, \phi(\xi+\eta)\rangle\rangle = \langle T_\eta, \phi(a+\eta)\rangle = \langle T_{x-a}, \phi\rangle.$$

Hence (3.) $\delta_{(a)} * \delta_{(b)} = \delta_{(a+b)}$

because
$\langle \delta_{(a)\xi} \otimes \delta_{(b)\eta}, \phi(\xi+\eta)\rangle = \langle \delta_{(a)\eta}, \langle \delta_{(b)\xi}, \phi(\xi+\eta)\rangle\rangle = \langle \delta_{(a)\eta}, \phi(b+\eta)\rangle = \phi(a+b).$

(4.) $\partial_x\delta_x * T = \partial_x T$
from

$$\langle \partial_x\delta_x * T, \phi\rangle = \langle \partial_x\delta_x \otimes T_\eta, \phi(x+\eta)\rangle =$$
$$\langle T_\eta, \langle \partial_x\delta_x, \phi(x+\eta)\rangle\rangle = -\langle T_\eta, \partial_x\phi(\eta)\rangle = \langle \partial_\eta T_\eta, \phi(\eta)\rangle.$$

This can be extended to:

(5.) $\partial_x^{\,n}\delta_x * T = \partial_x^{\,n}T.$

So **derivatives** (**OD** type) of a **distribution** equate to **convolution** of the **distribution** with the corresponding **derivative** of δ. So for any **derivative operator** D,

$$D\delta * T = DT.$$

Thus for example the **Laplacian operator** $\Delta := \partial_x^{\,2} + \partial_y^{\,2} + \partial_z^{\,2}$ obeys:

$$\Delta T = (\partial_x^{\,2}\delta_x + \partial_y^{\,2}\delta_y + \partial_z^{\,2}\delta_z) * T.$$

It can be shown that D, acting on a **convolution** itself, goes down onto *one or other* member, *not* distributing in the manner of **axiom II** for **differentiation** across an ordinary product:

$$D(S * T) = DS * T = S * DT.$$

This can be applied to *potential theory* in physics. The **Newtonian potential** of a charge distribution $\rho(\boldsymbol{x})$, where $\boldsymbol{x}$ is a point in 3-space, is given by:

$$U(\boldsymbol{x}) = (\rho(\boldsymbol{t})/|\mathbf{x} - \boldsymbol{t}|).\mathbf{d}x\mathbf{d}y\mathbf{d}z\,\big|\,\mathbf{V}$$

where **V** is all of 3-space. This is simply $U = \rho * 1/|\mathbf{x}| = \rho * 1/r$, where $r := \sqrt{(x^2 + y^2 + z^2)}$. Here $\rho(\boldsymbol{x})$ is a **function** over 3-space, but the definition of **potential** could now be extended to any **distribution** as:

$$U_T(\boldsymbol{x}) = T * 1/r.$$

Thus, applying the **Laplacian**, $\Delta U = \Delta(T * 1/r)$, which from the *one or other* member rule above, gives:

$$\Delta U = T * \Delta(1/r).$$

Now $\Delta(1/r)$ can be evaluated from the argument in Section 7.7 as being simply $-4\pi\delta(r)$, and this gives the result:

$$\Delta U = T * (-4\pi\delta(r))$$

but by $\delta * T = T * \delta = T$, this reduces to **Poisson's equation**:

$$\Delta U = -4\pi T.$$

Schwartz, whom we are now following closely, goes on to introduce a theorem to the following effect.

If D_x is a **linear differential operator** of the form:

$$D_x = \partial_x{}^m + a_1.\partial_x{}^{m-1} + \ldots + a_{m-1}.\partial_x + a_m,$$

then $D_x\delta(x)$ has an inverse of the form $H(x).Z(x)$ where $H(x)$ is the **Heaviside step function** and $Z(x)$ is the solution of the **homogeneous ordinary differential equation** $D_xZ(x) = 0$ subject to the **initial conditions**:

$$Z(0) = \partial_xZ(0) = \partial_x{}^2Z(0) = \ldots = \partial_x{}^{m-2}Z(0) = 0, \text{ but } \partial_x{}^{m-1}(0) = 1.$$

Since $\delta(x)$ is the **unit element** in this **convolution algebra**, this inverse must satisfy:

$$D_x\delta(x) * (H(x).Z(x)) = \delta(x).$$

The proof is really very easy. Since $D_x\delta(x) * H(x).Z(x) = D_x(H(x).Z(x))$, and

$$\mathbf{d}(H(x).Z(x)) = \mathbf{d}H(x).Z(x) + H(x).\mathbf{d}Z(x) = (\delta(x).Z(x) + H(x)\partial_xZ(x)).\mathbf{d}x,$$

and this equals $\partial_x(H(x).Z(x)).\mathbf{d}x$, we can evaluate all the derivatives quickly, using

$\delta(x).Z(x) = \delta(x).Z(0) = 0$, and so on for the higher **derivatives**,

leaving only:

$$\partial_x{}^k(H(x).Z(x)) = H(x).\partial_x{}^kZ(x) \qquad \text{for } k < m,$$

and

$$\partial_x{}^m(H(x).Z(x)) = H(x).\partial_x{}^mZ(x) + \delta(x),$$

so that:

$$D_x\delta(x) * (H(x).Z(x)) = H(x).D_xZ(x) + \delta(x) = \delta(x).$$

$H(x).Z(x)$ is called the **elementary solution** of the **convolution equation**:

$$D_x\delta(x) * X(x) = \delta(x).$$

In this sense, $X(x) = H(x).Z(x) = [D_x\delta(x)]^{-1}$.

As an example, the **elementary solution** of $D_x = (\partial_x - \lambda)$ is:

$$(\partial_x\delta(x) - \lambda.\delta(x))^{-1} = (\delta' - \lambda\delta)^{-1} = H(x).\mathbf{e}^{\lambda x}$$

and we shall return to this result in the next Section.

8.6 Applications

The practicality of all these techniques arises principally because they transform **derivative** operations into *algebraic* ones. So we could consider our RLC circuit of Section 3.5, obeying equations:

$$E + v_R + v_C + v_L = 0$$

$$v_R = -i.R$$

$$C.\mathbf{d}v_C = -i.\mathbf{d}t$$

$$L.\mathbf{d}i = -v_L.\mathbf{d}t.$$

The equations in **differentials** need to be switched into equations in v_C and v_L to use them in the first equation, which is easy enough. For the v_C equation, either we put

$$v_C(t) = \mathbf{d}v_C \,|\, [0, t] = -(i/C).\mathbf{d}\tau \,|\, [0, t]$$

or we could use

$$C.\partial_t v_C = -i, \qquad \text{with } v_C(0) = 0,$$

(both of which would give the **Laplace transform** $\mathfrak{L}(v_C) = -\mathfrak{L}(i)/pC$), and for the v_L,

$$L.\partial_t i = -v_L.$$

We then proceed to plug everything into the first equation to get:

$$E - i.R - (i/C).\mathbf{d}\tau \,|\, [0, t] - L.\partial_t i = 0.$$

The most useful transform here is the **Laplace** where, noting that

$$E.(\mathbf{e}^{-pt}.\mathbf{d}t \,|\, [0, \infty]) = -(E/p)\mathbf{d}(\mathbf{e}^{-pt}) \,|\, [0, \infty] = -(E/p)(0 - 1) = E/p,$$

and writing $\mathfrak{L}(i(t))$ as $I(p)$, we get, since $\mathfrak{L}(\partial_t i) = p.\mathfrak{L}(i(t)) - i(0) = p.I(p) - i(0)$:

$$E/p - I(p).R - I(p)/(pC) - p.L.I(p) + i(0) = 0$$

so that ($\times\, p \longrightarrow$):

$$E + p.i(0) = I(p).(Lp^2 + pR + 1/C),$$

so that:

$$I(p) = = \frac{p.i(0) + E}{Lp^2 + p.R + \frac{1}{C}}.$$

We now need to expand the awkward **denominator** by the technique of **partial fractions**.

First we break it into factors of the form:

$$(p - a_1)(p - a_2) = Lp^2 + p.R + 1/C.$$

It should be apparent from Section 1.6 that a_1 and a_2 are the two **roots** of the **quadratic equation** $Lp^2 + p.R + 1/C = 0$, for if $p = a_1$ or $p = a_2$, we will have:

$$(p - a_1)(p - a_2) = Lp^2 + p.R + 1/C = 0.$$

So, using the quadratic formula from Section 1.6, $a_{1,2} = (-R \pm \surd(R^2 - 4L/C))/2L$, where a_1 takes the + of the $\pm$ and a_2 takes the $-$. We can put:

$$\alpha := -R/(2.L)$$

and

$$\beta := +\surd(R^2 - 4L/C))/2L = +\surd(R^2 - 4L/C))/\surd(4L^2) = +\surd(R^2/4L^2 - 1/LC),$$

to give $a_1 = \alpha + \beta$, $a_2 = \alpha - \beta$, where it should be remembered that β will be **imaginary** if it happens that $R^2/4L^2 < 1/LC$. So we now have:

$$I(p) = \frac{p.i(0) + E}{Lp^2 + p.R + \frac{1}{C}} = \frac{p.i(0) + E}{(p - \alpha - \beta)(p - \alpha + \beta)}$$

Partial fractions are the reverse of evaluating a sum of two fractions by putting them over a common **denominator**. We need to find A and B such that:

$$\frac{A}{(p - \alpha - \beta)} + \frac{B}{(p - \alpha + \beta)} = \frac{p.i(0) + E}{(p - \alpha - \beta)(p - \alpha + \beta)}.$$

Cross-multiplication (multiplying through by $(p - \alpha - \beta)(p - \alpha + \beta)$) shows that this will be satisfied if:

$$A(p - \alpha + \beta) + B(p - \alpha - \beta) = p.i(0) + E.$$

Since p and 1 are **linearly independent**, this will be satisfied for arbitrary p if:

$$A + B = i(0) \qquad \text{and} \qquad A(\beta - \alpha) - B(\alpha + \beta) = E.$$

So $B = i(0) - A$, and $A(\beta - \alpha) - (i(0) - A)(\alpha + \beta) = E$. This latter expression gives:

$$A(\beta - \alpha + \alpha + \beta) - i(0)(\alpha + \beta) = E, \text{ or } A = (E + i(0).(\alpha + \beta))/2\beta.$$

So we have

$$A = (E + i(0)(\alpha + \beta))/2\beta$$

and

$$B = (2\beta.i(0) - E - i(0)(\alpha + \beta))/2\beta = (-E - i(0)(\alpha - \beta))/2\beta.$$

Check: $A + B = (E + i(0)\alpha + i(0)\beta - E - i(0)\alpha + i(0)\beta)/2\beta = 2\beta i(0)/2\beta = i(0)$.

So all told we have:

$$I(p) = \frac{(E + i(0)(\alpha + \beta))/2\beta}{(p - \alpha - \beta)} + \frac{(-E - i(0)(\alpha - \beta))/2\beta}{(p - \alpha + \beta)}.$$

Now only the **denominators** involve p and they take a recognisable form, because:

$$\mathfrak{L}(\mathbf{e}^{at}) = \mathbf{e}^{(a-p)t}.\mathbf{d}t \,|\, [0, \infty] = 1/(p - a).$$

So we have that $\mathfrak{L}^{-1}(1/(p - a)) = \mathbf{e}^{at}$. So we can evaluate these two terms immediately, remembering that $p - \alpha - \beta = p - (\alpha + \beta)$ and $p - \alpha + \beta = p - (\alpha - \beta)$, as:

$$i(t) = (1/2\beta)\{ (E + i(0).(\alpha + \beta)).\mathbf{e}^{(\alpha + \beta)t} + (-E - i(0).(\alpha - \beta)).\mathbf{e}^{(\alpha - \beta)t} \}.$$

Here $\alpha = -R/(2L)$ and $\beta = +\sqrt{(R^2/4L^2 - 1/LC)}$.

So if the **discriminant** $R^2/4L^2 - 1/LC$ is *negative*, β is **imaginary**, $= jb$, and because, if b is **real**,

$$\mathbf{e}^{\pm jbt} = \mathbf{cos}(bt) \pm j.\mathbf{sin}(bt),$$

we will get a **sinusoidal** or *oscillatory* solution. *But not otherwise.*

α is always **real** and we can factor this out to give:

$$i(t) = (\mathbf{e}^{\alpha t}/2\beta)\{\, (E + i(0).(\alpha + \beta)).\mathbf{e}^{\beta t} + (-E - i(0).(\alpha - \beta)).\mathbf{e}^{-\beta t}\,\}$$

and if β is **imaginary**, this equals:

$$(\mathbf{e}^{\alpha t}/2\beta)\{\, (E + i(0).(\alpha + jb)).(\mathbf{cos}(bt) + j.\mathbf{sin}(bt)) - (E + i(0).(\alpha - jb)).(\mathbf{cos}(bt) - j.\mathbf{sin}(bt))\}.$$

But this equals

$$(\mathbf{e}^{\alpha t}/2\beta)\{(E + i(0).\alpha + i(0).\, jb - E - i(0).\alpha + i(0).jb).\mathbf{cos}(bt)$$

$$+ (E + i(0).\alpha + i(0).\, jb + E + i(0).\alpha - i(0).jb)j.\mathbf{sin}(bt)\}$$

$$= (\mathbf{e}^{\alpha t}/2\beta)\{(2i(0).\, jb).\mathbf{cos}(bt) + (2E + 2\, i(0).\alpha)j.\mathbf{sin}(bt)\}$$

$$= 2j.(\mathbf{e}^{\alpha t}/2\beta)\{(i(0).b).\mathbf{cos}(bt) + (E + i(0).\alpha).\mathbf{sin}(bt)\}$$

This is **real** because β is **imaginary**, and setting $\beta = jb$,

$$i(t) = (\mathbf{e}^{\alpha t}/b)\{(i(0).b).\mathbf{cos}(bt) + (E + i(0).\alpha).\mathbf{sin}(bt)\}.$$

Laurent Schwartz uses this very same example to demonstrate his **convolution algebra**, which is closely analogous to the **Laplace transform**. Indeed, the **Laplace transform** can be seen as a formalization of this very technique.

Schwartz puts the equation of the *RLC*-circuit, treating *E* now as a time-dependent variable $e(t)$:

$$e(t) - i.R - (i/C).\mathbf{d}\tau\,|\,[0, t] - L.\partial_t i = 0.$$

To get rid of the awkward **integral, differentiate** through by ∂_t to get:

$$\partial_t e(t) - \partial_t i(t).R - (i(t)/C) - L.\partial_t^2 i(t) = 0.$$

As a **convolution equation**, this becomes:

$$\delta'(t) * e(t) = [L.\delta''(t) + R.\delta'(t) + (1/C)\delta(t)] * i(t),$$

where $\delta'(t) = \partial_t\delta(t)$ and $\delta''(t) = \partial_t^2\delta(t)$. To find the **elementary solution** of the **convolution** as defined at the end of the last Section, we simply put $e(t) = \delta(t)$, the **unit element** of the **convolution algebra**. Now since $\phi(t) * \delta(t) = \phi(t)$ whatever $\phi(t)$ is, this means the LHS reduces to $\delta'(t) * e(t) = \delta'(t) * \delta(t) = \delta'(t)$. So we need to solve for $i(t)$ in:

$$[L.\delta''(t) + R.\delta'(t) + (1/C)\delta(t)] * i(t) = \delta'(t).$$

To find the inverse, Schwartz uses much the same technique as I did above, putting $\delta \equiv 1$, and $\delta' \equiv p$ and $\delta'' \equiv p^2$ to give the form:

$$\text{“}i(t)\text{”} = p/[Lp^2 + Rp + (1/C)].$$

As above, this separates into:

$$p/[(p-\alpha-\beta)(p-\alpha+\beta)] = \frac{A}{(p-\alpha-\beta)} + \frac{B}{(p-\alpha+\beta)}.$$

So we must have $A(p-\alpha+\beta) + B(p-\alpha-\beta) = p$, so this time $A + B = 1$, and so $B = 1 - A$, and:

$$-A\alpha + A\beta + (1-A)(-\alpha-\beta) = -A\alpha + A\beta + (A-1)(\alpha+\beta)$$

$$= -A\alpha + A\beta + A\alpha + A\beta - \alpha - \beta = 2A\beta - \alpha - \beta = 0,$$

so $A = (\alpha + \beta)/2\beta$ and so $B = (2\beta - \alpha - \beta)/2\beta = -(\alpha - \beta)/2\beta$.

Now in the **convolution algebra**, we have from the end of the last Section:

$$(p-\lambda)^{-1} = H(t).\mathbf{e}^{\lambda t}$$

so again we get:

$$i(t) = (H(t)/2\beta)\{(\alpha+\beta).\mathbf{e}^{(\alpha+\beta)t} - (\alpha-\beta).\mathbf{e}^{(\alpha-\beta)t}\}.$$

With $E = 0$ and $i(0) = 1$, this is the same result as we obtained before with the **Laplace transform**.

Slightly obscured in that example is a concept that is important in applications. Very often, we have a system which can reasonably be divided into being “driven” by an **input** and responding with an **output** or **response**. The foregoing system could be described that way if we view $e(t)$ as the **input** and $i(t)$ as the **output**. Then we have, starting from the equation:

$$\partial_t e(t) - \partial_t i(t).R - (i(t)/C) - L.\partial_t^2 i(t) = 0.$$

Hence:

$$\partial_t e(t) = [\partial_t i(t).R + (i(t)/C) + L.\partial_t^2 i(t)].$$

Taking the **Laplace transform** and remembering that

$$\mathfrak{L}(\partial_t i(t)) = p.\mathfrak{L}(i(t)) - i(0) = p.I(p) - i(0)$$
$$\mathfrak{L}(\partial_t^2 i(t)) = p^2\mathfrak{L}(i(t)) - pi(0) - \partial_t i(0) = p^2\, I(p) - pi(0) - \partial_t i(0),$$

we get:

$$p.E(p) - e(0) = [Rp + 1/C + L.p^2].I(p) - (R + p.L).i(0) - L\partial_t i(0)$$

which gives:

$$I(p) = \frac{p.E(p) - e(0) + (R + pL).i(0) + L\,\partial_t i(0)}{[Rp + \frac{1}{C} + L.p^2]}.$$

If we ignore those irritating $e(0)$ and $i(0)$ terms, assuming $e(0) = i(0) = \partial_t i(0) = 0$, we get the simple form:

$$I(p) = \frac{p.E(p)}{[Rp + \frac{1}{C} + L.p^2]}. \qquad (8.6.1)$$

Now let $e(t)$ be approximated by a sharp pulse at $t = 0$ so that $e(t) = \delta(t)$, then:

$$E(p) = \mathfrak{L}(e(t)) = \delta(t).\mathbf{e}^{-pt}.\mathbf{d}t \,\big|\, [0, \infty] = \mathbf{e}^{-p.0} = 1$$

and if we assume that $e(0) = i(0) = \partial_t i(0) = 0$ when $e(t) = \delta(t)$, we do get:

$$I_\delta(p) = \frac{p}{\left[Rp + \frac{1}{C} + L.p^2\right]}.$$

Equally Schwartz's **convolution algebra** gives:

$$[L.\delta''(t) + R.\delta'(t) + (1/C)\delta(t)] * i(t) = \delta'(t) * e(t) = \delta'(t) * \delta(t) = \delta'(t),$$

which translates directly into:

$$[Lp^2 + Rp + 1/C].I_\delta(p) = p,$$

or

$$I_\delta(p) = \frac{p}{\left[Lp^2 + Rp + \frac{1}{C}\right]}.$$

Either way, $I_\delta(p)$ is called the **impulse response** (in p-space) of the circuit, the "**impulse**" in question being the **Dirac delta functional**, which in electronics parlance is often called the **impulse function**. The **impulse response** is the **output** that results when the **input** is a **Dirac impulse** $\delta(t)$.

The importance of this concept is that, as long as we can reasonably ignore those irritating **initial conditions** $e(0)$, $i(0)$, etc., the **response** or **output** for *any* **input** can be found using the **impulse response**. For in the general case where $e(t)$ is arbitrary, we get from the equation (8.6.1) above:

$$I(p) = \frac{p.E(p)}{[Rp + \frac{1}{C} + L.p^2]} = I_\delta(p).E(p).$$

or

$$i(t) = L^{-1}(I_\delta(p).E(p)) = i_\delta(t-\tau).e(\tau).\mathbf{d}\tau \,\big|\, [-\infty, \infty] = i_\delta(t) * e(t).$$

Here I've used the fact that products in p-space become **convolutions** in t-space, and I write $i_\delta(t)$ to indicate the *actual* **impulse response** in time.

In summary:

> The actual **response** of a circuit to a time-varying **input** is the **convolution** of the **impulse response** with the **input**.

You may feel a bit worried about those neglected **initial conditions**. They can be lumped together to give a generalised numerator as:

$$p.E(p) - e(0) + (R + p.L).i(0) + L\partial_t i(0) = p.E(p) + pL.i(0) + R.i(0) - e(0) + L\partial_t i(0)$$

$$= p.E(p) + pk_L + k_0.$$

Inductors are rather unusual anyway, so effectively we have to allow for just an extra term in:

$$\frac{k_0}{\left[Lp^2 + Rp + \frac{1}{C}\right]}$$

which can be handled by much the same methods.

To take this theory a little further, I'll introduce the concept of an **operational amplifier**, which is shown in *Figure 8.6.1*. In this circuit, we have an applied **input voltage** $e_i(t)$, an **amplifier** with **gain** A, which converts the **voltage** at *its* **input** e' to a **voltage** at its **output** e_o with $e_o = -A.e'$,[139] an **input impedance** Z_i and a **feedback loop** with an **impedance** Z_f. Here the "**impedances**" are just combinations of **resistors** and **capacitors**, the term **impedance** being a generalisation of **resistance** to the case where time-dependent elements are involved, and today that virtually always means **capacitors**. **Currents** i_i and i_f are through Z_i and Z_f as shown in the *Figure*.

[139] **Amplifiers** commonly invert the signal, hence the minus sign.

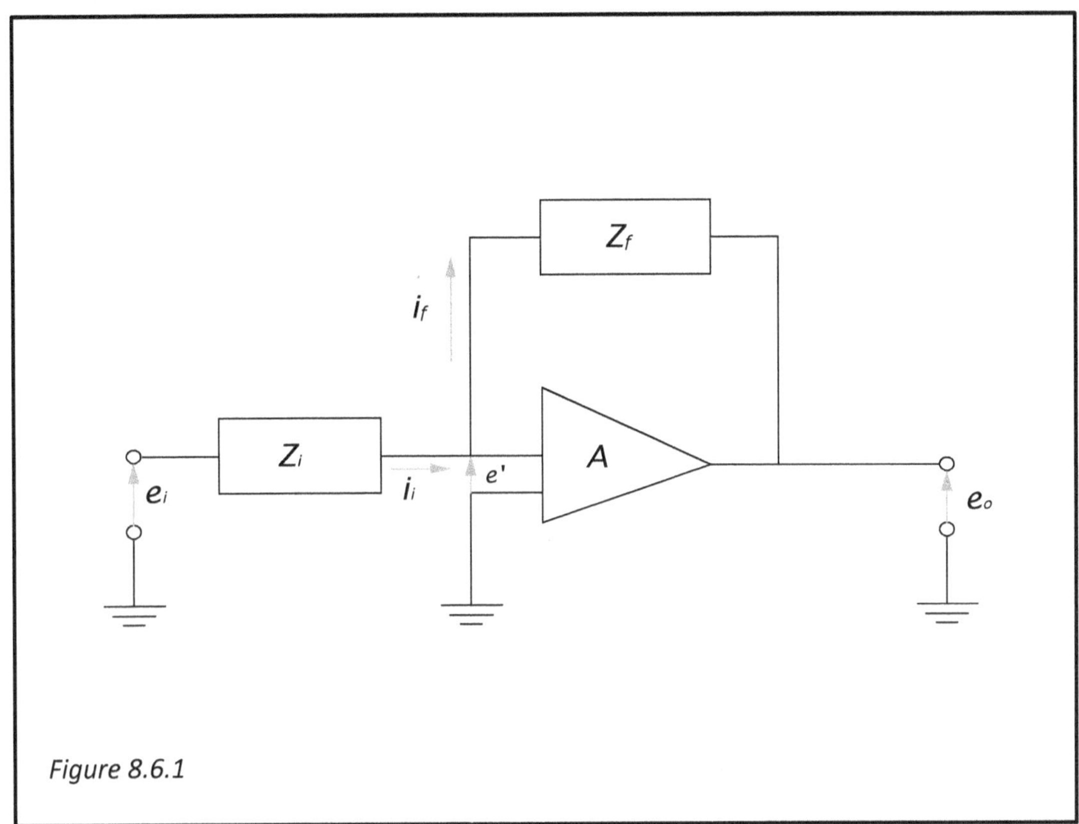

Figure 8.6.1

The **current** into the **amplifier** is usually reckoned as negligible. When such circuits are coupled or cascaded together, the addition of the second stage makes almost no difference to the **output voltage** of the first stage as calculated without the second stage being present, a great simplification. Under this hypothesis, by **Kirchhoff's current law** at the point just left of *A* we will have:

$$i_i = i_f \, .$$

We now just switch into *p*-space keeping the same *e*'s and *i*'s, and we can directly describe the individual elements from the standard equations thus:

For a **resistor**:	$e = -iR$	$\rightarrow$	$e(p) = -i(p).R$
For a **capacitor**:	$C.\partial_t e = -i$	$\rightarrow$	$e(p) = -i(p)/pC,$
For an **inductor**:	$L.\partial_t i = -e$	$\rightarrow$	$e(p) = -Lp.i(p).$

So we can express any combination of *R*'s and *C*'s in Z_i and Z_f as obeying:

$$e = -Z_i(p).i \qquad \text{and} \qquad e = -Z_f(p).i.$$

For example, if Z_i consists of a **resistor** *R* and a **capacitor** *C* in series, we have:

$$e = -[R + 1/pC].i$$

which you should recognise as our *RLC* case above without the *L*. This equation appeared there as (8.6.1)(ignoring the **initial conditions** as usual):

$$p.E(p) = [Rp + 1/C + L.p\,].I(p)$$

so $$E(p) = [Rp + 1/C].I(p)/p = [R + 1/pC].I(p)$$

with the input voltage $E(p)$ replacing the element voltage e, so the − sign disappears.

From the *Figure* the **voltage** across Z_f is $e_o - e'$, and that across Z_i is $e' - e_i$, so we get:

$$i_i = i_f \quad \rightarrow \quad (e' - e_i)/Z_i\,(p) = (e_o - e')/Z_f(p).$$

But $e_o = -A.e'$, so $e' = -\,e_o/A$, so:

$$(-e_o/A - e_i).Z_f\,(p) = (e_o + e_o/A).Z_i\,(p)$$

or

$$e_o[1 + (1/A) + (1/A).Z_f\,(p)/Z_i\,(p)] = -(\,Z_f\,(p)/Z_i\,(p)).e_i.$$

Again we approximate, assuming A is very large, so $(1/A)[1 + Z_f\,(p)/Z_i\,(p)] \approx 0$, and we get the basic equation of an **operational amplifier**:

$$e_o = -(Z_f\,(p)/Z_i\,(p)).e_i.$$

Judicious choice of the two **impedances** enables all sorts of clever behaviour to be obtained from this same basic concept.

Firstly, I'll just clear up one question: what is the combined **impedance** for a **resistor** and a **capacitor** in *parallel*? Remember from Section 1.14 that in this case the **voltages** across the two elements are equal, so:

$$e_R = -\,i_R R = e_C = -\,i_C/pC =: -e.$$

The total **impedance** equals that **voltage** divided by the combined **currents** $i_R + i_C$. So:

$$Z_{RC\|} = e/(i_R + i_C) = e/(e/R + e.pC) = 1/(1/R + pC).$$

Typical **operational amplifiers** are:

- Where $Z_f = 1/pC$ (a **capacitor**) and $Z_i = R$ (a **resistor**), then $e_o = -(Z_f(p)/Z_i\,(p)).e_i = (-1/pRC).e_i$. This circuit is an **integrator**, since × $1/p$ in the p-domain corresponds to **integration** in the *time* domain.

- Where $Z_f = R$ and $Z_i = 1/pC$, then $e_o = -(Z_f(p)/Z_i(p)).e_i = -pRC.e_i$, and this multiplication by p in the p-space corresponds to **differentiation** in t.
- Where $Z_f = R_1$ and $Z_i = R_2$, we get a *scaler*: $e_o = -R_1/R_2.e_i$.

A very similar logic holds when e and i are **sinusoidal**, of which the general case is that of a **function** which can be represented as a **Fourier transform**:

$$e(t) = \langle e(f), \varepsilon(ft)\rangle = \langle e(f), \mathbf{e}\uparrow(2\pi jft)\rangle = e(f).\mathbf{e}\uparrow(2\pi jft).\mathbf{d}f \,|\, [-\infty, \infty].$$

Here f represents **frequency** in cycles per second or Hz. Making the change of variables $f = \omega/2\pi$, so $\mathbf{d}f = \mathbf{d}\omega/2\pi$, we get:

$$\langle e(f), \varepsilon(ft)\rangle = (1/2\pi)\langle e(\omega/2\pi), \mathbf{e}\uparrow(j\omega t)\rangle = (1/2\pi)e(\omega/2\pi).\mathbf{e}\uparrow(j\omega t).\mathbf{d}\omega \,|\, [-\infty, \infty].$$

Rather improperly, we'll just write $e(\omega/2\pi)$ as $e(\omega)$, even though the functional dependence of $e(f)$ and $e(\omega)$ are not strictly the same, and we'll drop the leading $(1/2\pi)$ as it will occur in every term as we'll assume $\varepsilon(ft)$ dependence throughout.[140] So we can write this as:

$$\langle e(f), \varepsilon(ft)\rangle = \langle e(\omega), \mathbf{e}\uparrow(j\omega t)\rangle = e(\omega).\mathbf{e}\uparrow(j\omega t).\mathbf{d}\omega \,|\, [-\infty, \infty].$$

Similarly we can write:

$$i(t) = \langle i(\omega), \mathbf{e}\uparrow(j\omega t)\rangle = i(\omega).\mathbf{e}\uparrow(j\omega t).\mathbf{d}\omega \,|\, [-\infty, \infty].$$

Here the **transform** is into the space of the **frequency** variables f and $\omega = 2\pi f$. The variable ω, the **frequency** in **radians** per second, is generally preferred here to f, the **frequency** in *cycles* per second, simply because we don't get a lot of 2π's when we **differentiate**.

Now the equation for the *RC* series circuit is:

$$e(t) = v_R + v_C$$

with

$$v_R = i(t).R \qquad \text{and} \qquad C.\partial_t v_C = i(t).$$ [141]

First look at $\partial_t v_C(t)$, which we can write as $\partial_t\mathfrak{F}^{-1}(v_C(\omega)) = \partial_t\langle v_C(\omega), \mathbf{e}\uparrow(j\omega t)\rangle$ and which expands to:

$$\partial_t v_C(t) = \langle v_C(\omega), j\omega.\mathbf{e}\uparrow(j\omega t)\rangle = \langle j\omega.v_C(\omega), \mathbf{e}\uparrow(j\omega t)\rangle.$$

This gives, by $\partial_t v_C = i(t)/C$:

---

[140] I'm afraid engineers do this sort of thing all the time!

[141] Written with $e(t)$ and the v's on opposite sides we don't need the – signs.

$$\langle j\omega.v_C(\omega), \mathbf{e}{\uparrow}(j\omega t)\rangle = (1/C).\langle i(\omega), \mathbf{e}{\uparrow}(j\omega t)\rangle.$$

So $$j\omega.v_C(\omega) = (1/C).i(\omega)$$

or

$$v_C(\omega) = (1/j\omega C).i(\omega).$$

But

$$e(t) = \langle e(\omega), \mathbf{e}{\uparrow}(j\omega t)\rangle = \langle i(\omega), \mathbf{e}{\uparrow}(j\omega t)\rangle.R + \langle v_C(\omega), \mathbf{e}{\uparrow}(j\omega t)\rangle$$

$$= \langle i(\omega), \mathbf{e}{\uparrow}(j\omega t)\rangle.R + \langle i(\omega)/(j\omega C), \mathbf{e}{\uparrow}(j\omega t)\rangle$$

or

$$e(\omega) = i(\omega).R + i(\omega)/(j\omega C).$$

This is clearly an **impedance** in ω-space:

$$e(\omega)/i(\omega) = Z_{RC(\omega)} = [R + 1/j\omega C].$$

Since a huge amount of the theory of **signals** in electronics involves such generalised **sinusoidal signals**, vast tracts of the theory are simply written this way, where it is taken for granted that we are actually in ω-space. So electronics texts would commonly write such an equation as just:

$$e/i = Z_{RC} = [R + 1/j\omega C].$$

No warning is given that we are not in the time domain, and like so many other shorthand notations widely used in practice, this can occasion endless grief for the unsuspecting student!

8.X Exercises

1. Consider the function on [0, 1] defined by

    a. For $0 \leq \tau < ½$: $z = \{0, -1 + 4\tau\}$

    b. For $½ \leq \tau \leq 1$: $z = \{-3 + 4\tau, 0\}$

    These are a pair of crossed lines through the **origin**. Evaluate the general form of the nth term of the **Fourier series** for this function on τ:[0, 1]. You will need to use **integration by parts**, from $\mathbf{d}(ax.\varepsilon(-nx)) = a\varepsilon(-nx).\mathbf{dx} + ax.(-2\pi nj).\varepsilon(-nx).\mathbf{dx}$. More importantly, you *must* evaluate the integral over [0, ½] and over [½, 1] separately or you'll get the wrong answer! Try writing a quick program in Java or C# or your language of choice to graph the sum of the first few terms from $n = -N$ to $n = N$.

2. Work out the **Fourier series** for the off-centre "looping square" from (1, 1) to (2, 1) to (2, 2) to (1, 2) and back to (1, 1) for a full cycle.

3. Do the same for the off-center circle with centre at (1, 1) but radius ½.

4. The square pulse train in *Figure 8.2.3* evaluates to a **series** in **cosines**, as it should, being an **even** function. Yet the apparently analogous single square pulse in *Figure 8.3.1* has a **Fourier *transform*** in a **sine**. Why?

5. Evaluate the **Fourier series** for the symmetrical sawtooth waveform of **period** 2 running from (0, 0) to (1, 1) and back down to (2, 0). Evaluate the **Fourier transform** for a single "tooth" (i.e. one **period**) of this.

6. A PID controller is an operational amplifier with Z_i of *Figure 8.7.1* consisting of a **capacitor** C_1 in **parallel** withe a **resistor** R_1, and Z_f a **resistor** R_2 and a **capacitor** C_2 in **series**. Show that the **transfer function**, which is the **Laplace transform** of the ratio of the output to the input voltages for *zero initial conditions*, is $(\mathbf{e}_o/\mathbf{e}_i)(s) = -(R_2 + 1/C_2s)/(1/(C_1s + 1/R_1))$.

7. Evaluate the **Laplace transforms** of $\mathbf{e}^{at}$, **sin**(ωt), **cos**(ωt), **sinh**(at) and **cosh**(at), and also $t.\mathbf{e}^{at}$, $t.$**sin**(ωt), and $t.$**cos**(ωt). (These are the ones you end up using all the time!)

8. Take the Laplace transform of the equation $\partial_t^{\,2}y(t) + \omega_0^{\,2}y(t) = C.\mathbf{sin}(\Omega t)$. From this, inverting the result using **partial fractions** and the results of the previous question, demonstrate the **resonance** condition that when $\omega_0 = \Omega$, $y(t) = (C/2\omega_0^{\,2})(\mathbf{sin}(\omega_0 t) - \omega_0 t.\mathbf{cos}(\omega_0 t))$ where the $\omega_0 t$ factor in the second term causes it to increase without limit. (If you set $\omega_0 = \Omega$ you get a *double root*, a condition you should have discovered in the previous exercise.) (E. Kreyszig, *Advanced Engineering Mathematics*, Wiley 2nd Ed, p.211.)

9. The **discrete convolution** of two sequences of numbers $\{g_n\}$ and $\{h_m\}$ is defined as $\{g* h\} = \{(g * h)_n\} := \sum_{-\infty}^{\infty} g_{n-m}\, h_m$. A sequence $\{g_n\}$ is said to have **bounded support** if $g_n = 0$ if $n < n_1$ or $n > n_2$. Sequences may have elements for $n < 0$ or $n = 0$. If $\{h\}$ here has **bounded support,** $(g * h)_n$ reduces to $\sum_{m_1}^{m_2} g_{n-m}\, h_m$. Show that:

    a. For sequences of **bounded support** the sum of the elements in $\{g * h\}$ equals the product of the sum of the elements in $\{g_n\}$ and the sum of the elements in $\{h_m\}$.

    The sequence $\{J\}$ defined with $J_0 = 1$ but $J_n = 0$ for all $n \neq 0$ has the property analogous to the **delta function** that $\{J\} * \{g\} = \{g\}$. (R.N.Bracewell, *The Fourier Transform and its Applications*, McGraw-Hill 3rd Ed, Ch. 3)

10. Given a function $\phi(t)$ that is zero for $t < t_0$ and for $t > t_1$ and so of **bounded support**, for an arbitrary number $N > 0$, define a **sampling interval** $\tau := |t_1 - t_0|/N$. Then the **Discrete Fourier Transform** or **DFT** $\Phi(\omega)$ of $\phi(t)$ is defined by $\Phi(\omega) :=$ (**DFT**(ϕ))$(\omega) := N^{-1}\sum_{\tau=0}^{N-1} \varphi(\tau).\, \varepsilon(-(\omega/N)\tau)$. Show the **DFT Inversion Theorem** that $\phi(\tau) = \sum_{\omega=0}^{N-1} \Phi(\omega).\, \varepsilon((\omega/N)\tau)$.

11. Prove **Parseval's formula** that if $\{\varphi_1, \varphi_2, \ldots \varphi_n, \ldots\}$ constitute a **complete orthonormal basis** (i.e. $\varphi_m(\xi).\varphi_n(\xi).\mathbf{d}\xi \,|\, [a, b] = \delta_{mn}$) for functions over an **interval** $[a, b]$, then for $\lim_{N\to\infty} \sum_{n=1}^{N} c_n g_n\,(x) = f(x)$, then $\sum_{n=1}^{N} c_n^2 \to f^2(x).\mathbf{d}x \,|\, [a, b]$. (I've used two distinct notations for **limit** here!)

12. Prove the **autocorrelation theorem**: that if $f(x)$ has the **Fourier transform** $\varphi(\omega)$, then its **autocorrelation function** $f^*(\xi).f(\xi + a).\mathrm{d}\xi \,|\, [-\infty, \infty]$ has the **Fourier transform** $|\varphi(\omega)|^2$. Show how this could suggest an interpretation for $[\delta(x)]^2$. (Bracewell, *The Fourier Transform op.cit*, McGraw-Hill 2000, p.122.)

Chapter 9

PROJECTIVE GEOMETRY

9.1 Introduction

Projective Geometry has its origins in the theory of **perspective**. To put this into context, I'll give a brief résumé of general ideas involved in the **projection** of a three-dimensional scene or object onto a two-dimensional surface.

In architecture and engineering, a rather limited set of geometrical projections are used, which fall into two main classes: **parallel** and **perspective**. This may seem to be in marked contrast to geographers, who select from a very large range of projections to accommodate the problem of mapping a spherical earth onto plane paper. But there's a big difference: **projections** in engineering are mapping a three-dimensional object onto a plane two-dimensional surface; the **projections** used in cartography are mapping a curved two-dimensional surface onto a plane two-dimensional surface.

As there's a certain amount of confusion about terminology here, I will follow the terminology used in F. D. K. Ching's *Architectural Graphics*, a standard U.S. textbook, which gives the terms familiar to most American architects and which are those used in the almost universally American software employed in CAD work.

The **projections** are most easily distinguished by the use of two auxiliary concepts. The first is that of **projector lines** or **projectors**, which are simply straight lines so defined that precisely one alone passes through each point of the three-dimensional object being depicted. The second is that of the plane on which the two-dimensional image will appear, which is called the **picture plane**. Exactly how the **projectors** are defined defines the **projection**: for where the unique **projector** through any point in space P with **coordinates** (x, y, z) passes through the **picture plane** defines the **image** of P. Prof. Ching distinguishes three main classes:

- **Orthographic Projection**

  The **projectors** are parallel to each other and perpendicular to the **picture plane**.

- **Oblique Projection**

  The **projectors** are parallel to each other and at an oblique angle (i.e. an angle other than a right angle) to the **picture plane**.

- **Perspective Projection**

  The **projectors** all pass through a single point in space SP unique to the **projection** that represents a single eye of the observer. This unique point is called the **eye point** or **station point**. I will use both terms interchangeably.

The first two are both **parallel** projections, which is why I refer to only two main classes, **parallel** and **perspective**. Of these projections, **perspective** is much the most interesting, both mathematically and artistically, and its understanding by Brunelleschi in 1425 was one of the turning points of the modern era. There's a widespread belief today that perspective is some sort of Western convention, but it isn't: it's a true scientific discovery, because in perspective the **image** in the **picture plane** of any point in space lies in exactly the direction, as seen by the observer's eye, in which the point itself does. So a **perspective image** is a true realization of what the observer sees.

Parallel projections are sometimes referred to simply as "**paraline**" projections in the USA, but this is an informal term. **Orthographic projections** are sometimes mistakenly called **orthogonal**, but because this term has established usage in mathematics it should be avoided.

The three **projections** can be defined very easily algebraically. To do this, we need to choose **coordinates** specific to each of the three **projections,** and if we also define **picture plane coordinates** ξ and ψ measuring distances in the x and y directions *within the* **picture plane**, we can look at the three main **projection** types in detail, using diagrams showing just the xz-plane. The logic for the yz-plane is similar in each case.

The **picture plane**, which we assume lies at right angles to the xz-plane, and indeed, except for the **oblique** case, actually is parallel to the xy-plane, now appears as a line, since it's seen edge on. That line is marked PP in the three diagrams which follow, and for clarity, I've chosen to put each on a separate page.

- For **Orthographic Projection**

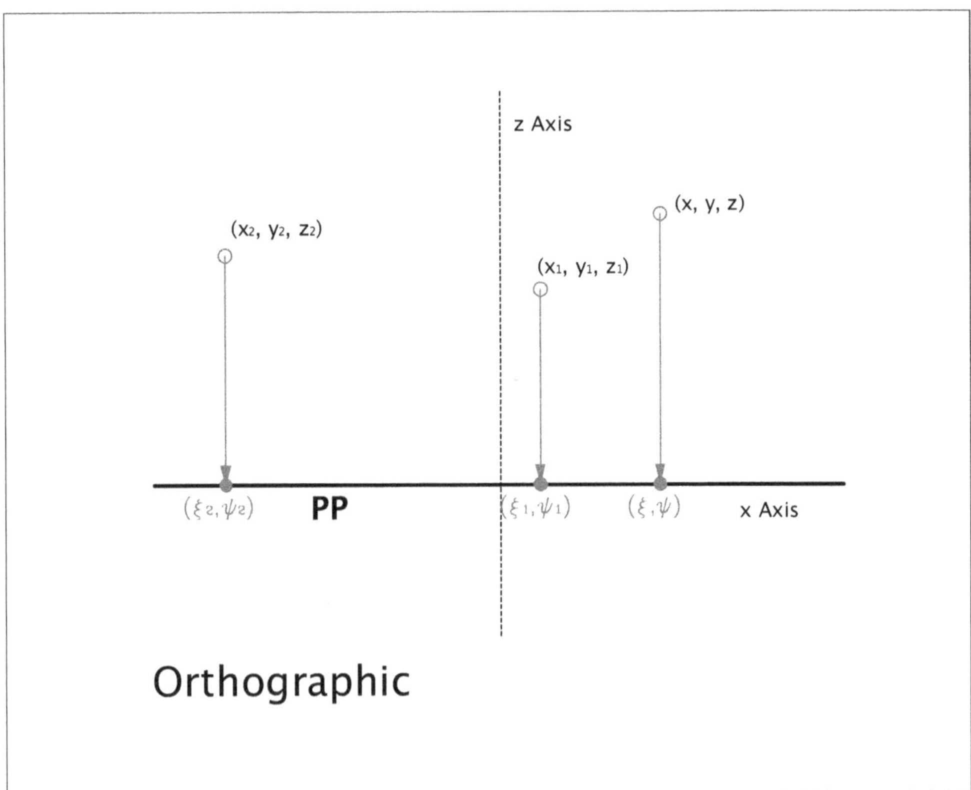

Figure 9.1.1

Choose **coordinates** such that the **projectors** are parallel to the *z* axis and the *x* and *y* axes are parallel to the **picture plane**. Then the **image** of (*x*, *y*, *z*) in the **picture plane** is at the point (ξ, ψ) = (*x*, *y*), independent of the value of *z*. In other words, all points with the same value of *z* map into the same **image point**.

- For **Oblique Projection**

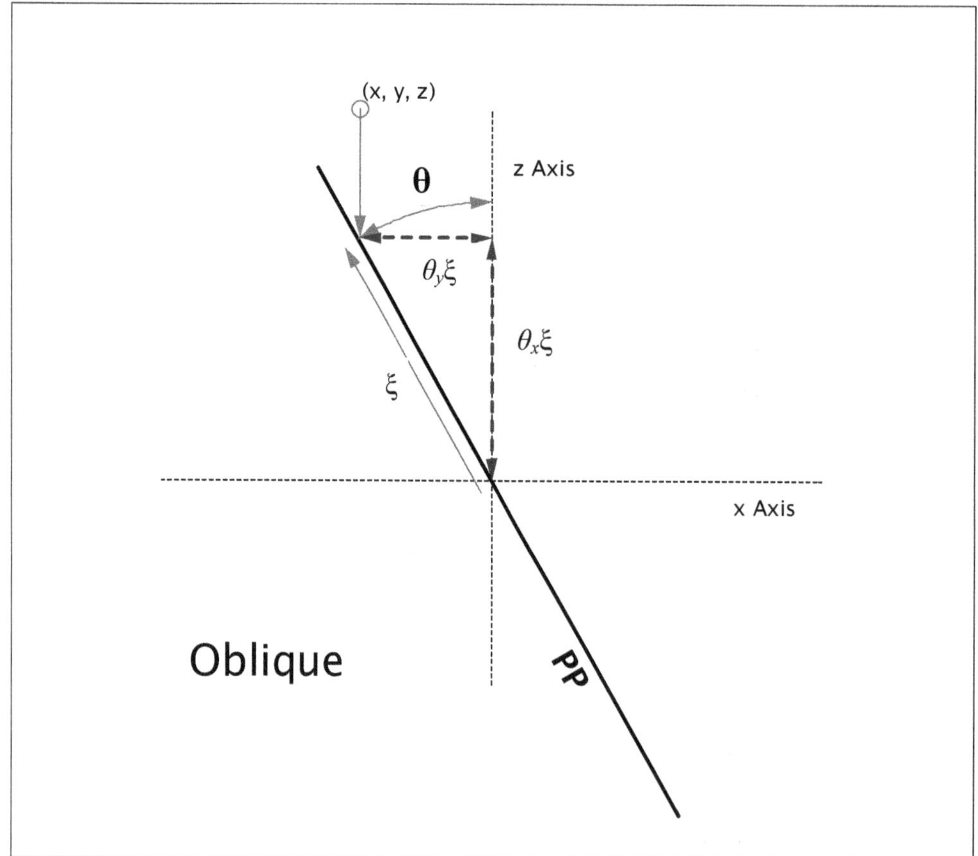

Figure 9.1.2

Choose **coordinates** such that the **projectors** are parallel to the *z* axis and at an angle **θ** = (θ_x, θ_z) to the **picture plane** in the *xz*-plane, but at right angles to the **picture plane** in the *yz*-plane. In other words, rotate the *x* and *y* axes about the *z* until the obliquity falls in the *xz*-plane. Then the **image** of (*x*, *y*, *z*) in the **picture plane** will fall at the point (ξ, ψ) = (x/θ_y, *y*) as can be seen in *Figure 9.1.2*. In this diagram, θ_x, the **cosine** part of **θ**, runs along the *z* axis, and θ_y, the **sine** part, runs parallel to the *x* axis. This is because the angle here is being measured anticlockwise from the *z* axis in this diagram. Note that if **θ** = **j** ≡ 0 + *j*.1, so the angle between the **picture plane** PP and the *z* axis is 90°, then (ξ, ψ) = (*x*/1, *y*) = (*x*, *y*) and the projection reduces to the ordinary **orthographic projection**.

• For **Perspective Projection**

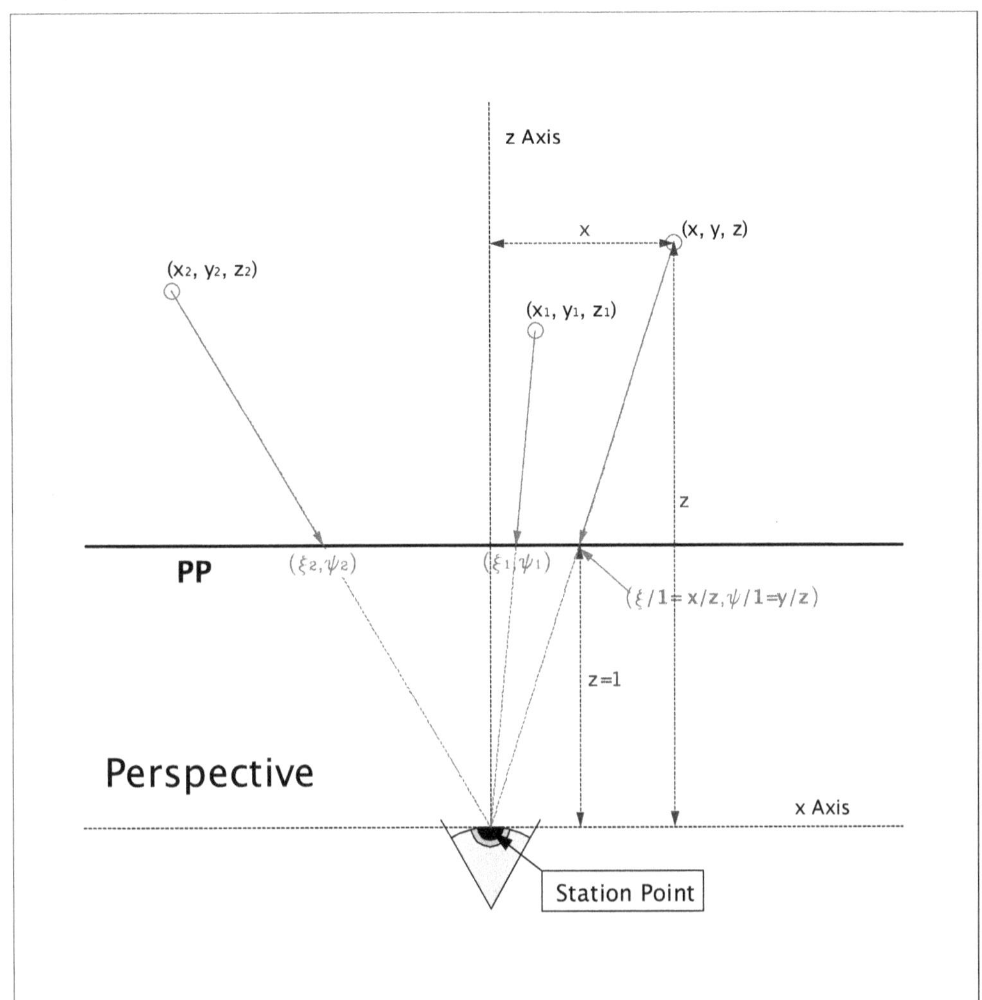

Figure 9.1.3

Choose **coordinates** so that the *z* axis is at right angles to the **picture plane**, and so the *x* and *y* axes are parallel to the **picture plane**. Let the unique **eye point** or **station point** be at the **origin** (0, 0, 0). Now scale the **coordinates** so that the **picture plane** coincides with the plane *z* = 1. Then, by similar triangles, as in *Figure 9.1.3*, we can see that the **image** of (*x*, *y*, *z*) in the **picture plane** will be at the point (ξ, ψ) = (*x*/*z*, *y*/*z*). This division by the third **coordinate** will be hugely important in what follows.

Perspective is unique in that all points in a given *direction* have **images** closer and closer to a specific **limiting point** in the **picture plane** as the points are selected further and further from the **eye point**. This **limiting point** is unique to that direction. We can see this easily enough.

Take any point in the $z = 0$ plane $(x_0, y_0, 0)$, and take any direction defined as $(\theta_x, \theta_y, 1)$. Then all points along the line through $(x_0, y_0, 0)$ running in the direction $(\theta_x, \theta_y, 1)$ will obey the equation:

$$P = (x, y, z) = (x_0, y_0, 0) + t.(\theta_x, \theta_y, 1),$$

where we're assuming ordinary **vector algebra**. Then as $t \to \infty$, the **image** of P is:

$$p = (x/z, y/z) = ((x_0 + t\theta_x)/t, (y_0 + t\theta_y)/t) = (x_0/t + \theta_x, y_0/t + \theta_y) \to (\theta_x, \theta_y).$$

So the **limiting point** of the **images** is *independent of* (x_0, y_0). *All* points lying far enough away in the direction defined by (θ_x, θ_y) have their **images** in the neighbourhood of *the same* **image point** given by precisely that direction itself (θ_x, θ_y).

This unique **image point** for any given direction from the **eye point** is called the **vanishing point** for that direction. By taking $(x_0, y_0) = (0, 0)$, or the **eye point** itself (we had assumed $z = 0$ here) we can see that this **image point** is precisely the point where the unique line through the **eye point** in the direction $(\theta_x, \theta_y, 1)$ intersects the **picture plane**.

In this I've assumed that "θ_z" = 1 always to avoid a notorious case. If we take "θ_z" = 0, so we run in a direction $(\theta_x, \theta_y, 0)$ from the **eye point**, our line *never* intersects the **picture plane** but simply runs off to infinity *parallel* to the **picture plane**. So *directions parallel to the* **picture plane** do not have **vanishing points** *lying in the* **picture plane**. This anomaly was the entire reason for the introduction of **projective geometry**.

Algebraically we now have:

$$P = (x, y, z) = (x_0, y_0, 0) + t.(\theta_x, \theta_y, 0),$$

so as $t \to \infty$, the **image point** would be undefined as the z **coordinate** of P is consistently zero:

$$p = (x/z, y/z) = ((x_0 + t\theta_x)/0, (y_0 + t\theta_y)/0) \to (\infty, \infty).$$

So, in the sense that $x/0 = \infty$, *any* point in a direction parallel to the **picture plane**, *however near to the* **eye point**, has its **image** at infinity.

This issue is so important that I will devote the next Section to the consideration of **vanishing points**.

9.2 On Vanishing Points

Architectural students are introduced to perspective drawing with the idea of one-point, two-point and three-point perspectives. I'm not sure who introduced these concepts, but they go back a long way. Although a valuable teaching aid, these ideas have occasioned some misunderstanding, such as the often repeated remark that Renaissance art is based on the concept of one-point perspective.

I'll repeat the definition, paraphrasing the form of it given by Prof. Ching again (p.92):

As two parallel lines recede into the distance, the space between them appears to diminish progressively until they appear to meet at a particular point on the **picture plane PP**. This point is called the **vanishing point** for that particular pair of lines and all other lines parallel to them.

Every direction in space that is not itself parallel to the **picture plane** has a **vanishing point** which is the point where a line drawn from the station point (SP) in the direction in question intersects the **picture plane PP**.

I want to stress that there is nothing inherently one- two- or three-point about any **perspective** projection. The "n-pointedness" is *simply an aspect of the orientation to the* **picture plane** *of the object being depicted*. This is such a source of misunderstanding that I'll go into it here at some length.

Buildings are most commonly rectilinear structures, for which there is a clear choice of **Cartesian axes** as defined in Section 1.2 which will make virtually all the walls parallel to either the x or the y axis if we assume that the z axis is chosen to represent the vertical. These preferred axes are often called **body axes** in mechanics. Prof. Ching calls them the **principal axes** of the building, a term that is also used in mechanics. I will write these **body axes** as **i**, **j**, **k**.

These are not the same as the axes introduced for the three kinds of projection in the last Section. Those are known as **camera axes** or **eye axes**, even though the concept of a **camera** or **eye point** is strictly only valid for **perspective**.

Most of the subsequent classification of **projections** depends *solely on the orientation of the object's* **body axes** *to the* **picture plane**. So . . .

- an **isometric projection** is an **orthographic projection** wherein the three **body axes** are all at the same angle to the **picture plane**. So if the **normal** to the **picture plane** is **n**, $\mathbf{i} \cdot \mathbf{n} = \mathbf{j} \cdot \mathbf{n} = \mathbf{k} \cdot \mathbf{n}$ so that $\mathbf{n} = (1/\sqrt{3})(\mathbf{i} + \mathbf{j} + \mathbf{k})$, and **n** is at an angle of **arcos**$(1/\sqrt{3})$ or about 54.7356° to each of the **body axes**, and *none* of the **body axes** are parallel to the **picture plane**,
- a **dimetric projection** is a **orthographic projection** in which two of the **body axes** are at the same angle to the **picture plane**, but not the third,
- an **axonometric projection** is *either* an **orthographic projection** in which none of the three **body axes** is parallel to the **picture plane**, *or* it can be used to refer to an **oblique parallel projection** in which two of the **body axes** *are* parallel to the **picture plane**, a form much used in the nineteenth century by Auguste Choisy,
- a one-point **perspective** is a **perspective** in which two of the three **body axes** are parallel to the **picture plane** so the third *must* be at right angles to it,
- a two-point **perspective** is one in which one of the **body axes** (most commonly the vertical **k**) is parallel to the **picture plane** but neither of the others is, and finally,
- a three-point **perspective** is one in which none of the **body axes** are parallel to the **picture plane**.

I don't raise the last three to the status of formal terms by putting them in bold. The two kinds of "axonometric" are confused because they look vaguely similar, but because of the ambiguity in the use of this term, it's probably best avoided.

The reason why the various kinds of one-point, two-point or three-point **perspective** are so called is because that is *the number of* **vanishing points** *of the three* **body axes** *that are actually present in the* **picture plane**. So in a two-point **perspective** the third **vanishing point** is at infinity because that third axis (generally **k**, the z axis) is parallel to the **picture plane**.

All this reflects simply the positioning of the **body axes** and is nothing intrinsic to the projection itself. So, for example, **perspective projection** *is in no way limited to Cartesian axes and there is no essential connection between Cartesian axes and perspective.*

Just to clear away some of this "1- 2- 3-point" misunderstanding, I will state Point No.1:

- *Every single point in the* **picture plane** *in any* **perspective** *is a* **vanishing point**. It is the **vanishing point** for the infinite *bundle* of lines parallel to the line from the **station point** (the "eye") to the point in question.

So our one to three **vanishing points** are not *the* **vanishing points** of the perspective, *but simply the three we happen to be interested in when drawing a particular rectilinear structure orientated in a particular way.*

Nevertheless, Cartesian axes are obviously exceptionally important and this leads to an interesting question. Instead of looking at how particular orientations of the Cartesian axes appear in the picture plane, let's turn the question inside out. Given *any* set of three points in the picture plane, chosen at random, do they define a possible orientation of the Cartesian axes that would make these three points the **i**, **j** and **k body axes' vanishing points**? I think the problem is quite illuminating and I will give the answer by a somewhat heuristic argument that is amplified in Appendix A.

In this discussion, I'll be dealing primarily with the **body axes**, so when I refer to the *x* axis, I'll mean **i**, when the *y* axis **j**, and when the *z*, **k**, and I'll refer to their respective **vanishing points** in the **picture plane** as VP_x, VP_y and VP_z. We won't need to refer to **camera space** again for a while.

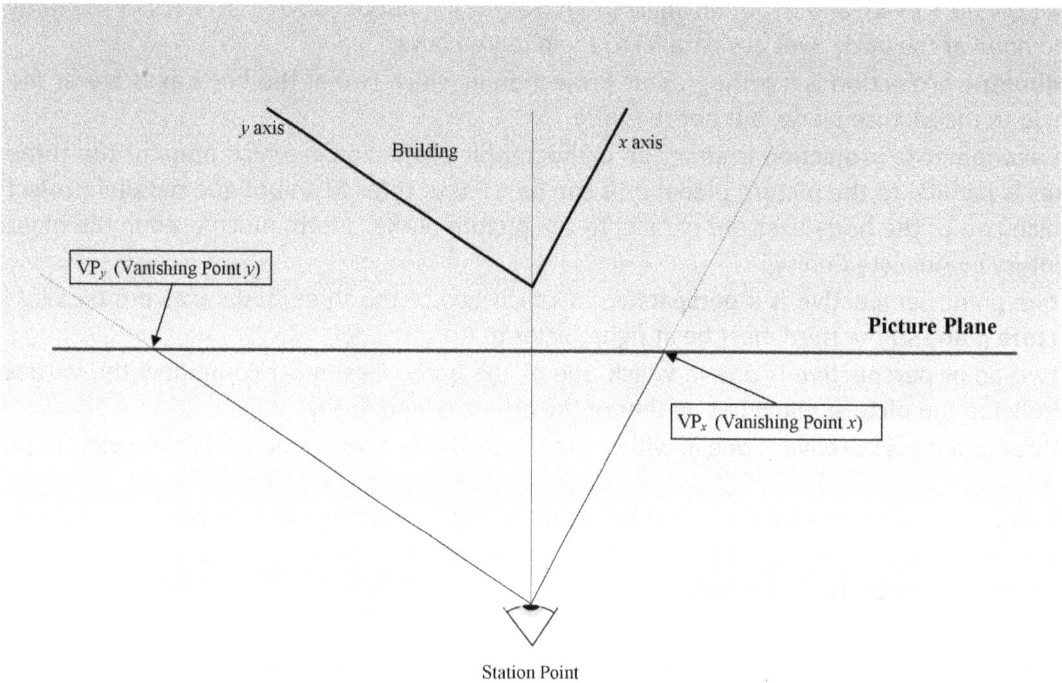

Figure 9.2.1

First look at *Figure 9.2.1*, which shows the standard construction for Cartesian **vanishing points** in two dimensions. Here SP is the **station point**, PP the **picture plane**, VP_x the **x- vanishing point** and VP_y the **y-** one. The placing of VP_x and VP_y clearly depends on the orientation of the **x-** and **y-** axes of the object being depicted. In particular, if either of these axes is parallel to the **picture plane**, its corresponding VP goes off "to infinity" and the "two-point" perspective becomes a "one-point". Now suppose that we are *given* VP_x and VP_y arbitrarily, and we want to see what orientations of the **x-** and **y-** axes of the object scene are compatible with this choice of VP_x and VP_y. Assume first of all that both **vanishing points** are at finite positions, as shown in *Figure 9.2.2*. Since the angle at the **station point** <VP_x-SP-VP_y> must be a right angle for the **vanishing points** to be Cartesian, we see immediately from school geometry that *the SP must lie on the semi-circle whose diameter is the line VP_x-VP_y.* (The angle in a semi-circle is a right angle.) So this determines the possible axes consistent with this placing of VP_x and VP_y, three of which are shown in *Figure 9.2.2*. (The dashed lines dropping from these points in the figure refer to the argument in Appendix A.) *Furthermore, the placing of the* **station point** *uniquely determines the orientation of the axes and vice versa.*

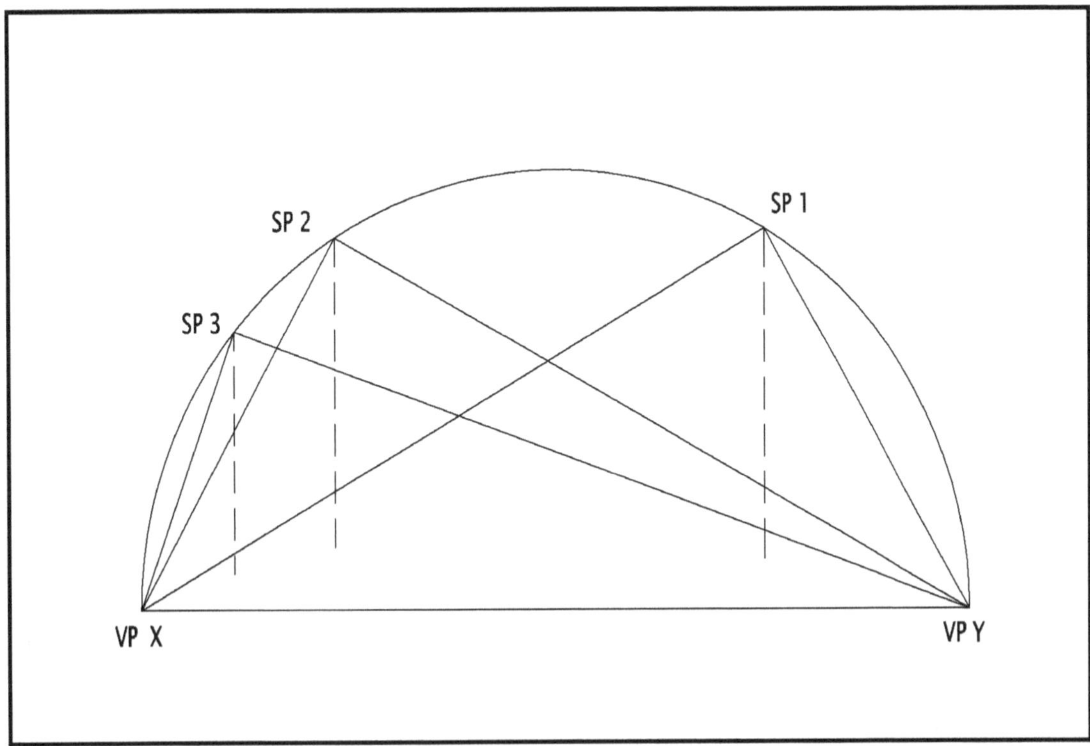

Figure 9.2.2

If one VP, say VP_y, is at infinity (when it can be thought of as having a *direction* in the **picture plane** but no position), then the angle <VP_x-SP-VP_y> must still be a right angle but now the line SP-VP_y will be parallel to the **picture plane** and so the line SP-VP_x must be perpendicular to the **picture plane** and *any* position for SP on the line perpendicular to the PP passing through VP_x is compatible with this.

(VP_x and VP_y cannot both be at infinity in the two-dimensional case.)

The corresponding construction in three dimensions is shown in *Figure 9.2.3*. This is most easily visualised as the corner of a cube poking through our PP with vertex SP and edges intersecting our arbitrarily given VP_x, VP_y and VP_z. The direction of these edges determines the orientation of our depicted scene's Cartesian axes, the line SP-VP_x giving the direction of the **x-** axis, SP-VP_y the **y-** and SP-VP_z the **z-** axis. The question now – and this was my original query – is: "Is any arbitrary assignment of

VP_x, VP_y and VP_z compatible with this construction?" The answer is **no**. Extending the angle in a semi-circle argument (Appendix 1), it turns out that VP_x VP_y and VP_z *must form a triangle in which all three angles are acute-angled.*

There are two exceptional, limiting, special cases:

1. The angle at one VP, say VP_z, is zero. Then both the others must be right-angles for if either were less than a right-angle, the other would have to be more. This case in fact corresponds to VP_z being off at infinity and is our familiar "two-point" perspective, for which we have thus shown that the line VP_x-VP_y must be at right-angles to the direction of VP_z.
2. Two of our VPs, say VP_z and VP_y, are at infinity. In this case *both* the lines SP-VP_z and SP-VP_y must be parallel to the picture plane, so by our "cube corner" construction the third (SP-VP_x) must be perpendicular to the **picture plane** and SP can be at any point along this line. This corresponds to the single VP at infinity for the two-dimensional case and is our familiar "one-point" perspective.

These are the only cases in which a right-angle is admissible.

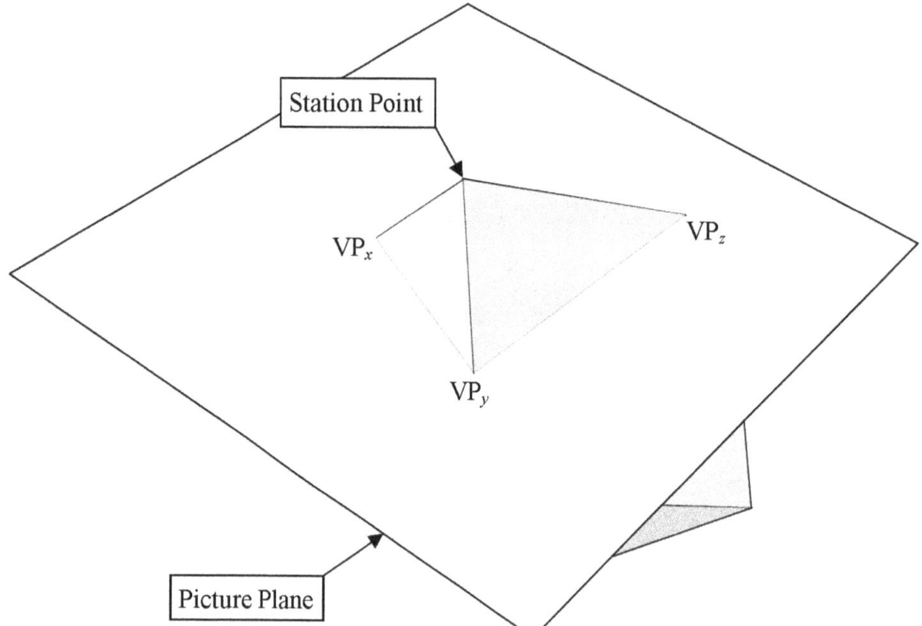

Figure 9.2.3

Note that the position of the SP is uniquely determined by the placing of the three VPs, except in the two special non-finite cases above. Case (1.) corresponds to our earlier two-dimensional case where the SP can lie on a semi-circular arc, and case (2.) admits of the SP being at an arbitrary distance from the PP, which can be regarded as its lying on an arc of infinite radius.

But always, as in two dimensions, the placing of the **station point** uniquely determines the orientation of the axes and vice versa.

An important aspect of perspectives that is often overlooked is that the *viewing position* must match the **station point**. For example, a picture taken with a 105mm lens on a 35mm film printed up to A3 size, should be viewed from about four feet away. The ratios of distance to picture plane width should remain the same. We commonly fail to do this and this accounts for much of the distortion experienced when viewing wide-angle or long-focus images and in turn explains much of the belief that perspective projections are not always "true". They are, but they must be viewed from the right position.

9.3 Homogeneous Coordinates

Jean Victor Poncelet was a French mathematician and engineer who was sent on Napoleon's expedition to Moscow. In the retreat of 1812 he was taken prisoner. To pass through the rigours of both cold and imprisonment, he dabbled in the theory of **projection** that we've just been reviewing.

It occurred to Poncelet that the awkward case of the **vanishing points** lying "at infinity" for any line parallel to the **picture plane** didn't apply to *the lines themselves*. The *line* through the **eye point** remained perfectly well-defined whether it was parallel to the **picture plane** or not. From this almost absurdly simple idea, the entire subject of **projective geometry** evolved.

In **projective geometry** the *lines* through the **origin** (with which we can now formally replace the **eye point**) are called **projective points**, which I will also refer to as **P-points,**[142] but because they strictly constitute a two-dimensional **manifold** I may also refer to them as **2-P-points**. Any point in 3-space uniquely defines such a **P-point** but not the other way round: there is a multiplicity of points in 3-space corresponding to the same **P-point**.

For if we take the 3-point $P_0 = (x_0, y_0, z_0)$, then it defines a line through the **origin** by:

$$(x, y, z) = t.(x_0, y_0, z_0), \qquad \text{or} \qquad P = t.P_0$$

and so this *line* is a **P-point**. But any other 3-point on the same line, say $P_1 = t_1.P_0$, also defines the same line or **P-point** by:

$$P = s.P_1 = st_1.P_0.$$

We call $[x_0, y_0, z_0]$ the **homogeneous coordinates** of the **P-point**, which I'll write as [P]. The totality of all **P-points** is called the **projective plane**. We can say that:

$$[x_1, y_1, z_1] \equiv [x_0, y_0, z_0]$$

or that

$[x_1, y_1, z_1]$ and $[x_0, y_0, z_0]$ represent the same **P-point**

[142] This is my version of the convention used in Brannan, Esplen and Gray or writing these as Points with a capital P. They also writes Lines with a capital L which I write as P-lines.

if and only if there is some number λ such that $x_1 = \lambda x_0$, $y_1 = \lambda y_0$ and $z_1 = \lambda z_0$. We can of course also write this as $(x_1, y_1, z_1) = \lambda.(x_0, y_0, z_0)$ or as $P_1 = \lambda P_0$ for the 3-points that the same triplets represent. But note the difference:

$$(x_1, y_1, z_1) \neq (x_0, y_0, z_0) \text{ and } P_1 \neq P_0$$

but $[x_1, y_1, z_1] = [x_0, y_0, z_0]$ and $[P_1] = [P_0]$ if $P_1 = \lambda P_0$ for some λ.

Now we can **normalize** the **homogeneous coordinates** of any **P-point** by taking the unique representative of $[P_0] \equiv [x_0, y_0, z_0]$ which has **unit magnitude** or:

$$x_0^2 + y_0^2 + z_0^2 = 1.$$

These special 3-points, which are not the same as the lines or **P-points** but are in **one-to-one correspondence** with them, are clearly the direct analogue of **angles** as I defined them in Chapter 4. Instead of representing (or indeed defining) **angles** as points on the **unit circle**, we are now identifying points on the **unit sphere** as corresponding directly to **P-points:**

- To every point on the **unit sphere** there corresponds a unique **P-point**.
- But I should emphasize that it's conventional to say that to every **P-point** in the **projective plane** there corresponds a unique *pair* of **antipodal** points on the **unit sphere**, because the same line of **P-point** passes through both sides.

This distinction of **P-points** matching *pairs* of **antipodal** points will not be important in what follows, but it is the way that mathematicians prefer to see the **projective plane**.

I will call these points on the **unit sphere** simply **directions** as I did informally in Section 9.1, or more specifically **3-directions** to distinguish them from "**2-directions**" or **angles**. When I do not need to distinguish between **P-points** and **directions**, I'll commonly use the term **direction** and may well use **homogeneous coordinates** that are *not* **normalized**, precisely as I did in Section 9.1.

You can now see that Poncelet's device is somewhat analogous to my device of treating **angles** as points on the **unit circle**, and again we have one **coordinate** more than it appears we need, but similar advantages will accrue from this.

But, and it is a big BUT, the subsequent development of the subject is quite different. A possible reason, although this is only my guess, is that there is no **geometric algebra** that is **closed** in three dimensions, where **closure**, as in Section 1.5, means that the product of multiplying two entities is another of the same kind. So in two dimensions,

$$(x + y\mathbf{e}_1\mathbf{e}_2) \times (u + v\mathbf{e}_1\mathbf{e}_2) = (xu - yv) + (xv + yu)\mathbf{e}_1\mathbf{e}_2$$

and in four,

$$(s + x\,\mathbf{e}_3\mathbf{e}_2 + y\,\mathbf{e}_1\mathbf{e}_3 + z\,\mathbf{e}_2\mathbf{e}_1) \times (t + u\,\mathbf{e}_3\mathbf{e}_2 + v\,\mathbf{e}_1\mathbf{e}_3 + w\,\mathbf{e}_2\mathbf{e}_1)$$

$$= (st - xu - yv - zw) + (su + xt + yw - zv)\mathbf{e}_{32}$$

$$+ (sv + yt + zu - xw)\mathbf{e}_{13} + (sw + zt + xv - yu)\mathbf{e}_{21},$$

writing $\mathbf{e}_3\mathbf{e}_2$ as $\mathbf{e}_{32}$ as in Section 5.9. The first is the algebra of **complex numbers**, the second that of **quaternions**. There is an algebra of **octonions** in eight dimensions, but that's all. In three dimensions, we always end up with terms of **grades** that didn't appear in the arguments of the product.[143]

The immediate significance of **homogeneous coordinates** is that if the underlying 3-space is the **camera space** of Section 9.2, one particular representation of any **P-point** gives the **image** of the original 3-point on the **picture plane**. So if we take an arbitrary point (x, y, z), then:

$$[x, y, z] \equiv (1/z)[x, y, z] = [x/z, y/z, 1].$$

$(x/z, y/z, 1)$ are the **coordinates** of the **image** of (x, y, z) on the **picture plane** at $z = 1$. So,

> Points in space with equivalent **homogeneous coordinates**, and so belonging to the same **P-point**, have the same **image** in the **picture plane**.

So the preferred **homogeneous coordinates** of any **P-point** are *not* x, y, z on the **unit sphere** such that $x^2 + y^2 + z^2 = 1$, but the **coordinates** $x, y, 1$ where the third **coordinate** is unity.

This emphasis leads to a wholly different development of the subject from the style seen in Chapter 4.

There is one special case. If the line corresponding to the **P-point** runs *parallel* to the plane $z = 1$, then its z **coordinate** must be identically zero: whatever point on the line we choose to give us our **homogeneous coordinates** we cannot have $z = 1$. *All* **homogeneous coordinates** for such a **P-point** are of the form $[x, y, 0]$.

Such **P-points** are called **ideal P-points** or just **ideal points**, and they correspond to the "points at infinity" that we were trying to get away from in the first place. So the "singularity" of such points still appears in the new theory as a special case.

Most introductions to **homogeneous coordinates**, particularly in the field of computer graphics, draw attention to the fact that these **coordinates** enable us to incorporate **translation** into **matrix multiplication** as well as **scaling** and **rotation**. We've already looked at the nastiest of these three, **rotations**, in Section 5.10. A **rotation** in two dimensions through an angle $\boldsymbol{\theta} = (\theta_x, \theta_y)$ is given the result of multiplying (x, y), written as a *column* **vector** with the x above the y, by the **matrix**:

$$\begin{pmatrix} \theta_x & -\theta_y \\ \theta_y & \theta_x \end{pmatrix}$$

[143] A statement often made but rarely proved. R.J.Brown, *Elements of Modern Topology*, McGraw-Hill 1968 p.133 gives two references.

which we can derive as $(\theta_x + j\theta_y)(x + jy) = (\theta_x x - \theta_y y) + j(\theta_x y + \theta_y x)$, giving $x\dagger = \theta_x x - \theta_y y$ and $y\dagger = \theta_x y + \theta_y x$ as the new **coordinates**, with the minus sign swapping down into the bottom-left corner of the **matrix** for **rotation** the other way where we'd use $\theta_x - j\theta_y$.

A **scaling** is even easier: **scaling** all x values by λ_x and all y values by λ_y can be done by multiplying (x, y), again seen as a *column* **vector**, by the *diagonal* **matrix**:

$$\begin{pmatrix} \lambda_x & 0 \\ 0 & \lambda_y \end{pmatrix}$$

to give $x\dagger = \lambda_x x$ and $y\dagger = \lambda_y y$. These two **matrices** can be multiplied together to give a combined **scaling** and **rotation matrix**.

But the much more trivial case of **translation** cannot be done this way. Here we have:

$$x\dagger = x + \mu_x \qquad y\dagger = y + \mu_y$$

and this cannot be done by a **matrix multiplication**. But if we take our (x, y) point to be the first two **homogeneous coordinates** of a 3-point $(x, y, 1)$, we *can*. For the 3×3 **matrix** given by:

1	0	μ_x
0	1	μ_y
0	0	1

acting on the column **vector**:

x
y
1

does give $x\dagger = 1.x + \mu_x$, and $y\dagger = 1.y + \mu_y$ with the third ("z") **coordinate** $z\dagger = 1$. Raising our original **2-matrices** to the enlarged forms:

θ_z	$-\theta_y$	0
θ_y	θ_x	0
0	0	1

and

λ_x	0	0
0	λ_y	0
0	0	1

we have three 3×3 **matrices** which we can multiply them all together to give **scaling**, **rotation** *and* **translation** all by a single **matrix multiplication**.

The logic extends smoothly to the original points being in 3-space, when we represent them as being **homogeneous coordinates** [*x*, *y*, *z*, 1] of a **P-point** now deriving from an imaginary original *four-dimensional* space.

The **translation matrix** now has a final column of (μ_x, μ_y, μ_z, 1) in *column* form, where μ_z is the **translation** in *z*, and this **matrix** is otherwise all zeroes except for 1's along the diagonal. The **scaling matrix** is all zeroes except for the values (λ_x, λ_y, λ_z, 1) along the *diagonal* where λ_z is the scaling factor for *z*, and the **rotation matrix** is simply the 3×3 original defining the **rotation** (a rather complicated form which I won't give here) with a fourth row and fourth column added taking just the values (0, 0, 0, 1).

The final complete transformation **matrix** obtained by multiplying all these together will still have the values (μ_x, μ_y, μ_z, 1) in the fourth *column* and (0, 0, 0, 1) in the fourth *row* and so looks like this:

(9.3.1)

a_{11}	a_{12}	a_{13}	μ_x
a_{21}	a_{22}	a_{23}	μ_y
a_{31}	a_{32}	a_{33}	μ_z
0	0	0	1

Because computer graphics aficionados are crazy about **matrices**, these **homogeneous coordinate matrices** are used everywhere. The fact that it would obviously be much easier to do the **translation** by the simple addition by which it is defined is simply ignored.

Consistently with my original terminology at the start of this Section, I will also call **P-points** with four **coordinate** values [*x*, *y*, *z*, *w*] as **3-P-points** and these will typically be represented as [*x*, *y*, *z*, 1] with the fourth **coordinate** unity, just as **2-P-points** are typically written as [*x*, *y*, 1] with the *third* unity.

Another interpretation comes in here. Suppose we take the difference (in the usual **vector** way) between two **3-P-points**:

$[x, y, z, 1] - \{p, q, r, 1] = [x - p, y - q, z - r, 0]$.

We've ended up with an **ideal point** having the fourth **coordinate** zero.

Again, if we apply our **matrix transformations** to an **ideal point**, there will be no effect from the **translation.** We can see this easily in the two-dimensional case:

$$\begin{bmatrix} 1 & 0 & \mu_x \\ 0 & 1 & \mu_y \\ 0 & 0 & 1 \end{bmatrix} \times \begin{bmatrix} x \\ y \\ 0 \end{bmatrix}$$

will give $x^\dagger = 1.x + 0.\mu_x = x$, and $x^\dagger = 1.y + 0.\mu_y = y$.

In computer graphics it is customary to regard "**real**" **points** with a *w* **coordinate** of 1, and so having *real* **images** in the **picture plane** at $z = 1$, as representing actual "points" (3-points) of space, and **ideal points**, with a *w* **coordinate** necessarily zero, as *directions* pure and simple. So "*directions*" in this sense are always the result of differences between "points" and are unaffected by **translations**. You will commonly see this convention:

- **Real points** with a *w* **coordinate** of 1 are called **Point3d** objects,
- **Ideal points** with a *w* **coordinate** of zero are called **Vector3d** objects.

The terminology I've used here is that in the Java3D system. The "**d**" on the end each time refers to "double-precision" and is nothing to do with the geometrical dimension. That only appears in the "**3**". This is the formal implementation of the idea I mentioned in Section 5.3.

9.4 *z*-Interpolation and Cross Ratio

I'll step aside from the main line of development to give a couple of illustrations of the fact that the ratio of distances is not preserved under perspective, something in a sense obvious from the very fact that as the **image** of a line in space approaches its **vanishing point**, equal distances along the original line map into **image points** ever closer and closer.

z-**interpolation** works out how changes in the **image** position in the **picture plane** of a straight line in space map into changes in the *z* distance of the original points. It's important in computer graphics for making sure that such things as texture mappings "shrink" correctly as the **vanishing point** of the line is approached.

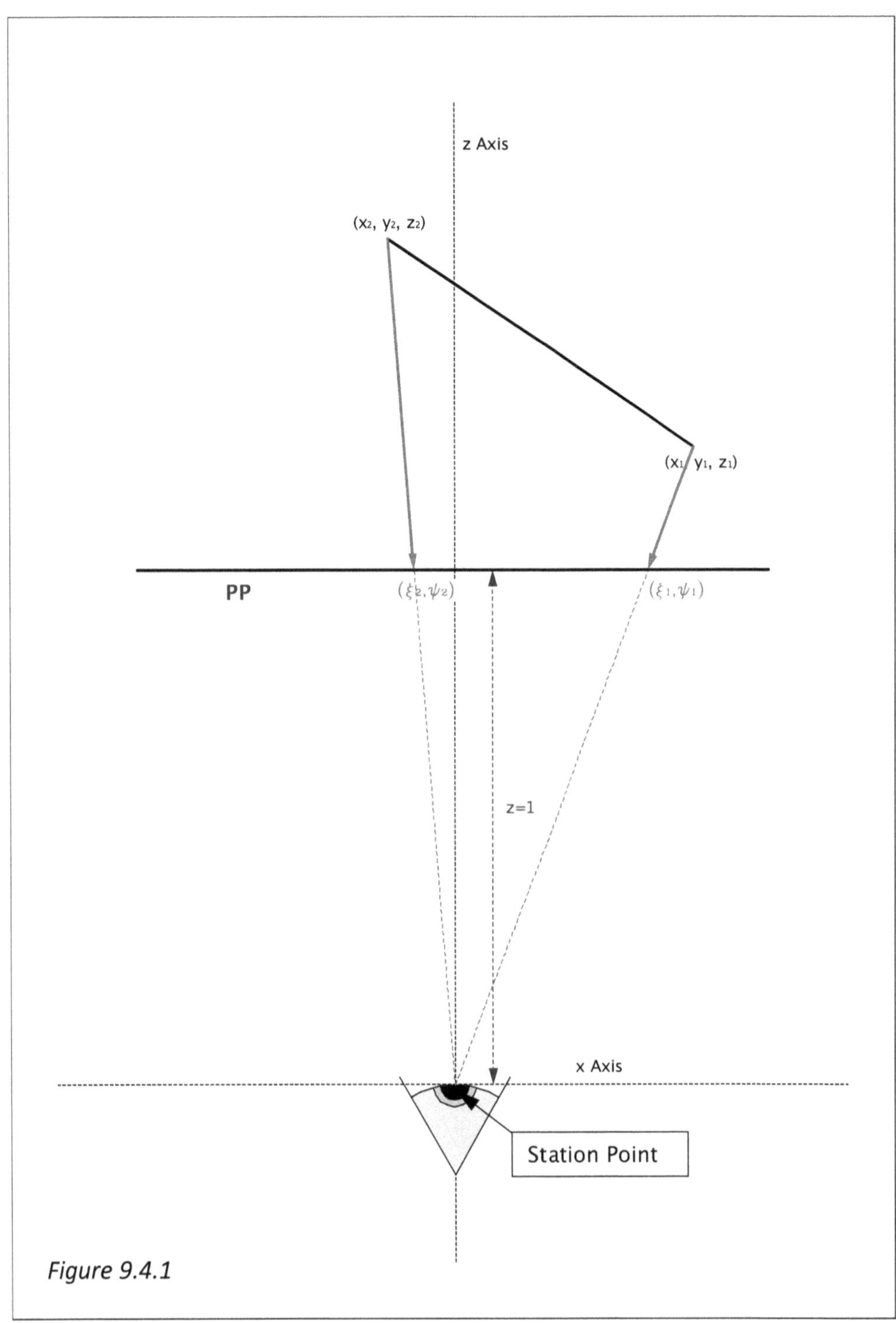

Figure 9.4.1

As shown in *Figure 9.4.1*, we work in just *xz*-space.[144] In this we can assume the equation of the original line in space is $ax + bz = c$. So the **image** of any point on the line appears at $\xi = x/z$, which is, dividing the equation through by az, $x/z = c/(az) - b/a$. It's convenient to transform this:

[144] I take this analysis largely from Eric Lengyel, *Mathematics for 3D Game Programming and Computer Graphics*, 2nd Ed. Charles River Media 2004.

$$\xi = c/(az) - b/a \longrightarrow \qquad 1/z = (a\xi + b)/c.$$

Now map the **image** of the line between ξ_1 and ξ_2 as $\xi = \xi_1(1 - t) + \xi_2.t$. Substitute this ξ into the equation in $1/z$ to give:

$$1/z = (a[\xi_1(1 - t) + \xi_2.t] + b)/c = (a\xi_1(1 - t))/c + (a\xi_2.t)/c + b/c.$$

But $b/c = (b/c)(1 - t) + (b/c)t$, so this becomes:

$$1/z = [(a\xi_1 + b)/c](1 - t) + [(a\xi_2.t + b)/c]t$$

but this equals

$$1/z = (1/z_1)(1 - t) + (1/z_2).t.$$

So it's the *reciprocals* of z that vary as the distance across the **image plane** or **picture plane**. Of course as $z \longrightarrow \infty$, $\xi = c/(az) - b/a \longrightarrow -b/a$, the x position of the **vanishing point** of the line.

Cross Ratio is much more important. Again we work just in xz-space, but rather than assuming a **picture plane** at $z = 1$, we'll imagine *two* arbitrary lines on the xz-plane, which correspond to arbitrarily positioned **picture planes** PP_1 and PP_2 if we imagine them to extend infinitely in the y direction. Again assume the **station point** or **eye point** is the **origin** with $x = 0$ and $z = 0$. This configuration is shown in *Figure 9.4.2*, which shows a line ABCD and two **picture planes** PP_1 and PP_2. This is a slightly more general configuration than we will need.

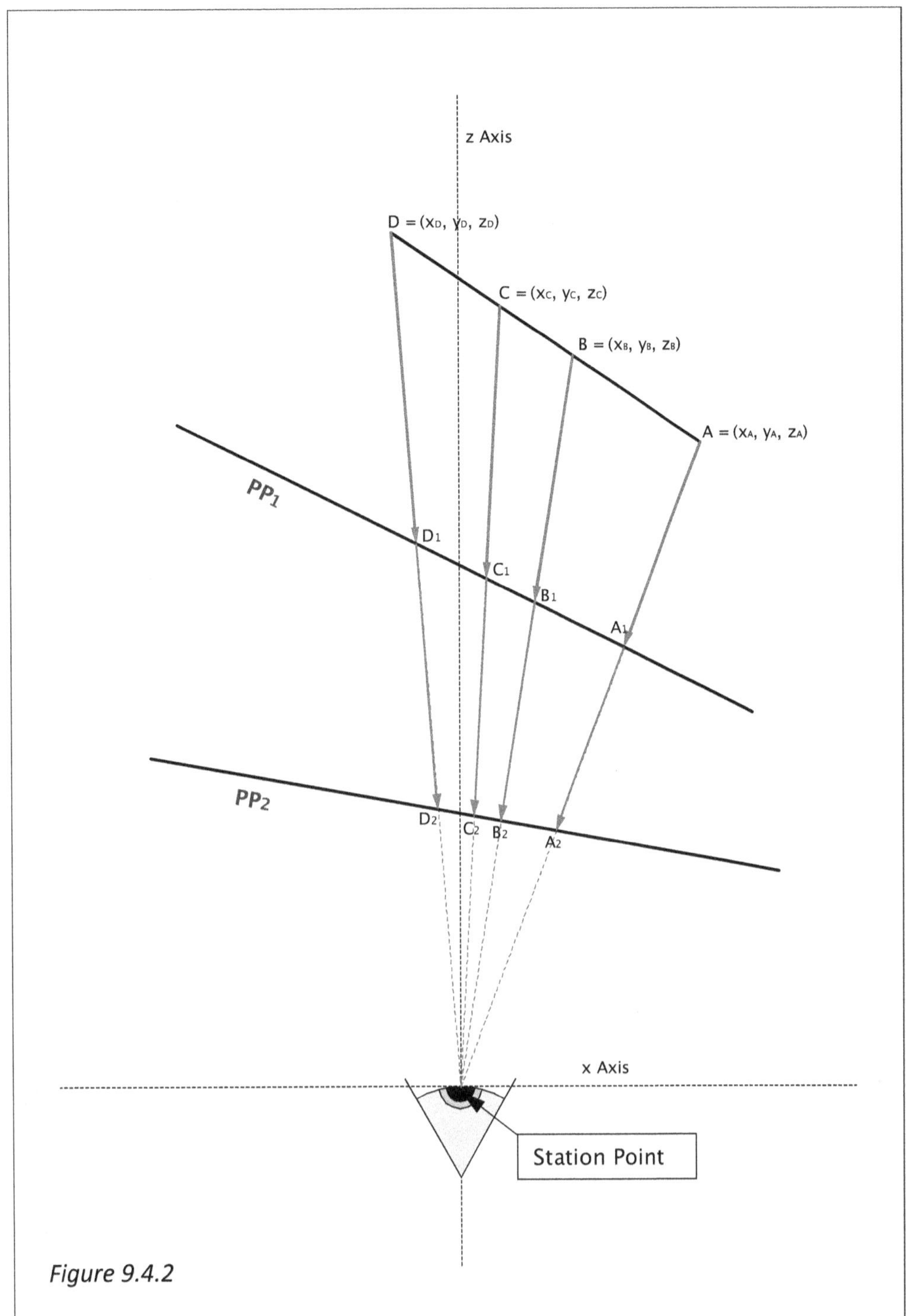

Figure 9.4.2

Let PP_1 obey the equation $a_1x + b_1z = c_1$, and PP_2 the equation $a_2x + b_2z = c_2$.

Take four points along the line in continuing order of *z*, and denoted A, B, C and D. Let the **P-point** lines radiating from O, the **origin**, and passing through A, B, C and D intersect PP_1 in A_1, B_1, C_1 and D_1 and PP_2 in A_2, B_2, C_2 and D_2 respectively. Then writing the distance between A and B in the line as AB and so with the distances within their respective planes of all other points, then:

$$(AB \times CD)/(BC \times DA) = (A_1B_1 \times C_1D_1)/(B_1C_1 \times D_1A_1)$$
$$= (A_2B_2 \times C_2D_2)/(B_2C_2 \times D_2A_2),$$

and so for the distances within *any* "**picture planes**" stretched across the path of the **projector lines** from the **station point** or **origin** O to A, to B, to C and to D. The ratio appearing for each of the three planes above, which we may write in general in the form $(A^*B^* \times C^*D^*)/(B^*C^* \times D^*A^*)$, is *always the same*.

Again:

$$(AB \times CD)/(BC \times DA) = (A^*B^* \times C^*D^*)/(B^*C^* \times D^*A^*).$$

The ratios appearing on either side of this equation are the **cross ratios** between the four points in their respective planes. The equation above asserts that the **cross ratios** for points with the same **homogeneous coordinates** (and so along the same lines from the **origin**) are invariant in different **image planes**.

Now $AB = \sqrt{((x_B - x_A)^2 + (z_B - z_A)^2)}$, so the LHS must be:

$$\frac{\sqrt{((x_B - x_A)^2 + (z_B - z_A)^2)} \times \sqrt{((x_D - x_C)^2 + (z_D - z_C)^2)}}{\sqrt{((x_C - x_B)^2 + (z_C - z_B)^2)} \times \sqrt{((x_D - x_A)^2 + (z_D - z_A)^2)}}.$$

Since $\sqrt{(xy)} = \sqrt{x}.\sqrt{y}$ we can take the roots out and we will have that the **cross ratios** are constant if the expressions *without the roots* are. Putting $\xi_A = x_A/z_A$, we need to show that the value of the **cross ratio** depends only on ξ_A, ξ_B, ξ_C and ξ_D.

It isn't easy from first principles, but let's try:

on PP_1, $a_1x + b_1z = c_1$, so if $\xi = x/z$, then $z = x/\xi \longrightarrow a_1x + b_1x/\xi = c_1$

and this gives $x = c_1/(a_1 + b_1/\xi)$. Substituting for $x = z\xi$, we get $a_1z\xi + b_1z = c_1$, so

$$z = c_1/(a_1\xi + b_1).$$

Now for the expression for the distance AB^2, we have:

$$(x_B - x_A)^2 + (z_B - z_A)^2 = ((c_1/(a_1 + b_1/\xi_B)) - (c_1/(a_1 + b_1/\xi_A)))^2$$
$$+ ((c_1/(a_1\xi_B + b_1)) - (c_1/(a_1\xi_A + b_1)))^2.$$

Take $(c_1/(a_1 + b_1/\xi_B)) - (c_1/(a_1 + b_1/\xi_A))$ first. This gives:

$$\frac{(c_1(a_1 + b_1/\xi_A)) - c_1(a_1 + b_1/\xi_B))}{((a_1 + b_1/\xi_B)(a_1 + b_1/\xi_A))} = \frac{(c_1b_1(1/\xi_A - 1/\xi_B))}{((a_1 + b_1/\xi_B)(a_1 + b_1/\xi_A))}.$$

Now multiply top and bottom by $\xi_A\xi_B$ and we get:

$$\frac{(c_1b_1(\xi_B - \xi_A))}{((a_1\xi_A\xi_B + b_1\xi_A)\,(a_1 + b_1/\xi_A))} = \frac{(\,c_1b_1(\xi_B - \xi_A))}{((a_1\xi_B + b_1)(a_1\xi_A + b_1))}\,.$$

This term is still all squared.

Much more easily, the $(z_B - z_A)$ part gives:

$$(c_1/(a_1\xi_B + b_1)) - (c_1/(a_1\xi_A + b_1)) = \frac{(c_1(a_1\xi_A + b_1) - c_1(a_1\xi_B + b_1))}{((a_1\xi_B + b_1)(a_1\xi_A + b_1))}$$

$$= \frac{(c_1a_1(\xi_A - \xi_B))}{((a_1\xi_B + b_1)(a_1\xi_A + b_1))}\,.$$

Again this term is squared, but now *both terms have the same denominator*, so we get all told:

$$\frac{(c_1^2a_1^2(\xi_A - \xi_B)^2 + c_1^2b_1^2(\xi_B - \xi_A)^2)}{((a_1\xi_B + b_1)(a_1\xi_A + b_1))^2} = \frac{(\,c_1^2(a_1^2 + b_1^2)(\xi_A - \xi_B)^2)}{((a_1\xi_B + b_1)(a_1\xi_A + b_1))^2}.$$

This is AB^2. So CD^2 will take just the same form with $\xi_A \rightarrow \xi_C$ and $\xi_B \rightarrow \xi_D$. Taking the roots ($\surd$) we'll get:

$$AB \times CD = \frac{(c_1\surd(\,a_1^2 + b_1^2)(\xi_A - \xi_B) \times c_1\surd(\,a_1^2 + b_1^2)(\xi_C - \xi_D))}{(a_1\xi_B + b_1)(a_1\xi_A + b_1)(a_1\xi_D + b_1)(a_1\xi_C + b_1)}.$$

Similarly

$$BC \times AD = \frac{(c_1\surd(\,a_1^2 + b_1^2)(\xi_B - \xi_C) \times c_1\surd(\,a_1^2 + b_1^2)(\xi_A - \xi_D))}{(a_1\xi_C + b_1)(a_1\xi_B + b_1)(a_1\xi_D + b_1)(a_1\xi_A + b_1)}\,.$$

But the denominators are identical, so:

$$\frac{(AB \times CD)}{(BC \times AD)} = \frac{(c_1^2(a_1^2 + b_1^2)(\xi_A - \xi_B)(\xi_C - \xi_D))}{(c_1^2(a_1^2 + b_1^2)(\xi_B - \xi_C)(\xi_A - \xi_D))} = \frac{(\xi_A - \xi_B)(\xi_C - \xi_D)}{(\xi_B - \xi_C)(\xi_A - \xi_D)}.$$

And *yes*, this *does* depend only on ξ_A, ξ_B, ξ_C and ξ_D.

The signs I've chosen here are intended to give the positive square roots for the configuration shown in *Figure 9.4.2* where $\xi_A > \xi_B > \xi_C > \xi_D$ and so to give a *positive* **cross ratio**. A more common definition corresponds to:

$$\frac{(\mathrm{AB} \times \mathrm{CD})}{(\mathrm{BC} \times \mathrm{DA})} = \frac{(c_1^{\;2}(a_1^{\;2} + b_1^{\;2})(\xi_A - \xi_B)(\xi_C - \xi_D))}{(c_1^{\;2}(a_1^{\;2} + b_1^{\;2})(\xi_B - \xi_C)(\xi_D - \xi_A))} = \frac{(\xi_A - \xi_B)(\xi_C - \xi_D)}{(\xi_B - \xi_C)(\xi_D - \xi_A)} .$$

In general, **cross ratios** are signed rather than just positive, but there is in fact quite a lot of flexibility in how we define them. We could have chosen the product:

((AC × BD)/(BC × DA)).

We'd now get the form for the numerator:

$$\mathrm{AC} \times \mathrm{BD} \;= \frac{(c_1\sqrt{(\,a_1^{\;2} + b_1^{\;2})}(\xi_A - \xi_C) \times c_1\sqrt{(\,a_1^{\;2} + b_1^{\;2})}(\xi_B - \xi_D))}{(a_1\xi_B + b_1)(a_1\xi_A + b_1)(a_1\xi_D + b_1)(a_1\xi_C + b_1)} ,$$

and this would reduce to:

$$\frac{(\mathrm{AC} \times \mathrm{BD})}{(\mathrm{BC} \times \mathrm{DA})} = \frac{(c_1^{\;2}(a_1^{\;2} + b_1^{\;2})(\xi_A - \xi_C)(\xi_B - \xi_D))}{(c_1^{\;2}(a_1^{\;2} + b_1^{\;2})(\xi_B - \xi_C)(\xi_D - \xi_A))} = \frac{(\xi_A - \xi_C)(\xi_B - \xi_D)}{(\xi_B - \xi_C)(\xi_D - \xi_A)} .$$

This is the form used in the excellent standard text by Brannan, Esplen and Gray.[145] This may not be immediately obvious from their definition. To obtain the form I'm using here in arbitrary 3-space (so not restricted to the *xz*-plane) you need to express C and D in terms of A and B. Now if these are all **collinear** then both C and D must fit the equation:

$\mathrm{X} = (1 - t_\mathrm{X})\mathrm{A} + t_\mathrm{X}\mathrm{B}$ for some t_X.

So we will have:

$\mathrm{C} = (1 - t_\mathrm{C})\mathrm{A} + t_\mathrm{C}\mathrm{B}$ for some t_C,
$\mathrm{D} = (1 - t_\mathrm{D})\mathrm{A} + t_\mathrm{D}\mathrm{B}$ for some t_D.

We also have:

$\mathrm{A} = (1 - t_\mathrm{A})\mathrm{A} + t_\mathrm{A}\mathrm{B}$ so $t_\mathrm{A} = 0$,
$\mathrm{B} = (1 - t_\mathrm{B})\mathrm{A} + t_\mathrm{B}\mathrm{B}$ so $t_\mathrm{B} = 1$.

Then the relevant **cross ratio** can be expressed in terms of the **parameter** *t* as:

$$\frac{(\mathrm{AC} \times \mathrm{BD})}{(\mathrm{BC} \times \mathrm{DA})} = \frac{(t_\mathrm{A} - t_\mathrm{C})(t_\mathrm{B} - t_\mathrm{D})}{(t_\mathrm{B} - t_\mathrm{C})(t_\mathrm{D} - t_\mathrm{A})} = \frac{-t_\mathrm{C}(1 - t_\mathrm{D})}{(1 - t_\mathrm{C})t_\mathrm{D}} .$$

[145] D.A.Brannan, M.F.Esplen, J.J.Gray, *Geometry*, CUP 1999.

Brannan, Esplen and Gray cleverly lay a trap for the unwary student into which I fell headlong. In their leading example, they choose four points A, B, C and D which are *not* in fact **collinear** in *space*, but *are* in the sense of projective geometry. And that is, that just as **P-points** correspond to *lines* in space, so we can think of **P-lines** which correspond to *planes* in space passing through the **origin**. **P-lines** will cut the **picture plane** in a *line*, hence *projectively* they can be thought to correspond to "lines". Furthermore, there is now *one unique* plane (or **P-line**) through the **origin** parallel to the **picture plane** which does *not* have an **image** in the **picture plane** and this is called the **ideal (P-)line**.

Two **P-points** are **collinear** (in the projective sense) if they lie in the same **P-line**. So their defining 3-points must lie in the same *plane*.

- In this case, the t equation won't necessarily hold for C and D, so we have two alternatives. Either we find another C or D belonging to the same **P-point** [C] or [D] and so with $C* = \lambda C$ or $D* = \mu D$, and solve for:

$$\lambda C = (1 - t_C)A + t_C B \qquad \text{for some } t_C,$$
$$\mu D = (1 - t_D)A + t_D B \qquad \text{for some } t_D.$$

- Or we use the more general substitution:

$$C = \alpha A + \beta B \qquad \text{for some } \alpha, \beta,$$
$$D = \gamma A + \delta B \qquad \text{for some } \gamma, \delta.$$

Brannan, Esplen and Gray use the second method, and indeed define the **cross ratio** as:

$$CR = \beta\gamma/\alpha\delta.$$

This is more consistent with the spirit of projective geometry, but I think I'll always prefer the intuitive:

$$CR = \frac{(AC \times BD)}{(BC \times DA)} = \frac{(t_A - t_C)(t_B - t_D)}{(t_B - t_C)(t_D - t_A)} = \frac{-t_C(1 - t_D)}{(1 - t_C)t_D}.$$

Cross ratios have applications. *Figure 9.4.3* shows the most common.

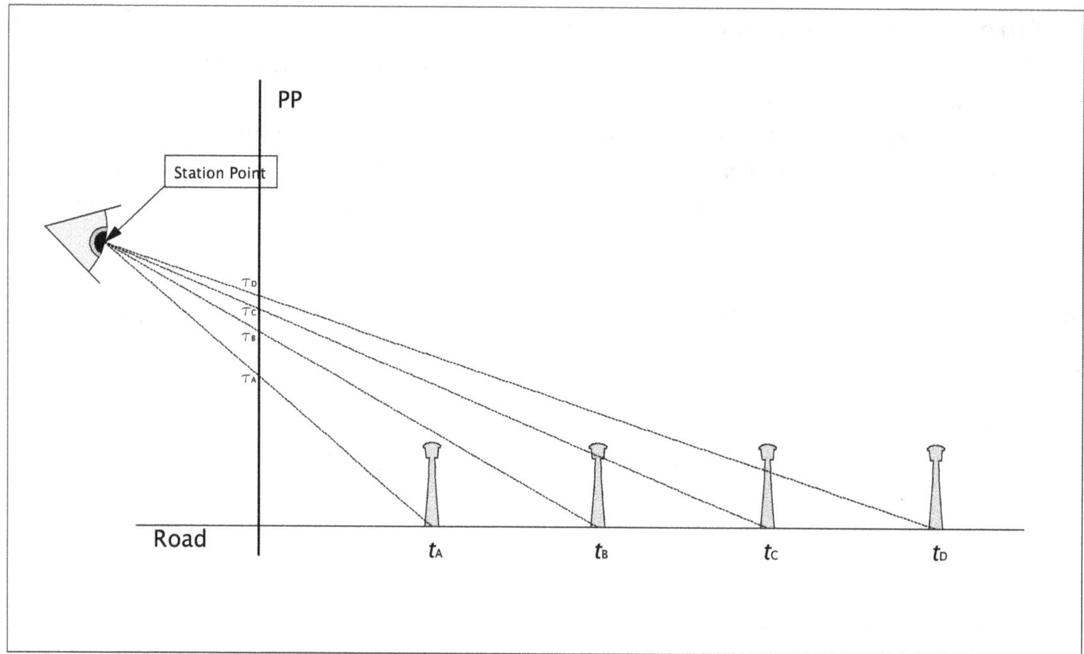

Figure 9.4.3

We have a **picture plane** PP which we can assume to be vertical, and a landscape running horizontally away from us. We assume that now the **eye point** is above ground level,[146] and so we have a **perspective** view of the landscape appearing on the **picture plane**. If there are four posts distributed in a straight line along a roadway as shown in the *Figure*, at t_A, t_B, t_C and t_D, their **cross ratio** on the ground (calculated from the *t*'s) must equal the **cross ratio** between their **images** in the **picture plane** by the basic result in this Section.

So if all four are equally spaced with $t_A = 0$, $t_B = 1$, $t_C = 2$ and $t_D = 3$, the CR above will be:

$$-2(1 - 3)/((1 - 2).3) = +4/-3 = -1⅓.$$

On the other definition, which henceforth I'll call CR*, where

$$CR^* = (AB \times CD)/(BC \times DA) = ((t_A - t_B)(t_C - t_D))/((t_B - t_C)(t_D - t_A))$$

$$= -1(2 - 3)/((1 - 2)\times 3) = -⅓.$$

This might appear in the **picture plane** as four **images** at $\tau_A = 0$, $\tau_B = 6$, $\tau_C = 8$ and $\tau_D = 9$,

$$((\tau_A - \tau_B)(\tau_C - \tau_D))/((\tau_B - \tau_C)(\tau_D - \tau_A))$$

[146] Historically, this was the distinction between the **eye point** and the **station point**. The **station point** meant the point on the *ground below* the observer's eye. Today, the two terms are used interchangeably, with **station point** being actually the more common, albeit historically the less correct.

$$= -6(8 - 9)/((6 - 8)\times 9) = +6/(-18) = -\tfrac{1}{3} \text{ again.}$$ [147]

Using CR here, we'd get

$$CR = (AC \times BD)/(BC \times DA) = ((\tau_A - \tau_C)(\tau_B - \tau_D))/((\tau_B - \tau_C)(\tau_D - \tau_A))$$

$$= -8(6 - 9)/((6 - 8)\times 9) = +24/(-18) = -1\tfrac{1}{3}.$$

So yet again the **cross ratio** is preserved whichever definition we use.

This Section has made use of the concept of *alternative* **picture planes** lying in different positions in space but always not passing through the **origin**. A mapping such as we saw in *Figure 9.4.2* from one such **picture plane** PP_1 to another PP_2 by rays emanating from the **origin** is sometimes called a **perspectivity**.[148]

9.5 Joins, Meets and Duality

We have seen in the previous Section that **P-points** that lie in the same **P-line** are defined to be **collinear** in the **projective plane**.

Now two **P-points** are two lines in space passing through the **origin**. A **P-line** is a *plane* in space passing through the **origin**. If the lines defining the **P-points** belong to *distinct* **P-points**, then they won't be collinear in space, and so they must define a plane in space, and since both lines pass through the **origin**, the plane they define must also pass through the **origin** and so must define a **P-line**.

The **P-line** defined by two distinct **P-points** P and Q is called the **join** of the two **P-points** and is written as $P \wedge Q$.

We have a remarkable symmetry here:

- **Collinearity Property**: any two *distinct* **P-points** P and Q, with $P \neq Q$, define a unique **P-line** called the **join** of P and Q,
- **Incidence Property**: any two *distinct* **P-lines** *p* and *q* intersect in a unique **P-point** called the **meet** of *p* and *q*.

This symmetry is of profound importance in **projective geometry**, where it is referred to as **duality**. It gives rise to the most attractive feature of **projective geometry**, its **duality principle**:

> **The Duality Principle of Projective Geometry**: To every theorem in terms of **P-points** and **meets**, there is a corresponding **dual theorem** in terms of **P-lines** and **joins**, and vice versa.

---

[147] This example comes from the charming little book by W.W.Sawyer, *A Prelude to Mathematics*, Pelican 1955, 1963.

[148] Brannan, Esplen and Gray, *op. cit.*, p.98.

In passing, note that any two **P-points** are always **collinear**: there must be a **P-line**, their **join**, that contains them both. Only *three or more* **P-points** may not be **collinear** in the **projective plane.**[149]

It's clearly useful to refer to **P-points** simply as P and Q, a form I've reserved so far for points in the underlying 3-space that define the **P-points** themselves as OP or OQ. If we're to do that, we'll need some way of getting *back* to the 3-space entities, and here it becomes useful to use **vector** algebra.

Any line in space can be defined by an equation like:

$$\mathbf{x} = \mathbf{p}_0 + t.\mathbf{v}.$$

In **projective geometry**, if the line represents a **P-point**, $\mathbf{p}_0$ can always be chosen as the **origin** O = (0, 0, 0), so this equation simplifies to:

$$\mathbf{x} = t.\mathbf{v}.$$

The line or **P-point** is completely defined by **v** alone. Because any point $t\mathbf{v}$, whatever the value of t, lies on the same line, the **coordinates** (tv_x, tv_y, tv_z) define the same line for any t. This is the basic idea behind **homogeneous coordinates** (which are also sometimes called **projective coordinates**).

A *plane* in space is defined by an equation like:

$$(\mathbf{x} - \mathbf{p}_0) \cdot \mathbf{n} = 0$$

where $\mathbf{p}_0$ is again an arbitrary point in the plane, and **n** is the **normal vector** to the plane – the **vector** in space that is perpendicular to the plane.

But yet again, in **projective geometry**, if the plane represents a **P-line**, $\mathbf{p}_0$ can always be chosen to be the **origin O**, and since $\mathbf{O} \cdot \mathbf{n} = 0$ always, the equation for the **p-line** reduces to:

$$\mathbf{x} \cdot \mathbf{n} = 0.$$

So a **P-point** is uniquely defined by a **3-vector v**, but a **P-line** is also uniquely defined by a **vector n**, and since $\mathbf{x} \cdot \lambda\mathbf{n} = \lambda(\mathbf{x} \cdot \mathbf{n}) = \lambda.0 = 0$, again **n** and $\lambda\mathbf{n}$ both define the same plane through the **origin** or the same **P-line**.

So we could write **P-points** as $[\mathbf{v}] \equiv [t.\mathbf{v}]$ for arbitrary t, but we could also write **P-lines** as $[\mathbf{n}] \equiv [\lambda\mathbf{n}]$ for arbitrary λ![150]

---

[149] **Collinearity** of **P-points** in the **projective plane** does not correspond to collinearity in 3-space of their defining lines, because collinearity in 3-space of the lines through P and Q is *identity* of their corresponding **P-points**. To make the distinction clear, I'll use bold, as in **collinearity**, only for **P-points** in the **projective plane**.

[150] Note that **P-points** and **P-lines** are not signed because t or λ may be negative, so there's no **chirality** here.

This is obviously a prescription for endless confusion, but there's another interpretation that works for planes.

A plane through the **origin** is also uniquely defined by any two **linearly independent vectors u** and **v** that lie in the plane. It can be shown that the **normal vector** to the plane is given by:

$$*(\mathbf{u} \wedge \mathbf{v})$$

where $*$ is the **pseudoscalar dual** operation between **vectors** and **bivectors** defined in Section 5.9. This is easy to see for the planes defined by the **orthogonal coordinate axes** themselves, because for example the *xy*-plane is given by the **bivector**:

$$\mathbf{e}_x \wedge \mathbf{e}_y$$

and $*(\mathbf{e}_x \wedge \mathbf{e}_y) = \mathbf{e}_z$, which clearly *is* the **normal vector** to the *xy*-plane. So planes (through the **origin**) are unambiguously defined by **bivectors** and we can write:

$$p = [\mathbf{u} \wedge \mathbf{v}],$$

with just the single pair of square brackets, to indicate a **P-line** *unambiguously*.

As before $[\mathbf{u} \wedge \mathbf{v}] = [\lambda(\mathbf{u} \wedge \mathbf{v})]$ for any λ: both define the same **P-line**.

We can now redefine the [] notation for **P-lines** to give instead a **bivector B**:

$$p = [\lambda\mathbf{B}]$$

where $\mathbf{B} = *\mathbf{n}$ or $*\mathbf{B} = \mathbf{n}$, which is to say the **bivector** and the **normal vector** are **pseudoscalar dual**. Remember that **pseudoscalar duality** doesn't affect the components of the **vector** or **bivector**:

$$*(x\mathbf{e}_{yz} + y\mathbf{e}_{zx} + z\mathbf{e}_{xy}) = x\mathbf{e}_x + y\mathbf{e}_y + z\mathbf{e}_z,$$

and

$$*(x\mathbf{e}_x + y\mathbf{e}_y + z\mathbf{e}_z) = x\mathbf{e}_{yz} + y\mathbf{e}_{zx} + z\mathbf{e}_{xy}.$$

So **n** and **B** are readily identifiable.

If $\mathbf{u} = (x_u, y_u, z_u)$ is seen as a point in space P, and $\mathbf{v} = (x_v, y_v, z_v)$ is seen as another point Q, then:

$$p = [\mathbf{u} \wedge \mathbf{v}]$$

is the unique **P-line** containing both **P-points** P and Q, and so is clearly the **join** of P and Q. So *p*, which we can also write as $P \wedge Q$, is the **join** of P and Q, and so we can define:

$$P \wedge Q = \mathbf{join}(P, Q) := [\mathbf{u} \wedge \mathbf{v}].$$

With **P-points** represented by **vectors** and **P-lines** represented by **bivectors**, we can use **pseudoscalar duality** to establish a **duality** between **P-points** and **P-lines** and say for a **P-point** P = [**p**] and a **P-line** *p*:

$*\mathrm{P} = p$ if $p = [*\mathbf{p}]$

and

$*p = \mathrm{P}$,

or, looking at it the other way, for a **vector v**, if $[\mathbf{v}] = \mathrm{P}$, then $[*\mathbf{v}] = *\mathrm{P}$, a **P-line**.

So every **P-point** is **dual** to some **P-line** and every **P-line** is **dual** to some **P-point**. So this defines the **projective duality** in terms of the **pseudoscalar duality**.

Unfortunately, the conventions used in ordinary geometry are in direct conflict with those used in **geometric algebra**.
In geometry, points are written in upper-case as P, Q, etc., but lines are generally written in italicised lower-case as *p*, *q*, etc.
In **geometric algebra, vectors** (corresponding to **P-points**) are written in lower-case and **bivectors** (corresponding to **P-lines**) are written in upper-case.
Beware!

When working in the plane, either Euclidean or **projective**, it is usual to omit any **join** operator $\wedge$ between points or **P-points**, and just write PQ for the line or **P-line** joining P and Q.

Having established these preliminaries, things get interesting.

Assume $\mathrm{P} = p_x\mathbf{e}_x + p_y\mathbf{e}_y + p_z\mathbf{e}_z$ and $\mathrm{Q} = q_x\mathbf{e}_x + q_y\mathbf{e}_y + q_z\mathbf{e}_z$, both ordinary points in 3-space, then we have well-defined:

$$\mathrm{P} \wedge \mathrm{Q} = (p_xq_y - p_yq_x)\mathbf{e}_{xy} + (p_yq_z - p_zq_y)\mathbf{e}_{yz} + (p_zq_x - p_xq_z)\mathbf{e}_{zx}.$$

again using the shorthand notation $\mathbf{e}_{xy} \equiv \mathbf{e}_x\mathbf{e}_y$.

Now suppose we also have $\mathrm{R} = r_x\mathbf{e}_x + r_y\mathbf{e}_y + r_z\mathbf{e}_z$, then:

$$\mathrm{P} \wedge \mathrm{R} = (p_xr_y - p_yr_x)\mathbf{e}_{xy} + (p_yr_z - p_zr_y)\mathbf{e}_{yz} + (p_zr_x - p_xr_z)\mathbf{e}_{zx}.$$

Take the **pseudoscalar duals** of each of these to get:

$$*(\mathrm{P} \wedge \mathrm{Q}) = (p_yq_z - p_zq_y)\mathbf{e}_x + (p_zq_x - p_xq_z)\mathbf{e}_y + (p_xq_y - p_yq_x)\mathbf{e}_z$$
$$*(\mathrm{P} \wedge \mathrm{R}) = (p_yr_z - p_zr_y)\mathbf{e}_x + (p_zr_x - p_xr_z)\mathbf{e}_y + (p_xr_y - p_yr_x)\mathbf{e}_z.$$

Now form:

$$\begin{aligned} *(\mathrm{P} \wedge \mathrm{Q}) \wedge *(\mathrm{P} \wedge \mathrm{R}) = {} & ((p_yq_z - p_zq_y)(p_zr_x - p_xr_z) - (p_zq_x - p_xq_z)(p_yr_z - p_zr_y))\mathbf{e}_x\mathbf{e}_y + \\ & ((p_zq_x - p_xq_z)(p_xr_y - p_yr_x) - (p_xq_y - p_yq_x)(p_zr_x - p_xr_z))\mathbf{e}_y\mathbf{e}_z + \\ & ((p_xq_y - p_yq_x)(p_yr_z - p_zr_y) - (p_yq_z - p_zq_y)(p_xr_y - p_yr_x))\mathbf{e}_z\mathbf{e}_x \end{aligned}$$

$$= (p_yq_zp_zr_x - p_yq_zp_xr_z - p_zq_yp_zr_x + p_zq_yp_xr_z - p_zq_xp_yr_z + p_zq_xp_zr_y + p_xq_zp_yr_z - p_xq_zp_zr_y)\mathbf{e}_x\mathbf{e}_y +$$
$$(p_zq_xp_xr_y - p_zq_xp_yr_x - p_xq_zp_xr_y + p_xq_zp_yr_x - p_xq_yp_zr_x + p_xq_yp_xr_z + p_yq_xp_zr_x - p_yq_xp_xr_z)\mathbf{e}_y\mathbf{e}_z +$$
$$(p_xq_yp_yr_z - p_xq_yp_zr_y - p_yq_xp_yr_z + p_yq_xp_zr_y - p_yq_zp_xr_y + p_yq_zp_yr_x + p_zq_yp_xr_y - p_zq_yp_yr_x)\mathbf{e}_z\mathbf{e}_x$$

$$= (p_yp_zq_zr_x - \mathbf{p_xp_yq_zr_z} - p_z^2q_yr_x + p_xp_zq_yr_z - p_yp_zq_xr_z + p_z^2q_xr_y + \mathbf{p_xp_yq_zr_z} - p_xp_zq_zr_y)\mathbf{e}_x\mathbf{e}_y +$$
$$(p_xp_zq_xr_y - \mathbf{p_yp_zq_xr_x} - p_xp_xq_zr_y + p_xp_yq_zr_x - p_xp_zq_yr_x + p_xp_xq_yr_z + \mathbf{p_yp_zq_xr_x} - p_xp_yq_xr_z)\mathbf{e}_y\mathbf{e}_z +$$
$$(p_xp_yq_yr_z - \mathbf{p_xp_zq_yr_y} - p_yp_yq_xr_z + p_yp_zq_xr_y - p_yp_xq_zr_y + p_yp_yq_zr_x + \mathbf{p_xp_zq_yr_y} - p_yp_zq_yr_x)\mathbf{e}_z\mathbf{e}_x.$$

Now within each of these three main terms, the terms in bold cancel out, leaving:

$$= p_z \times (p_y(q_zr_x - q_xr_z) + p_x(q_yr_z - q_zr_y) + p_z(q_xr_y - q_yr_x))\mathbf{e}_{xy} +$$
$$p_x \times (p_z(q_xr_y - q_yr_x) + p_x(q_yr_z - q_zr_y) + p_y(q_zr_x - q_xr_z))\mathbf{e}_{yz} +$$
$$p_y \times (p_x(q_yr_z - q_zr_y) + p_y(q_zr_x - q_xr_z) + p_z(q_xr_y - q_yr_x))\mathbf{e}_{zx}$$

$$= (p_x(q_yr_z - q_zr_y) + p_y(q_zr_x - q_xr_z) + p_z(q_xr_y - q_yr_x)) \times (p_x\mathbf{e}_{yz} + p_y\mathbf{e}_{zx} + p_z\mathbf{e}_{xy}).$$

But $p_x\mathbf{e}_{yz} + p_y\mathbf{e}_{zx} + p_z\mathbf{e}_{xy} = *(p_x\mathbf{e}_x + p_y\mathbf{e}_y + p_z\mathbf{e}_z) = *\mathrm{P}$, and if we write

$$\lambda := (p_x(q_yr_z - q_zr_y) + p_y(q_zr_x - q_xr_z) + p_z(q_xr_y - q_yr_x))$$

then the whole expression is of the form:

$$\lambda \times *\mathrm{P}.$$

So we can say that:

$$*(*(\mathrm{P} \wedge \mathrm{Q}) \wedge *(\mathrm{P} \wedge \mathrm{R})) = \lambda\mathrm{P},$$

but $[\lambda\mathrm{P}] = [\mathrm{P}]$, and OP is the line that P ∧ Q and P ∧ R have in common, so:

$$[*(*(\mathrm{P} \wedge \mathrm{Q}) \wedge *(\mathrm{P} \wedge \mathrm{R}))] = [\mathrm{P}] = \mathbf{meet}([\mathrm{P} \wedge \mathrm{Q}], [\mathrm{P} \wedge \mathrm{R}]),$$

or in words, the **meet** of P ∧ Q and P ∧ R is the **P-point** corresponding to the **dual** of the **outer product** (or **alternating product**) of the **duals** of P ∧ Q and P ∧ R.

This turns out to be quite general, so that we can actually *define* the **meet** of two **P-lines** [A] and [B] as:

$$\mathbf{meet}([\mathrm{A}], [\mathrm{B}]) = [\mathrm{A}] \vee [\mathrm{B}] := [*(*\mathrm{A} \wedge *\mathrm{B})].$$ [151]

But we also have, for two **P-points** [P] and [Q], that their **join** corresponds to the **bivector** between P and Q:

---

[151] As in Doran and Lasenby, *op. cit.* p.346.

join([P], [Q]) = [P] ∧ [Q] := [P ∧ Q].

So, bringing all our **dual** definitions together, and putting them in terms of 3-points (or **3-vectors**) P and Q, and for A, B **bivectors**, we have these definitions:

(1) [P] ∧ [Q] := [P ∧ Q]
(2) ∗[P] := [∗P]
(3) ∗[A] := [∗A]

and we can write, using ∗∗ = 1:

∗([A] ∨ [B]) = [∗∗(∗A ∧ ∗B)] = [∗A ∧ ∗B]
∗([P] ∧ [Q]) = [∗P] ∨ [∗Q]

where [∗P] and [∗Q] are **P-lines** corresponding to the **bivectors** ∗P and ∗Q. The first equation, which is simply a restatement of the definition of [A] ∨ [B] above, is an equation between **P-points** on each side, the second is an equation between **P-lines**. To justify it, we expand its RHS, using the definition of [A] ∨ [B], as:

[∗P] ∨ [∗Q] = [∗(∗∗P ∧ ∗∗Q)] = [∗(P ∧ Q)]

but under our definitions,

[∗(P ∧ Q)] =: ∗[P ∧ Q] (definition 3)

and

∗[P ∧ Q] =: ∗([P] ∧ [Q]) (definition 1).

Here I've used the reversed definition operator =: to emphasize that the RHS is actually defined as the LHS here.

In passing, I'll point out that the factor λ which we obtained from expanding ∗(P ∧ Q) ∧ ∗(P ∧ R) is the **determinant** of the **matrix**:

$$\begin{pmatrix} p_x & p_y & p_z \\ q_x & q_y & q_z \\ r_x & r_y & r_z \end{pmatrix}.$$

The **determinant** is $p_x(q_y r_z - q_z r_y) + p_y(q_z r_x - q_x r_z) + p_z(q_x r_y - q_y r_x)$, which you can see easily by rewriting the middle term as $-p_y(q_x r_z - q_z r_x)$. It contains all the **permutations** of *xyz* with the *cyclic* ones (e.g. *yzx*) appearing as positive and the *anticyclic* (e.g. *xzy*) negative just as was explained in Section 1.8. In this context, we can replace the a_1 row in Section 1.8 with *p*, the a_2 row with *q*, and the a_3 row with *r* and the expansion of the **determinant** given in that Section then appears in the rather neat form:

$$\Sigma_{(ijk)}\, \varepsilon_{ijk}\, p_i\, q_j\, r_k$$

where $\Sigma_{(ijk)}$ means the sum over all **permutations** in 123, which are just the six terms (123), (132), (312), (321), (231) and (213) and where the sign ε_{ijk} is:

For **cyclic permutations:** $\varepsilon_{123} = \varepsilon_{312} = \varepsilon_{231} = +1$
For **anti-cyclic permutations:** $\varepsilon_{321} = \varepsilon_{213} = \varepsilon_{132} = -1$.

This concept of a **determinant** of a **matrix** formed from three **vectors** is so useful that I will introduce a notation for such **determinants**, writing the **determinant** above in the form:

$$\mathbf{det}\,(\mathbf{p}, \mathbf{q}, \mathbf{r}) = \quad \Sigma_{(ijk)}\, \varepsilon_{ijk}\, p_i\, q_j\, r_k.$$

Since the **pseudoscalar dual** of the **trivector** $\mathbf{e}_{xyz}$ is simply the **scalar** 1, this **determinant** also appears as $*(\mathrm{P} \wedge \mathrm{Q} \wedge \mathrm{R})$ wherein only terms in the **trivector** survive, and perhaps its most popular form is that written using the **Gibbs cross product** defined in Section 5.9 as:

$$\mathbf{a} \times \mathbf{b} := -\mathbf{I}(\mathbf{a} \wedge \mathbf{b})$$

which we can now see equals just $*(\mathbf{a} \wedge \mathbf{b})$. In terms of the **cross product** and the ordinary **dot product**, the **determinant** is simply:

$$\mathrm{P} \cdot (\mathrm{Q} \times \mathrm{R}).$$

Written this way, the **determinant** is known as the **scalar triple product** and is often abbreviated to (P, Q, R), which from $(\mathrm{P}, \mathrm{Q}, \mathrm{R}) = *(\mathrm{P} \wedge \mathrm{Q} \wedge \mathrm{R})$ can be seen to equal $-(\mathrm{Q}, \mathrm{P}, \mathrm{R}) = (\mathrm{Q}, \mathrm{R}, \mathrm{P})$, etc.

Now this **determinant** is a very important quantity for our present analysis. Because,

if and only if $\mathrm{P} \cdot (\mathrm{Q} \times \mathrm{R}) = 0$, the three points [P], [Q] and [R] are **collinear.**

The basic reason is most easily seen from the determinantal form itself. For if the **determinant**:

$$\begin{vmatrix} x & y & z \\ p_x & p_y & p_z \\ q_x & q_y & q_z \end{vmatrix}$$

is zero, it means that the three rows (x, y, z), (p_x, p_y, p_z) and (q_x, q_y, q_z) are **linearly dependent**, so there exist α and β such that:

$$\mathbf{x} = \alpha\mathrm{P} + \beta\mathrm{Q}$$

so [**x**] must be a point in the **P-line** determined by P and Q, much as our foregoing analysis showed, so $\mathbf{x} \cdot (\mathrm{P} \times \mathrm{R}) = -\,\mathrm{P} \cdot (\mathbf{x} \times \mathrm{R}) = \mathrm{P} \cdot (\mathrm{R} \times \mathbf{x}) = 0$ is the equation of the **P-line** through P and Q. If R lies on that line, so that (P, Q, R) = 0, it must by definition be **collinear** with P and Q.

The determinantal form, incidentally, also gives a neat expression for the simple **cross product** itself, which, for $\mathbf{e}_1$, $\mathbf{e}_2$ and $\mathbf{e}_3$ (or $\mathbf{e}_x$, $\mathbf{e}_y$ and $\mathbf{e}_z$) an **orthonormal basis** (i.e. $\mathbf{e}_i \cdot \mathbf{e}_j = \delta_{ij}$) enables us to write $\mathbf{p} \times \mathbf{q}$ as the "**determinant**":

$$\begin{vmatrix} p_x & p_y & p_z \\ q_x & q_y & q_z \\ \mathbf{e}_x & \mathbf{e}_y & \mathbf{e}_z \end{vmatrix}$$

or in the **permutation** expansion:

$$\mathbf{p} \times \mathbf{q} = \Sigma_{(ijk)}\, p_i\, q_j\, \mathbf{e}_k.$$

To convert one into the other, we interpret p_1 as p_x, p_2 as p_y and so on.

As a final point, it may be noted that the old **cross product** actually gives a cleaner interpretation of the **meet**. As shown earlier in this Section, a **P-line** in the **projective plane** corresponds to a plane through the **origin**, and if this plane has a **normal p**, the equation of that plane is:

$$\mathbf{p} \cdot \mathbf{x} = p_x x + p_y y + p_z z = 0.$$

Now suppose we have a second plane through the **origin** with **normal a**, so that a point **x** is in this plane if $\mathbf{a} \cdot \mathbf{x} = 0$. The **meet** of the two is the line (or **P-point**) where simultaneously $\mathbf{p} \cdot \mathbf{x} = 0$ and $\mathbf{a} \cdot \mathbf{x} = 0$. But that is the line through the **origin O** in the direction $\mathbf{p} \times \mathbf{a}$:

$$\mathbf{x} = \mathbf{O} + (\mathbf{p} \times \mathbf{a}).t.$$

To see this, we take the **dot product** of this equation with **p**, to get:

$$\mathbf{p} \cdot \mathbf{x} = \mathbf{p} \cdot \mathbf{O} + \mathbf{p} \cdot (\mathbf{p} \times \mathbf{a}).t = 0 + 0.t = 0$$

because, as is easily shown, **det** (**p**, **p**, **a**) = 0, i.e. a **determinant** of a **matrix** with two rows equal is zero. Similarly $\mathbf{a} \cdot \mathbf{x} = 0$, so [**p** × **a**] is the **meet** of the two **P-lines** [[**p**]] and [[**a**]], which we may also write as [∗**p**] and [∗**a**], since ∗**p** and ∗**a** are **bivectors**.

That this is equivalent to the earlier definition is clear because that definition, in terms of the **bivectors** ∗**p** and ∗**a** would be:

$$*\mathbf{p}\,.\,*\mathbf{a} = *(*(*\mathbf{p}) \wedge *(*\mathbf{a})) = *(\mathbf{p} \wedge \mathbf{a}),$$

but $*(\mathbf{p} \wedge \mathbf{a}) = \mathbf{p} \times \mathbf{a}$, so we have $*\mathbf{p}\,.\,*\mathbf{a} = \mathbf{p} \times \mathbf{a}$.

9.6 Desargues's Theorem

Desargues's Theorem is the most famous, and was really the seminal, result of **projective geometry**. It was discovered by a mathematical architect and engineer called Girard Desargues (1591-1661) in the early Seventeenth Century, long before Poncelet.

It's a most surprising result.

Look at *Figure 9.6.1*. In it we have two triangles ABC and $A^\dagger B^\dagger C^\dagger$ (written as A'B'C' in the *Figure*) which are related by a **perspectivity** such that the same point Ω lies on the line $AA^\dagger$, the line $BB^\dagger$ and the line $CC^\dagger$. Assume that AB is not parallel to $A^\dagger B^\dagger$ ($AB \nparallel A^\dagger B^\dagger$), and similarly $BC \nparallel B^\dagger C^\dagger$ and $AC \nparallel A^\dagger C^\dagger$, so that all these three pairs of lines have an intersection. Then call P the intersection of BC and $B^\dagger C^\dagger$ (the side *opposite* A), call Q the intersection of AC and $A^\dagger C^\dagger$, and R the intersection of AB and $A^\dagger B^\dagger$. Then P, Q and R lie on the same line.

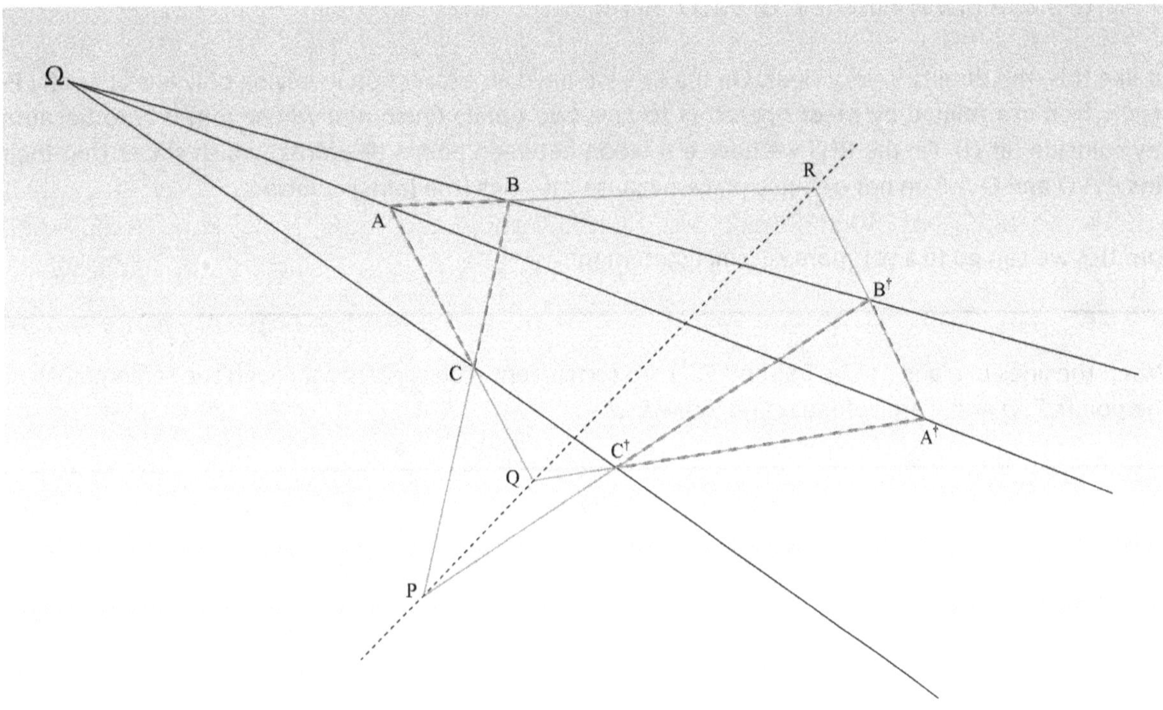

Figure 9.6.1

In terms of **meets** and **joins**, and omitting any square brackets as we assume we're working in the plane throughout, the theorem is:

if (**meet**($AA^\dagger$, $BB^\dagger$) = **meet**($BB^\dagger$, $CC^\dagger$) = Ω above,

then **join**(P, Q) = **join**(Q, R).

We can write this using the notation from the previous Section, using V to mean the intersection (**meet**) in the plane, and Λ to indicate the **join** between two points in the plane, when the theorem takes the form:

if ($AA^† \vee BB^† = BB^† \vee CC^†$),

then ($BC \vee B^†C^†) \wedge (AC \vee A^†C^†) = (AC \vee A^†C^†) \wedge (AB \vee A^†B^†$).

The converse is also true: if P, Q and R lie along a common line, $AA^†$, $BB^†$ and $CC^†$ will meet in a common point.

But we can phrase it more succinctly than that, because if two **P-points** are the same, their **join** is zero, and if two **P-lines** are collinear i.e. their defining planes are coplanar), they have no point of intersection so their **meet** is zero. This, using P, Q and R, and defining $p = AA^†$, $q = BB^†$ and $r = CC^†$, and using the equivalence symbol ≡, gives us the neat form:

$(p \vee q) \wedge (q \vee r) = 0 \equiv (P \wedge Q) \vee (Q \wedge R) = 0.$

Put like this, the **duality** is very clear. On the LHS we have an expression involving only lines (strictly **P-lines**) which are related by **meet** operators to give two points (**P-points**) whose **join** is zero because they coincide (at Ω). On the RHS we have a relation between points (**P-points**) which states that their **joins** $P \wedge Q$ and $Q \wedge R$ do not define a plane because the lines (the **joins**) coincide.

From this we can go to a yet more succinct statement:

When the lines *p*, *q* and *r* ($AA^†$, $BB^†$ and $CC^†$) are **concurrent** (run together through the same point) the points P, Q and R are **collinear** and *vice versa*.

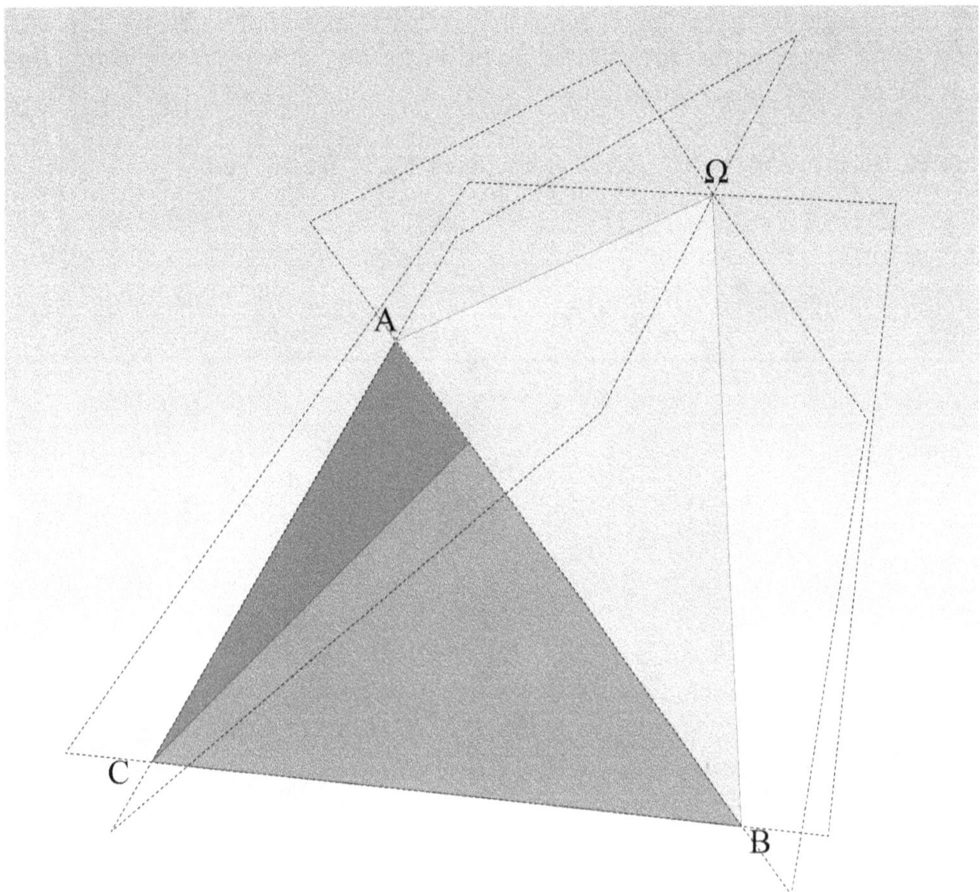

Figure 9.6.2

The key to the proof lies in looking at the model extended into 3-space, as shown in *Figures 9.6.2* and *9.6.3*. First look at *9.6.2*. This shows three planes in space coming towards the viewer which all intersect in a common point Ω. As they move away from Ω, they diverge to create a triangular "hole" between them. Now imagine that two "**picture planes**" PP_1 and PP_2 pass through this diverging fan of the three planes at two distinct positions, but both on the same side of Ω. This is the configuration seen in *Figure 9.6.3,* which is the definitive picture for the rest of this Section and reflects the two-dimensional configuration of *Figure 9.6.1*.

A compelling way to visualize this diagram is to imagine that Ω is an actual **eye point**, and that the planes in which the two triangles lie are two *alternative* **picture planes** PP_1 and PP_2. Seen from Ω, the two triangles ABC and $A^+B^+C^+$ are "in perspective" *exactly* by the defining hypothesis of the theorem. So ABC precisely covers $A^+B^+C^+$ and from Ω the two triangles cannot be distinguished at all. From Ω the critical pair of lines AB and A^+B^+ (and the pairs BC and B^+C^+, and AC and A^+C^+) lie precisely on top of each other and don't actually appear to have an intersection at all. But because they're actually running down the distinct planes PP_1 and PP_2, they must in fact cross *along the line where these two planes cross*. This is the key idea behind this intuitive proof which I take from Brannan, Esplen and Gray.[152]

[152] Brannan, Esplen and Gray, *op. cit.* p.101.

So from Ω the triangle in space $\Omega B^{\dagger}A^{\dagger}$ actually includes ΩAB and thereby we can see that the quadrilateral $ABB^{\dagger}A^{\dagger}$ is planar, because all four points lie in the plane which we might write as $\Omega A \wedge \Omega B = \lambda \Omega A^{\dagger} \wedge \Omega B^{\dagger}$ (which *is* well-defined!)

Similarly $ACC^{\dagger}A^{\dagger}$ lies in one plane ($\Omega A \wedge \Omega C$), and so does $BCC^{\dagger}B^{\dagger}$ (in $\Omega B \wedge \Omega C$).

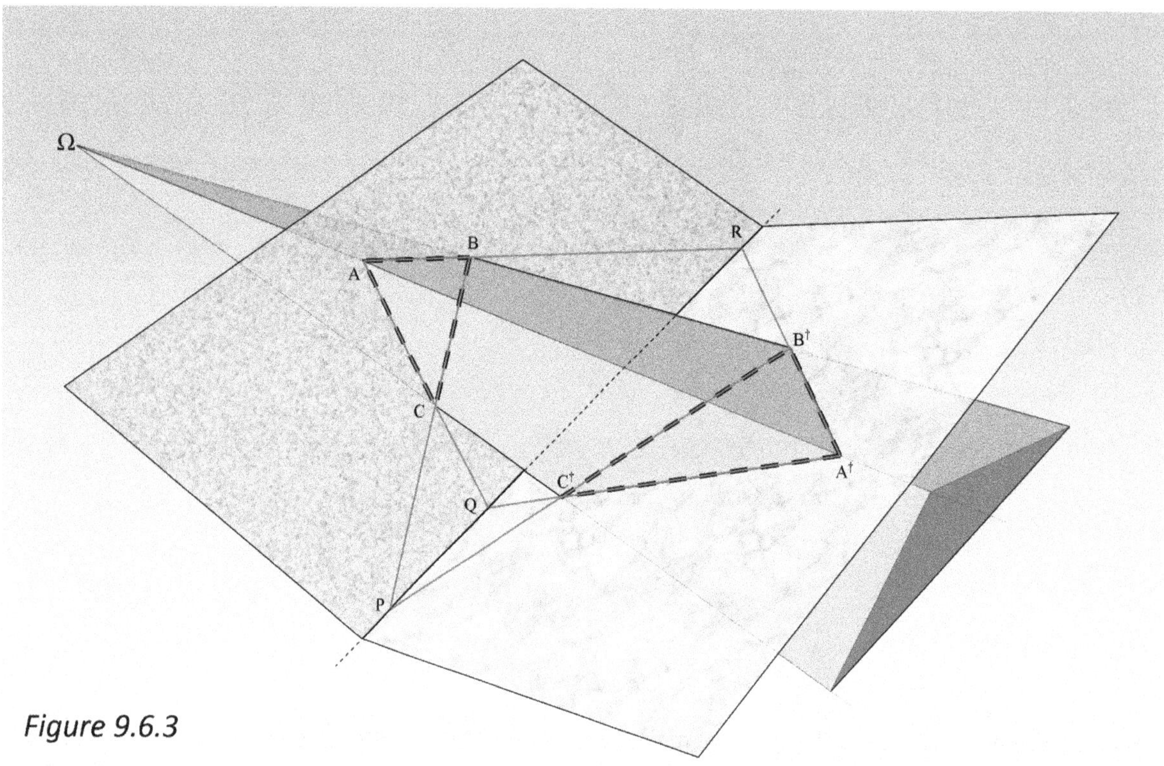

Figure 9.6.3

Now consider the triangle ABC formed by the *actual* lines in space AB, AC, BC which lie in the plane PP_1, and similarly the triangle $A^{\dagger}B^{\dagger}C^{\dagger}$ lies in the plane PP_2. The intersection of these two planes is the *line* in space, which we might write as:

$$PP_1 \vee PP_2.$$

AB lies in PP_1, and $A^{\dagger}B^{\dagger}$ lies in PP_2, so where the three plane $OAA^{\dagger}B^{\dagger}B$, PP_1 and PP_2 meet, our point R, now interpreted as an *actual* point in 3-space as shown in *Figure 9.6.3*, must lie in the *three* planes:

$\Omega AA^{\dagger}B^{\dagger}B$
PP_1
PP_2

Similarly the meeting of AC and $A^{\dagger}C^{\dagger}$, the *actual* point Q, must lie in

$\Omega AA^{\dagger}C^{\dagger}C$

PP_1
PP_2,

and of BC and $B^{\dagger}C^{\dagger}$, the *actual* point P, in

$\Omega BB^{\dagger}C^{\dagger}C$
PP_1
PP_2.

In all three cases they lie on the intersection of the two **picture planes** containing the triangles:

PP_1
PP_2,

so they must be collinear in space. It can trivially be shown that then [P], [Q] and [R] are **collinear** in the **projective plane**.

So also, the two **join** planes $\Omega R \wedge \Omega P$ and $\Omega P \wedge \Omega Q$, using Ω itself as an arbitrary **origin**, must be equal, which can be seen as the 3-space version of our original **P-line** equation:

$$(AB \vee A^{\dagger}B^{\dagger}) \wedge (BC \vee B^{\dagger}C^{\dagger}) = (BC \vee B^{\dagger}C^{\dagger}) \wedge (AC \vee A^{\dagger}C^{\dagger}).$$

The theorem is rarely proved *as it stands*, as a theorem in plane geometry. The attractions of the **projective** interpretation usually prove irresistible. But this can be done, and I'll outline such a proof first.

To do this we work from the plane figure, *Figure 9.6.1*, and forget completely for now all about **projective geometry**!

Firstly, note that $A^{\dagger}$ lies on the line ΩA and so must obey an equation:

$$A^{\dagger} = \Omega + \lambda(A - \Omega) \qquad \text{for some value } \lambda \neq 1.$$

Also

$$B^{\dagger} = \Omega + \mu(B - \Omega) \qquad \text{for some } \mu \neq 1,$$
$$C^{\dagger} = \Omega + \delta(C - \Omega) \qquad \text{for some } \delta \neq 1.$$

The equation of the line through AB is $X = (1 - t)A + tB$, and through $A^{\dagger}B^{\dagger}$ is $X = (1 - s)A^{\dagger} + sB^{\dagger}$, so the intersect or **meet** of these two lines must be at:

$$(1 - t)A + tB = R = (1 - s)A^{\dagger} + sB^{\dagger}$$

and substituting for $A^{\dagger}$ and $B^{\dagger}$,

$$(1 - t)A + tB = (1 - s)[\Omega + \lambda(A - \Omega)] + s[\Omega + \mu(B - \Omega)].$$

We can rearrange this to get an equation in t and s as:

$$t(B - A) + s(\lambda A - \mu B - (\lambda - \mu)\Omega) = (1 - \lambda)(\Omega - A).$$

Similar equations for the Q = AC V $A^{\dagger}C^{\dagger}$ and P = BC V $B^{\dagger}C^{\dagger}$ intersects, using:

$$(1 - p)A + pC = Q = (1 - q)A^{\dagger} + qC^{\dagger}$$
$$(1 - u)B + uC = P = (1 - v)B^{\dagger} + vC^{\dagger}$$

give:

$$p(C - A) + q(\lambda A - \delta C - (\lambda - \delta)\Omega) = (1 - \lambda)(\Omega - A)$$
$$u(C - B) + v(\mu B - \delta C - (\mu - \delta)\Omega) = (1 - \mu)(\Omega - B).$$

Now choose Ω to be the **origin** of the plane, so that Ω = (0, 0), and we end up with the simplified form (for the point R = AB V $A^{\dagger}B^{\dagger}$:

$$t.(B - A)_x + s.(\lambda A_x - \mu B_x) = (\lambda - 1)A_x$$
$$t.(B - A)_y + s.(\lambda A_y - \mu B_y) = (\lambda - 1)A_y$$

where I've split the **2-vector** equation into its x and y parts. The **determinant** here is

$$(B - A)_x(\lambda A_y - \mu B_y) - (B - A)_y(\lambda A_x - \mu B_x)$$

which simplifies to:

$$(\mu - \lambda)(A_xB_y - A_yB_x).$$

Using Cramer's rule, the numerator **determinant** is

$$(\lambda - 1)A_x(\lambda A_y - \mu B_y) - (\lambda - 1)A_y(\lambda A_x - \mu B_x) = -\mu(\lambda - 1)(A_xB_y - A_yB_x)$$

so we obtain $t = -\mu(\lambda - 1)/(\mu - \lambda)$.

So $\quad R = A - [\mu(\lambda - 1)/(\mu - \lambda)].(B - A)$,

and we also get:

$$Q = A - [\delta(\lambda - 1)/(\delta - \lambda)].(C - A)$$
$$P = B - [\delta(\mu - 1)/(\delta - \mu)].(C - B).$$

Now to establish whether these are collinear, we need to verify that:

$$P = (1 - \tau)R + \tau.Q$$

for some value τ, and so that P lies on the line connecting R and Q regardless of the choice of A, B and C. So we expand P − (1 − τ)R − τ.Q as:

$$B - [\delta(\mu - 1)/(\delta - \mu)].(C - B) - (1 - \tau)(A - [\mu(\lambda - 1)/(\mu - \lambda)].(B - A))$$

$$- \tau.(A - [\delta(\lambda - 1)/(\delta - \lambda)].(C - A))$$

and collect terms in A, B and C. After a few slips in the algebra, I finally get:

$$[1 + (1 - \tau)\mu(\lambda - 1)/(\mu - \lambda) + \tau.\delta(\lambda - 1)/(\delta - \lambda)] \times A$$

$$+ [\mu(\delta - 1)/(\delta - \mu) + (1 - \tau)\mu(\lambda - 1)/(\mu - \lambda)] \times B$$

$$+ [\delta(1 - \mu)/(\delta - \mu) + \tau.\delta(\lambda - 1)/(\delta - \lambda)] \times C$$

and for P − (1 − τ)R − τ.Q = 0, we want this expression to be zero for the same τ regardless of the values of A, B and C.

So by **linear independence**, we want to show that the value of τ that zeroes the value of the coefficient of A also zeroes those of B and C.

So $\quad 1 + (1 - \tau)\mu(\lambda - 1)/(\mu - \lambda) + \tau.\delta(\lambda - 1)/(\delta - \lambda) = 0$

gives $\quad (\mu - \lambda)(\delta - \lambda - \mu(\lambda - 1)(\delta - \lambda) = \tau[\mu(\lambda - 1)(\delta - \lambda) - \delta(\lambda - 1)(\mu - \lambda)]$

which gives $\tau = (1 - \mu)(\delta - \lambda)/[(1 - \lambda)(\delta - \mu)]$

and indeed this same value of τ gives:

$$\mu(\delta - 1)/(\delta - \mu) + (1 - \tau)\mu(\lambda - 1)/(\mu - \lambda) = 0$$

and

$$\delta(1 - \mu)/(\delta - \mu) + \tau.\delta(\lambda - 1)/(\delta - \lambda) = 0$$

so making P = (1 − τ)R − τ.Q for all values of A, B, C, λ, μ and δ. So P, Q and R are indeed always collinear in the plane. **QED**

The theorem can be proved easily in 3-space too, using a device that is a favourite in **projective geometry**. This ploy is to transform our **coordinates** into ones that will make the statement of the problem much simpler. We can do it in various ways, but I'll choose to do it using Ω as the new **origin**.

So first we **translate coordinates** by $x -= \Omega_x$, $y -= \Omega_y$, $z -= \Omega_z$, using computer notation which I expect will be very familiar to any reader of this Chapter.[153]

Now in these new **coordinates** we have $\Omega = (0, 0, 0)$, a big simplification. But we can go further. Knowing (A_x, A_y, A_z), (B_x, B_y, B_z) and (C_x, C_y, C_z) we can solve for the nine $\{a_{ij}\}$ values in the **matrix A** that will give:

$$a_{11}A_x + a_{12}A_y + a_{13}A_z = 1$$
$$a_{21}A_x + a_{22}A_y + a_{23}A_z = 0$$
$$a_{31}A_x + a_{32}A_y + a_{33}A_z = 0$$

$$a_{11}B_x + a_{12}B_y + a_{13}B_z = 0$$
$$a_{21}B_x + a_{22}B_y + a_{23}B_z = 1$$
$$a_{31}B_x + a_{32}B_y + a_{33}B_z = 0.$$

$$a_{11}C_x + a_{12}C_y + a_{13}C_z = 0$$
$$a_{21}C_x + a_{22}C_y + a_{23}C_z = 0$$
$$a_{31}C_x + a_{32}C_y + a_{33}C_z = 1.$$

For example,

$$a_{21}A_x + a_{22}A_y + a_{23}A_z = 0$$
$$a_{21}B_x + a_{22}B_y + a_{23}B_z = 1$$
$$a_{21}C_x + a_{22}C_y + a_{23}C_z = 0$$

solves this set for a_{21}, a_{22} and a_{23}.

If we apply *this* change of **coordinates** as well, by multiplying any point in the 3-space by the **matrix A**, we end up with **coordinates** in which:

$$\Omega = (0, 0, 0),\ A = (1, 0, 0),\ B = (0, 1, 0) \text{ and } C = (0, 0, 1).$$

Now ΩA, ΩB and ΩC are new **basis vectors** for the space, and any **vector** or point in the 3-space can now be defined in terms of them (they must be **linearly independent** by the premises of the theorem).

But $A^\dagger$ lies along ΩA, $B^\dagger$ lies along ΩB, and $C^\dagger$ lies along ΩC, so these can be expressed as:

$$A^\dagger = a.A = a(1, 0, 0) = (a, 0, 0),$$
$$B^\dagger = b.B = b(0, 1, 0) = (0, b, 0),$$
$$C^\dagger = c.C = a(0, 0, 1) = (0, 0, c),$$

for suitable values *a*, *b*, *c* which will place the three points on the second **picture plane** PP_2, which can anyway be regarded as *defined* purely by the points $A^\dagger$, $B^\dagger$ and $C^\dagger$.

---

153 In conventional notation, we define $x' = x - \Omega_x$, $y' = y - \Omega_y$, $z' = z - \Omega_z$, so that in terms of the new **coordinates** (x', y', z'), the **coordinates** of Ω are (0, 0, 0).

So now all our key points have easy **coordinates**. We can now work out where AB and $A^{\dagger}B^{\dagger}$ intersect in the ΩAB plane in which they both lie:

$$R = AB \vee A^{\dagger}B^{\dagger}: \quad A + s.(B - A) = R = A^{\dagger} + t.(B^{\dagger} - A^{\dagger}).$$

But $B - A = (0, 1, 0) - (1, 0, 0) = (-1, 1, 0)$, and similarly $B^{\dagger} - A^{\dagger} = (-a, b, 0)$. So our equation defining R becomes:

$$(1, 0, 0) + s.(-1, 1, 0) = (a, 0, 0) + t.(-a, b, 0).$$

The z part gives merely $0 = 0$, but from the x and y we get:

$$1 - s = a - ta$$
$$s = bt.$$

We can solve this for s to obtain:

$$s = b(a - 1)/(a - b) = bt,$$

so we have:

$$R = (1 - b(a - 1)/(a - b), b(a - 1)/(a - b), 0)$$

which simplifies to

$$R = (a(1 - b)/(a - b), b(a - 1)/(a - b), 0).$$

In a similar way we can solve the equation for Q:

$$Q = AC \vee A^{\dagger}C^{\dagger}: \quad A + u.(C - A) = Q = A^{\dagger} + v.(C^{\dagger} - A^{\dagger}),$$

which gives:

$$1 - u = a - av$$
$$u = vc.$$

This gives $u - c(a - 1)/(a - c)$, and so:

$$Q = (a(1 - c)/(a - c), 0, c(a - 1)/(a - c)).$$

Finally, since

$$P = BC \vee B^{\dagger}C^{\dagger}: \quad B + p.(C - B) = P = B^{\dagger} + q.(C^{\dagger} - B^{\dagger}),$$

we solve for P as:

$$P = (0, b(1-c)/(b-c), c(b-1)/(b-c)).$$

Now we could proceed by using the condition that the **determinant**:

$$\begin{vmatrix} P_x & P_y & P_z \\ Q_x & Q_y & Q_z \\ R_x & R_y & R_z \end{vmatrix}$$

is zero, for then P, Q and R are **linearly dependent** and there must be α and β such that:

$$P = \alpha Q + \beta R.$$

This would mean that P lies in the plane defined by ΩQ and ΩR, because it can be so expressed, and so as **P-points** [P], [Q] and [R] are **collinear** in the **projective plane** because the **P-points** [P], [Q] and [R] all lie in the same **P-line** or the same plane in 3-space through the new **origin Ω**.

But we can do better than that. We can easily show that the *actual* 3-points P, Q and R are collinear in 3-space. To do this we need to establish that there is a unique value of t such that:

$$P = Q + t.(R - Q).$$

This equation now expands to:

$$(0, b(1-c)/(b-c), c(b-1)/(b-c)) = (a(1-c)/(a-c), 0, c(a-1)/(a-c)) +$$

$$t.[\ (a(1-b)/(a-b), b(a-1)/(a-b), 0) - (a(1-c)/(a-c), 0, c(a-1)/(a-c))\ ].$$

Much as in the foregoing two-dimensional proof, *the same* t must give identity in the x, the y, and the z components of the equation.

The x part is:

$$0 = a(1-c)/(a-c) + t.[\ a(1-b)/(a-b) - a(1-c)/(a-c)\ ].$$

This solves for t as:

$$t = (c-1)(a-b)/((1-a)(b-c))$$

and this value of t also gives identity in the y and in the z components of the equation, so it is indeed true that:

$$P = Q + t.(R - Q) = Q + [(c - 1)(a - b)/((1 - a)(b - c))](R - Q),$$

and so P, Q and R are strictly collinear in 3-space too.

We have now three proofs of the Desargues theorem, one a "heuristic proof" in 3-space, then a strict proof in the plane, and finally again a strict proof in 3-space. Can we find a version that actually uses the **duality** with which we began?

If we could establish that the line $p = \Omega A = \Omega A^{\dagger} = AA^{\dagger}$ actually was **dual** to the point defined by the intersection of the opposite sides of the triangles, $P = BC \vee B^{\dagger}C^{\dagger}$, which is the **meet** of BC and $B^{\dagger}C^{\dagger}$, then we could proceed, working in the underlying 3-space, by taking the **dual** of:

$$(P \wedge Q) \vee (Q \wedge R) = 0$$

to give

$$*(P \wedge Q) \wedge *(Q \wedge R) = (*P \vee *Q) \wedge (*Q \vee *R) = (p \vee q) \wedge (q \vee r) = 0,$$

and conversely,

$$*[(p \vee q) \wedge (q \vee r) = 0] = *(p \vee q) \vee *(q \vee r) = 0$$

$$= (*p \wedge *q) \vee (*q \wedge *r) = (P \wedge Q) \vee (Q \wedge R) = 0.$$

This would prove the theorem and its converse very trivially.

It turns out that this isn't directly possible. This led me to suggest in the first edition of this book that the **pseudoscalar duality** did not actually yield the **projective principle of duality**. But Hestenes himself makes it very clear that it does, and to see this we need to look at his own papers, particularly D. Hestenes and R. Ziegler, *Projective Geometry with Clifford Algebra*, in *Acta Applicandae Mathematicae*, Vol. **23**, (1991) 25-63, although another illuminating paper is D. Hestenes, *Universal Geometric Algebra*, in the quarterly journal *Simon Stevin*, Vol. **62**, (1988), Number 3-4. Doran and Lasenby merely sketch the argument here (their p.348).

Hestenes uses the identity:

$$\langle \mathbf{AB}(\mathbf{c} \wedge \mathbf{c}') \rangle = (\mathbf{A} \wedge \mathbf{c})\cdot(\mathbf{B} \wedge \mathbf{c}') - (\mathbf{A} \wedge \mathbf{c}')\cdot(\mathbf{B} \wedge \mathbf{c}), \tag{9.6.1}$$

where **A** and **B** are **bivectors** and **c**, **c**' **vectors**. I'll need to explain the angle-bracket notation on the left, which I've somewhat avoided until now.

Since **geometric products** throw up terms of different **grades** (which Hestenes sometimes refers to as **steps**), he commonly uses a notation such as:

$$\mathbf{A}_r\mathbf{B}_s = \langle\mathbf{AB}\rangle_{|r-s|} + \langle\mathbf{AB}\rangle_{|r-s|+2} + \ldots + \langle\mathbf{AB}\rangle_{r+s}$$

where the **blades A** and **B** are explicitly labelled as being of **grades** r and s respectively, and the parts on the RHS are the resulting parts of the product, of **grade** $|r-s|$, $|r-s|+2, \ldots$ through to the $r+s$ part (this is the general case – note that they go up in steps of 2).

So in general he *defines* $\mathbf{A}_r \cdot \mathbf{B}_s$ as the lowest **grade** part $\langle\mathbf{AB}\rangle_{|r-s|}$, and $\mathbf{A}_r \wedge \mathbf{B}_s$ as the *highest* **grade** part $\langle\mathbf{AB}\rangle_{r+s}$, a concept which initially caused me much confusion when I first encountered it in Doran and Lasenby's text as it suggests an uncontrolled explosion in algebraic complexity!

If $|r-s| = 0$, there will be a **scalar** part (so this only happens when multiplying objects of the same **grade**)

$$\mathbf{A}_r \cdot \mathbf{B}_r = \langle\mathbf{AB}\rangle_{|r-r|} = \langle\mathbf{AB}\rangle_0 .$$

Because we're often multiplying objects of the same **grade**, this comes up rather a lot, so Hestenes uses the shorthand of *omitting* the subsript 0 here and just writing

$$\mathbf{A}_r \cdot \mathbf{B}_r = \langle\mathbf{AB}\rangle.$$

This is the notation we see above.

To prove (9.6.1), I'll cheat and use a **basis** $\{\mathbf{e}_1, \mathbf{e}_2, \mathbf{e}_3\}$ as I did in Chapter 5. $\mathbf{A} \wedge \mathbf{c}$ on the RHS then expands to

$$(A_1\mathbf{e}_{23} + A_2\mathbf{e}_{31} + A_3\mathbf{e}_{12}) \wedge (c_1\mathbf{e}_1 + c_2\mathbf{e}_2 + c_3\mathbf{e}_3).$$

Since this is a strict **alternating product** we can't have any repeating terms like $\mathbf{e}_{232}$ in the result, so it evaluates to:

$$(A_1c_1\mathbf{e}_{231} + A_2c_2\mathbf{e}_{312} + A_3c_3\mathbf{e}_{123}) = (A_1c_1 + A_2c_2 + A_3c_3)\mathbf{e}_{123} = (A_1c_1 + A_2c_2 + A_3c_3)\mathbf{I}.$$

So

$$(\mathbf{A} \wedge \mathbf{c})\cdot(\mathbf{B} \wedge \mathbf{c}') - (\mathbf{A} \wedge \mathbf{c}')\cdot(\mathbf{B} \wedge \mathbf{c})$$

$$= (A_1c_1 + A_2c_2 + A_3c_3)(B_1c'_1 + B_2c'_2 + B_3c'_3)\mathbf{I}^2 - (A_1c'_1 + A_2'c_2 + A_3c'_3)(B_1c_1 + B_2c_2 + B_3c_3)\mathbf{I}^2$$

$$= [A_1B_2(c_1c'_2 - c'_1c_2) + A_1B_3(c_1c'_3 - c'_1c_3) + A_2B_1(c_2c'_1 - c'_2c_1) + A_2B_3(c_2c'_3 - c'_2c_3) +$$

$$A_3B_1(c_3c'_1 - c'_3c_1) + A_3B_2(c_3c'_2 - c'_3c_2)]\mathbf{I}^2.$$

Here all the $A_iB_ic_ic'_i$ terms have cancelled out. Closer examination, and remembering $\mathbf{I}^2 = -1$, reduces this back to:

$$-[(A_1B_2 - A_2B_1)(c_1c'_2 - c'_1c_2) + (A_1B_3 - A_3B_1)(c_1c'_3 - c'_1c_3) + (A_2B_3 - A_3B_2)(c_2c'_3 - c'_2c_3)],$$

and it is a **scalar**. Call this result (9.6.2).

On the LHS,

$$\mathbf{AB} = (A_1\mathbf{e}_{23} + A_2\mathbf{e}_{31} + A_3\mathbf{e}_{12})(B_1\mathbf{e}_{23} + B_2\mathbf{e}_{31} + B_3\mathbf{e}_{12})$$

$$= A_1B_1(\mathbf{e}_{23})^2 + A_1B_2\mathbf{e}_{2331} + A_1B_3\mathbf{e}_{2312} + A_2B_1\mathbf{e}_{3123} + A_2B_2(\mathbf{e}_{31})^2 + A_2B_3\mathbf{e}_{3112}$$

$$+ A_3B_1\mathbf{e}_{1223} + A_3B_2\mathbf{e}_{1231} + A_3B_3(\mathbf{e}_{12})^2$$

$$= -(\Sigma A_iB_i) + (A_2B_1 - A_1B_2)\mathbf{e}_{12} + (A_3B_2 - A_2B_3)\mathbf{e}_{23} + (A_1B_3 - A_3B_1)\mathbf{e}_{31}. \qquad (9.6.3)$$

Now $\mathbf{c} \wedge \mathbf{c}' = (c_1\mathbf{e}_1 + c_2\mathbf{e}_2 + c_3\mathbf{e}_3) \wedge (c'_1\mathbf{e}_1 + c'_2\mathbf{e}_2 + c'_3\mathbf{e}_3)$

$$= (c_1c'_2 - c_2c'_1)\mathbf{e}_{12} + (c_2c'_3 - c_3c'_2)\mathbf{e}_{23} + (c_3c'_1 - c_1c'_3)\mathbf{e}_{31}. \qquad (9.6.4)$$

Multiplying (9.6.3) and (9.6.4) together, only terms like $\mathbf{e}_{12}\mathbf{e}_{12}$ will give **scalar** results, and since $\mathbf{e}_{12}\mathbf{e}_{12} = \mathbf{e}_{23}\mathbf{e}_{23} = \mathbf{e}_{31}\mathbf{e}_{31} = -1$, we get

$$\langle \mathbf{AB}(\mathbf{c} \wedge \mathbf{c}')\rangle = -(A_2B_1 - A_1B_2)(c_1c'_2 - c_2c'_1) - (A_3B_2 - A_2B_3)(c_2c'_3 - c_3c'_2)$$

$$-(A_1B_3 - A_3B_1)(c_3c'_1 - c_1c'_3).$$

This equals (9.6.2) exactly but for the sign. Since we are only concerned with when both sides are zero, this will not affect the argument.

Returning to the Desargues' theorem, we run headlong into the conflict, boxed in the last Section, between **geometric algebra** and geometry!

We have **P-points** A, B and C and A', B' and C' defining the two triangles, and the **P-points** that are to be proven **collinear** are P, Q and R. These must be represented by **vectors** in the **geometric algebra** which we expect to be *lower-case*, so I'll adopt the convention that the corresponding **vectors** *are* lower-case, but **bold**.

So A ≡ **a**, or A = [**a**], P ≡ **p**, so P = [**p**], etc. The *lines* or **P-lines** along the sides of the triangles are represented by **bivectors** and Hestenes puts them as $A = b \wedge c$, $B = a \wedge c$, and $C = a \wedge b$. But this is just too confusing, so I'll again write these in **bold** when I mean the **algebraic** entities as:

$\mathbf{A} = \mathbf{b} \wedge \mathbf{c}$, $\mathbf{B} = \mathbf{a} \wedge \mathbf{c}$, and $\mathbf{C} = \mathbf{a} \wedge \mathbf{b}$.

So we'll have for side BC, opposite A, of triangle ABC,

a = BC = [**A**] = [$\mathbf{b} \wedge \mathbf{c}$].

It is confusing but I want to keep fairly close to Hestenes's notation as far as I can.

The theorem states that (*Figure 9.6.1*):

> When the lines p, q and r ($AA^{\dagger}$, $BB^{\dagger}$ and $CC^{\dagger}$) are **concurrent** (run together through the same point) the points P, Q and R are **collinear** and *vice versa.*

From the argument at the end of the previous Section, the **collinearity** of P, Q and R is given in terms of their defining **vectors** as:

$$\mathbf{p} \wedge \mathbf{q} \wedge \mathbf{r} = 0.$$

The condition for **concurrence** of lines is rather more clumsy; Hestenes shows that for lines p, q and r represented by **bivectors P**, **Q** and **R**, it is given in terms of the **meet** of **P** and **Q** as:

$$(\mathbf{P} \vee \mathbf{Q}) \wedge \mathbf{R} = 0.$$

In the two-dimensional **projective plane** of **Desargues's theorem**, this reduces to:

$$\langle \mathbf{PQR} \rangle = 0,$$

using Hestenes' notation for the **scalar** part of the **geometric product**, and we can see that this is going to be where (9.6.1) comes in. This condition actually corresponds to my

$$(p \vee q) \wedge (q \vee r) = 0,$$

the condition that p, q and r all meet in the same point, or are **concurrent**.

Hestenes then defines two auxiliary products for each triangle:

for ABC: $\mathbf{J} = \mathbf{a} \wedge \mathbf{b} \wedge \mathbf{c}$
for $A^{\dagger}B^{\dagger}C^{\dagger}$: $\mathbf{J}' = \mathbf{a}' \wedge \mathbf{b}' \wedge \mathbf{c}'$.

He now does a curious thing: he expands the sides of the triangles in a way that seems clearly incorrect.

He takes $\mathbf{A} = \mathbf{b} \wedge \mathbf{c}$ and identifies this with $^*\mathbf{AJ} = \mathbf{A}.\mathbf{I}^{-1}\mathbf{J}$, and so with all the other sides. But

$$\mathbf{J} = \mathbf{det}(\mathbf{a}, \mathbf{b}, \mathbf{c}).\mathbf{I} = \Sigma\varepsilon_{ijk}a_i b_j c_k.\mathbf{I},$$

so this gives

$$^*\mathbf{AJ} = \mathbf{A}.\mathbf{I}^{-1}.\mathbf{det}(\mathbf{a},\mathbf{b}, \mathbf{c}).\mathbf{I} = \mathbf{A}.\mathbf{det}(\mathbf{a},\mathbf{b}, \mathbf{c}).$$

This *doesn't* equal **A** unless **det(a,b, c)** = 1 which there's no reason to suppose. Hestenes explicitly refers to the **determinantal** expansion of **J** and **J'**, so we seem to be agreed on that. It could be that he is saying that the two are **projectively equivalent**, a term he uses to refer to any **vector** $\lambda\mathbf{v}$ that represents the same **P-point** as **v**:

$V = [\mathbf{v}] = [\lambda\mathbf{v}]$.

So we need to prove that:

$$\langle(\mathbf{a} \wedge \mathbf{a}')(\mathbf{b} \wedge \mathbf{b}')(\mathbf{c} \wedge \mathbf{c}')\rangle = 0 \qquad \equiv \qquad \mathbf{p} \wedge \mathbf{q} \wedge \mathbf{r} = 0.$$

i.e. when the lines $\mathbf{a} \wedge \mathbf{a}'$, *q* $\mathbf{b} \wedge \mathbf{b}'$ and $\mathbf{c} \wedge \mathbf{c}'$ ($AA^\dagger$, $BB^\dagger$ and $CC^\dagger$) are **concurrent**, the points **p, q** and **r** (P, Q and R) are **collinear** and *vice versa*.

We use (9.6.1) to expand $\langle(\mathbf{a} \wedge \mathbf{a}')(\mathbf{b} \wedge \mathbf{b}')(\mathbf{c} \wedge \mathbf{c}')\rangle$ as (with a bit of switching of order):

$$(\mathbf{b}' \wedge \mathbf{b} \wedge \mathbf{c})\cdot(\mathbf{a} \wedge \mathbf{c}' \wedge \mathbf{a}') - (\mathbf{b} \wedge \mathbf{b}' \wedge \mathbf{c}')\cdot(\mathbf{c} \wedge \mathbf{a} \wedge \mathbf{a}'),$$

but we already have e.g. $\mathbf{b} \wedge \mathbf{c} = \mathbf{A} = {}^*\mathbf{A}\mathbf{J}$, and $\mathbf{b}' = \mathbf{A}' \vee \mathbf{C}'$ (i.e. $\mathbf{b}'$ is the **meet** of the lines $\mathbf{A}'$ and $\mathbf{C}'$, that is to say of $\mathbf{b}' \wedge \mathbf{c}'$ and $\mathbf{a}' \wedge \mathbf{b}'$), so

$$\mathbf{b}' \wedge \mathbf{b} \wedge \mathbf{c} = (\mathbf{A}' \vee \mathbf{C}') \wedge {}^*\mathbf{A}\mathbf{J}.$$

But

$$^*(\mathbf{A}' \vee \mathbf{C}') = -(\mathbf{A}' \vee \mathbf{C}')\mathbf{I} = -({}^*\mathbf{A}' \wedge {}^*\mathbf{C}'),$$

so $$(\mathbf{A}' \vee \mathbf{C}') = -(\mathbf{A}' \vee \mathbf{C}')\mathbf{I}\,\mathbf{I} = -({}^*\mathbf{A}' \wedge {}^*\mathbf{C}')\mathbf{I},$$

but again using the device of **projective equivalence** we can write this as:

$$(\mathbf{A}' \vee \mathbf{C}') = -(\mathbf{A}' \vee \mathbf{C}')\mathbf{I}\mathbf{I} \equiv ({}^*\mathbf{A}' \wedge {}^*\mathbf{C}')\mathbf{det}(\mathbf{a}', \mathbf{b}', \mathbf{c}')\mathbf{I} = ({}^*\mathbf{A}' \wedge {}^*\mathbf{C}')\mathbf{J}',$$

so all told,

$$\mathbf{b}' \wedge \mathbf{b} \wedge \mathbf{c} = ({}^*\mathbf{A}' \wedge {}^*\mathbf{C}')\mathbf{J}' \wedge {}^*\mathbf{A}\mathbf{J} = ({}^*\mathbf{A}' \wedge {}^*\mathbf{C}' \wedge {}^*\mathbf{A})\mathbf{J}\mathbf{J}'.$$

Similarly, looking at this as $\mathbf{b}' \wedge (\mathbf{b} \wedge \mathbf{c}) = (({}^*\mathbf{A}' \wedge {}^*\mathbf{C}') \wedge {}^*\mathbf{A})\mathbf{J}\mathbf{J}'$.

$$\mathbf{a} \wedge (\mathbf{c}' \wedge \mathbf{a}') = (({}^*\mathbf{C} \wedge {}^*\mathbf{B}) \wedge {}^*\mathbf{B}')\mathbf{J}\mathbf{J}',$$
$$\mathbf{b} \wedge (\mathbf{b}' \wedge \mathbf{c}') = (({}^*\mathbf{A} \wedge {}^*\mathbf{C}) \wedge {}^*\mathbf{A}')\mathbf{J}\mathbf{J}',$$
$$(\mathbf{c} \wedge \mathbf{a}) \wedge \mathbf{a}' = ({}^*\mathbf{B} \wedge ({}^*\mathbf{C}' \wedge {}^*\mathbf{B}'))\mathbf{J}\mathbf{J}',$$

so

$$\langle(\mathbf{a} \wedge \mathbf{a}')(\mathbf{b} \wedge \mathbf{b}')(\mathbf{c} \wedge \mathbf{c}')\rangle = -\mathbf{J}\mathbf{J}'[({}^*\mathbf{B} \wedge {}^*\mathbf{C}' \wedge {}^*\mathbf{B}')\cdot({}^*\mathbf{A} \wedge {}^*\mathbf{C} \wedge {}^*\mathbf{A}')$$
$$-({}^*\mathbf{C} \wedge {}^*\mathbf{B} \wedge {}^*\mathbf{B}')\cdot({}^*\mathbf{A} \wedge {}^*\mathbf{C}' \wedge {}^*\mathbf{A}')]$$

where Hestenes's has reversed the two terms. This equals:

$$-\mathbf{J}\mathbf{J}'[({}^*\mathbf{B} \wedge {}^*\mathbf{B}' \wedge {}^*\mathbf{C}')\cdot({}^*\mathbf{A} \wedge {}^*\mathbf{A}' \wedge {}^*\mathbf{C}) - ({}^*\mathbf{B} \wedge {}^*\mathbf{B}' \wedge {}^*\mathbf{C})\cdot({}^*\mathbf{A} \wedge {}^*\mathbf{A}' \wedge {}^*\mathbf{C}')]$$

We now use (9.6.1) again, identifying **A** as ***A** ∧ ***A'**, and **B** as ***B** ∧ ***B'** to switch this back to:

$$\mathbf{JJ'}\langle(^*\mathbf{A}\wedge{}^*\mathbf{A'})(^*\mathbf{B}\wedge{}^*\mathbf{B'})(^*\mathbf{C}\wedge{}^*\mathbf{C'})\rangle.$$

But (***A** ∧***A'**) is the **meet** of **A** = (**b** ∧ **c**) and **A'** = (**b'** ∧ **c'**), but this is the meet of the corresponding lines BC and B'C' in the two triangles, and is, from *Figure 9.6.1*, the point P. Similarly (***B** ∧ ***B'**) is Q and (***C** ∧ ***C'**) is R, so we've proved that when the lines (**a** ∧ **a'**) and (**b** ∧ **b'**) and (**c** ∧ **c'**) are **concurrent**, or

$$\langle(\mathbf{a}\wedge\mathbf{a'})(\mathbf{b}\wedge\mathbf{b'})(\mathbf{c}\wedge\mathbf{c'})\rangle = 0$$

then the points P, Q, R (or **p**, **q** and **r** as **vectors**) are **collinear** or

$$\mathbf{p}\wedge\mathbf{q}\wedge\mathbf{r} = \langle\mathbf{pqr}\rangle = 0.$$

So that completes our fourth proof of **Desargues's theorem**.

9.7 Projective Transformations

As we saw in Section 9.3, the device of doing transformations in a space one dimension higher than the space we're actually interested in enables us to incorporate **translations** as ordinary transformations about the **origin** in the underlying space.

This means we can define a generalized transformation of the **projective plane**, as a simple invertible transform about the **origin** in 3-space, or a generalized transformation of **projective space** as an invertible transform about the **origin** in 4-space, and so on.

Christian Felix Klein (1849-1925), a pure mathematician teaching at Erlangen in Germany in the Nineteenth Century, proposed what came to be called the "Erlangen Programme", of which the key idea was that different geometries could best be compared by defining the kinds of transformation that do not affect the critical properties of that geometry. So Euclidean geometry's critical properties are lengths and angles, and transformations that preserve these define the geometry. Such transformations are simple **rotations** about any axis, and **translations**. The **cross ratio** can be seen as the critical property of **projective geometry** and since **invertible linear transformations** in the n+1-space preserve that and so, under Klein's model, define **projective geometry**.

Klein's idea, coming as it did at the time when the discovery of non-Euclidean geometries was the mathematical sensation of the day, has become so influential that the entire development of academic geometry has been based on it ever since.

For the present purposes the concept is not that essential, and so I will just review it very briefly.

A **projective transformation** in the **projective plane** is a **function** τ defined as:

$\tau: [\mathbf{x}] \rightarrow [\mathbf{Ax}]$

where **A** is an **invertible 3×3 matrix.**[154]

Two successive **projective transformations** Σ and τ defined by **matrices Σ** and ***T*** can be "multiplied together" to give a composite **transformation** simply by using the product of their **matrices**:

$\Sigma\tau: [\mathbf{x}] \rightarrow [\boldsymbol{\Sigma} \boldsymbol{T}\mathbf{x}].$

This operation is called **composition** of **functions**.

Under **composition** of functions **projective transformations** form a **group**, as defined in Section 1.5. So:

1. If Σ and τ are **projective transformations**, Στ is, (**closure**)
2. There is an **identity transformation** *I* given by $I: [\mathbf{x}] \rightarrow [\mathbf{Ix}]$ where **I** is the **identity matrix**,
3. τ has an **inverse** τ^{-1} given by the **inverse matrix** $\mathbf{A}^{-1}$
4. $(\tau\Sigma)\psi \equiv \tau(\Sigma\psi)$ (**associativity**).

Projective transformations do *not* correspond to the **perspectivities** mentioned in Section 9.4. In fact they don't very obviously relate to any of the main concepts of **projective geometry** *per se*. They just happen to preserve the so-called **projective properties** such as:

- **Collinearity**
- Incidence (the crossing of two lines)
- **Cross ratio** (as of course do **perspectivities**)
- quadrilateralism – i.e. quadrilaterals stay quadrilaterals
- **Desargues's Theorem**
- the property of being a non-degenerate **projective conic.**[155]

I always have a vague feeling there's a certain circularity in these ideas: we define the kind of transformation that preserves the properties we're most interested in, and then define a given property as a **projective** (or Euclidean or inversive or affine or whatever) property if it's preserved under the chosen kind of transformation. The key distinguishing traits of a given geometry – as, surely, defined in its axiomatic foundations – are in danger of getting lost along the way.

Projective transformations for the **projective plane** are uniquely defined by how they transform just *four* distinct points in the **projective plane**. So given two quadrilaterals ABCD and $A^{\#}B^{\#}C^{\#}D^{\#}$, there is a unique **projective transformation** that maps $A \rightarrow A^{\#}$, $B \rightarrow B^{\#}$, $C \rightarrow C^{\#}$ and $D \rightarrow D^{\#}$.

---

[154] Brannan, Esplen and Gray, *op.cit.* p.114.

[155] A non-degenerate **projective conic** is an ellipse, or a parabola or a hyperbola that doesn't reduce to a straight line. Because these are all sections through a cone, and the apex of the cone clearly can be interpreted as an **eye point** from which they're all indistinguishable, all **projective conics** are in a sense equivalent in **projective geometry.**

The theorem that establishes this is called the **Fundamental Theorem of Projective Geometry**.

Brannan, Esplen and Gray use this theorem to prove Desargues's theorem in yet another way, but one that is closely related to the 3-space proof I gave towards the end of the last Section. They choose to transform the **projective plane** into **coordinates** where Ω = (1, 1, 1), A = (1, 0, 0), B = {0, 1, 0) and C= (0, 0, 1). So again they are choosing an **origin** O ≠ Ω. By logic similar to that which I used in the last Section, this gives $A^{\dagger}$ = (*a*, *b*, *b*) for some *a* and *b*, which is equivalent, as **homogeneous coordinates**, to (*a*/*b*, *b*/*b*, *b*/*b*) = (*p*, 1, 1), and similarly $B^{\dagger}$ = (1, *q*, 1) and $C^{\dagger}$ = (1, 1, *r*).

From this they easily find P = BC V $B^{\dagger}C^{\dagger}$, and similarly Q and R, but note that these are not necessarily the *original* or *actual* P, Q and R shown in *Figure 9.6.3*, but points that have *the same* **homogeneous coordinates**, and so which lie in the same direction from O, or are the same *seen as* **P-points** P, Q and R, which we might now interpret as **vectors p**, **q**, **r** or, better, as **vector classes** [**p**], [**q**] and [**r**], where we identify [**p**] ≡ [λ**p**].

They then evaluate **det**(P, Q, R) for these three **P-points**, and this is easily shown to equal zero, so the **P-points** P, Q and R are **collinear** in the **projective plane**, a slightly weaker result than I obtained in the last Section working strictly in 3-space, but a form more sympathetic to the concepts of **projective geometry**.

9.A Appendix A: Demonstration of Acute-angled Triangle Property for Three Dimensions

First of all return to *Figure 9.2.2*, but imagine now that we are in the three-dimensional case. Then this construction must now refer to the plane formed by VP_x and VP_y in the **picture plane** and the **station point** SP. In other words we are now looking at the VP_x-VP_y-SP *face* of the cube seen in *Figure 9.2.3*. Then because the angle VP_x-SP-VP_y is a right angle, the line to VP_z – the *third* axis from SP – must run straight towards us (for right-handed co-ordinates) or straight away from us (for left-handed co-ordinates). It must run perpendicular to the plane of the paper. Now rotate the diagram about the line VP_x–VP_y, either towards us or away from us. The VP_z line will now appear as the dashed line shown in *Figure 9.A.2*, always rotating in the *plane* perpendicular to the VP_x–VP_y line defined by SP. The line shown dashed in *Figure 9.2.2* is really a suggestion of how this would appear to us as the VP_x-VP_y-SP plane is rotated out of the plane of the picture, when our semi-circle would of course actually become an ellipse. The true geometry is best seen in *Figure 9.A.1* of this appendix, which shows an oblique view of the construction (actually for the left-handed case, VP_x being the nearer to us of VP_x and VP_y in this view).

Now here is the key point: the only possible VP_z positions are those which lie on the intersection of one of these dashed lines and the **picture plane**, which is the base plane of *Figure 9.A.1. No other positions are accessible consistent with the cube-corner construction.* That places bounds on the accessible area defined by the line in the **picture plane** passing through VP_x perpendicular to the VP_x–VP_y line and the line though VP_y perpendicular to the VP_x–VP_y line. No points outside these two lines can be reached, and by the perpendicular construction of these lines, the angles at VP_x, VP_z-VP_x-VP_y and at VP_y, VP_x-VP_y-VP_z must both be less than 90°. But more than that, the *nearest* position to the VP_x–VP_y line that can be accessed is *when the semi-circle rolls right over to lie flat in the picture plane*, so points that lie *within* the semi-circles defined on either side of the VP_x–VP_y line are again inaccessible. Only such points Q would have the angle VP_x-Q-VP_y more than a right-angle. All points lying in the picture plane *outside* these semi-circles and so all points accessible to this construction – call them P – must have their VP_x-P-VP_y angle *less than a right-angle*. So the angle VP_x-VP_z-VP_y is also an acute angle. QED!

(The total set of accessible VP_z positions in the picture plane for a given VP_x–VP_y line are shown in the shaded area of *Figure 9.A.2*.)

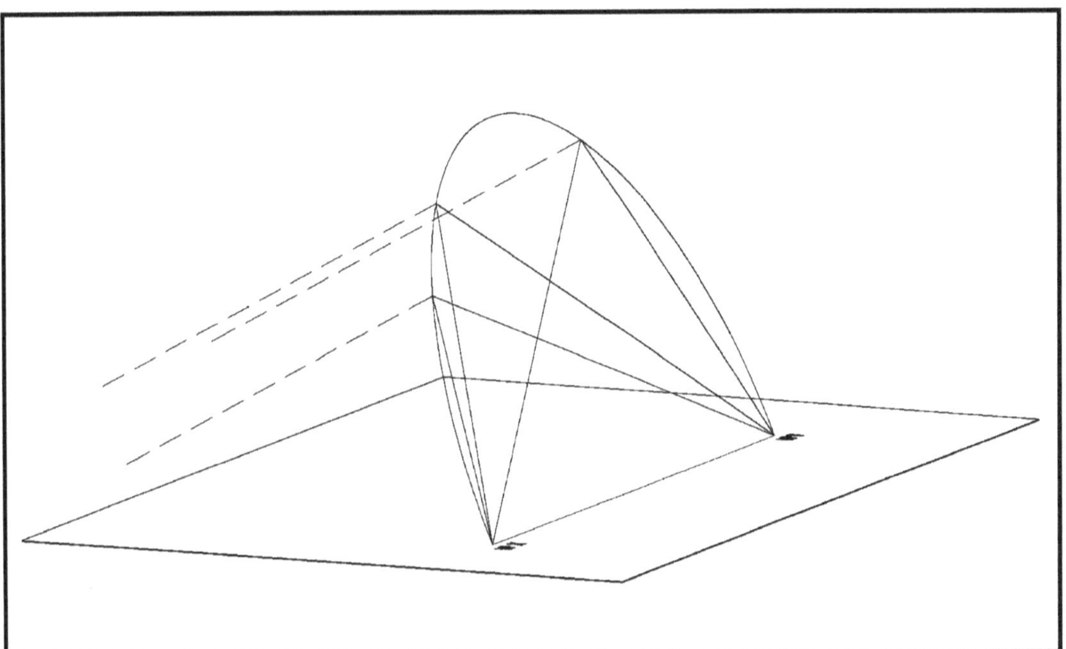

Figure 9.A.1

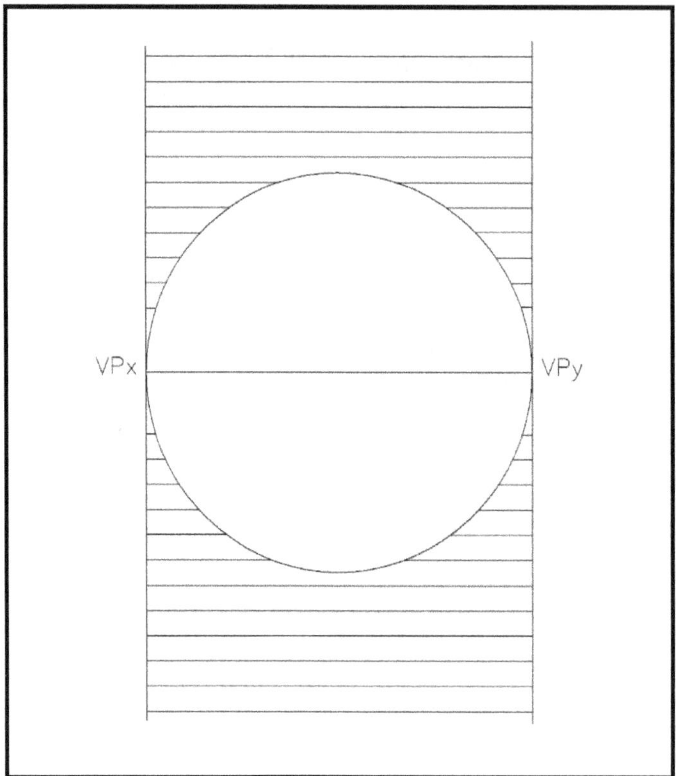

Figure 9.A.2

9.X Exercises

1. The discovery of **perspective** by Brunelleschi was one of the seminal discoveries of modern science. Write a critical essay on the beginnings of the **perspective** revolution in art, and why Alberti chose to simplify the concept by the introduction of the simplifying notions of one-point, two-point and three-point perspective. David Hockney has written a book espousing the idea that it was impossible that painters like Caravaggio could have created the realism they did by the understanding of perspective alone, but must have used some form of *camera obscura*. Comment.

2. Taking the 4×4 **matrix** shown in Equation 9.3.1, which I'll call **Φ**, and assuming it acts on a point (*x*, *y*, *z*,1) where the 1 differentiates as **point** from a **vector** (the difference between two **points** and so not subject to **translation** changes, but ony to **rotations** and **scaling**), show that the **inverse** transformation is given by $\begin{pmatrix} \mathbf{M}^{-1} & -\mathbf{M}^{-1}\mathbf{T} \\ \mathbf{0} & 1 \end{pmatrix}$, where **M** is the original 3×3 **matrix** constituting the upper left part of **Φ**, the **0** indicates three zeroes in a row, and the $-\mathbf{M}^{-1}\mathbf{T}$ indicates the column 3-**vector** obtained by applying $-\mathbf{M}^{-1}$ to **T** = $\begin{pmatrix} \mu_x \\ \mu_y \\ \mu_z \end{pmatrix}$, the **translation column vector** in the original **matrix** in (9.3.1). (Lengyel,

Mathematics for 3D Game Programming and Computer Graphics, Charles River Media, 2004.)

3. A plane in ordinary space is generally described by an equation $Ax + By + Cz + D = 0$. Putting $\mathbf{P} = (x, y, z)$ and $\underline{\mathbf{n}} = (A, B, C)/\mathbf{norm}\ (A, B, C)$, this becomes $\mathbf{P} \cdot \underline{\mathbf{n}} + D' = 0$, where $D' = D/\ \mathbf{norm}\ (A, B, C)$. Then $d = \mathbf{Q} \cdot \underline{\mathbf{n}} + D'$ gives the distance, signed according to the direction of $\underline{\mathbf{n}}$, of the arbitrary point $\mathbf{Q}$ from the plane. Show that the intersection of a line $\mathbf{p}(t) = \mathbf{S} + t\mathbf{V}$ with the plane is given by $t = -(\underline{\mathbf{n}}\cdot\mathbf{S} + D')/\mathbf{N}\cdot\mathbf{V}$, or writing the plane as the **4-vector** $\mathbf{L} = (\underline{\mathbf{n}}, D')$, $t = -\mathbf{L}\cdot\mathbf{S}/\mathbf{L}\cdot\mathbf{V}$, where $\mathbf{V}$, as a **4-vector**, is taken to have $w = 0$. (Lengyel, *op.cit.*)

4. Show that the **4-matrix** $(1/z)\times \begin{pmatrix} \frac{2n}{r-l} & 0 & \frac{r+l}{r-l} & 0 \\ 0 & \frac{2n}{t-b} & \frac{t+b}{t-b} & 0 \\ 0\ \ 0 & & \frac{-(f+n)}{f-n} & \frac{2nf}{f-n} \\ 0\ \ 0 & & -1 & 0 \end{pmatrix}$ acting on $\begin{pmatrix} x \\ y \\ z \\ 1 \end{pmatrix}$, this last representing an original point of 3-space in **homogeneous coordinates** with $w = 1$ to indicate a point, maps points with $z = f$ to $z' = -1$, with $z = n$ to $z' = +1$, with $x/z = l/n$ to $x' = -1$, with $x/z = r/n$ to $x' = +1$, and with $y/z = b/n$ to $y' = -1$, with $y/z = t/n$ to $y' = +1$, where n is the near plane and is *negative* and f is the far plane and is *negative* as camera space points towards $-z$. Show that this matrix transforms all points in the original space within the **homogeneous** values $-l/n < -x/z < -r/n$, and $-b/n < y/z < -t/n$, and $-n < -z < -f$ into a cube of dimensions $[-1, 1] \times [-1, 1] \times [-1, 1]$ whilst preserving depth information. (All the inequalities seem crazy unless you remember all depth values are negative!) (Lengyel, *op.cit.*)

5. The device used in the third proof given of **Desargues' s theorem** I described as a favourite in **projective geometry**. By assigning arbitrary points in the plane figure as being [1, 0, 0], [0, 1, 0], and [0, 0, 1] we define new $\mathbf{e}_1$, $\mathbf{e}_2$ and $\mathbf{e}_3$ for any given **origin**. In that proof, I chose the **origin** [0, 0, 0] to be Ω so that I worked in 3-space from the start. But often when applying **projective** methods to plane geometry we use what is called the **Fundamental Theorem of Projective Geometry** (Brannan, Esplen and Gray) by which, as long as no three of A, B, C and D are **collinear**, we can always find an **origin** that assigns **homogeneous coordinates** of [1, 0, 0] to A, [0, 1, 0] to B, [0, 0, 1] to C, *and* [1, 1, 1] to D. Use this to prove **Pappus's theorem** that given two lines Λ_1 and Λ_2 in the plane (the ordinary plane) with A, B, C lying on Λ_1 and A', B' and C' along Λ_2, and defining $AB' \vee A'B =: P$, $AC' \vee A'C =: Q$, and $BC' \vee B'C =: R$, then P, Q and R are **collinear**. (Set A = [1, 0, 0], A' = [0, 1, 0], P = [0, 0, 1] and R = [1, 1, 1]).

6. Prove the **unique fourth point theorem** of **cross-ratios**: if $CR(A, B, C, X) = CR(A, B, C, Y)$ for **P-points** A, B, C, X and Y, then $X = Y$. Hence prove that if $CR(A, B, C, D) = CR(A, E, F, G)$ then BE, CF and DG are concurrent. (Brannan, Esplen and Gray)

7. Prove **Pappus's theorem** using **cross-ratios**: ley V be $\Lambda_1 \vee \Lambda_2$. Define $S := AC' \vee BA'$, $T := CA' \vee BC'$. Then *from A* as "eye-point" or "origin", V and B are on the same **P-Line**, as are A' with itself, and B' with R and C' with S. So $CR(V, A', B', C') = CR(B, A', R, S)$. Similarly, from C as "eye-point" we have $CR(V, A', B', C') = CR(B, T, P, C')$. Using this and the last result, prove **Pappus's theorem**. Note that the first equation here uses V and B as being collinear, *even*

though they're on opposite sides of A. "Between-ness" is not a projective property, and the **cross-ratio** still holds. (Brannan, Esplen and Gray)

8. Using a simple limiting argument, show that **cross-ratio** still holds even when one of the points involved is an **ideal point** with respect to one of the **picture planes** involved – i.e. it lies in a direction from the origin parallel to this **picture plane**. Work out the simplified **cross-ratio** formulae for all four such cases: *A* **ideal**, *B* **ideal** or *C* or *D*.

9. The **commutator product** between two **multivectors** (elements of a **geometric algebra**) is defined as **M** × **N** = ½(**MN** – **NM**) where the products on the right are **geometric products**. Show that this obeys the **Jacobi identity**: **L** × (**M** × **N**) = (**L** × **M**) × **N** + **M** × (**L** × **N**).

10. Show that Pappus's Theorem can be proved as the identity ⟨(**a** ∧ **a**')(**b** ∧ **b**')(**c** ∧ **c**')⟩ + ⟨(**a** ∧ **b**')(**b** ∧ **c**')(**c** ∧ **a**')⟩ + ⟨(**a** ∧ **c**')(**b** ∧ **a**')(**c** ∧ **b**')⟩ = 0, using the same identity we used to prove **Desargues's Theorem**.

Appendix

METHODS OF INTEGRATION

A.1 Introduction

Because **integration** is not what might be called a deductive process but one requiring a certain amount of trial and error and inspired guesswork, it is difficult for students first coming to the subject.

Equally, because it doesn't involve anything new *conceptually*, it doesn't really fit into the framework of a volume such as this one which is devoted to the conceptual foundations of its subject.

So I've decided to relegate to this Appendix a brief summary, somewhat in note form, of some of the basic tricks used to find the **antiderivatives** of a given **differential form**.

Many of these methods involve a change in the variable of integration, and although this Appendix will only briefly touch on such matters it's important to remember this rule:

> Whenever you change a variable of integration, you must make a matching change in the **limits of integration** on the right-hand side of the **integration** symbol " | ".

So if you have an **integral** of the form $\phi(x).\mathbf{d}x \mid [a, b]$, if you replace x by, say, $y = x^2$, you must also replace $\mathbf{d}x$ by $\frac{1}{2}y^{-\frac{1}{2}}.\mathbf{d}y$ (by: $\mathbf{d}y = 2x.\mathbf{d}x \rightarrow \mathbf{d}x = (1/2x).\mathbf{d}y = \frac{1}{2}(1/\sqrt{y}).\mathbf{d}y$) but *more importantly* remember to replace $[a, b]$ by $[a^2, b^2]$ because if $x = a$, <u>y will now be a^2</u>.

This is very easily forgotten!

In this example the integral is changed to:

$$\phi(\sqrt{y}).\ \tfrac{1}{2}y^{-\frac{1}{2}}.\mathbf{d}y \mid [a^2, b^2].$$

The subject of "Methods of Integration" is very standard and appears in a very similar way in every basic calculus textbook. This Appendix simply skims through the key topics as covered in an early edition of the excellent standard text by G.B.Thomas, *Calculus and Analytic Geometry*, published by

Addison-Wesley, albeit with some modifications, for example where my ε notation is able to give new insights. You should refer to this or another standard text for more details.

A.2 Use of Inverse Trigonometric Functions

I'll start with the case which came up in Section 4.4 – the integral needed to get the **arc length** around the circle. In that Section I tried to show how the answer is developed from first principles, but in practice tables of the **inverse trigonometric functions** such as $\mathbf{sin}^{-1}(y)$ and $\mathbf{cos}^{-1}(x)$, giving the **arc length** θ in terms of the **sine** or **cosine** are readily available.

I'll first consider the $\mathbf{cos}^{-1}$ integral using my ε notation.

Suppose $u = \varepsilon_x(\tau)$. Then $\mathbf{d}u = -2\pi.\varepsilon_y(\tau).\mathbf{d}\tau = -2\pi.\sqrt{(1 - \varepsilon_x^2(\tau))}.\mathbf{d}\tau = -2\pi.\sqrt{(1 - u^2)}.\mathbf{d}\tau$.

But this means that:

$$2\pi.\mathbf{d}\tau = = 2\pi.\mathbf{d}\varepsilon_x^{-1}(u) = \frac{-\mathbf{d}u}{\sqrt{(1-u^2)}}.$$

This means that if we have an integral like $\dfrac{-\mathbf{d}u}{\sqrt{(1-u^2)}}\Big|\,[0, ½]$, we can evaluate it as:

$$-2\pi.\mathbf{d}\varepsilon_x^{-1}(u)\ \big|\ [0, ½] = -2\pi.(\varepsilon_x^{-1}(½) - \varepsilon_x^{-1}(0)).$$

The problem here is not the one boxed above. Instead it's that these $\varepsilon_x^{-1}(u)$ values are ambiguous, reflecting the ± nature of the square root in the $\dfrac{-\mathbf{d}u}{\sqrt{(1-u^2)}}$ **integrand** (which means the **differential form** being integrated).

On the **interval** [0, 1], $\varepsilon_x(\tau)$ takes the value of ½ at 60°, or 1/6 of the way round the **unit circle**, but also at 300° (−60°), or 5/6. It takes the value 0 at *either* ¼ or ¾.

So

$\varepsilon_x^{-1}(½) = 1/6, 5/6,$
$\varepsilon_x^{-1}(0) = ¼, ¾.$

If we assume the positive square root, we should then get:

$$-2\pi(1/6 - ¼) = 2\pi(¼ - 1/6) = (1/6)\pi.$$

Using 5/6 and ¾ would give the opposite value: $-2\pi(5/6 - ¾) = -2\pi((10 - 9)/12) = -(1/6)\pi$.

As should be apparent in *Figure A.2.1* below, these are in fact the only permissible answers, because if we paired, say, 1/6 and ¾ as end points, any **arc** connecting these passes either through (1, 0) = 1 or through (−1, 0) = −1, inevitably having a **cosine** projection *outside* the highlighted band between 0 and ½, and so *outside* the [0, ½] **interval**. This is a subtle point but important.

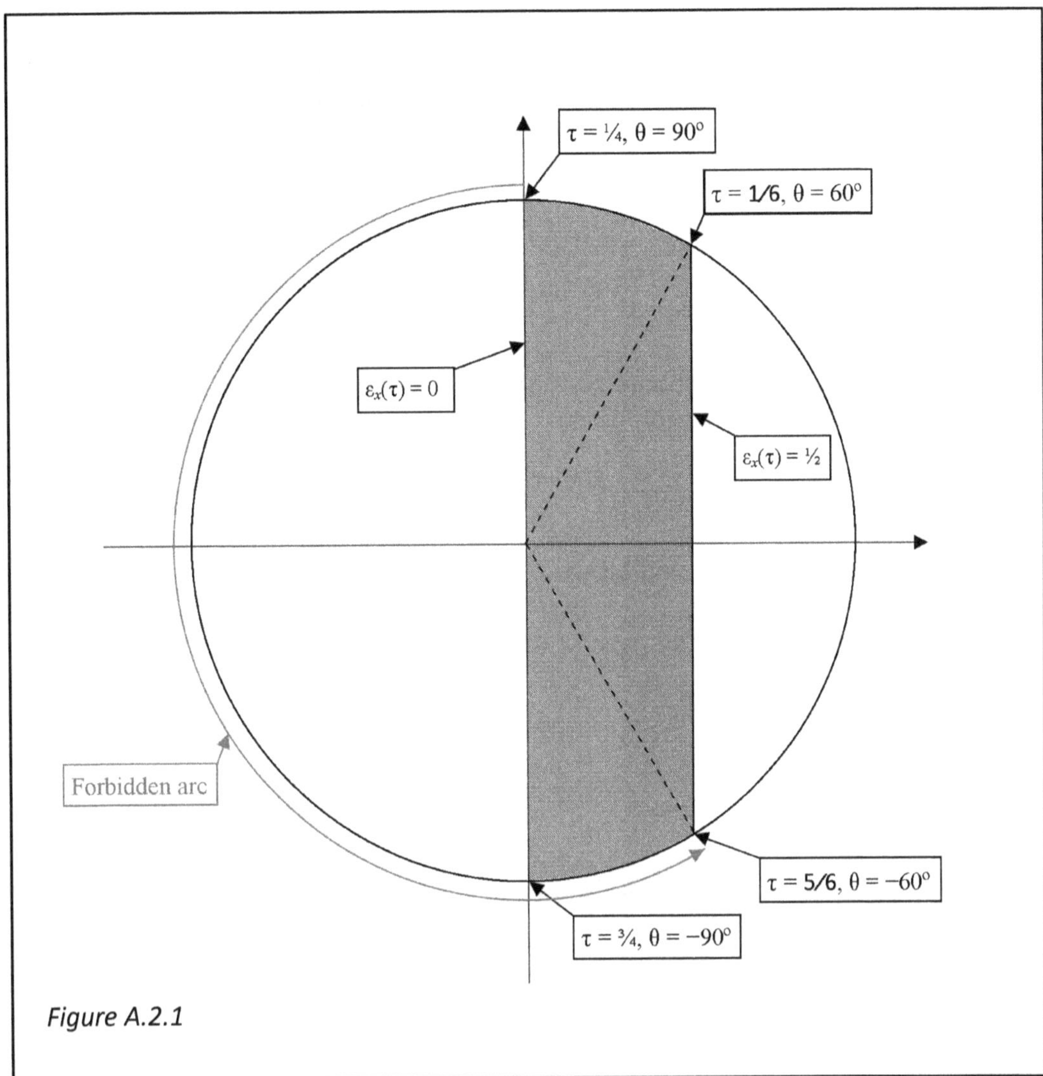

Figure A.2.1

Note that the assumption that the **integrand** be **real** limits the permissible values of u to the **interval** [−1, 1] because outside that $\sqrt{(1-u^2)}$ becomes **imaginary**.

As it happens, because the **differential** formulae for $\varepsilon_x(\tau)$ and $\varepsilon_y(\tau)$ involve a 2π which isn't present in the "straight" functions **sin**(θ) and **cos**(θ), this is one place where the traditional notation is actually slightly simpler.

Taking the $\mathbf{sin}^{-1}$ case this time,

$$u = \mathbf{sin}(\theta) \longrightarrow \mathbf{d}u = \mathbf{cos}(\theta).\mathbf{d}\theta = \sqrt{(1 - \mathbf{sin}^2(\theta))}.\mathbf{d}\theta,$$

so

$$\frac{\mathrm{d}u}{\surd(1-u^2)} = \mathbf{d}(\mathbf{sin}^{-1}(u)),$$

without the − sign in the **sin**/ε_y case. The analogous interval to that above would be, in terms of values of the **sine**, [0, ½] would correspond to an **arc** from (1, 0) (the *unique* point where **sin**(θ) = 0) to ± π/6, or to 30° or 330°, because 12 × π/6 = 2π and 1/12th of the circle is 30°.

So $$\frac{\mathrm{d}u}{\surd(1-u^2)} \,|\, [0, ½] = \mathbf{d}(\mathbf{sin}^{-1}(u)) \,|\, [0, ½] = \mathbf{sin}^{-1}(½) - \mathbf{sin}^{-1}(0) = \pm \pi/6 - 0 = \pm \pi/6.$$

The third most commonly used **trigonometric function** after **sin**() and **cos**() is the ratio of the two, called the **tangent function** (just to be confusing):

$$\mathbf{tan}(\theta) = \mathbf{sin}(\theta)/\mathbf{cos}(\theta) = \varepsilon_y(\tau)/\varepsilon_x(\tau).$$

So $$\mathbf{d}(\mathbf{tan}(\theta)) = (\mathbf{cos}(\theta).\mathbf{d}(\mathbf{sin}(\theta)) - \mathbf{sin}(\theta).\mathbf{d}(\mathbf{cos}(\theta)))/\mathbf{cos}^2(\theta)$$

$$= (\mathbf{cos}^2(\theta).\mathbf{d}\theta + \mathbf{sin}^2(\theta).\mathbf{d}\theta)/\mathbf{cos}^2(\theta) = (1 + \mathbf{tan}^2(\theta)).\mathbf{d}\theta$$

so we have, setting u = **tan**(θ),

$$\frac{\mathrm{d}u}{(1+u^2)} = \mathbf{d}\theta = \mathbf{d}(\mathbf{tan}^{-1}(u))$$

a particularly neat result. Again tables of **tan**$^{-1}$ are readily available.

Being the ratio of **sin** to **cos**, **tan**(θ) can take on any values over the entire **real line** [−∞, ∞], so, looking at this the other way, the argument u in this integral can also span any part of the **real line** so

$$\frac{\mathrm{d}u}{(1+u^2)} \,|\, [u_1, u_2] = \mathbf{d}(\mathbf{tan}^{-1}(u)) \,|\, [u_1, u_2]$$

is well-defined for *any* u_1 and u_2, as should also be apparent from the nature of the expression $(1 + u^2)$ without any limiting square root.

There are three other basic **trigonometric functions**:

sec(θ) := 1/**cos**(θ) **csc**(θ) := 1/**sin**(θ) **cot**(θ) = 1/**tan**(θ) = **cos**(θ)/**sin**(θ).

They are called respectively the **secant**, the **cosecant** and the **cotangent** and all appeared in the early Renaissance, heralding the development of modern mathematics, but still nearly two centuries before the **calculus**.

Note these must be written as above or as, e.g. **sec**(θ) = (**cos**(θ))$^{-1}$, *not* as **cos**$^{-1}$(x) which means the **inverse** of **cos**(θ) – the θ value that has a **cosine** of x, not the **reciprocal** of **cos**(θ).

A useful identity is:

$$\mathbf{tan}^2(\theta) + 1 = \mathbf{sec}^2(\theta)$$

which shows that above I could have written

$$\mathbf{d}(\mathbf{tan}(\theta)) = (1 + \mathbf{tan}^2(\theta)).\mathbf{d}\theta = \mathbf{sec}^2(\theta).\mathbf{d}\theta.$$

We have

$$\mathbf{d}(\mathbf{sec}(\theta)) = \mathbf{d}((\mathbf{cos}(\theta))^{-1}) = -((\mathbf{cos}(\theta))^{-2}).\mathbf{d}(\mathbf{cos}(\theta)).\mathbf{d}\theta = +((\mathbf{cos}(\theta))^{-2}).\mathbf{sin}(\theta).\mathbf{d}\theta$$

$$= \mathbf{sec}(\theta).\mathbf{tan}(\theta).\mathbf{d}\theta.$$

If here we set u = **sec**(θ), we have, using the identity above:

$$\mathbf{d}u = u.\sqrt{}(u^2 - 1).\mathbf{d}\theta$$

or

$$\mathbf{d}\theta = \frac{\mathbf{d}u}{u.\sqrt{}(u^2 - 1)}.$$

$\sqrt{}(u^2 - 1)$ here is a representation of **tan**(θ), and the ambiguous root is + if $\theta \in [0, ½\pi]$, and – (*negative*) if $\theta \in [½\pi, \pi]$. But in these intervals u itself (**cos**(θ))$^{-1}$) is also + and – respectively, so

$$\mathbf{d}\theta = \frac{\mathbf{d}u}{u.\sqrt{}(u^2 - 1)} \text{ if } u > 0$$

$$\mathbf{d}\theta = \frac{\mathbf{d}u}{-u.\sqrt{}(u^2 - 1)} \text{ if } u < 0$$

and this is customarily summarized as

$$\mathbf{d}\theta = \frac{\mathbf{d}u}{|u|.\sqrt{}(u^2 - 1)}$$ using the expression for absolute value.

This gives us the first of two extra formulae:

$$\frac{\mathbf{d}u}{|u|.\sqrt{}(u^2 - 1)} = \mathbf{d}(\mathbf{sec}^{-1}(u))$$

$$\frac{-\mathbf{d}u}{|u|.\sqrt{}(u^2 - 1)} = \mathbf{d}(\mathbf{csc}^{-1}(u))$$

There's a form for **d(cot**$^{-1}$(u)) too, taking the same form as that for **d(tan**$^{-1}$(u)) except **d**u → −**d**u much as we see with the pair above.

The closely similar algebra of the **hyperbolic functions** means that they too give such formulae, which I express both in conventional noitation and my η-notation of Section 4.10:

$$\frac{\mathbf{d}u}{\sqrt{(1+u^2)}} = \mathbf{d}(\sinh^{-1}(u)) = \mathbf{d}(\eta_y^{-1}(u)) \qquad \frac{\mathbf{d}u}{\sqrt{(u^2-1)}} = \mathbf{d}(\cosh^{-1}(u)) = \mathbf{d}(\eta_x^{-1}(u)).$$

Defining **tanh**(θ) = **sinh**(θ)/**cosh**(θ), we get:

$$\frac{\mathbf{d}u}{(1-u^2)} = \mathbf{d}(\tanh^{-1}(u))$$

which only holds for integrations within the **interval** $u \in [-1, 1]$, but outside that this same integral appears as:

$$\frac{\mathbf{d}u}{(1-u^2)} = \mathbf{d}(\coth^{-1}(u)) \text{ for } \quad |u| > 1$$

where **coth**(θ) = 1/**tanh**(θ).

We also have:

$$\frac{-\mathbf{d}u}{u\sqrt{(1-u^2)}} = \mathbf{d}(\operatorname{sech}^{-1}(u))$$

$$\frac{-\mathbf{d}u}{|u|.\sqrt{(1+u^2)}} = \mathbf{d}(\operatorname{csch}^{-1}(u)),$$

using **sech**(θ) = 1/**cosh**(θ), **csch**(θ) = 1/**sinh**(θ).

A.3 Other Trigonometric Integrals

For any n, setting u = **sin**(ax) so that **d**u = a.**cos**(ax).**d**x makes

$$\sin^n(ax).\cos(ax).\mathbf{d}x = a^{-1}u^n.\mathbf{d}u$$

and this is easily integrated as $(a.(n+1))^{-1}\mathbf{d}(u^{n+1})$, except in the special case where $n = -1$, when this becomes

$$a^{-1}.\mathbf{d}{\downarrow}|u| = a^{-1}.\mathbf{d}(\ln|u|).$$

This gives:

$$\mathbf{sin}^n(ax).\mathbf{cos}(ax).\mathbf{d}x = (a.(n + 1))^{-1}\mathbf{d}(\mathbf{sin}^{n+1}(ax))$$

and

$$\mathbf{cot}(ax).\mathbf{d}x = a^{-1}.\mathbf{d}(\mathbf{ln}|\mathbf{sin}(ax)|),$$

results you might not expect.

Another in the same family is

$$\mathbf{sin}^3(x).\mathbf{d}x$$

which we can expand to $(1 - \mathbf{cos}^2(x)).\mathbf{sin}(x).\mathbf{d}x = -(1 - u^2).\mathbf{d}u$ using $u = \mathbf{cos}(x)$, giving

$$\mathbf{d}(⅓u^3 - u) = \mathbf{d}(⅓\mathbf{cos}^3(x) - \mathbf{cos}(x)).$$

$\mathbf{tan}^2(x).\mathbf{d}x$ can be solved by $u = \mathbf{tan}(x) \rightarrow \mathbf{d}u = \mathbf{sec}^2(x).\mathbf{d}x = (1 + u^2).\mathbf{d}x$,
so

$$\mathbf{tan}^2(x).\mathbf{d}x \rightarrow u^2.\mathbf{d}u/(1 + u^2).$$

But it's easier just to put:

$$\mathbf{tan}^2(x).\mathbf{d}x = (\mathbf{sec}^2(x) - 1).\mathbf{d}x = \mathbf{sec}^2(x).\mathbf{d}x - \mathbf{d}x = \mathbf{d}(\mathbf{tan}^2(x)) - \mathbf{d}x,$$

both of which are directly integrable, e.g.:

$$\mathbf{d}(\mathbf{tan}^2(x)) - \mathbf{d}x\,|\,[a, b] = \mathbf{tan}^2(b) - b - \mathbf{tan}^2(a) + a.$$

Talking of **tan**,

$$\mathbf{tan}(x).\mathbf{d}x = (\mathbf{sin}(x)/\mathbf{cos}(x)).\mathbf{d}x = -\mathbf{d}(\mathbf{cos}(x))/\mathbf{cos}(x) = -\mathbf{d}(\downarrow\mathbf{cos}(x)) = -\,\mathbf{d}(\mathbf{ln}(\mathbf{cos}(x))),$$

but here we need to keep in mind that ↓ or **ln** is only defined for *positive* numbers.

From $(x + jy)^2 = x^2 - y^2 + 2jxy$, for (x, y) on the **unit circle**, we have:

$$\mathbf{cos}(2\theta) = \mathbf{cos}^2(\theta) - \mathbf{sin}^2(\theta)$$

and

$$\mathbf{sin}(2\theta) = 2.\mathbf{cos}(\theta).\mathbf{sin}(\theta)$$

because squaring the **complex number** form of an **angle** corresponds to doubling the **angle** itself.

So from the $\mathbf{cos}(2\theta)$ formula above,

$$1 + \mathbf{cos}(2\theta) = \mathbf{cos}^2(\theta) + \mathbf{sin}^2(\theta) + \mathbf{cos}^2(\theta) - \mathbf{sin}^2(\theta) = 2\mathbf{cos}^2(\theta)$$

and

$$1 - \cos(2\theta) = \cos^2(\theta) + \sin^2(\theta) - \cos^2(\theta) + \sin^2(\theta) = 2\sin^2(\theta).$$

These formulae enable us to evaluate, for example,

$$\cos^4(x).dx = (\cos^2(x))^2.dx = (½(1 + \cos(2x))^2.dx = ¼(1 + 2\cos(2x) + \cos^2(2x)).dx$$

$$= ¼(1 + 2\cos(2x) + ½(1 + \cos(4x))).dx$$

$$= ⅜.dx + ¼.d(\sin(2x)) + (1/32).d(\sin(4x)).$$

The **inverse trigonometric functions** can also be used to evaluate integrals containing forms like $(a^2 \pm u^2)$ rather than just $(1 \pm u^2)$.

For example, in

$$\frac{du}{\sqrt{(a^2 - u^2)}}$$

set $u = a.\sin(x)$

$\rightarrow$ 1. $du = a.\cos(x).dx$
$\rightarrow$ 2. $a^2 - u^2 = a^2(1 - \sin^2(x)) = a^2\cos^2(x)$.

So $\quad (a^2 - u^2)^{-½}du = (a^2\cos^2(x))^{-½}\ a.\cos(x).dx = \pm(a.\cos(x)/a.\cos(x)).dx = \pm dx$

and this equals $\pm d(\sin^{-1}\left(\frac{u}{a}\right))$ from the original $u = a.\sin(x)$.

Another example is

$$\frac{du}{(a^2+ u^2)} = \left(\frac{1}{a}\right)d(\tan^{-1}\left(\frac{u}{a}\right)).$$

and another is:

$$\frac{du}{\sqrt{(a^2 - u^2)}} = \pm\sec(x).dx = d(\ln|\sec(x) + \tan(x)|)$$

using $u = a.\tan(x)$, and this can be worked back to give:

$$d(\ln(\sqrt{(a^2 + u^2)} + u))$$

from $\tan(x) = u/a$ and $\sec(x) = \sqrt{(1 + \tan^2(x))} = \sqrt{(1 + u^2/a^2)}$, because dividing through by a merely adds $-\ln(a)$ to the result and $d(\ln(a)) = 0$.

Finally a rather more exotic substitution can be obtained this way:

$$1 + \mathbf{cos}(2\theta) = 2\mathbf{cos}^2(\theta) \rightarrow \mathbf{cos}(u) = 2.\mathbf{cos}^2(u/2) - 1 = 2/\mathbf{sec}^2(u/2) - 1$$

$$= 2/(1 + \mathbf{tan}^2(u/2)) - 1.$$

So if we set $z = \mathbf{tan}(u/2)$, we get

$$\mathbf{cos}(u) = 2/(1 + z^2) - 1 = (2 - (1 + z^2))/(1 + z^2) = (1 - z^2)/(1 + z^2).$$

$\mathbf{d}(\mathbf{tan}(\theta)) = (1 + \mathbf{tan}^2(\theta)).\mathbf{d}\theta$ now gives, using $\theta = u/2$:

$$\mathbf{d}z = (1 + z^2).\mathbf{d}\theta = \tfrac{1}{2}(1 + z^2).\mathbf{d}u \rightarrow \mathbf{d}u = 2.\mathbf{d}z/(1 + z^2).$$

A simple example of the application of this is:

$$(1 + \mathbf{cos}(u))^{-1}.\mathbf{d}u = \{2/(1 + z^2)\}^{-1}.\{2.\mathbf{d}z/(1 + z^2)\}$$

using $\mathbf{cos}(u) = 2/(1 + z^2) - 1$ and the formula for $\mathbf{d}u$, and this is:

$$\{(1 + z^2)/2\}\{2.\mathbf{d}z/(1 + z^2)\} = \mathbf{d}z = \mathbf{d}(\mathbf{tan}(u/2))!$$

Everything else cancels out.

A.4 Use of Partial Fractions

I've used **partial fractions** in Section 8.6. Given an **integrand** $[\phi(x)/\psi(x)].\mathbf{d}x$, we work on the factor $\phi(x)/\psi(x)$:

- if the **degree** (highest power) of $\phi(x)$ is greater than the degree of $\psi(x)$, perform a **long division** as in Section 1.9, and work with the **remainder**

then

- if $(x - r)$ is a linear factor of $\psi(x)$, then if $(x - r)^n$ is the highest power of $(x - r)$ that divides $\psi(x)$ assign n **partial fractions** for $(x - r)$ as:

$$\frac{A_1}{(x-r)} + \frac{A_2}{(x-r)^2} + \ldots + \frac{A_n}{(x-r)^n}.$$

- if $(x^2 + px + q)$ is a quadratic factor of $\psi(x)$, then if $(x^2 + px + q)^m$ is the highest power of $(x^2 + px + q)$ that divides $\psi(x)$ assign n **partial fractions** for $(x^2 + px + q)$ as:

$$\frac{B_1x+C_1}{(x^2+px+q)} + \frac{B_2x+C_2}{(x^2+px+q)^2} + \ldots + \frac{B_mx+C_m}{(x^2+px+q)^m}.$$

So, for example, suppose we have:

$$\psi(x) = x^4 - 2dx^3 + (c + d^2)x^2 - 2cdx + cd^2.$$

Remarkably, this can be factorised as:[156]

$$((x^2 + c)(x - d)^2) = (x^2 + c)(x^2 - 2dx + d^2) = x^4 - 2dx^3 + d^2x^2 + cx^2 - 2cdx + cd^2.$$

So an expression like

$$(px + q)/(x^4 - 2dx^3 + (c + d^2)x^2 - 2cdx + cd^2) = (px + q)/((x^2 + c)(x - d)^2).$$

This can be expanded as the set of **partial fractions**:

$$(px + q)/((x^2 + c)(x - d)^2) = \frac{B_1x+C_1}{(x^2 + c)} + \frac{A_1}{(x-d)} + \frac{A_2}{(x-d)^2}.$$

Now multiply through by $((x^2 + c)(x - d)^2)$:

$$px + q = (B_1x + C_1)(x - d)^2 + A_1(x^2 + c)(x - d) + A_2(x^2 + c)$$

$$= (B_1x + C_1)(x^2 - 2dx + d^2) + A_1(x^3 - dx^2 + cx - cd) + A_2(x^2 + c)$$

$$= B_1x^3 - 2B_1dx^2 + B_1d^2x + A_1x^3 + (C_1 - A_1d + A_2)x^2 + (-2dC_1 + A_1c)x$$

$$+ (C_1d^2 - A_1cd + A_2c).$$

Equating like powers of x:

x^3: $B_1 + A_1 = 0$
x^2: $-2B_1d + C_1 - A_1d + A_2 = 0$
x: $p = B_1d^2 + (-2dC_1 + A_1c)$
1: $q = C_1d^2 - A_1cd + A_2c.$

We have four equations in four unknowns $\{A_1, A_2, B_1, C_1\}$, all linear. Eliminate the first as $A_1 = -B_1 =: -B$ (i.e. writing B_1 as B) to give:

$$-2Bd + C_1 + Bd + A_2 = 0$$
$$p = Bd^2 + (-2dC_1 - Bc)$$
$$q = C_1d^2 + Bcd + A_2c.$$

Now similarly write A_2 as A and C_1 as C:

---

156 Obviously this is how I worked it out!

$$- Bd + C + A = 0$$
$$p = Bd^2 + (-2dC - Bc)$$
$$q = Cd^2 + Bcd + Ac.$$

The first gives $C = Bd - A$ which gives:

$$p = Bd^2 + (-2d(Bd - A) - Bc) = (B - 2B)d^2 + 2Ad - Bc = 2Ad - Bd^2 - Bc$$
$$q = (Bd - A)d^2 + Bcd + Ac.$$

From the first,

$$p + B(d^2 - c) = 2Ad \rightarrow A = (p + B(d^2 - c))/2d$$

so finally

$$q = (Bd - (p + B(d^2 - c))/2d)d^2 + Bcd +((p + B(d^2 - c))/2d)c$$

from which we could solve for B! I think × $2d$ gives:

$$2dq = 2Bd^4 - pd^2 - Bd^2(d^2 - c) + 2Bd^2 + pc + Bc(d^2 - c)$$

so

$$B = (2dq + pd^2 - pc)/(2d^4 - d^2(d^2 - c) + 2d^2 + c(d^2 - c)).$$

The denominator expands to:

$$2d^4 - d^4 + d^2c + 2d^2 + cd^2 - c^2 = d^4 + 2d^2 + 2cd^2 - c^2 = d^4 + 2d^2(1 + c) - c^2.$$

The **partial fractions** that result lead to **integrands** of two forms:

- $\dfrac{\mathbf{d}x}{(x-r)^n}$
- $\dfrac{(ax+b)\mathbf{d}x}{(x^2 + px + q)^m}$

The first is easily handled by setting $u = (x - r) \rightarrow$ **d**u = **d**$x \rightarrow u^{-n}$.**d**u, a standard form.
The second is a little more tricky and again I take the analysis from G.B.Thomas, *Calculus and Analytic Geometry*.

First complete the square in the denominator:

$$x^2 + px + q = (x + ½p)^2 + q - ¼p^2.$$

Then set $u = x + ½p$, $c^2 = q - ¼p^2$. This gives $ax + b = a(u - ½p) + b = au + b'$. so we now have:

$\frac{(au+b')\mathbf{d}u}{(u^2+c^2)^m}$ = ½$a.(u^2 + c^2)^{-m}(2u.\mathbf{d}u) + b'.(u^2 + c^2)^{-m}.\mathbf{d}u$.

In the first term here set $z = u^2 + c^2 \rightarrow \mathbf{d}z = 2u.\mathbf{d}u \rightarrow$ ½$a.(u^2 + c^2)^{-m}(2u.\mathbf{d}u)$ = ½$a.z^{-m}.\mathbf{d}z$.

The second is reminiscent of the $\mathbf{tan}^{-1}$ formula from Section A.2, so we try:

$$u = c.\mathbf{tan}(\theta) \rightarrow \mathbf{d}u = c.\mathbf{sec}^2(\theta).\mathbf{d}\theta$$

$$\rightarrow u^2 + c^2 = c^2.\mathbf{sec}^2(\theta)$$

which gives:

$$b'.(u^2 + c^2)^{-m}.\mathbf{d}u = c^{1-2m}.\mathbf{cos}^{2m-2}(\theta).\mathbf{d}\theta.$$

This is evaluated by **integration by parts**.

A.5 Integration by Parts

Let's take the foregoing example:

$$\mathbf{cos}^m(\theta).\mathbf{d}\theta.$$

Express it as

$$\mathbf{cos}^{m-1}(\theta).\mathbf{cos}(\theta).\mathbf{d}\theta = \mathbf{cos}^{m-1}(\theta).(\mathbf{cos}(\theta).\mathbf{d}\theta).$$

Now $\mathbf{cos}(\theta).\mathbf{d}\theta = \mathbf{d}(\mathbf{sin}(\theta))$, so this is:

$$\mathbf{cos}^{m-1}(\theta).\mathbf{d}(\mathbf{sin}(\theta)) = \mathbf{d}(\mathbf{cos}^{m-1}(\theta).\mathbf{sin}(\theta)) - \mathbf{sin}(\theta).\mathbf{d}(\mathbf{cos}^{m-1}(\theta))$$

by

$$\mathbf{d}(xy) = y.\mathbf{d}x + x.\mathbf{d}y \qquad \rightarrow \qquad x.\mathbf{d}y = \mathbf{d}(xy) - y.\mathbf{d}x.$$

$\mathbf{d}(\mathbf{cos}^{m-1}(\theta).\mathbf{sin}(\theta)) \mid [a, b] = \mathbf{cos}^{m-1}(b).\mathbf{sin}(b) - \mathbf{cos}^{m-1}(a).\mathbf{sin}(a)$ directly by the **adjoint rule** so this term drops out, and

$$-\mathbf{sin}(\theta).\mathbf{d}(\mathbf{cos}^{m-1}(\theta)) = -\mathbf{sin}(\theta).(m-1).\mathbf{cos}^{m-2}(\theta).\mathbf{d}(\mathbf{cos}(\theta))$$

$$= (m-1).\mathbf{sin}^2(\theta).\ \mathbf{cos}^{m-2}(\theta).\mathbf{d}\theta.$$

We can use $\mathbf{sin}^2 = 1 - \mathbf{cos}^2$ to convert this to:

$$(m-1).(1 - \mathbf{cos}^2(\theta)).\mathbf{cos}^{m-2}(\theta).\mathbf{d}\theta,$$

and it looks like we're getting nowhere! The highest term is still of degree m. But looking at the expression as a whole (greying out the **d**() term), we have:

$$\mathbf{cos}^m(\theta).\mathbf{d}\theta = \mathrm{d}(\cos^{m-1}(\theta).\sin(\theta)) + (m-1).(1 - \mathbf{cos}^2(\theta)).\mathbf{cos}^{m-2}(\theta).\mathbf{d}\theta$$

$$= \mathrm{d}(\cos^{m-1}(\theta).\sin(\theta)) + (m-1).\mathbf{cos}^{m-2}(\theta).\mathbf{d}\theta - (m-1).\mathbf{cos}^2(\theta).\mathbf{cos}^{m-2}(\theta).\mathbf{d}\theta$$

$$= \mathrm{d}(\cos^{m-1}(\theta).\sin(\theta)) + (m-1).\mathbf{cos}^{m-2}(\theta).\mathbf{d}\theta - (m-1).\mathbf{cos}^m(\theta).\mathbf{d}\theta$$

so, bringing together the two underlined terms in $\mathbf{cos}^m$:

$$m.\mathbf{cos}^m(\theta).\mathbf{d}\theta = \mathrm{d}(\cos^{m-1}(\theta).\sin(\theta)) + (m-1).\mathbf{cos}^{m-2}(\theta).\mathbf{d}\theta.$$

This constitutes a **reduction formula** bringing the degree down by two at each step.

A.6 Contour Integration

A favourite device for evaluating **improper integrals** is to imagine that they are "really" **line integrals** in the **complex plane** that just happen to be along the **real line**.

We start with **Cauchy's integral theorem** that if a function $\phi(z)$ is **analytic** over an area Σ in the **complex plane**, or $\partial_z{*}\phi(z) = 0$ over Σ, then $\phi(z).\mathbf{dz} \mid \mathbf{b}\Sigma = 0$.

Looking at *Figure A.6.1*, we can see that if the integral of $\phi(z)$ around the closed loop $\mathbf{b}\Sigma$ is zero, then taking any two points along $\mathbf{b}\Sigma$, z_1 and z_2, we must have that:

$$\phi(z).\mathbf{dz} \mid \mathbf{b}\Sigma_{z1\to z2} + \phi(z).\mathbf{dz} \mid \mathbf{b}\Sigma_{z2\to z1} = 0$$

or that

$$\phi(z).\mathbf{dz} \mid \mathbf{b}\Sigma_{z1\to z2} = -\ \phi(z).\mathbf{dz} \mid \mathbf{b}\Sigma_{z2\to z1} = 0$$

where I write $\mathbf{b}\Sigma_{z1\to z2}$ to indicate the arc of $\mathbf{b}\Sigma$ *anticlockwise* from z_1 to z_2, and $\mathbf{b}\Sigma_{z2\to z1}$ to indicate the arc, still *anticlockwise*, form z_2 back to z_1 that completes the closed curve $\mathbf{b}\Sigma$.

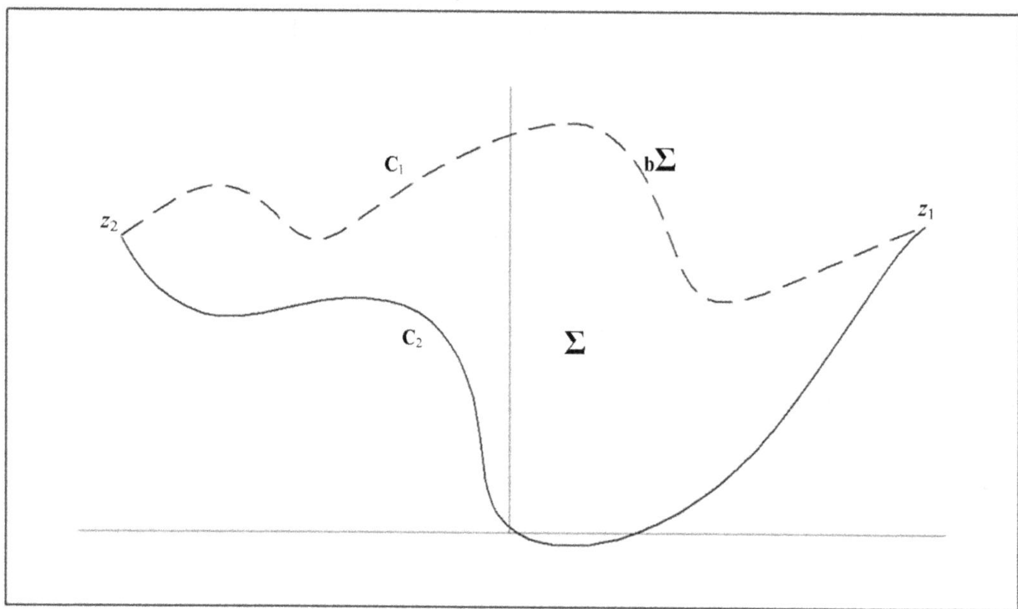

Figure A.6.1

Let's call $\mathbf{C}_1 := \mathbf{b\Sigma}_{z1\to z2}$ and $\mathbf{C}_2 := \mathbf{b\Sigma}_{z2\to z1}$. so

$$\phi(z).\mathbf{d}z \,|\, \mathbf{C}_1 = -\ \phi(z).\mathbf{d}z \,|\, \mathbf{C}_2.$$

Now *reverse* $\mathbf{C}_2$ to $-\mathbf{C}_2$, so that it too now runs from z_1 to z_2 in the *opposite* sense of this same partial arc in the original $\mathbf{b\Sigma}$.
Then we have:

$$\phi(z).\mathbf{d}z \,|\, \mathbf{C}_1 = \phi(z).\mathbf{d}z \,|\, (-\mathbf{C}_2).$$

So now

The integrals over *both* arcs from z_1 to z_2 are equal if $\phi(z)$ is **analytic** ($\partial_z{*}\phi(z) = 0$) over the area enclosed between them (the original Σ).

This is sometimes called the **principle of deformation of path**.

The second part of the argument follows from *Figure A.6.2*. Here we have a representation of an **improper integral** along the **real line**, now depicted in the **complex plane**. The integral is taken over the **interval** $[-a, a]$, but we now *close* the path by the semicircular arc in the upper half-plane Ω_a, subscripted a because the actual arc is defined by the value of a.

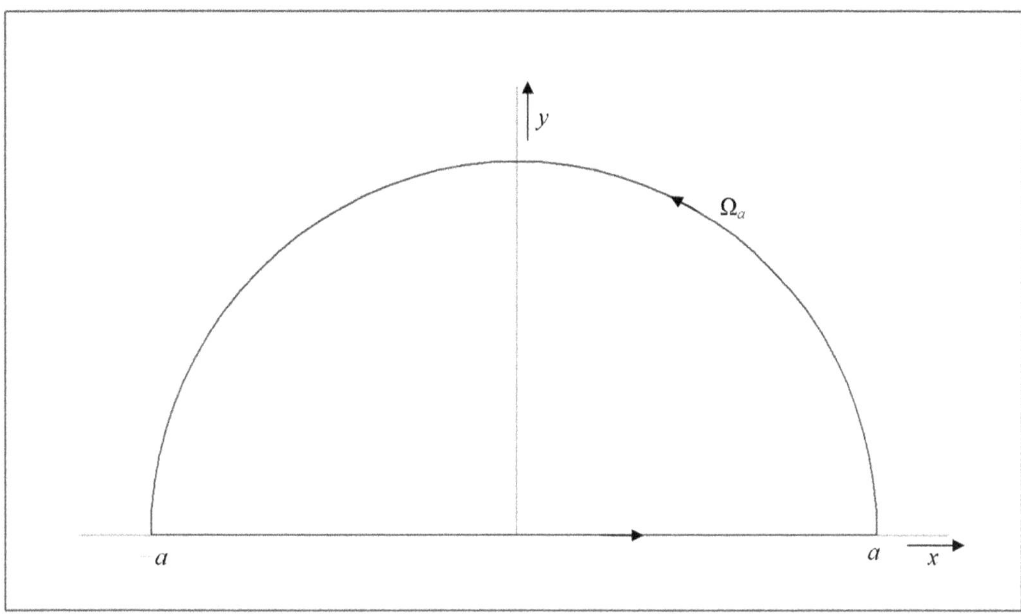

Figure A.6.2

This transforms the path of integration into a *closed curve*, which is also known, from the cartographic analogy, as a **contour**.

> If we can show that as $a \to \infty$, the value of the **integrand** $\phi(z)$.**d**z over the semicircular arc Ω_a tends to zero, then the contribution to the integral over this part is zero, and the integral over the entire **contour** equals the integral over $[-a, a] \to [-\infty, \infty]$.

Then we can *replace* the **improper integral** over the **real line *R*** by the **contour integral** over $[\boldsymbol{R} + \Omega_{a\to\infty}]$ and use the **Cauchy integral formula** to evaluate it.

An example is if we find that $\phi(z) \sim \mathbf{e}^{-a}$ over Ω_a, which will tend to zero as $a \to \infty$.[157]

NOTE that it may be that we need to use a semicircle in the *lower* half-plane instead. If over Ω_a, $\phi(z) \sim \mathbf{e}^{+a}$, it may well be that $\phi(z) \sim \mathbf{e}^{-a}$ over a circle in the lower half-plane Ω^-_a, as shown in *Figure A.6.3*.

[157] I'm using "~" here to indicate "is of the order of".

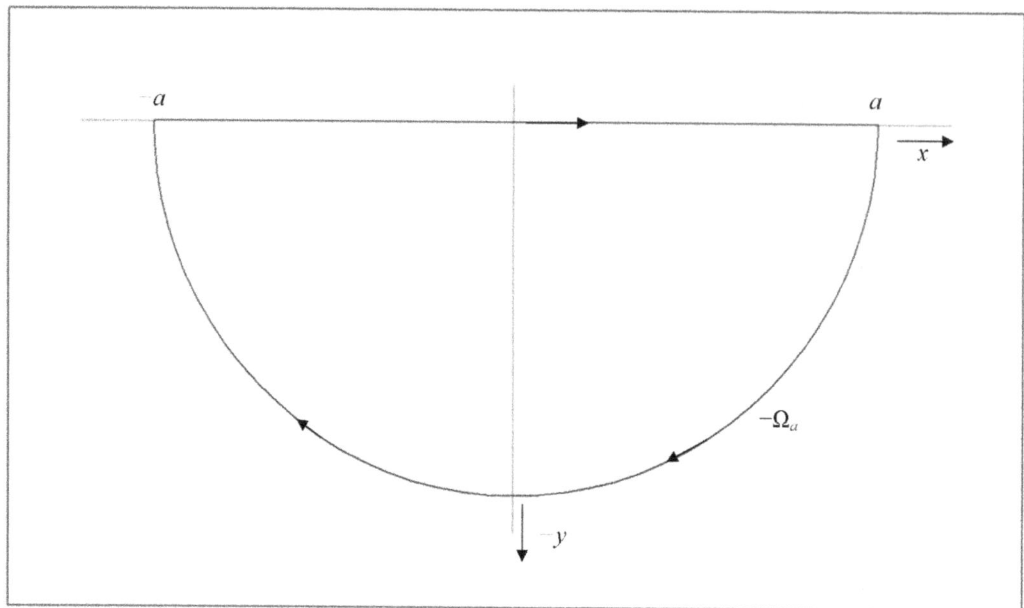

Figure A.6.3

In this case we complete the **improper integral** by the arc $-\ \Omega^{-}_{a}$, taking the **contour integral** in the *clockwise* direction.

The normal procedure then is to find all the **singular points** enclosed in whichever half-plane has needed to be chosen, and use the **Cauchy integral formula** to evaluate the **residues** at each **pole** by:

$$[\phi(z)/(z - z_0)].\mathbf{d}z\,|\,\mathbf{C} = 2\pi j.\phi(z_0),$$

which for several **poles** becomes:

$$\Sigma_k[\phi(z)/(z - z_k)].\mathbf{d}z\,|\,\mathbf{C}_k = \Sigma_k\, 2\pi j.\phi(z_k).$$

This is all much as described in Section 7.2. In passing, I'll mention that for a **pole** of higher **order** such as

$$f(z) = \phi(z)/(z - z_0)^n$$

we use the value $2\pi j.\mathbf{lim}_{z\to z0}(1/(n-1)!).\partial_z^{\,n-1}((z - z_0)^n f(z))$.

For a **simple pole**, $n = 1$, and this formula simplifies to the usual **Cauchy integral formula**:

$$2\pi j.\mathbf{lim}_{z\to z0}(1/(0)!).\partial_z^{\,0}((z - z_0)^1 f(z)) = 2\pi j.(z - z_0)^1 f(z) = 2\pi j.\phi(z_0).$$

The actual "**residue**" is the expression without the $2\pi j$, so the usual **simple pole** version is just $\phi(z_0)$. The general expression is $\mathbf{lim}_{z\to z0}(1/(n-1)!).\partial_z^{\,n-1}((z - z_0)^n f(z))$.

Sometimes (indeed more often than not) things are a little less straightforward, and the actual **singularities** or **poles** may in fact be *outside* the area enclosed by the **contour of integration**.

As it's directly relevant to Section 8.3, I'll sketch the evaluation of the **sinc** function as described in E. Kreyszig, *Advanced Engineering Mathematics*, Wiley 1968.

We want to find **(sin**(x)/x)**.d**x | [$-\infty$, ∞].

We work with (**e**^{jz}/z).**d**z | [$-\infty$, ∞] as on the **real line** (where z is **real**) the **imaginary part** of this is the same integral.

We use the **contour** shown in *Figure A.6.4*:

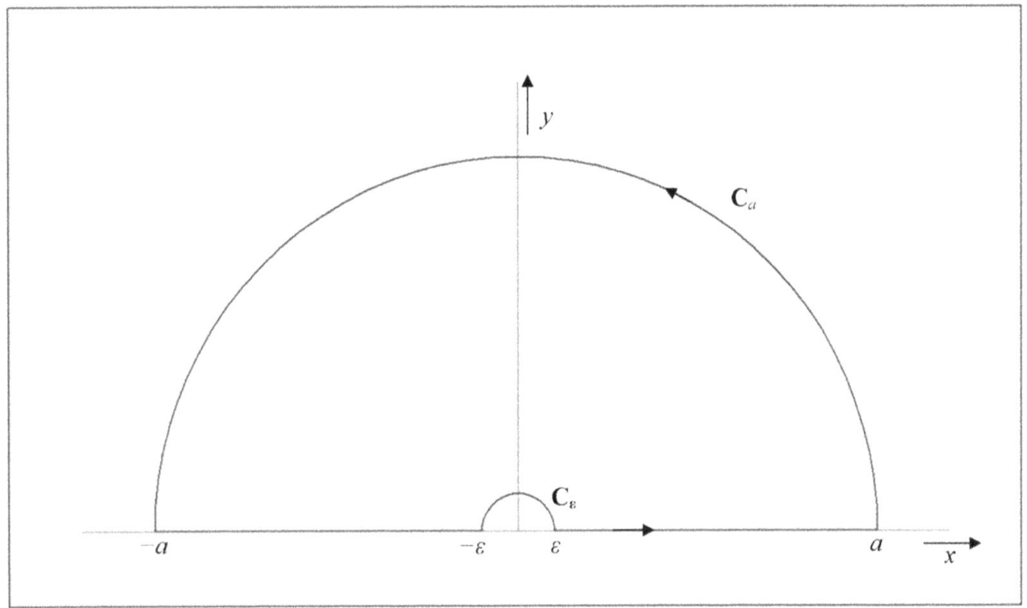

Figure A.6.4

The complete **contour** here is [$-a$, $-\varepsilon$] + C_ε + [ε, a] + C_a, and within the contained area (**e**^{jz}/z) is **analytic** – its one **singularity** is at z = 0, which we "avoid" by the little loop C_ε.

So (**e**^{jz}/z).**d**z | {[$-a$, $-\varepsilon$] + C_ε + [ε, a] + C_a } = 0

by the **Cauchy integral theorem**.

First we want to establish that over the large semicircle C_a the integral tends to zero as $a \to \infty$. Here z = a.**e**$^{j\theta}$, so **d**z = aj.**e**$^{j\theta}$.**d**θ.

So $(e^{jz}/z).dz \,|\, C_a = (e^{jz}/a.e^{j\theta}).aj.e^{j\theta}.d\theta \,|\, [0, \pi] = j.e^{jz}.d\theta \,|\, [0, \pi]$.

The value of this is $\leq |e^{jz}|.d\theta \,|\, [0, \pi]$.

$$|e^{jz}| = |e\uparrow j(a.(\cos(\theta) + j\sin(\theta))| = |e\uparrow(ja.\cos(\theta))| \times |e\uparrow(-a.\sin(\theta))| = 1.e\uparrow(-a.\sin(\theta)).$$

So $|e^{jz}|.d\theta \,|\, [0, \pi] = e\uparrow(-a.\sin(\theta)).d\theta \,|\, [0, \pi]$.

The problem part of this integral is where $\theta \sim 0$ where we might expect significant non-zero contributions however large a may be.

To deal with this, Kreyszig splits the integral in half first, at $\pi/2$:

$$e^{-a.\sin(\theta)}.d\theta \,|\, [0, \pi] = e^{-a.\sin(\theta)}.d\theta \,|\, [0, \tfrac{1}{2}\pi] + e^{-a.\sin(\theta)}.d\theta \,|\, [\tfrac{1}{2}\pi, \pi].$$

In the second integral, set $\theta = \pi - \theta^* \rightarrow d\theta = -\, d\theta^*$ and $[\tfrac{1}{2}\pi, \pi]_\theta \rightarrow [\tfrac{1}{2}\pi, 0]_{\theta^*}$, so

$$e^{-a.\sin(\theta)}.d\theta \,|\, [\tfrac{1}{2}\pi, \pi] = -\, e^{-a.\sin(\pi-\theta^*)}.d\theta^* \,|\, [\tfrac{1}{2}\pi, 0] = +\, e^{-a.\sin(\pi-\theta^*)}.d\theta^* \,|\, [0, \tfrac{1}{2}\pi].$$

But $\sin(\pi-\theta^*) = \sin(\theta^*)$, so renaming back to θ, we see that

$$e^{-a.\sin(\theta)}.d\theta \,|\, [0, \pi] = 2.e^{-a.\sin(\theta)}.d\theta \,|\, [0, \tfrac{1}{2}\pi].$$

He then splits it at a small value ε:

$$|e^{jz}|.d\theta \,|\, [0, \pi/2] = e\uparrow(-a.\sin(\theta)).d\theta \,|\, [0, \varepsilon] + e\uparrow(-a.\sin(\theta)).d\theta \,|\, [\varepsilon, \pi/2].$$

Because the **integrand** is monotone decreasing from 0 to $\pi/2$ as $\sin(\theta)$ gets larger, making $e\uparrow(-a.\sin(\theta))$ steadily smaller, the maximum possible value of the first integral is:

$$e^0.d\theta \,|\, [0, \varepsilon] = \varepsilon$$

and of the second is:

$$e\uparrow(-a.\sin(\varepsilon)).d\theta \,|\, [\varepsilon, \pi/2] = e\uparrow(-a.\sin(\varepsilon)).(\pi/2 - \varepsilon).$$

So

$$|e^{jz}|.d\theta \,|\, [0, \pi] = 2.|e^{jz}|.d\theta \,|\, [0, \pi/2] \leq 2\{\varepsilon + e\uparrow(-a.\sin(\varepsilon)).(\pi/2 - \varepsilon)\}.$$

Choosing ε as small as we like, and then choosing a large enough to make $e\uparrow(-a.\sin(\varepsilon))$ as small as we like, we can infer that as $a \rightarrow \infty$:

$$|e^{jz}|.d\theta \,|\, [0, \pi] = 2.|e^{jz}|.d\theta \,|\, [0, \pi/2] = 0$$

so in the **limit** as $a \to \infty$, the integral over the large semicircle C_a is zero.

Having eliminated the large semicircle, we must have:

$$(\mathbf{e}^{jz}/z).\mathbf{d}z \mid \{[-a, -\varepsilon] + C_\varepsilon + [\varepsilon, a]\} = 0,$$

so that

$$(\mathbf{e}^{jz}/z).\mathbf{d}z \mid \{[-a, -\varepsilon] + [\varepsilon, a]\} = -(\mathbf{e}^{jz}/z).\mathbf{d}z \mid C_\varepsilon.$$

So we just want to evaluate $(\mathbf{e}^{jz}/z).\mathbf{d}z \mid C_\varepsilon$.

Expand $\mathbf{e}^{jz}$ in a **Taylor expansion** about $z = 0$:

$$\mathbf{e}^{jz} = 1 + jz + ½j^2z^2 + O(z^3)$$

to give

$$(\mathbf{e}^{jz}/z).\mathbf{d}z = ((1 + jz + ½j^2z^2 + O(z^3))/z).\mathbf{d}z$$

$$= \mathbf{d}z/z + jz.\mathbf{d}z/z - ½z.\mathbf{d}z + O(z^2).\mathbf{d}z$$

and as before set $z = \varepsilon.\mathbf{e}^{j\theta} \to \mathbf{d}z = j\varepsilon.\mathbf{e}^{j\theta}\mathbf{d}\theta$, and $C_\varepsilon \to [\pi, 0]$ in θ, to give:

$$(\mathbf{e}^{jz}/z).\mathbf{d}z = (j\varepsilon.\mathbf{e}^{j\theta}\mathbf{d}\theta)/(\varepsilon.\mathbf{e}^{j\theta}) + j(j\varepsilon.\mathbf{e}^{j\theta}\mathbf{d}\theta) - ½\varepsilon.\mathbf{e}^{j\theta}.\ j\varepsilon.\mathbf{e}^{j\theta}\mathbf{d}\theta + O(\varepsilon^3)$$

$$= j.\mathbf{d}\theta - \varepsilon.\mathbf{e}^{j\theta}\mathbf{d}\theta + O(\varepsilon^2).$$

So only the first term contributes as $\varepsilon \to 0$, and we get:

$$\mathbf{lim}_{\varepsilon\to 0}(\mathbf{e}^{jz}/z).\mathbf{d}z \mid C_\varepsilon = j.\mathbf{d}\theta \mid [\pi, 0] = j.0 - j.\pi = -j\pi.$$

So

$$\mathbf{lim}_{\varepsilon\to 0}(\mathbf{e}^{jz}/z).\mathbf{d}z \mid \{[-a, -\varepsilon] + [\varepsilon, a]\} = -(-j\pi) = j\pi.$$

Since $(\mathbf{sin}(x)/x).\mathbf{d}x \mid [-\infty, \infty]$ is the **imaginary part** of this result, we have:

$$(\mathbf{sin}(x)/x).\mathbf{d}x \mid [-\infty, \infty] = \pi.$$

For the integral in Section 8.3 we need $(\mathbf{sin}(\pi x)/\pi x).\mathbf{d}x \mid [-\infty, \infty]$, so we'd use

$$(\mathbf{e}^{j\pi z}/\pi z).\mathbf{d}z = ((1 + j\pi z + ½j^2\pi^2z^2 + O(z^3))/\pi z).\mathbf{d}z$$

the $z = \varepsilon.\mathbf{e}^{j\theta} \rightarrow \mathbf{d}z = j\varepsilon.\mathbf{e}^{j\theta}\mathbf{d}\theta$ substitution is unaffected, and the surviving term is now:

$$\mathbf{d}z/\pi z = (j\varepsilon.\mathbf{e}^{j\theta}\mathbf{d}\theta)/(\pi\varepsilon.\mathbf{e}^{j\theta}) = j.\mathbf{d}\theta/\pi,$$

giving

$$(\mathbf{sin}(\pi x)/\pi x).\mathbf{d}x \,|\, [-\infty, \infty] = 1.$$

GLOSSARY

This final section gives informal definitions of most of the key terms that you may encounter in the text. Generally a fuller definition is available in the main text, and this can usually be located by the first reference to the term in the Index which follows, but sometimes I give a particularly pertinent Section number here in **bold**. The definitions given here are not intended to be rigorous but more to serve as a brief reminder of the substance of a concept.

0-form, 1-form, 2-form, *n*-form See **Differential Form**

2-space The space of a plane.

3-space Ordinary three-dimensional space. The concept extends to "*n*-space".

2-vector, 3-vector, *n*-vector See **standard vector**.

Absolute Value The magnitude of a number regardless of its sign. So 3 and −3 both have **absolute value** 3. Generally written $|x|$.

Adjoint Two **operators** are **adjoint** if the action of one on the left-hand **argument** of a **product** gives the same result as the action of the other on the right-hand **argument**. If the two **operators** are one and the same, it is called **self-adjoint**. **2.6**

Adjoint Co-Derivative (my terminology)The definition of an artificial operator **adjoint** to the **exterior derivative** under the **integral scalar product** between **forms**. Strictly there is no such actual **operator**, but the definition can be accepted as a kind of shorthand. **7.5**

Adjoint Definition The definition of **integration** as a product between a **differential form** and a **simplex** where the **differential operator** is **adjoint** to the **boundary operator** of the **simplex**. **2.6, 6.1**

Adjoint Differentiation The use of the **adjoint exterior derivative** or **adjoint co-derivative** to define the concept of **differentiation in the sense of distributions**, which is Laurent Schwartz's full name for what I call **adjoint differentiation**.

Adjoint Exterior Derivative (my terminology)The definition of an artificial operator **adjoint** to the **co-differential operator** under the **integral scalar product** between **forms**. Strictly there is no such actual **operator**, but the definition can be accepted as a kind of shorthand. **7.5**

Algorithm A rule or procedure for the calculation of a **function**. The term is commonly used to describe the underlying logic of a computer programme. **1.5**

Alternating (product) A product of two **expressions** whose value changes sign if the two **arguments** are interchanged. Also called an **antisymmetric** or **anticommutative product** or property, and sometimes the **Outer Product**.

Analytic Function Also known as a **holomorphic function**, this is a function from the **complex plane** into the **complex plane**, for which the derivative with respect to $z*$ is zero everywhere. **7.2**

Angle An **angle** is either a point on the **unit circle** or the distance between two such points, which itself can be treated as a point on the **unit circle**. **4.1**

Anticommutative (product) See **Alternating (product)** (The term is used independently in 5.4, 5.9)

Anti-cyclic Permutation See **Permutation**.

Antiderivative The **(n−1)-form** from which an **exact n-form** can be **derived** by application of the **differential operator**.

Antisymmetric See **Alternating (product)** (term is used independently in 5.4, 5.9)

Antisymmetric Graded Algebra A **graded algebra** based on an **alternating** or **antisymmetric product. 5.6**

Arc Length A defined measure of the distance along a curve in the plane or in space. **4.4**

Area A single **real number** which conventionally gives a **measure** to any surface. The historical importance of **area** in the inception of the **calculus** cannot be overestimated.

Argument An **argument** of a **function** is a placeholder for one of the values that must be supplied to evaluate that **function** at a given point in its **domain**. **1.1**

Associative An **operator** between two **arguments** is **associative** if it can be evaluated pair by pair in either order to give the same result. So, for example: $a + (b + c) = (a + b) + c$. **1.5**

Axiom A fundamental rule defining the operation of a mathematical **operator** or principle. **1.5**

Axis The representation of a **coordinate** either on a two-dimensional graph or in three-dimensional space. The **axis** of a **coordinate** is strictly the line defined by all the other **coordinates** having a value of zero. **1.2**

Basic Space The underlying space of **vectors** on which elements of the **dual space** act to give a simple numerical result.

Basis A set of **linearly independent vectors** in a **vector space** such that any **vector** in the **vector space** can be defined as a **linear combination** of elements in the **basis**. **5.1**

Bilinear A **function** of two **arguments** is **bilinear** if it is **linear** in both **arguments**. **5.4**

Bivector A **bivector** is a **linear combination** of **alternating products** of two **vectors**. So a **grade-2 vector**. **5.9**

Boolean Operator A **function** whose **arguments** can only take the two values 0 and 1, and whose result likewise has the **range** 0 and 1. **1.4**

Bound Variable Also called a **dummy variable**, one which is used in the calculation of a sum (as an **index**) or an **integral**, when it appears with its **differential**, and whose value is undefined outside the expression defining the sum or **integral**.

Boundary An **$(n-1)$-simplex** associated with any given ***n*-simplex** which gives a precise meaning to the "edge" of the ***n*-simplex**. **1.13**

Boundary Operator An **operator** which generates the **boundary** of a **simplex**. **1.13**

Bromwich integral The **inverse** of a **Laplace transform**. The line in the **complex plane** along which it is defined is called the **Bromwich line**.

Calculus The theory of the **differential operator** and the associated theory of **integration**. **2.1**

Camera Space The set of **coordinates** used in **projective geometry** or computer graphics to generate a **perspective**, and in which the **eye point** is at the **origin** and the z **axis** intersects the **picture plane** at right angles. Also **camera axes**. **9.1**

Capacitor An electrical circuit element obeying a particular differential law and which can store electrical charge. **3.5**

Cartesian Product The **Cartesian product** of two sets or lists is the set of all **ordered pairs** (x, y) where the first element comes from the first list and the second from the second. Normally used where the "sets" or "lists" are **coordinate axes**, so giving an infinite number of possibilities. **6.5**

Cauchy Integral Formula The formula that for an **analytic function** in the **complex plane** $\phi(z)$, any **integral** around a ring **C** about z_0 of $(\phi(z)/(z - z_0)).\mathbf{d}z$ equals $2\pi j\phi(z_0)$. **7.2**

Cauchy Integral Theorem The assertion that the **integral** of an **analytic function** over any subset of the **complex plane** is zero. **7.2**

Cauchy Principal Value A way of evaluating an **improper integral** around a **singular point** on the **real line** to return a finite value.

Cauchy-Riemann Equations Auxiliary equations in the x and y parts of a function in the **complex plane** that establish that the function is **analytic**. **7.2**

Chain Strictly a signed sum of any **simplexes** of the same **dimension**. I tend to use the term preferentially for **1-simplexes**. **1.13**

Chain Rule The substitution of an **expression** or **expressions** for one or more **differentials** into a **differential form** of **grade** one (a **1-form**). **2.2**

Characteristic Equation The zero **determinant** condition that must be satisfied for an **eigenvalue equation** to have a solution. Also used to indicate the algebraic equation corresponding to a **Linear Differential Equation.**

Chirality Left-handedness versus right-handedness. **1.2**

Clifford Algebra A synonym for **Geometric Algebra**.

Closed Of an **operator** in a **group, closure** means its **range** is still in the **group**. Of a **differential form** ω, being **closed** means **d**ω = **0**. Of a **mesh**, being **closed** means all its **edges** are **shared**. **1.5**, **6.2**, **6.6**

Codifferential Operator A lowering operator that acts on a **form** to give a **form** of **grade** one lower, and in conjunction with the **differential operator** returns a **Laplacian**. Basically ±∗**d**∗, where ∗ is the **Hodge star operator**. **7.5**

Coefficient A factor multiplying one of the defining elements in a **linear combination**. The term is not used very precisely. **1.7**

Coefficient Function A **coefficient function** is a **scalar function** appearing as a **coefficient** to a **differential** or to a product of **differentials** in a **differential form**. The term therefore subsumes **ordinary** and **partial derivatives**. **2.2**

Cogredient An adjective used to describe objects attached to a **vector space** that transform in the same way under a change of **basis**, like the **basis vectors** themselves and the **components** of the **reciprocal basis** elements. Two things that transform under **dual transforms** are called **contragredient**.

Collinear In this text three **P-points** in the **projective plane** are **collinear** (in **bold**) if they belong to the same **P-line**. I also use the word as "collinear", *not* **bold**, to indicate that three points in space lie in the same line.

Commutative A product of two **arguments** whose value is unchanged if the two **arguments** are interchanged. Also called a **symmetric** property. **1.5**

Complex Analysis The theory of functions from the **complex plane** into the **complex plane**.

Complex Conjugate The **complex number** $x - jy$ is the **complex conjugate** of the **complex number** $x + jy$. For **complex numbers** on the **unit circle**, corresponding to **angles**, the two are multiplicative **inverse elements** of each other. **4.2**

Complex Number A **complex number** is an element of the specific algebra with one **imaginary**. **4.5**

Complex Plane The ordinary two dimensional xy-plane treated as a space of **complex numbers**.

Concurrent Three **P-lines** in the **projective plane** are **concurrent** if they all intersect in the same **P-point**.

Constant A placeholder for a number whose value does not change as algebraic or differential operations are performed on any **expression** containing it.

Constant of Integration In a generic expression for all possible **original scalar functions** from which a given **differential form** of **grade** one (i.e. in simple **differentials**) has **derived**, a **constant term** must be included as this will disappear in the **differentiation. 2.4**

Contraction The elimination of an **index** by summing over it, or of a **real number variable** by **partial integration. 6.C**

Contragredient See **Cogredient**

Convergence The condition that as the value of one of its **arguments** approaches a certain value, often $\pm\infty$, the value of the entire **expression** becomes ever closer to one precise number. this number is the **limit** to which the **expression converges**.

Convex A set of points in space is said to be **convex** if any line joining any two points of the space lies entirely within the set. **Simplexes** are **convex**.

Convolution The particular form of an **integral** derived from two functions $\phi(x)$ and $\psi(x)$ which has the **differential form** $\phi(x)\psi(y-x)$.**d**x. The **Fourier Transform** of a **convolution** equals the product of the **Fourier Transforms** of the two constituent functions. Also called a **convolution integral. 8.6**

Convolution Algebra An algebra in which the basic product operation is the **convolution** between two functions to form a new function. **8.6**

Coordinate A **variable** that is one of those contributing to the specification of the position of a point in space. "Space" here may also be a line or a plane, or if more than three **coordinates** are needed, a "hyperspace". **1.2**

Coordinate Differential A **differential** of one of the **coordinates** of a space, as **d**x, **d**y or **d**z in 3-space or xyz-space. **6.6**

Cosine The **cosine** of a point on the **unit circle** is its x **coordinate. 4.2**

Cosine Function The **cosine** of a point on the **unit circle** expressed as a function of the **arc length parameter. 4.4**

Covector An element of the **dual space** of a **vector space**.

Cramer's Rule An **algorithm** for solving a set of **linear equations** using **determinants. 1.8**

Cross Product The traditional product of two **vectors** to give a third **vector** as result, as opposed to the **scalar product** which gives a *number* result. **5.9, 9.5**

Cross Ratio A product between the distances between the **images** of points in a **perspective** that is always the same in all **picture planes**, however orientated. **9.4**

Cyclic Permutation See **Permutation**.

Definite Integration True **integration**, forming the **integral** of a **differential form** acting on a **simplex chain** or a **manifold**.
Degree The **degree** of a **term** is the number of **factor variables** multiplied together in it. In this sense, $x^3 = x \times x \times x$ is of **degree** 3, as is x^2y. **1.9**

Delta Function also called the **delta distribution, delta functional** or **Dirac delta function, Dirac delta distribution** or **Dirac delta functional**. An element of the space **dual** to a space of **functions** treated as **vectors** that abstracts the value of any **function** in the **basic space** at a specific point. **7.3**

Denominator In a division of one **expression** by another, the one *doing the dividing*. So in a/b, or $a \div b$, the **denominator** is b, the second **argument** of a division **operator**.

Derivative (also verb: **Derive**) In this text, a **derivative** is a **variable**, or an **expression**, or an **equation** that is generated from another by the application of the **differential operator**. This is also called the **exterior derivative**. More conventionally, the term is restricted to the **coefficient functions** of a **differential form. 2.1**

Desargues's Theorem The seminal theorem of **projective geometry. 9.6**

Determinant A single number defined for a square **matrix** that has an **alternating** property, that interchanging rows or interchanging columns of the **matrix** changes the value of the **determinant** to its negative. **1.7**

Diatonic Scales The true musical scales of seven notes only, which are defined by particular ratios of the frequencies or pitches of notes in the scale. **1.12**

Differential A **differential** is the entity that results from the application of the **differential operator** to a single **variable. 2.1**

Differential Form A **linear combination** in either **differentials** or products of **differentials**. If products are involved, they must all be of the same **grade**. If the **linear combination** is in simple **differentials** (*not* products) it is a **(grade) 1-form**. Otherwise, a **(grade) n-form** has only products of **grade *n***. A **0-form** is a **scalar function. 2.4, 6.4**

Differential Geometry The subject of the **graded algebra** of **differentials** seen from a geometrical standpoint.

Differential Operator The fundamental defining **operator** of the **differential calculus. 2.1**

Differentiation Application of the **differential operator. 2.1**

Dimension The number of **coordinates** required to specify the position of a point in a given space. A special case is the number of **basis vectors** of a **vector space**.

Dirac Comb A set of **Dirac delta functions** whose **Fourier Transform** represents either a **periodic function** or a **function** over a finite **domain. 8.4**

Dirac Delta Function or **distribution** See **Delta Function**.

Discontinuity In a simple function on the **real line**, a point where the value of the function jumps abruptly. **7.3**

Discriminant The **discriminant** is the **expression** $b^2 - 4ac$ in the solution of a **quadratic equation** which determines whether its values are **real** or **complex numbers. 1.6**

Distribution A function, or more strictly a **functional** that is a member of the **dual space** of a **basic space** of **functions.** So a **distribution** acts **linearly** on **functions** in the **basic space**, and returns a single numerical value. **7.1**

Divergence Theorem The assertion that the **adjoint definition** of **integration** extends to the **integration** of **3-forms** on **3-manifolds. 6.8**

Divergent Integral An **integral** that doesn't return a finite value, either because its **domain of integration** is infinite or because its **differential form** has **singular points** where some **coefficient function** of the **form** runs off to infinity.

Domain The range of allowed values of the **arguments** of a **function. 1.1**

Domain of Integration The **simplex** or **manifold** which is the second argument in an **integral**, the first being a **differential form. 6.5**

Dot Product The ordinary **scalar product** of two **vectors**. Also called the **inner product. 5.4**

Duality Principle of Projective Geometry The principle that any theorem in **projective geometry** expressed in terms of **P-points** and **joins** will also have an analogous theorem expressed in terms of **P-lines** and **meets**, and therefore *vice versa*. **9.5**

Dual Space The **vector space** consisting of all **linear functions** on a given **vector space** with **real number** values, i.e. **scalar linear functions** on the given **vector space**. A **vector space** and its **dual space** have the same **dimension. 5.4**

Dual Transforms My term for the relationship between the transforms of **vectors** under a change of **basis** and those of the corresponding **dual space**, sometimes called **covectors**. The term is symmetrical: the transform of **covectors** is **dual** to that of **vectors**, but the transform of **vectors** is also **dual** to that of **covectors. 5.5**

Edge A **1-simplex**, especially one occurring in the **boundary** of a **2-simplex. 1.13, 1.15**

Eigenfunction A **function** in a **vector space** of **functions** that, acted on by a specific **linear operator**, returns a result proportional to itself, a **scalar multiple** of itself. **8.1**

Eigenvalue The value of the **scalar** multiplier which, multiplied by a particular **vector**, which may be a **function**, returns the same result as a specific **linear operator** does when it acts on the same **vector** or **function. 5.11**

Eigenvector A **vector** that, acted on by a specific **linear operator**, returns a result proportional to itself, a **scalar multiple** of itself. **5.11**

Elementary Solution In a **convolution algebra**, the **elementary solution** of a particular differential operator (not *the* **differential operator**) is the **function** which, acted upon by that differential operator, returns a **delta function. 8.6**

Equal Temperament Scale A compromise musical scale which approximates both the **major** and **minor diatonic scales.** It was introduced to reduce the cost of instruments like church organs which have resonators (sound generators) of fixed frequency. **1.12**

Equation A statement that two **expressions** are equal. The term is usually reserved for the case that they are equal only for certain values of the **variables** appearing in the **equation**. Otherwise it is an **identity. 1.6**

Euler Characteristic (sometimes **Euler Index**) A number which distinguishes certain aspects of the **topology** of **simplexes** and **manifolds** in space. **1.15**

Euler's Formula The formula showing the **exponential** property of **complex numbers. 4.6**

Even Permutation The same thing as a **cyclic permutation.**

Exact A **differential form** of any **grade** is **exact** if it can be **derived** by the application of the **differential operator** to a **differential form** of one **grade** less. **2.4, 6.2**

Exponential Function The **exponential function** is the **inverse** of the **natural logarithm** and, as a **function** of x, equals the number **e** (= 2.71828182845...) raised to the power x, *i.e.* $\mathbf{e}^x$. **3.2**

Expression A mathematical formula, generally, but not necessarily, involving just +,−, × and / **operators** and powers like x^3, which represents a **function.**

Exterior Derivative The application of the **differential operator** to any **scalar function** or any **differential form** returns the **differential form** one **grade** higher that is its **exterior derivative**. The actual operation of the **differential operator** and indeed the **operator** itself, are sometimes also known by this name. **2.1, 6.6**

Exterior Product Another name for the **alternating product** $\wedge$, or **outer product.**

Extremum A **minimum** or a **maximum** of a **function.**

Eye Point The specific point in space where all the **projectors** of a **perspective** intersect. In practice it corresponds to the position of the observer's eye, and is the **origin** of **camera space**, and the **origin** in **projective geometry**. **9.1**

Face A **2-simplex** or, more generally, an interconnected set of **2-simplexes** all lying in the same plane. **1.15**

Facet A **2-simplex**, so a triangular element of a **mesh**. **1.13, 1.15**

Factor One of a list of **expressions** or **variables** or **constants** multiplied together.

Factorial Function The **function** written $n!$ which equals $n \times (n - 1) \times (n - 2) \times \ldots 1$. So $5! = 5\times4\times3\times2\times1$, for example.

Field See **Ordered Field**

Formal Dual The actual theorem **dual** to a given theorem in **projective geometry** whose existence is implied by the **Duality Principle**. **9.5**

Formal Duality of Projective Geometry The **Dual Principle of Projective Geometry. 9.5**

Fourier Analysis The general theory of the **Fourier** and **Laplace transforms** and **Fourier Series**, or of specifically the **Fourier Transform** alone. **8.3**

Fourier Coefficient The **coefficient** or **scalar** multiplier of a **basis vector** in the representation of a **vector** in terms of a given **basis**. Also used for the **coefficient** or **scalar** multiplier of an **eigenfunction** **e**↑$2\pi jnx$ of a **Fourier Series**.

Fourier Inversion Theorem The theorem that **functions** obeying certain **convergence** criteria can be represented as an **iterated integral** of themselves with two **Fourier eigenfunctions** of the form **e**↑$(\pm 2\pi jyx)$. **8.2**

Fourier Series The **Fourier Transform** of a **periodic function** or one over a finite **domain**, which can be represented as a *sum* of **Fourier eigenfunctions** of the form **e**↑$(\pm 2\pi jnx)$ for **integer** values n. **8.4**

Fourier Transform The **integral scalar product** over an infinite **domain** of a **function** with a **Fourier eigenfunction** of the form **e**↑$(-2\pi jyx)$. **8.3**

Fourier-Laplace Transform Another name for the **Laplace transform** or for both the **Fourier** and **Laplace Transforms** considered as variant forms of the same general transform. **8.5**

Fubini's Theorem The assertion that evaluating **multiple integrals** by **partial integration** is independent of the order in which the **variables** are chosen. Commonly thought to be analogous to **Young's Theorem** in **differentiation. 6.C**

Function A **function** is a rule, generally defined by an algebraical **expression**, from which a precise and unique value can be calculated if values are given for any of its inputs or **arguments. 1.1**

Functional Most generally, any **function** that takes a **function** as an **argument**. More specifically, a member of the **dual space** of a **basic space** of **functions**, and so a **distribution**. **7.1**

Fundamental Period The longest **interval** over which a **periodic function** precisely repeats. **8.4**

Fundamental Theorem of the Calculus The **adjoint** relation between the **differential operator** and the **boundary operator** in an **integral**, expressed not as a definition, but as a theorem depending on a distinct definition of **integration**. **2.6**

Fundamental Theorem of Projective Geometry The assertion that there is a unique **projective transformation** defined by the transformation of any four **P-points** in the **projective plane**. **9.7**

Gaussian Elimination An **algorithm** for solving a set of **linear equations** *not* using **determinants**. **1.8**

Generalized Function Another name for a **distribution**¸ or a member of the **dual space** of a **basic vector space** of **functions**. **7.1**

Geometric Algebra The most powerful **graded algebra** on **vector spaces**. **5.8**

Geometric Product An **associative** product between **vectors** that forms the foundation of **Geometric Algebra**. **5.8**

Geometric Series An **infinite series** in powers of a single **variable**. **1.10**

Gibbs Cross Product The ordinary **cross product** between two **vectors** which gives a third **vector orthogonal** to the two we started with. **5.9, 9.5**

Good Function A **function** in a **vector space** of **functions** satisfying certain **convergence** criteria such that it has a well-defined **Fourier transform**.

Grade The **grade** of a **vector space** is the number of **vectors** that are multiplied together to define the **basis elements** of the **vector space**. So any **vector space** of **grade** higher than *one* is derived from a **vector space** of **grade** one, also called a.**grade-1 vector space**, as a **vector space** of **grade** *n* is also called a **grade-n vector space**. **5.6**

Graded Algebra A mathematical structure with an underlying **vector space**, and a product of **vectors**, usually, but not necessarily, an **alternating product**, that gives rise to several tiers of vector spaces of different **grades**. **5.6**

Gradient (**vector**) The **gradient** of a **scalar function** is the **vector** whose components are those of the **differential** of the **scalar function**: $\mathbf{grad}(\phi) = \partial_x\phi.\mathbf{i} + \partial_y\phi.\mathbf{j} + \partial_z\phi.\mathbf{k} = \partial_x\phi.\mathbf{e}_1 + \partial_y\phi.\mathbf{e}_2 + \partial_z\phi.\mathbf{e}_3$. **6.6**

Gram-Schmidt Orthogonalization A strategy for developing an **orthogonal basis** for a **vector space** given a **basis** that is merely **linearly independent** (which any **basis** must be).

Graph The representation of a **function** as a two-dimensional image on paper. **1.2**

Green's Formula One of a range of **identities** between **integrals** that correspond to the extension of **integration by parts** (itself deriving from our Axiom II of the **differential operator**) to higher **grade differential forms**. **7.5**

Green's Function A **function** that acts through an **integral scalar product** with the result of an **elliptic differential operator** itself acting on an arbitrary **function**, and returns the original arbitrary **function**. So the **Green's Function inverts** the **elliptic operator** through an **integral scalar product**. **7.7**

Green's Theorem The theorem that an **exact 2-form** evaluated as an **integral** over an area in the plane gives the same value as the **integral** of the **1-form** from which it **derives** does over the **boundary** of that area. **6.8**

Group A minimal axiomatic system that describes properties common to addition and multiplication, as well as many other mathematical operations. **1.5**

Heaviside Function or **Heaviside Step Function** Any of a family of functions that change abruptly from a value of 0 to a value of 1 at a given **point of discontinuity**. The different members of the family differ in the choice of that point, and on which side the value goes to 1. **7.3**

Hermitian Operator A rather elaborate kind of **self-adjoint operator** used in **vector spaces** with **complex scalars** or **Hilbert spaces**. **8.1**

Hilbert Space A **vector space** with **complex numbers** as **scalars.**

Hodge Duality In an **antisymmetric graded algebra** whose highest **grade** is n, a correspondence can be established between elements in **grade** i and **grade** $(n - i)$. **5.9**, **7.4**

Hodge $*$ Operator or **Hodge Star Operator** The **operator** that switches between an element of a **graded algebra** and its **Hodge dual**. **5.9**, **7.4**

Holomorphic Function An **analytic function** in the **complex plane**.

Homogeneous Coordinates The **coordinates** of any point on a line through the **origin** representing a **P-point** in the **projective plane**. Any point on the line is considered as good as any other, so **P-points** have an infinite range of **homogeneous coordinates**. **9.3**

Homogeneous Equation A **linear equation** is **homogeneous** if it equates a **linear combination** to zero. **1.7**

Ideal Point A **P-point** in the **projective plane** that corresponds to a line that is parallel to the putative **Picture Plane**. **9.3**

Identity A statement that two **expressions** are equal for all values of the **variables** appearing in the **identity**. **1.7**

Identity Element The unique member of a **group** that, multiplying any member of the **group**, leaves that member unchanged as the result. **1.5**

Identity Matrix The unique **matrix** – unique for any given **dimension** – that, multiplying any **matrix**, leaves that **matrix** unchanged as the result. Also called the **unit matrix. 1.8**

Imaginary An **imaginary** is an element of an algebra – a generalized number – whose square has the value −1. No ordinary number squares to this value. **4.5**

Impedance In **Fourier-Laplace transform** theory, time **derivatives** are replaced by algebraic operations, so **capacitors** and **inductors** can both be represented as modified **resistances** with a frequency **parameter** ω or a real **exponential parameter** p. Such generalized **resistances** are called **impedances. 8.7**

Implicit Differentiation Application of the **differential operator** to both sides of an **equation. 2.2**

Improper Integral An **integral** which either has a **pole singularity** somewhere in the **domain of integration** where a **coefficient function** of a **differential** goes to infinity (**of the second kind**) or the **domain of integration** itself is unbounded (**of the first kind**). **7.3**

Indefinite Integration An operation purely on a **differential form** that involves finding its **antiderivative** or the **form** (which may be a **scalar function**) from which it is **derived**.

Index A whole number **argument** to a **function. 1.3**

Induction A method of proof, that establishes a statement for the value of some **index** of 1, and then shows that if the statement holds for **index** value n, it must hold for **index** value $n + 1$. **1.10**

Inductor An electrical circuit element obeying a particular differential law and which represents the storage of **electrical current** energy in a magnetic field. It gives a measure of the reluctance of the circuit to accept a change in **current. 3.5**

Inequality A statement that two **expressions** are *not* equal. Generally it also states an **ordering relation** between them. **1.6**

Infinite Series A sum with an infinite number of **terms. 1.10**

Infinitesimal A nasty thing from the early days of the **calculus. 2.3**

Inner Product Another name for the **scalar product**.

Integer A whole number, not necessarily positive, like 3 or 77,929 or −395.

Integral An **integral** is either the defining **expression**, or the resulting value, of an **integration** between a **differential form** and a **simplex** or **manifold. 2.6**

Integral Operator The **operator**, which I write as "|", between a **differential form** and a **simplex** defining an **integral. 2.6**

Integral Product The evaluation of **integrating** between a **differential form** and a **simplex** or **manifold**. The **integral operator** defines the **product**.

Integral Scalar Product An **integral** over a specified **domain of integration** where two **functions** over that **domain** are multiplied together, so that the operation obeys the axioms defining a **scalar product**. For **differential forms**, we take the **outer product** of one **form** with the **Hodge dual** of the other. **7.4**

Integration is a specific definition of a method of assigning a value to a product of a **differential form** and a **simplex**. The **simplex argument** can be extended to arbitrary ***n*-dimensional manifolds**. **2.6**

Integration by Parts The use of Axiom II or the **Leibnitz rule** on one of the pairs of products resulting from applying the **differential operator** to give a new expression in the **exact differential** of the **original expression** and a new **integral** involving the other member of the pair. **2.4**, **7.5**

Integration by the Method of Residues Put in simplest terms, the device of **integrating** by the use of the **Cauchy Integral Formula**. The **residue** at a **pole** is the result of applying the **Cauchy Integral Formula** at that **pole**. **7.2**

Interval A **1-simplex** lying on an **axis** of a **coordinate**, commonly written $[a, b]$, where a and b are the two **boundary** points of the **simplex**. **1.13**

Inverse A **function** that exactly cancels a given **function**, so that if f is the given **function** and f^{-1} denotes its **inverse**, $f^{-1}f(x) = ff^{-1}(x) = x$, where x is any **argument** or list of **arguments**. The term is also used to describe the **inverse element** of a **group**. **1.4**, **1.5**

Inverse Fourier Transform The operation that restores an original **function** from its **Fourier Transform function**.

Inverse Matrix The **matrix** whose product with a given **matrix** returns the **identity matrix**.

Irrational Number A number that cannot be represented as the **quotient** of two **integers**. $\sqrt{2}$ and π are examples.

Isometric Projection An **orthographic projection** in which all three **Cartesian body axes** of the structure being viewed are at the same angle to the **picture plane**. **9.2**

Iterated Integral A **multiple integral** evaluated by eliminating one variable at a time through successive **integrations**. As each variable is eliminated we obtain an **integral** one **grade** lower. The process is also called **partial integration**.

Jacobian Determinant When a **differential form** of **grade** one undergoes a **transformation of coordinates**, the result is the **chain rule**. For **forms** of higher **grade**, because of the **alternating products** multiplying the **differentials** a new **factor** appears that has the form of a **determinant**. This more complex combination of **partial derivatives** than the simple ones appearing in the **chain rule** is called a **Jacobian determinant**.

Join The unique **P-line** in the **projective plane** between two **P-points**. It is actually a plane in the underlying space, defined by the lines defining the **P-points. 9.5**

Kirchhoff's Laws The two fundamental laws governing electrical circuits. **1.14**

Kronecker Delta The **function** of two **integers** written δ_{ij} which is *one* if both **arguments** are the same, otherwise zero. **5.4**

Kronecker Property The property that the **dot product** or **inner product** between any two **vectors** in a **basis** has the value 0 if the two **vectors** are different or the value 1 if they're the same. Such a **basis** is **orthonormal.**

Kronecker Scalar Product In this text, the standard **scalar product** or **dot product** whose **metric tensor** equals the **Kronecker Delta. 5.4**

Laplace Inversion Formula The specialization of the **Fourier Inversion Formula** to a **Laplace transform. 8.5**

Laplace-Beltrami Operator The generalization of the **Laplacian operator** to arbitrary **grade forms**, using the **exterior derivative** or **differential operator** and the **co-differential operator. 7.5**

Laplacian Most simply, the **scalar function** derived by the application of the **Laplacian operator** to a simple function ϕ. In 3 dimensions it takes the form $\partial_x{}^2\phi + \partial_y{}^2\phi + \partial_z{}^2\phi$.

Laplacian Operator The **operator** that generates the **Laplacian:** $\partial_x{}^2 + \partial_y{}^2 + \partial_z{}^2$.

LDE My shorthand for **Linear Differential Equation.**

Leibnitz Property The property $\mathbf{d}(x \times y) = x \times \mathbf{d}y + y \times \mathbf{d}x$ that distinguishes the action of the **differential operator** acting on a **scalar function** form other **linear operators**. My "**axiom II**" of the **differential operator. 2.1**

Limit An infinite sequence of **expressions**, indexed by an **index** *i*, converges to a **limit** if the values of the **expressions** become ever closer to one particular value as *i* increases. So an **infinite series** converges to a **limit** if the magnitude of the individual **terms** becomes ever closer to zero.

Logarithm The special function **ln**(*x*) which is the value of the **integral** $\mathbf{d}\xi/\xi \mid [1, x]$. **3.1**

Line Integral The **integral** of a **1-form**, generally in a space of **dimension** *more than one*. **6.2**

Line Segment (also sometimes **Line Interval**, but see **Interval**) A less formal name for a **1-simplex. 1.13**

Linear The term **linear** is used in a variety of contexts. The archetypal one is a **linear term**, which, in *x*, is a **term** involving only *x* itself, and no higher powers like x^2 or x^3. An **equation** is **linear** if it involves only **linear** and **constant terms**, like $ax + by + c = 0$, whose graph is a straight line, and it is from this that the term **linear** comes. A **function** $g(x,\{\xi_i\})$, where $\{\xi_i\}$ represents various other **arguments**, is **linear** in the **argument** *x* if $g(ax + by,\{\xi_i\}) = ag(x,\{\xi_i\}) + bg(y,\{\xi_i\})$. Axiom I of Section 2.1 shows this form for the **linearity** of an **operator**, where it is the functional *form* that is significant, not the *value*. A **linear space** is another name for a **vector space**.

Linear Combination A **linear combination** is an **expression** of the form $a_1x_1 + a_2x_2 + a_3x_3 + \ldots a_nx_n$ where each term consists of a **coefficient**, here the a_i's, multiplying some more fundamental entity, commonly a number or a **vector** or a **differential**, sometimes a **function**.

Linear Differential Equations (also **LDE**) As used here, this term refers to a set of **equations** in **1-forms**, or **differential forms** of **grade** one. It is conventionally used to refer to **equations** in **ordinary** or **partial derivatives** with **constant coefficients**. The two concepts closely correspond to each other. **3.5**

Linear Equation An **equation** involving only **linear combinations** and **constant terms**. **1.7**

Linear Function See **Linear**.

Linear Independence A set of entities, generally **vectors** or **functions**, is **linearly independent** if any **linear combination** of them involving numerical **coefficients** is zero only if all the **coefficients** are zero. In this case, no one element can be expressed in terms of the others. **1.7**

Linear Operator A **linear operator** really only differs from a **linear function** in that the result of application of the **operator** is something more than a simple numerical value. Thus L might operate on **functions** ϕ and ψ to generate new **functions** $L(\phi)$ and $L(\psi)$, and L is a **linear operator** if $L(a\phi + b\psi) = aL(\phi) + bL(\psi)$.

Linear Transformation a change of **coordinates** in which the new **coordinates** are **linear functions** of the old.

List Notation In this book, I may write $\{a_i\}$ to indicate a list of things a_1, a_2, a_3, etc.

Logarithm The **natural logarithm** is the **function** defined as the **integral** $(1/t)$.**d**$t\,|\,[1, x]$. It can only be evaluated as an **infinite series. 3.1**

Long Division An **algorithm** for the division of two **polynomials**. A specialization of the **algorithm** to powers of ten gives a general method for dividing numbers. **1.9**

Magnitude The **magnitude** of a **vector** is the square root of the **scalar product** of the **vector** with itself. The concept is closely related to the **absolute value** of a number, and the same notation is often used, as in: $|\mathbf{v}|$. **5.4**

Major Arc Two points on the **unit circle** are either diametrically opposite across the circle or are joined around the circle by two arcs, one of which is longer than the other. The longer one is the **major arc**, the shorter the **minor arc**.

Major Scale One of the two types of **diatonic musical scales. 1.12**

Manifold A **manifold** is any part of a space of any number of **dimensions** each point of which can be precisely located by giving a value to a certain number of **arguments**. The number of **arguments** is the

dimension of the **manifold**, and is always less than or equal to the **dimension** of the space in which the **manifold** lies. The **arguments** are said to **parameterize** the **manifold**. A **manifold** with *n* **dimensions** is called an **n-manifold**. **6.3**

Map, Mapping A **map** is a **function** seen in aggregate, such as going from one **manifold** to another, or from a **simplex** to a **manifold**. **1.1**

Mathematical Induction See **Induction**.

Matrix A **function** of two whole number **arguments** which can generally take only a very limited range of values like 1 to 3. **Matrices** can be written on the page as a rectangular array of numbers. **1.8**

Matrix Multiplication An operation defined between two **matrices** that gives another **matrix**. **1.8**

Maximum A set of the values for the **arguments** of a numerical **function** that gives a result greater than that for any nearby sets of values. **2.5**

Meet In **projective geometry** the **meet** of two **P-lines** is the **P-point** where they intersect. **9.5**

Mesh An interconnected set of **2-simplexes** used to approximate an arbitrary surface. **1.13, 1.15**

Metric A **function** whose **arguments** are any two points in a space or **manifold** and which gives a measure of a *distance* between them.

Metric Tensor The **matrix** of all the **scalar products** of the elements of a **basis** of a **vector space**. **5.4**

Minimum A set of the values for the **arguments** of a numerical **function** that gives a result less than that for any nearby sets of values. **2.5**

Minor Arc See **Major Arc**.

Minor Scale One of the two types of **diatonic musical scales**. **1.12**

Multiple Integral The **integral** of an ***n*-form** over an ***n*-manifold** or ***n*-simplex** where the integer $n > 1$. **6.5**

Natural Basis In this text, A **basis** for a **vector space** of **standard vectors** where each member of the **basis** has just one non-zero component, which equals one. So [1, 0, 0], [0, 1, 0] and [0, 0, 1] give the **natural basis** for **standard 3-vectors**. **5.1**

Natural Logarithm See **Logarithm**.

Negative The **inverse element** of an addition. See **Group**.

Norm The **magnitude** of a **vector**. A **vector** is **normalized** when it is **scalar multiplied** with the reciprocal of its **norm**, as it then becomes a **unit vector**.

Normalization The operation of multiplying a **vector** by the reciprocal of its **magnitude** or **norm** to obtain a **unit vector**.

Numerator In a division of one **expression** by another, the one *being divided into*. See also **denominator**. So in a/b, or $a \div b$, the **numerator** is a, the first **argument** of the division **operator**.

Oblique Projection A **projection** of a three-dimensional object onto a flat surface wherein the **projector lines** run parallel to each other but at an angle to the surface. **9.1**

Octave Two musical notes differ by an **octave** if the frequency or pitch of one is twice that of the other. **1.12**

OD My shorthand for an **ordinary derivative**, the **coefficient** of the unique **differential** in the result of applying the **differential operator** to a **scalar function** of a single variable.

Ohm's Law The electrical circuit law that applies to a component the **current** through which is proportional to the **voltage** across it. The statement that this is so. **1.14**

Operator Either of two things: a **function** of two (sometimes just one) **arguments** represented by a special symbol placed *between* the **arguments** like +, ×, ∧; or else a syntactical agent, such as the **differential operator**, that transforms an entire **expression** or **equation** into a derived **expression** or **equation**.

Ordered Field The formal algebraic structure corresponding to the **real numbers**. **1.5**

Ordered Pair Two items whose order is important, as in the components of a **2-vector**: $[x, y] \neq [y, x]$.

Ordering Relation A **predicate** that states whether of two **expressions** one is greater than the other.

Ordinary Derivative The **coefficient function** of a **differential form** in one and only one **variable**. **2.2**

Oriented or **Orientated** Having **chiral** variants, being left and right-handed. **1.2**

Origin The centre of a **graph** or indeed of any space, where all the **coordinates** have the value zero. **1.2**

Original In this text, the **variable**, **expression** or **equation** from which a **derivative** is obtained by application of the **differential operator**. **2.1**

Orthogonal Two **vectors** are **orthogonal** if their **scalar product** is zero. **5.5**

Orthonormal Basis A **basis** for a **vector space** which has all its elements **orthogonal** to each other and of unit **magnitude**. **5.4, 5.5**

Outer Product Another name for the **alternating product** ∧.

Parallel Electrical circuit elements are in **parallel** if they are connected together in such a way that the voltages across them are all the same. The term is also used in its usual sense from Euclidean geometry. **1.14**

Parallel Projection Any **projection** of a three-dimensional object onto a flat surface in which all the **projector lines** are parallel. **9.1**

Parallelepiped The three-dimensional analogue of a parallelogram, in which opposite **faces** are parallel. **6.4**

Parameter (also **Parametric Variable** and vb. **parameterize**) A **parameter** is one of a set of n **real numbers** that specifies a point on an **n-manifold**.

Parametric Function A function giving a point on an ***n*-manifold** lying in an m-space, where $n < m$. **6.2, 6.3**

Partial Derivative A **coefficient function** of a **differential form** in more than one **variable**. **2.2**

Partial Fractions The representation of a **quotient** of **polynomials**, also called a **rational function**, as a sum of **terms** whose **denominators** are the **factors** of the **denominators** of the original **quotient**. **1.9, 8.7**

Partial Integration The traditional method for evaluating **multiple integrals** by **integrating** one **variable** at a time, temporarily treating the other **variables** as **constants**. The true justification of this procedure comes from the fact that any **form** on the **pseudoscalar** of n-space has n **antiderivatives**. The term is also used, by a form of synecdoche, for the **inverse** of **differentiation** treating all but one **variable** as **constant**. The reason is that this operation is involved in true **partial integration**.

Pascal's Triangle An array of **coefficients** having the property that each number in each row is the sum of the two numbers immediately above it. **1.11**

PD My shorthand for a **Partial Derivative**.

Periodic Function A **function** of a single **argument** all of whose values repeat after a fixed change in the value of that **argument**, generally time, so $\phi(t + T) = \phi(t)$ for all t and for a fixed T, called the **period** of the **function**. **8.4**

Periodicity The property of being **periodic**.

Permutation A re-ordering of a list of items. Thus 4321 and 4213 are **permutations** of 1234. A **permutation** is **cyclic** (or **even**) if, arranging the list around a circle, the **permutation** can be obtained by choosing a different starting point. Thus 3412 is a **cyclic permutation** of 1234, but 4321 is not. If the **permutation** can be obtained as a **cyclic permutation** of the elements in the list in *reversed order*, it is **anti-cyclic** (or **odd**). Thus 4321 and 2143 are **anti-cyclic permutations** of 1234. **1.8**

Perspective Any **projection** of a three-dimensional object onto a flat surface in which all the **projector lines** pass through one unique point in space defining the **projection**, which point is called the **station point** or **eye point**. **9.1**

Picture Plane In any **projection** of a three-dimensional object onto a flat plane, the **picture plane** is the formal name of the flat plane in question. **9.1**

P-line In **projective geometry** a **P-line** is actually a *plane* passing through the **origin**. Since such a plane intersects the **picture plane** in a line, every line in the **picture plane** corresponds to a **P-line**, but there is also a **P-line** that doesn't intersect the **picture plane**, which is called the **ideal line**. **9.5**

Poincaré's Lemma The assertion that applying the **differential operator** twice gives zero, and so that an **exact form** is **closed**. The converse – that a **closed form** is **exact** – is also true, at least "locally". **6.6**

Polar Coordinates or **Polar Representation** The use of the distance from the **origin** together with an **angle**, in the "single **real**" form, as **coordinates** for points in the plane.

Pole a point in the **domain** of **arguments** of a **function** or **differential form** where the value of the **function** or one of the **coefficient functions** in the **form** becomes infinite, usually because of a division by zero. **7.3**

Polynomial An **expression** in powers of one or several **variables**. **1.9**

P-point In **projective geometry** a **P-point** is actually a *line* passing through the **origin**. Since such a line intersects the **picture plane** in a point, every point in the **picture plane** corresponds to a **P-point**, but there are also **P-points** that don't intersect the **picture plane**, which are called **ideal points**. **9.3**

Predicate A **function** whose **range** is simply the two numbers 0 and 1, or any pair of entities analogous to these, such as **T**rue and **F**alse. **1.4**

Primary Colour A colour that stimulates only one of the human eye's three colour receptors. **1.11**

Projective Geometry Most simply, the theoretical device of analyzing the geometry of points and lines in a plane as if they were *lines* and *planes* respectively, passing through the **origin** in *space*. The two are in (almost) 1:1 correspondence, but the spatial geometry, perhaps surprisingly, is "cleaner" because points at infinity now become regular "points". The concept can be extended to higher dimensions, always using a representation one dimension higher still. **9.3**

Projectors or **Projector Lines** The lines in space that define any **projection** by assigning to any point in 3-space the **image point** where the unique **projector** through the original point intersects the **picture plane**. **9.1**

Projective Coordinates See **Homogeneous Coordinates**.

Pseudoscalar The unique **basis vector** of highest **grade** in an (**alternating**) **graded algebra**. The **grade** of the **pseudoscalar** equals the **dimension** of the underlying **vector space** of **grade** one. All members of this highest **grade** of the algebra are **linearly dependent**. **5.8**

Pullback The transformation of, or the transformed value of, a **differential form** when **mapped** into **coordinates** differing from those in which it was originally defined. Also, and perhaps more

importantly, the associated **mapping** of the **simplex** or **manifold** on which the **form** was to be **integrated. 6.2**
Quad A quadrilateral in the plane. So two **2-simplexes sharing** an **edge** lying in the same plane. The term refers especially to planar quadrilaterals used in a **mesh. 1.15, 6.C**

Quadratic An **expression** all of whose **terms** are of **degree** two, involving two **factor variables.** Such terms are x^2, xy, $(a + b)xz$. **1.6**

Quadratic Equation An **equation** which has at least one **term** of **degree** two, but none higher. **1.6**

Quaternion A **quaternion** is an element of the specific algebra which is defined with three **imaginaries** as opposed to a **complex number**, which is an element of the specific algebra with one **imaginary. 5.2**

Quotient The result of the division of one number or **polynomial** into another. **1.9**

Radian The unit of measure of an **angle** in terms of the length of the arc around the **unit circle. 4.4**

Range The range of values that can appear as the results of a **function.**

Real Line The set of all **real numbers** visualized as running along a line in space, which is a geometrical structure that has the same **topology.**

Real Number An ordinary number like 12 or −3.14156 or −0.000034521, as opposed to a **complex number** which has an **imaginary part. 1.5, 4.4**

Reciprocal Basis The **basis** of a **dual space** whose members, operating on the **basis** of the **vector space** whose **dual** this is, give the **Kronecker Delta**: $\varepsilon^i(\mathbf{e}_j) = \delta^i_j$. **5.5**

Regular Distributions or **Regular Functionals Linear functions** that take other **functions** as **arguments** and return a **real** or **complex number** and which *can* be represented by the action of the **integral scalar product** of some **function** with the **argument function. 7.1**

Remainder The part still undivided at any stage of a numerical or **polynomial** division. **1.9**

Resistor An electrical circuit element obeying Ohm's law. **1.14, 3.5**

Rotation Most simply, a transformation or **mapping** of **vectors** that preserves angles and distances, because it preserves **scalar products**: $(T(\mathbf{u})) \cdot (T(\mathbf{v})) = \mathbf{u} \cdot \mathbf{v}$. I wouldn't swear to it that *only* **rotations** have this property. **5.10**

Rotor A **bivector** or **quaternion** that describes a **rotation. 5.10**

Scalar Just a single **real** or **complex number** as opposed to a **vector**, although **complex numbers** can be interpreted as either **scalars** or **vectors.**

Scalar Function A **function** of one or more **arguments** that returns a *number* result.

Scalar Multiplication One of the two basic operations defining **vectors. 5.1**

Scalar Product A **function** of two **vectors** that returns a single number result. **5.4**, **7.4**

Scaling A **transformation** of a **vector space** in which each **vector**, represented as a **linear combination** of a certain set of **basis vectors**, has the **coefficient** of each **basis vector** multiplied by a fixed number. A **scaling** corresponds to a stretching of the **vectors** to different extents in different directions.

Schwartz Rule The definition of a generalized **Fourier Transform** as being **self-adjoint** *by definition*. **8.3**

Second Derivative Any of the **coefficient functions** of the **2-form** obtained by applying the **differential operator** to a **1-form**, or more generally, of the **1-form** obtained from applying the **differential operator** to a **coefficient function** of a **1-form**. The first definition only allows for expressions like $\partial_x\partial_y\phi(x, y)$ involving different ∂ operators, but the latter includes $\partial_x\partial_x\phi(x, y) = \partial_x^2\phi(x, y)$.

Secondary Colour A colour that stimulates two of the human eye's three colour receptors. **1.11**

Self-adjoint See **Adjoint**.

Series Electrical circuit elements are in **series** if they are connected together in such a way that the currents through them are all the same. **1.14**

Set The postulated **inverse** of a **predicate**. Thus we may talk of the "set" of everything that has, or that has not, the property of being round, or red, or square, or made of green cheese.

Shared Of an **edge** in a **mesh**, being **shared** means it occurs in the **boundaries** of two **faces**. **1.15**

Signal A **function** of time, most commonly **periodic**.

Signature The sign of something, + or −.

Simplex A **simplex** is a formal definition of the simplest **convex** structure in any number of **dimensions**. So a triangle is the **simplex** for two **dimensions**. A **simplex** with *n* **dimensions** is called an **n-simplex**. A point is a **0-simplex**, of **dimension** zero. **1.13**

Sine The **sine** of a point on the **unit circle** is its *y* **coordinate**. **4.2**

Sine Function The **sine** of a point on the **unit circle** expressed as a function of the **arc length parameter**. **4.4**

Singular Distributions or **Singular Functionals Linear functions** that take other **functions** as **arguments** and return a **real** or **complex number** and which *cannot* be represented by the action of the **integral scalar product** of some **function** with the **argument function**. **7.1**

Singularity or **Singular Point** A point in the **domain** of a **function** or, by extension, a **form**, where its value is "undefined", generally ±∞ but not necessarily: the step of the **Heaviside step function** has an undefined but not an infinite value. **7.3**

Sinusoidal Having the form of either the **sine function** or the **cosine function**, which differ only by **translation** along the **axis**, and are obtained as the **sine** or **cosine** of an **angle** as a **function** of the **arc-length parameter**.

Standard *n*-Simplex The **simplex** in *n* dimensions whose defining points are the **origin** with **coordinates** (0, 0, 0, 0, . .), and the points (1, 0, 0, 0, . . .), (0, 1, 0, 0, . . .), (0, 0, 1, 0, . . .) etc.

Standard Vector In this text, an ordered list of numbers treated as a **vector**. If the list has two numbers in it, we have a **2-vector**, if three a **3-vector** and so on. **5.1**

Station Point (also **Eye Point**) The unique point in space that defines a **perspective projection**. **9.1**

Stokes's Theorem The theorem that an **exact 2-form** evaluated as an **integral** over a surface has the same value as the **1-form** from which it is **derived** evaluated as an **integral** over the **boundary** of that surface. **6.8**

Strict Chain or **strict 1-chain** A **chain** of **1-simplexes** where the start point of each is the end point of the last. The concept can be extended to higher dimensions. **1.13**

Strict Duality In **projective geometry** the **duality** between a **P-line** defined as a plane through the **origin** in space, and a **P-point** defined by the line in space perpendicular to that plane. Useful algebraically, but not so significant as **Formal Duality**. **9.5**

Subspace A part of a **vector space** that has a **basis** in terms of which every member of the **subspace** can be expressed, but *not* every member of the whole space.

Surface Integral Specifically, an **integral** of a **2-form** over a **2-manifold** lying in 3-space or *xyz*-space, but **surface integrals** are the archetype of any **integration** between an ***n*-form** and an ***n*-manifold** lying in *m*-space, where $n < m$. **6.7**

Summation Symbol The Greek capital letter sigma Σ, used to indicate the sum of a large number of **terms**. **1.3**

Symmetric A term used in a variety of contexts, usually suggesting the sense of **commutative**.

Symmetric Matrix A **matrix** where $a_{km} = a_{mk}$ for all k, m. The sense of **commutativity** should be clear in this example.

Tangent A line or plane that touches, but does not intersect, a curve or surface.

Tangent vector An element of the **vector space** defined at a point on a **manifold** with **basis** derived from the **coordinates** $\{x^1, x^2, \ldots x^n\}$ as $\{\partial_{x^1}, \partial_{x^2}, \ldots \partial_{x^n}\}$. **6.6**

Taylor's Expansion or **Taylor's Theorem** A **polynomial** or an **infinite series** which defines a **function** in terms of its **ordinary derivatives** (if it has only one **argument**) or its **partial derivatives** (otherwise) evaluated at a single point. **2.7**

Tensor Product A product between two members of the **dual space** of a **vector space** that gives a **bilinear function** of two **vectors** in the original **vector space**. The concept can also be extended to products on the original **vector space** by regarding this as the **dual** of the **dual space. 5.5**

Term One of a list of **expressions** added together.

Tet My shorthand term for a **3-simplex**, especially in computer work. A **tet** differs from a **quad** in that a **quad** has all its **vertices** in the same plane. **1.15**

Topology The branch of mathematics that treats of the *connections* between things – generally, but not necessarily, **manifolds** – without involving a **metric**.

Translation A **translation** of points in space is simply a **mapping** or **transformation** that adds a constant **vector** to every point: so $(x, y, z) \longrightarrow (x + a, y + b, z + c)$. **5.10, 9.3**

Transpose The **transpose** of a **matrix** is the **matrix** whose columns were the rows of the original matrix, and vice versa. **1.8**

Trivector A **trivector** is a **linear combination** of **alternating products** of three **vectors**. So this is another name for a **grade-3 vector**. **5.9**

Unit Angle The **unit angle** is the point on the **unit circle** from which all other angles are measured. It is normally the point $(x, y) = (1, 0)$ or due East. North $(0, 1)$ might have been a better choice. I sometimes refer to this as the **zero angle** to stress that addition and multiplication are different interpretations of the same **group** property here. **4.2**

Unit Circle The **unit circle** is the circle on the xy plane centred at the **origin** and with radius = 1. **4.2**

Unit Matrix Another name for the **identity matrix**.

Unit Vector A **vector** of **magnitude** one. **5.4**

Vanishing Point The **image** in the **picture plane** in a **perspective** of any point in space infinitely far away. A **vanishing point** may be thought of as the **image** of a direction in space. **9.1, 9.2**

Variable A placeholder for a number which can take on any value, or any value within a specified range, in an **expression**.

Vector A **vector** is, formally, an entity to which the axioms of a **vector space** apply. **5.3**

Vector Addition One of the two basic operations defining **vectors**. **5.1**

Vector Bundle The set of **tangent vectors** at a point on a **manifold**. The set of **differentials** or **covectors** is called the **covector bundle**. **6.6**

Vector Space A **vector space** is an algebraic system that obeys the axioms of **vector addition** and **scalar multiplication**. **5.3**, **5.4**

Vertex A **0-simplex** or, in other words, a point in space. **1.13**, **1.15**

Volume The three-dimensional or higher-dimensional analogue of **area**, and so a single **real number** that gives a **measure** of a **simplex** or **manifold** of three or more **dimensions. 6.4**

Young's Theorem The theorem that the "cross" **partial derivatives** with respect to the same two **variables** of a **function** of two or more **arguments** are equal: $\partial_x\partial_y g(x,y) = \partial_y\partial_x g(x,y)$. **2.4**

Zero Angle See **Unit Angle**.

Zeroes of a Function The points in the **domain** of a **function** where it evaluates to 0.

SELECT BIBLIOGRAPHY

Authors	Title	Publishing
C.B.Allendoerfer and C.E.Oakley	*Principles of Mathematics*	McGraw-Hill, 2nd.Edn, 1963
R.L.Bishop and S.I.Goldberg	*Tensor Analysis on Manifolds*	Macmillan, 1968
R.N.Bracewell	*The Fourier Transform and its Applications*	McGraw-Hill, 3rd. Edn., 2000
R.Brown	*Elements of Modern Topology*	McGraw-Hill,1968
D.A.Brannan, M.F.Esplen and J.J.Gray	*Geometry*	Cambridge Univ. Press 1999
C.Doran and A.Lasenby	*Geometric Algebra for Physicists*	Cambridge Univ. Press 2003
S.Feferman	*The Number Systems*	Addison-Wesley, 1964
W.Franz	*Algebraic Topology*	Frederick Ungar Publishing, 1968
B. Friedman	*Principles and Techniques of Applied Mathematics*	Wiley, 1956, Dover 1990
H.Goldstein	*Classical Mechanics*	Addison-Wesley, 1950
D. Hestenes and G. Sobczyk	*Clifford Algebra to Geometric Calculus*	D. Reidel Publishing, 1984
R.E.Johnson and F.L.Kiokemeister	*Calculus with Analytic Geometry*	Allyn and Bacon, 1964
C.Kittel and H.Kroemer	*Thermal Physics*	W.H.Freeman and Co.,1980
S.Lang	*Linear Algebra*	Addison-Wesley, 1966
M.J.Lighthill	*An introduction to Fourier analysis and generalized functions*	Cambridge Univ. Press 1958
R.Lipsey and A.Chrystal	*Economics*	Oxford Univ.Press, 11th. Edn., 2007
S.MacLane and G.Birkhoff	*Algebra*	Macmillan, 1967
F.Mandl	*Statistical Physics*	Wiley, 2nd. Edn. 1988
J.Mathews and R.L.Walker	*Mathematical Methods of Physics*	W.A.Benjamin, 2nd. Edn., 1970
H.K.Nickerson, D.C.Spencer and N.E.Steenrod	*Advanced Calculus*	Van Nostrand, 1959, Dover 2011
J.R.Reitz and F.J.Milford	*Foundations of Electromagnetic Theory*	Addison-Wesley, 2nd. Edn., 1967
L.Schwartz	*Mathematics for the Physical Sciences*	Hermann, 1966 (English Edn. Addison-Wesley)

Select Bibliography

M.R.Spiegel	*Advanced Calculus*	**Schaum Publishing, 1963**
G.B.Thomas	*Calculus and Analytic Geometry*	**Addison-Wesley, 3rd. Edn., 1960**
J.Vince	*Geometric Algebra for Computer Graphics*	**Springer Verlag, 2008**

INDEX

0

1

2

3

4

A

E

F

G

H

I

J

K

L

M

N

O

P

Q

R

S

T

U

www.ingramcontent.com/pod-product-compliance
Lightning Source LLC
Chambersburg PA
CBHW082122210326
41599CB00031B/5848

* 9 7 8 0 9 9 3 5 4 8 3 0 7 *